Quantum Bounded Symmetric Domains

Translations of
MATHEMATICAL MONOGRAPHS

Volume 238

Quantum Bounded Symmetric Domains

Leonid L. Vaksman

Translated by Olga Bershtein
and Sergey D. Sinel'shchikov

American Mathematical Society
Providence, Rhode Island

Леонид Л. Ваксман

ОГРАНИЧЕННЫЕ КВАНТОВЫЕ СИММЕТРИЧЕСКИЕ ПРОСТРАНСТВА

Translated from a Russian manuscript. The present translation was created under license for the American Mathematical Society and is published by permission.

Translated from the Russian by Olga Bershtein and Sergey D. Sinel′shchikov.

2000 *Mathematics Subject Classification.* Primary 17B37, 20G42, 81R50; Secondary 22E47, 33D45, 43A85, 46L52.

For additional information and updates on this book, visit
www.ams.org/bookpages/mmono-238

Library of Congress Cataloging-in-Publication Data

Vaksman, L. L. (Leonid L'vovych), 1951–2007.
[Kvantovye ogranichennye simmetricheskie oblasti. English]
Quantum bounded symmetric domains / Leonid L. Vaksman ; translated by Olga Bershtein and Sergey Sinel′shchikov.
p. cm. — (Translations of mathematical monographs ; v. 238)
Includes bibliographical references and index.
ISBN 978-0-8218-4909-5 (alk. paper)
1. Symmetric domains. 2. Representations of quantum groups. 3. Quantum groups. 4. Noncommutative algebras. I. Title.
QC20.7.F84V3513 2010
512′.55—dc22

2010009598

∞ The paper used in this book is acid-free and falls within the guidelines established to ensure permanence and durability.
Visit the AMS home page at `http://www.ams.org/`

10 9 8 7 6 5 4 3 2 1 15 14 13 12 11 10

Contents

Foreword

This book presents basics of the theory of quantum bounded symmetric domains, which were developed by L. Vaksman and his team.

Our main goal in this foreword is to tell a little bit about the author, about the history of this book, and about its structure.

Leonid Vaksman (1951–2007) was a talented mathematician who loved teaching mathematics and explaining mathematical ideas to his colleagues. Starting from his undergraduate years, L. Vaksman was interested in many diverse mathematical fields, from mathematical physics to category theory. This is crucial in understanding why L. Vaksman was eager to cooperate fruitfully with very different kinds of students and colleagues, being able to suggest the right topics in each case.

Since the late 1980s L. Vaksman ran a seminar, first in Rostov (Russia), then in Kharkov (Ukraine). The range of topics discussed at the seminar was very wide and included such areas as the theory of quantum groups and their representations, the theory of $*$-algebras, noncommutative complex and harmonic analysis, special functions, etc. All related methods and tools were extensively explained and used. Among the participants of the seminar were Ya. Soibelman, L. Korogodski, S. Sinel'shchikov, D. Shklyarov, O. Bershtein, Ye. Kolisnyk.

Later L. Vaksman started to write down the problems, results, and ideas discussed at the seminar. These notes resulted in a fundamental survey on the theory of quantum bounded symmetric domains, which also included both complete results discussed earlier and a number of unsolved problems, together with related ideas and hints. Currently, this survey is available only in Russian. Simultaneously, L. Vaksman prepared a smaller version of that survey and supplied it with a preliminary chapter explaining his approach to the main topics of the book. After his death this text was prepared for publication by O. Bershtein, Ye. Kolisnyk, and S. Sinel'shchikov. This is the book you are looking at now.

The core of this book is the exposition of basic results on quantum bounded symmetric domains. Some of these results have already been published as articles and/or as preprints, some others are presented here for the first time. Regretably, it was impossible to include all the nice results in this shorter version.

On the other hand, the large introductory Chapter 1 is devoted to the simplest quantum bounded symmetric domain, namely, the quantum disc. The idea of this chapter is to explain the problems, results, and methods of the book in the simplest case (which is already quite nontrivial).

We are grateful to H. P. Jakobsen for helpful discussions on some parts of the book and his warm hospitality during our visit to Copenhagen in September 2008, where the translation work started. We also acknowledge the dedicated work of Ye. Kolisnyk in preparing this book for publication.

O. Bershtein, S. Sinel'shchikov
Kharkov, 2009

Preface

This book presents a study of quantum analogs of bounded symmetric domains. The latter domains are steady subjects of attraction for specialists in geometry, algebra, and analysis, basically as sources of exactly solvable problems of complex analysis, noncommutative harmonic analysis, and classical mathematical physics.

The simplest bounded symmetric domain is the unit disc $\mathbb{D} = \{z \in \mathbb{C} \mid |z| < 1\}$. Its quantum analog was introduced by S. Klimek and A. Lesniewski [**173**]. We find it reasonable to start with Chapter 1 on the quantum disc. Our motive here is to avoid distracting the reader by algebraic details, but instead to produce an outline of the problems of noncommutative complex and harmonic analysis in which we are interested. This chapter can be used as background material to a seminar for university students in mathematics.

The problems discussed in Chapter 1 admit a reformulation (many of them, even a solution) in a much more general context, namely within the framework of quantum theory of bounded symmetric domains [**289, 290**], whose basics are expounded in Chapter 2. This demonstrates an interplay between the theory of quantum groups and noncommutative complex analysis.

A more detailed view of Chapters 1 and 2 is given in the table of contents.

The author's intention for subsequent chapters was to present results of O. Bershtein, Ye. Kolisnyk, D. Proskurin, D. Shklyarov, S. Sinel'shchikov, A. Stolin, L. Turowska, L. Vaksman, and G. Zhang, [**33, 36, 34, 308, 258, 284, 320, 279, 318, 281, 287**], together with unpublished results on quantum bounded symmetric domains (some of those are already present in `xxx.lanl.gov`, see, e.g., [**317**], or will appear therein in the nearest future).

Regretably, this plan is not accessible due to some nonmathematical reasons. Instead, we are going to form a Web page dedicated to quantum bounded symmetric domains, which will contain a draft of the conjectured full version of the book. It is expected to be twice as large as the present volume; it will, in particular, contain a discussion on unsolved problems.

I am deeply grateful to my students O. Bershtein, Ye. Kolisnyk, L. Korogodski, D. Shklyarov, and coauthors Ya. Soibelman and A. Stolin. Also, my special thanks to H. Jakobsen, A. Klimyk, E. Koelink, S. Kolb, Yu. Samoilenko, K. Schmüdgen, L. Turowska, and G. Zhang for numerous helpful discussions on the results of this book.

A special role in my life was played by Vladimir Drinfeld. In the mid-1980s he taught me the basics of the theory of quantum groups and helped me to return to mathematics after an involuntary break I had to take from it for many years.

Here is some historical background. In the late 1970s, a study of exactly solvable problems of statistical mechanics and quantum field theory led L. Faddeev and his team to the creation of the quantum method of inverse scattering problems [**293,**

294, 305]. They introduced the quantum Yang-Baxter equation and associated its solutions to series of exactly solvable problems of mathematical physics. Their work [**206**] presents the solutions of this equation which were known before 1980; the authors mention that "its deep relations to the mathematical fields like group theory and algebraic geometry are coming into the picture". In the early 1980s plenty of literature was dedicated to bringing to light those relations and studying solutions of the quantum Yang-Baxter equation. In this context, we mention the works of E. Sklyanin, P. Kulish, V. Drinfeld, and N. Reshetikhin [**291, 204, 77, 292, 205**].

In 1984 V. Drinfeld introduced quantum analogs for universal enveloping algebras (see Subsection 2.1.1), and his talk on quantum groups at the Gelfand seminar became a crucial point in the theory of quantum groups. Independently, in 1985 M. Jimbo came to his version of quantum analogs of universal enveloping algebras. The crucial works in quantum group theory are [**78, 140**] as well as survey reviews [**79, 141, 94**].

A different approach to quantum groups is due to S. Woronowicz [**341, 342**]. This approach was used later in the theory of compact quantum groups [**344, 72, 66**], where the simplest example is the quantum group $SU(2)$ [**343, 251, 321, 228, 239, 240, 229, 191, 193, 170, 316, 37, 152, 153, 184, 189, 182, 180, 252, 254, 213, 1**].

Applications of the theory of quantum groups in low-dimensional topology and category theory as well as in conformal quantum field theory have been found later; see [**262, 307, 160, 150, 223, 222, 5, 247**]. In this book we do not consider these and many other applications, and we restrict ourselves to quantum analogs of homogeneous spaces of noncompact real Lie groups.

The initial results in this direction were obtained in late 1980s in [**319**]. In this work the quantum group of motions of the Euclidean plane was introduced; then it became a tool for solving a number of problems in noncommutative harmonic analysis and special functions [**56, 346, 347, 59, 217, 194, 183, 181, 187, 188, 186, 301, 38, 2**]. The quantum group of motions of plane can be derived from the quantum group $SU(2)$ by an application of the Inonu-Wigner contraction [**326**, p. 234]; this trick is also applicable to some other "nonhomogeneous" quantum groups [**58, 57, 75**].

P. Podleš and S. Woronowicz managed to produce a quantum analog for the Lorentz group [**255**] by applying another method to real forms of complex semisimple Lie groups. Their research has been succeeded in [**259, 253, 244, 243, 52, 51**].

Essential obstacles have been encountered on the way to a quantum analog of the group of motions of the Lobachevski plane (more precisely, to a quantum analog of some locally isomorphic group). It even happened that S. Woronowicz denied the very existence of this quantum group [**345**]. Nevertheless, he changed his opinion later, possibly after reading the work of L. Korogodsky [**200**]. Several years later, a construction of the required quantum group was completed by E. Koelink and J. Kustermans [**179**] within the Kustermans-Vaes axiomatics [**207, 208**].

The above references are not exhaustive; however, they provide a general idea about the development of noncompact quantum group theory at its earliest stage.

An apparent feature of [**200**] is a sharp discrepancy between the simplicity of the classical subject and the complexity of its quantum analog. This inspired an idea of not using function algebras on quantum groups when studying function algebras on quantum homogeneous spaces. Under this approach, the methods of representation theory of real reductive Lie groups could be the principal tools of research [**332, 333**]. In [**114**] Harish-Chandra introduced the holomorphic discrete series of representation for groups of Hermitian type. These representations are realized in weighted Bergman spaces of holomorphic functions on bounded symmetric domains [**197**]. The spaces of the associated action of Lie algebras are polynomial algebras on prehomogeneous complex vector spaces of commutative parabolic type (see Subsection 2.3.8). In this context it was natural to ask if there exist quantum analogs of bounded symmetric domains, weighted Bergman spaces, holomorphic discrete series, and the Sato–Bernstein polynomials of prehomogeneous vector spaces of commutative parabolic type (see [**234, 268**] for a background on Sato–Bernstein polynomials).

The positive answer to this question was supposed to lead to substantial progress in the theory of quantum prehomogeneous vector spaces and Harish-Chandra modules [**270**] over quantum universal enveloping algebras. This could result in a breakthrough in noncommutative complex analysis, a field coming back to a work by W. Arveson [**17**]. It became clear later that it was the right idea.

In late 1990s three teams of mathematicians obtained initial results in quantum theory of bounded symmetric domains. These teams acted independently, being unaware of each other and thinking of their methods as self-sufficient.

T. Tanisaki and his collaborators introduced q-analogs of prehomogeneous vector spaces of commutative parabolic type and found an explicit form of the associated Sato–Bernstein polynomials [**156, 155, 233, 154**].

H. Jakobsen then suggested an easier way of producing these quantum vector spaces; he also realized that he was working toward quantum Hermitian symmetric spaces of noncompact type [**136, 138**].

The authors of the papers listed above overlooked the so-called hidden symmetry of the quantum prehomogeneous vector spaces in question [**288, 282**]; perhaps, it was the reason why they made no crucial step on the way to quantum bounded symmetric domains.

Foundations of the quantum theory of bounded symmetric domains were laid in [**289**]. The compatibility of the approaches used in [**306, 136, 289**] has been established by D. Shklyarov [**276**].

CHAPTER 1

Quantum Disc

1.1. A caution to pedants

In some cases we found it inappropriate to include definitions to be used in the book, and we replaced them by suitable references.

One should be aware that some textbooks contain close notions which have the same names and could be hardly distinguished. The use of some conventional notation sometimes appears to be imprecise. Nevertheless, the context always brings to light what the author means. Here are some examples.

A basis of a vector space V normally means a subset $B \subset V$ of linearly independent vectors that span V [**211, 102**]. However, a matrix being associated to a linear map assumes a basis to be linearly ordered. This way the initial definition becomes implicitly altered, as the same subset $B \subset V$ with two different linear orders now determines two bases.

A completion of a metric space X is normally referred to as a complete metric space $\overline{X}$, together with an isometric embedding $i : X \to \overline{X}$ such that $i(X)$ is dense in $\overline{X}$ [**190**]. However, some authors [**237**], instead of referring to an arbitrary pair $(\overline{X}, i)$ with these properties, treat the completion as a single distinguished pair whose construction is present in all the textbooks on functional analysis.

The associativity of the Cartesian product of sets is written as $(X_1 \times X_2) \times X_3 = X_1 \times (X_2 \times X_3)$. However, it is not an equality, but a canonical isomorphism [**42**]: $(X_1 \times X_2) \times X_3 \cong X_1 \times (X_2 \times X_3)$.

Similarly, it would be reasonable to replace the equality $(X_1 \otimes X_2) \otimes X_3 = X_1 \otimes (X_2 \otimes X_3)$ with the relation $(X_1 \otimes X_2) \otimes X_3 \cong X_1 \otimes (X_2 \otimes X_3)$.

By the way, just like the notion of completion for metric spaces, there exists some ambiguity in treating the notion of a tensor product of vector spaces in the literature. Some authors [**161**] use the term *tensor product of vector spaces* V', V'' to refer to the vector space written as $V' \otimes V''$, together with a bilinear map $i : V' \times V'' \to V' \otimes V''$ which satisfies the following universality condition. Given a bilinear map $f : V' \otimes V'' \to W$, there exists a unique linear map $\widetilde{f} : V' \otimes V'' \to W$ with $f = \widetilde{f} i$. However, most authors [**41, 211, 202**], instead of referring to an arbitrary such pair with the universality property, call a tensor product just a single distinguished pair. We describe this construction below.

Just like groups and algebras, vector spaces can be defined in terms of generators and relations. To be more precise, associate to each set M the vector space $\mathcal{M}$ of formal linear combinations of elements of this set. With $N \subset \mathcal{M}$, one can consider the vector space defined as the quotient of $\mathcal{M}$ by the linear span of N. This quotient space will be called a vector space with the set of generators M and the set of relations N.

Consider the vector space $V' \otimes V''$ given by the set of generators $\{(v', v'') \in V' \times V''\}$ and the set of relations which is a union of the two subsets

$$\{(\lambda_1 v_1' + \lambda_2 v_2', v'') - \lambda_1(v_1', v'') - \lambda_2(v_2', v'')|\, \lambda_1, \lambda_2 \in \mathbb{C},\ v_1', v_2' \in V',\ v'' \in V''\},$$
$$\{(v', \lambda_1 v_1'' + \lambda_2 v_2'') - \lambda_1(v', v_1'') - \lambda_2(v', v_2'')|\, \lambda_1, \lambda_2 \in \mathbb{C},\ v' \in V',\ v_1'', v_2'' \in V''\}.$$

Denote by $v' \otimes v''$ the image of $(v', v'') \in V' \times V''$ under the canonical epimorphism of the linear span of all $(v', v'') \in V' \times V''$ onto $V' \otimes V''$. It is easy to demonstrate that the bilinear map

$$i : V' \times V'' \mapsto V' \otimes V'', \qquad i : v' \times v'' \mapsto v' \otimes v'',$$

possesses the universality property.

During the latest decades, the number of twins increase drastically due to a passage from a set-theoretic to category-theoretic approach to mathematics [**100**]. Of course, both viewpoints, as well as each of the twin notions, deserve existence.

1.2. Topology

1.2.1. Commutative C^*-algebras. In the late 1960s W. Arveson suggested in [**17**] a program of producing a noncommutative harmonic analysis. The main object of [**17**] is a pair of algebras such that one of them is a noncommutative analog of the algebra $C(K)$ of continuous functions on a compact space K and the other one is a noncommutative analog of a subalgebra of functions analytical in the interior of K.

Such an approach to the standard complex analysis appeared to be fruitful within the framework of the theory of (commutative) uniform algebras. The last theory is closely related to the problems of uniform approximation by rational functions in a complex domain [**98**]. In [**17**] Arveson essentially uses the principal notions of the theory of C^*-algebras. Let us recall some of them.

The ground field will be the field of complex numbers $\mathbb{C}$.

An algebra with unit will also be called unital. Now we consider the category of (not necessarily unital) algebras. In this category a morphism $F_1 \to F_2$ does not necessarily map the unit of F_1 to the unit of F_2, even if these units exist.

In a standard way we join a unit by applying a functor from the mentioned category to the category of unital algebras such that an algebra F is sent to the unital algebra $\widetilde{F} = \mathbb{C} \times F$ such that

$$(\alpha_1, f_1)(\alpha_2, f_2) = (\alpha_1\alpha_2, \alpha_1 f_2 + \alpha_2 f_1 + f_1 f_2), \qquad \alpha_1, \alpha_2 \in \mathbb{C},\ f_1, f_2 \in F.$$

The action of the introduced functor on morphisms is defined in an obvious way.

Recall that an algebra F is called a Banach algebra if it is a Banach space and

$$\|f_1 f_2\| \le \|f_1\| \cdot \|f_2\|, \qquad f_1, f_2 \in F.$$

EXAMPLE 1.1. Let K be a compact space. The algebra $C(K)$ of continuous functions on K endowed with the norm $\|f\| = \max_{x \in K} |f(x)|$ is a Banach algebra.

EXAMPLE 1.2. Let X be a locally compact topological space. Consider the disjoint union $\widetilde{X} = X \cup \{\infty\}$ of X and the one-point set $\{\infty\}$. We endow $\widetilde{X}$ with the weakest topology which contains all open subsets $U \subset X$ and the sets $\widetilde{X} \backslash K$ for all compact subsets $K \subset X$. The topological space $\widetilde{X}$ produced in this way is compact. The embedding $i : X \hookrightarrow \widetilde{X}$ is a continuous mapping and it is called the Tychonoff compactification of X. The Tychonoff compactification is a functor which assigns to a locally compact topological space X the compact space $\widetilde{X}$ with the marked point ∞. The algebra $C(\widetilde{X})$ of all continuous functions on $\widetilde{X}$ and its subalgebra $C_0(X) = \left\{ f \in C(\widetilde{X}) \middle| \ f(\infty) = 0 \right\}$ with the inherited norm $\|f\| = \max_{x\in X} f(x) = \max_{x\in\widetilde{X}} f(x)$ are Banach algebras. Note that one can obtain $C(\widetilde{X})$ by joining a unit to $C_0(X)$ once we forget the norms.

An involution $*$ is an antilinear involutive antiautomorphism of an algebra F:

$$(f_1 + f_2)^* = f_1^* + f_2^*, \quad (\lambda f)^* = \overline{\lambda} f^*, \qquad \lambda \in \mathbb{C},\ f_1, f_2, f \in F,$$
$$** = \mathrm{id}, \qquad (f_1 f_2)^* = f_2^* f_1^*, \qquad f_1, f_2 \in F.$$

Evidently, $1^* = 1$ since the unit is unique in F. An algebra with an involution is called a $*$-algebra.

A morphism of $*$-algebras is a homomorphism of algebras $j : F_1 \to F_2$ which intertwines the involutions $*j = j*$. As usual, an invertible morphism in the category of $*$-algebras is called an *isomorphism.*

A Banach algebra F with an involution $*$ is called a *C^*-algebra* if for each $f \in F$ one has

$$\|f\|^2 = \|f^* f\|. \tag{1.1}$$

A morphism of C^*-algebras is a continuous linear mapping of Banach spaces which is a morphism of $*$-algebras. Note that here we may skip the requirement of continuity; see Proposition 1.6.

The Banach algebra $C_0(X)$ with the involution $* : f \mapsto \overline{f}$ is, obviously, a *commutative C^*-algebra.*

Recall that a morphism of topological spaces is called *proper* if the preimage of each compact set is compact [**39**, p. 187]. Now we can consider the category with locally compact topological spaces as *objects* and continuous proper maps as *morphisms.* To each locally compact topological space X one can assign the C^*-algebra $C_0(X)$. Furthermore, to each continuous proper map $\varphi : X_1 \to X_2$ there corresponds the following morphism of C^*-algebras:

$$C_0(X_2) \to C_0(X_1), \qquad f(x) \mapsto f(\varphi(x)).$$

Now we obtain a *contravariant functor* from the category of locally compact topological spaces to the category of commutative C^*-algebras. It is known [**48**, p. 70] that these categories are antiequivalent. Now we recall the construction of the quasi-inverse functor [**100**, p. 94], the famous *Gelfand transform.*

Homomorphisms of a commutative C^*-algebra to the C^*-algebra $\mathbb{C}$ are called characters.[1] The set $\sigma(F)$ of all characters is called the *spectrum of the commutative C^*-algebra F.* To every element $f \in F$ there corresponds the function $\hat{f}(\chi) \stackrel{\text{def}}{=} \chi(f)$ on $\sigma(F)$. Now we can endow $\sigma(F)$ with the weakest topology in which all the functions $\{\hat{f} |\ f \in F\}$ are continuous. One can prove that the spectrum $\sigma(F)$ is locally compact and the Gelfand transform

$$F \to C_0(\sigma(F)), \qquad f \mapsto \hat{f},$$

is an isomorphism of C^*-algebras. The spectrum is compact if and only if the algebra F is unital [**48**, p. 70]. The action of the Gelfand transform on morphisms

[1]The requirements of continuity and involutivity might be skipped in the definition of a character [**48**, p. 70].

is determined in a natural way: every morphism $\varphi : F_1 \to F_2$ of commutative C^*-algebras corresponds to the continuous proper map of the spectra

$$\sigma(F_2) \to \sigma(F_1), \qquad \chi(f) \mapsto \chi(\varphi(f)).$$

EXAMPLE 1.3 (Stone–Čech compactification). We would like to consider a continuous embedding of the real axis to a compact such that each bounded continuous function on $\mathbb{R}$ admits a unique extension to a continuous function on the compact. Does this embedding exist? Consider the commutative C^*-algebra $C_b(\mathbb{R})$ of bounded continuous functions on $\mathbb{R}$ with the norm $\|f\| = \sup_{x\in\mathbb{R}} |f(x)|$. To every point $x_0 \in \mathbb{R}$, one can assign the following character:

$$f \mapsto f(x_0) \in \mathbb{R}, \qquad f \in C_b(\mathbb{R}).$$

One can easily show that the constructed embedding of $\mathbb{R}$ to the compact space $\sigma(C_b(\mathbb{R}))$ satisfies the required properties [**82**, p. 23].

EXAMPLE 1.4 (Quasi-periodic functions). Let $\gamma \in \mathbb{R}\setminus\mathbb{Q}$. Consider the smallest C^*-algebra $F \subset C_b(\mathbb{R})$ which contains all 2π-periodic functions and all $2\pi\gamma$-periodic functions. We would like to describe the spectrum $\sigma(F)$ of this commutative C^*-algebra. The elements $u_1 = e^{ix}$, $u_2 = e^{ix/\gamma}$ are unitary:

$$u_1 u_1^* = u_1^* u_1 = 1, \qquad u_2 u_2^* = u_2 u_2^* = 1.$$

We can introduce a continuous map of $\sigma(F)$ to the two-dimensional torus

$$\sigma(F) \to \mathbb{T}^2, \qquad \chi \mapsto (\chi(u_1), \chi(u_2)). \tag{1.2}$$

Now we show that this map is a homeomorphism. It is sufficient to verify its bijectivity. The Stone–Weierstrass theorem [**260**, p. 119] implies that F is the smallest C^*-subalgebra which contains u_1 and u_2. The last fact leads to the injectivity. The surjectivity follows from the fact that the dense subgroup of the torus $\{(e^{ix}, e^{ix/\gamma}) | \, x \in \mathbb{R}\} \subset \mathbb{T}^2$ (irrational rotation) belongs to the image of $\mathbb{R}$ under the natural embedding to $\sigma(F)$ and the subsequent map (1.2).

EXAMPLE 1.5 (Almost-periodic functions). Now we consider the smallest C^*-subalgebra $AP \subset C_b(\mathbb{R})$ which contains the set of exponential functions $\{e^{i\lambda x} | \, \lambda \in \mathbb{R}\}$. These functions are common eigenvectors of the shift operators

$$T_t : C_b(\mathbb{R}) \to C_b(\mathbb{R}), \qquad T_t : f(x) \mapsto f(x+t), \quad t \in \mathbb{R},$$

in $C_b(\mathbb{R})$. The well-known Bohr theorem claims that a function $f \in C_b(\mathbb{R})$ belongs to AP if and only if the set $\{T_t f | \, t \in \mathbb{R}\}$ is relatively compact in $C_b(\mathbb{R})$. Such functions are called *almost periodic*, and the compactification related to the C^*-algebra AP is called the *Bohr compactification.*

One of the elementary results of homotopy theory is that the unit circle $\mathbb{T}$ is not a retract of the closed unit disc $\overline{\mathbb{D}}$ [**39**, p. 143]. This can be rephrased by saying that there is no continuous map of $\overline{\mathbb{D}}$ to $\mathbb{T}$ which is identical on $\mathbb{T}$. We would like to translate this result to the language of the theory of Banach algebras. This will allow us to obtain its analog for a quantum disc.

Let us consider the exact sequence

$$0 \to C_0(\mathbb{D}) \to C(\overline{\mathbb{D}}) \xrightarrow{j} C(\mathbb{T}) \to 0 \tag{1.3}$$

in the category of commutative C^*-algebras, where j is the restriction of functions continuous in $\overline{\mathbb{D}}$ onto its boundary $\mathbb{T}$. This exact sequence is said to be *split* if there exists a homomorphism of commutative C^*-algebras $i : C(\mathbb{T}) \to C(\overline{\mathbb{D}})$ such that $ji = \mathrm{id}$. Using the antiequivalence of categories, we see that the above topological result means that the exact sequence (1.3) is not split.

1.2.2. Elements of the general theory of C^*-algebras. C^*-algebras are noncommutative analogs of algebras of continuous functions. Here we present some results from the theory of C^*-algebras. We follow [**48, 303**] and skip most of proofs.

An application of (1.1) to f and f^* yields the equality of the norms $\|f^*\| = \|f\|$. Indeed, $\|f\| \leq \|f^*\|$ since $\|f\|^2 = \|f^* f\| \leq \|f^*\| \cdot \|f\|$. The replacement of f by f^* leads to the opposite inequality. Hence a C^*-algebra F is a Banach $*$-algebra such that

$$\|f^*\| = \|f\|, \qquad \|f^* f\| = \|f^*\| \cdot \|f\| \tag{1.4}$$

for all $f \in F$. Some authors define a C^*-algebra by (1.4) instead of (1.1).

There are both unital and nonunital C^*-algebras. In the latter case the unital $*$-algebra $\widetilde{F} = \mathbb{C} \times F$ is endowed with a structure of C^*-algebra, as described in [**48**, p. 29]:

$$\|(\alpha, f)\| = \sup\{\|\alpha f_1 + f f_1\| | \, f_1 \in F, \ \|f_1\| = 1\}.$$

Given a unital algebra F, the spectrum of $f \in F$ is a set $\mathrm{spec}(f)$ of $\lambda \in \mathbb{C}$ such that the element $\lambda - f$ is not invertible in F. If F is nonunital, the spectrum of f is by definition the spectrum $\mathrm{spec}(f)$ in $\widetilde{F}$ [**45**, p. 10]. For each element f of the Banach algebra F there exists a limit

$$\rho(f) = \lim_{n\to\infty} \|f^n\|^{\frac{1}{n}} = \sup\{|\lambda| | \, \lambda \in \mathrm{spec}(f)\}$$

(see [**48**, p. 33], [**303**, p. 8]), which is called the spectral radius of f. If F is a C^*-algebra, then $\|f\| = \rho(f^* f)^{\frac{1}{2}}$ [**48**, p. 36], and we get the following result on automatic continuity [**303**, p. 21].

PROPOSITION 1.6. *Let F_1, F_2 be C^*-algebras and $\pi : F_1 \to F_2$ a morphism of $*$-algebras. Then*

$$\|\pi(f)\| \leq \|f\|, \qquad f \in F.$$

It is sufficient to prove the following statement for commutative C^*-algebras and a self-adjoint element f. In this special case the Gelfand transform [**303**, p. 22] yields the claim.

PROPOSITION 1.7. *Let F_1, F_2 be C^*-algebras and $\pi : F_1 \to F_2$ a monomorphism of $*$-algebras. Then $\|\pi(f)\| = \|f\|$ for all $f \in F$.*

Now we may exclude the isometry requirement from the definition of isomorphism of C^*-algebras since a Banach $*$-algebra has either no C^*-norm or a unique C^*-norm.

Let us present an example of a noncommutative C^*-algebra.

EXAMPLE 1.8. Consider the algebra $\mathcal{B}(H)$ of all bounded linear operators on a Hilbert space H. We can endow it with the operator norm $\|A\| = \sup_{\|v\|=1} \|Av\|$. It is easy to prove that this normed algebra is complete. Indeed, each fundamental sequence of operators is bounded, and its convergence on every vector follows from the completeness of H. Let us endow $\mathcal{B}(H)$ with the involution that assigns the adjoint operator A^* to an operator A. Now we would like to show that $\mathcal{B}(H)$ is a C^*-algebra:

$$\|A\|^2 = \|A^*A\|. \tag{1.5}$$

First, one can prove that given a bounded self-adjoint linear operator A, one gets

$$\|A\| = \sup_{\|v\|=1} |(Av, v)|. \tag{1.6}$$

The only thing that is not evident is the inequality $\|A\| \le \sup_{\|v\|=1} |(Av, v)|$. In the special case of a two-dimensional Hilbert space, the inequality is well known from the general courses on linear algebra. The general case reduces to this special case by the inequality

$$\|A\| \le \sup_{\dim L=2} \|P_L A|_L\|, \tag{1.7}$$

where L runs through the set of all two-dimensional subspaces, $A|_L$ is the restriction of A to L, and P_L is the orthogonal projection to L.[2] The proof of (1.5) is completed by (1.6), as in [**260**, p. 210]:

$$\|A^*A\| = \sup_{\|v\|=1} |(A^*Av, v)| = \sup_{\|v\|=1} (A^*Av, v) = \sup_{\|v\|=1} \|Av\|^2 = \|A\|^2.$$

The Gelfand–Naimark theorem [**48**, p. 32] claims that every C^*-algebra is isomorphic to a closed self-adjoint subalgebra of the C^*-algebra $\mathcal{B}(H)$.

Let J be a closed two-sided ideal of a C^*-algebra F. One can show that $J = J^*$ and the quotient space F/J with the norm

$$\|f + J\| = \inf_{x\in J} \|f + x\|, \qquad f \in F,$$

is a C^*-algebra [**48**, p. 48].

EXAMPLE 1.9. Let H be an infinite-dimensional separable Hilbert space. A linear operator A on H is called a finite rank operator if $\dim(AH) < \infty$. These operators form a two-sided ideal in $\mathcal{B}(H)$, and its closure $\mathcal{K}(H)$ coincides with the closed $*$-ideal of all compact operators [**260**, p. 223]. It is known [**103**, p. 90] that $\mathcal{K}(H)$ is a unique nontrivial closed two-sided ideal of the C^*-algebra $\mathcal{B}(H)$. The C^*-algebra $\mathcal{B}(H)/\mathcal{K}(H)$ is called the *Calkin algebra.*

Now we would like to present some results from operator theory that are closely related to the Calkin algebra. The essential spectrum $\operatorname{spec}_e(A)$ of a bounded linear operator $A \in \mathcal{B}(H)$ is the spectrum of the corresponding element of the Calkin algebra. A linear operator $A \in \mathcal{B}(H)$ is called *Fredholm operator* if the corresponding element of the Calkin algebra is invertible. Due to Atkinson's theorem [**113**, p. 97], it means that the kernel of A is finite dimensional, the image is closed and has a finite codimension. Evidently, the essential spectrum consists of $\lambda \in \mathbb{C}$ such that the operator $\lambda - A$ is not Fredholm.

The index of a Fredholm operator $\operatorname{ind}(A) \overset{\text{def}}{=} \dim \operatorname{Ker} A - \operatorname{codim} \operatorname{Im} A$ does not change under compact perturbations, and therefore it is a function on the group of invertible elements of the Calkin algebra [**162**, p. 300], [**76**, p. 138]. This function is continuous and its level sets are linearly connected [**76**, p. 136].

[2] In order to prove (1.7), one should consider two-dimensional subspaces that contain the pairs of vectors $\{v, Av\}$.

The simplest example of a Fredholm operator with index zero is the operator of one-sided shift S defined on an orthonormal basis $\{e_j\}_{j\in\mathbb{Z}_+}$ as follows: $S(e_j) = e_{j+1}$. The operator S cannot be decomposed into a sum of normal and compact operators since the index of a normal operator is 0. The latter fact follows from the spectral theorem [**82**, p. 50].

1.2.3. The Fock representation. Recall [**190**, p. 66] that a completion of a metric space X is its embedding $X \hookrightarrow \overline{X}$ to a complete metric space $\overline{X}$ with everywhere dense image. The completion exists and is essentially unique.

Let $\mathbb{D} = \{z \in \mathbb{C} |\, |z| < 1\}$. Consider the commutative C^*-algebra $C(\overline{\mathbb{D}})$ of continuous functions in the closed unit disc $\overline{\mathbb{D}}$ and the $*$-algebra of polynomial functions $\mathrm{Pol}(\mathbb{C})$ on the real plane $\mathbb{C} \simeq \mathbb{R}^2$. Due to the Weierstrass theorem on polynomial approximation [**260**, p. 120], the $*$-algebra $\mathrm{Pol}(\mathbb{C})$ is dense in $C(\overline{\mathbb{D}})$. Equivalently, the natural embedding $\mathrm{Pol}(\mathbb{C}) \hookrightarrow C(\overline{\mathbb{D}})$ is a completion of the $*$-algebra $\mathrm{Pol}(\mathbb{C})$ with respect to the corresponding norm. The choice of completion is not essential in many problems of functional analysis. This allows one to replace the particular C^*-algebra $C(\overline{\mathbb{D}})$ by any completion of $\mathrm{Pol}(\mathbb{C})$ with respect to this norm. Using this approach, we would like to construct a noncommutative C^*-algebra that will be called a *quantum analog of the commutative C^*-algebra $C(\overline{\mathbb{D}})$*.

Fix $q \in (0, 1)$ that will be referred to as a *deformation parameter*, i.e., as $q \to 1$ formulas of function theory in a quantum disc become usual formulas of the function theory in the disc. We discuss the role of this parameter and a quantization procedure in detail later in the book.

Let us start with a quantum polynomial algebra. Consider the free algebra $\mathbb{C}\langle z, z^*\rangle$ with two generators z, z^*. This algebra is usually identified with the algebra of formal linear combinations of words over the alphabet $\{z, z^*\}$. Evidently, there exists a unique involution $*$ in $\mathbb{C}\langle z, z^*\rangle$ which maps z to z^*. The element $(z^*z - q^2 z\, z^* - 1 + q^2) \in \mathbb{C}\langle z, z^*\rangle$ is self-adjoint. Hence it generates a two-sided self-adjoint ideal. Therefore an algebra $\mathrm{Pol}(\mathbb{C})_q$ with the generators z, z^* and the defining relations

$$z^*z = q^2 zz^* + 1 - q^2 \tag{1.8}$$

inherits the involution and becomes a $*$-algebra. This $*$-algebra will be considered as a q-analog of the algebra of polynomials in variables z and $\overline{z}$.

Now we should just complete $\mathrm{Pol}(\mathbb{C})_q$ to obtain an algebra of continuous functions in the quantum disc. How can one choose a norm? We follow the traditions of quantum mechanics. Substituting $q = e^{-\frac{h}{2}}$ and formally passing to the limit as $h \to 0$, we get

$$\{z^*, z\} \stackrel{\mathrm{def}}{=} \lim_{h\to 0} \frac{z^*z - zz^*}{ih} = \lim_{h\to 0} \frac{(1-e^{-h})(1-zz^*)}{ih} = \frac{1}{i}(1-|z|^2),$$

which is called a *quasi-classical limit*. Thus we endow the plane with a structure of Poisson manifold [**16**, p. 422]. The complement to the unit circle is endowed with the structure of a symplectic manifold [**16**, p. 175] with the symplectic form

$$\omega = \frac{2}{1-|z|^2} d\,\mathrm{Im}\, z \wedge d\,\mathrm{Re}\, z.$$

The unit disc is a *bounded* connected component of the complement to the unit circle. Experience of quantum mechanics [**31**] indicates that one should find a faithful irreducible representation of the $*$-algebra $\mathrm{Pol}(\mathbb{C})_q$ by *bounded* operators.[3] Given $f \in \mathrm{Pol}(\mathbb{C})_q$, we define its norm as the norm of a corresponding operator of the faithful representation. To realize this plan, it is important to know that the faithful irreducible $*$-representation exists and is essentially unique.

PROPOSITION 1.10. *The unital $*$-algebra $\mathrm{Pol}(\mathbb{C})_q$ has a unique (up to unitary equivalence)*[4] *faithful irreducible representation by bounded operators in a Hilbert space.*

PROOF. Consider a pre-Hilbert space $\mathcal{H}$ with an orthonormal basis $\{e_n\}_{n\in\mathbb{Z}_+}$ and the representation T_F of the $*$-algebra $\mathrm{Pol}(\mathbb{C})_q$ in $\mathcal{H}$ defined as follows:

$$T_F(z)e_n = \sqrt{1-q^{2(n+1)}}\,e_{n+1}, \qquad T_F(z^*)e_n = \begin{cases} \sqrt{1-q^{2n}}\,e_{n-1}, & n \in \mathbb{N}, \\ 0, & n = 0. \end{cases} \tag{1.9}$$

Let

$$y = 1 - zz^* = q^{-2}(1 - z^*z).$$

Then (1.8) leads to commutation relations

$$z^*y = q^2yz^*, \qquad zy = q^{-2}yz. \tag{1.10}$$

Hence every element f of $\mathrm{Pol}(\mathbb{C})_q$ has a decomposition

$$f = \sum_{j\in\mathbb{N}} z^j f_j(y) + f_0(y) + \sum_{j\in\mathbb{N}} f_{-j}(y)z^{*j}, \tag{1.11}$$

where f_j are polynomials in one variable. The uniqueness of the decomposition (1.11) and the faithfulness of T_F are easily checked by

$$T_F(p(y))e_n = p(q^{2n})e_n, \qquad n \in \mathbb{Z}_+,$$

for each polynomial p. Indeed, given $j \geq 0$, the polynomials f_j are restored from the matrix elements $(T_F(f)e_n, e_{n+j})$ of the operator $T_F(f)$, while the polynomials f_{-j} can be restored from the matrix elements $(T_F(f)e_{n+j}, e_n)$.

Operators of the representation T_F are bounded since $\|T_F(z)\| \leq 1$, $\|T_F(z^*)\| \leq 1$. Therefore these operators admit an extension by continuity from the pre-Hilbert space $\mathcal{H}$ to its completion H. Thus we obtain a faithful representation of $\mathrm{Pol}(\mathbb{C})_q$ in a Hilbert space that will be called the Fock representation. This representation, as well as its restriction to $\mathcal{H}$, is denoted by T_F.

Now we prove the irreducibility of T_F in H. Let e_{jk} be the linear operator in H defined by the formula $e_{jk}e_m = \delta_{k,m}e_j$, where $j,k,m \in \mathbb{Z}_+$ and $\delta_{k,m}$ is the Kronecker symbol. The operators e_{jk} are called matrix units. Consider the C^*-algebra $\mathcal{B}(H)$ of all bounded linear operators in H and the closure $\mathcal{A}$ of the algebra of operators $T_F(f)$ in $\mathcal{B}(H)$. To prove the irreducibility of T_F it suffices to show that $e_{jk} \in \mathcal{A}$ for all $j,k \in \mathbb{Z}_+$.

We apply functional calculus for continuous functions [**260**, p. 247] to the bounded self-adjoint linear operator $T_F(y)$. Fix $j \in \mathbb{Z}_+$, and let $\psi_j(\lambda)$ be a continuous function on the axis that equals 1 at $\lambda = q^{2j}$ and vanishes at all other points

---

[3]Here, "faithfulness" means that nonzero elements of an algebra correspond to nonzero operators of representation, and "irreducibility" means these operators have no nontrivial common invariant subspaces.

[4]Representations π_1, π_2 of an algebra F in Hilbert spaces H_1, H_2 are called *unitary equivalent* if $U\pi_1(f) = \pi_2(f)U$ for a certain unitary operator $U : H_1 \to H_2$ and all $f \in F$.

of the geometric series $q^{2\mathbb{Z}_+}$. Then $e_{jj} = \psi_j(T_F(y))$. Therefore, the diagonal matrix units e_{jj}, $j \in \mathbb{Z}_+$, belong to the C^*-algebra $\mathcal{A}$. Now (1.9) implies that

$$T_F(z)e_{jk} = \left(1 - q^{2(j+1)}\right)^{\frac{1}{2}} e_{(j+1)k}, \qquad e_{kj}T_F(z^*) = \left(1 - q^{2(j+1)}\right)^{\frac{1}{2}} e_{k(j+1)},$$

and all other matrix units belong to $\mathcal{A}$.

It remains to show that any faithful irreducible representation T' of $\mathrm{Pol}(\mathbb{C})_q$ by bounded operators in a Hilbert space H' is unitary equivalent to the Fock representation T_F.

LEMMA 1.11. *Let T' be an irreducible representation of the $*$-algebra $\mathrm{Pol}(\mathbb{C})_q$ by bounded operators in a Hilbert space H' and $T'(y) \neq 0$. Then there exists a real $a \neq 0$ such that*

$$\mathrm{spec}(T'(y)) \subset \{0\} \cup aq^{2\mathbb{Z}}.$$

The proof is given in the next subsection.

This lemma implies that nonzero points of the spectrum of $T'(y)$ belong to the geometric series $aq^{2\mathbb{Z}}$, $a \neq 0$, so they are its eigenvalues. Let λ_0 be the maximum modulus eigenvalue and v_0 the corresponding normed eigenvector: $T'(y)v_0 = \lambda_0 v_0$, $\|v_0\| = 1$. Then (1.10) implies that $T'(y)T'(z^*)v_0 = q^{-2}\lambda_0 T'(z^*)v_0$. But $q^{-2}\lambda_0$ is not an eigenvalue of $T'(y)$. Hence $T'(z^*)v_0 = 0$. Therefore

$$T'(y)v_0 = T'(1 - zz^*)v_0 = T'(1)v_0 = v_0,$$

and $\lambda_0 = 1$.

Similarly, one can use (1.10) to show that the relation $T'(y)v_n = q^{2n}v_n$ holds for the vectors $v_n = T'(z^n)v_0$, and

$$\|v_{n+1}\|^2 = \|T'(z)v_n\|^2 = (T'(z^*z)v_n, v_n) = (T'(1-q^2y)v_n, v_n) = (1-q^{2(n+1)})\|v_n\|^2.$$

Hence $v_n \neq 0$ are eigenvectors of the bounded self-adjoint operator $T'(y)$ corresponding to pairwise different eigenvalues. Therefore they are pairwise orthogonal. Using

$$T'(z)v_n = v_{n+1}, \qquad T'(z^*)v_{n+1} = (1 - q^{2(n+1)})v_n, \qquad T'(z^*)v_0 = 0,$$

one can show that the map

$$H_F \to H', \qquad e_n \mapsto \frac{e'_n}{\|e'_n\|}, \qquad n \in \mathbb{Z}_+,$$

is isometric and intertwines the representations (i.e., is a morphism). The map is surjective, so it accomplishes a unitary equivalence between T_F and T' since T' is irreducible. □

REMARK 1.12. The Fock representation was introduced by (1.9) although we have not discussed its origin. In fact, these formulas can be obtained by an argument similar to that in the last part of the proof of Proposition 1.10.

Here we consider one-dimensional representations of $\mathrm{Pol}(\mathbb{C})_q$ parameterized by points of the unit circle:

$$\rho_\varphi(z) = e^{i\varphi}, \quad \rho_\varphi(z^*) = e^{-i\varphi}, \qquad \varphi \in \mathbb{R}/(2\pi\mathbb{Z}). \tag{1.12}$$

These representations are pairwise nonequivalent.

COROLLARY 1.13. *Let T be an irreducible representation of the $*$-algebra $\mathrm{Pol}(\mathbb{C})_q$ by bounded operators in a Hilbert space. Then T is unitary equivalent either to a representation* (1.12) *if $T(y) = 0$ or to the Fock representation T_F if $T(y) \neq 0$.*

Therefore we obtain a complete list of irreducible representations of the $*$-algebra $\mathrm{Pol}(\mathbb{C})_q$ up to unitary equivalence.

1.2.4. Proof of Lemma 1.11. Let us suppose that the claim of the lemma is incorrect. Since $T'(y) \neq 0$, we have that the spectrum of $T'(y)$ intersects with the following two geometric series:

$$\mathrm{spec}(T'(y)) \cap a_1 q^{2\mathbb{Z}} \neq \varnothing, \qquad \mathrm{spec}(T'(y)) \cap a_2 q^{2\mathbb{Z}} \neq \varnothing,$$

where $a_1 q^{2\mathbb{Z}} \neq a_2 q^{2\mathbb{Z}}$.

Let us consider continuous functions $\Psi_1(\lambda)$, $\Psi_2(\lambda)$ that satisfy the following conditions:

- they are invariant under the homothety with coefficient q^2:
$$\Psi_1(q^2\lambda) = \Psi_1(\lambda), \qquad \Psi_2(q^2\lambda) = \Psi_2(\lambda);$$
- they take nonzero values on "their own" geometric series
$$\Psi_1(a_1 q^{2j}) \neq 0, \quad \Psi_2(a_2 q^{2j}) \neq 0, \qquad j \in \mathbb{Z};$$
- their product is 0.

Here it is implicit that $a_1, a_2 \neq 0$. Now the relation

$$T'(z^*)T'(y) = T'(z^*)T'(1 - zz^*) = T'(1 - z^*z)T'(z^*) = q^2 T'(y)T'(z^*)$$

implies that

$$T'(z^*)\, p(T'(y)) = p(q^2 T'(y))\, T'(z^*)$$

for each polynomial p in one variable. Hence,

$$T'(z^*)\Psi_j(T'(y)) = \Psi_j(q^2 T'(y))T'(z^*) = \Psi_j(T'(y))T'(z^*), \qquad j = 1, 2,$$

for both continuous functions Ψ_j on the real axis. The linear operators $A_1 = \Psi_1(T'(y))$, $A_2 = \Psi_2(T'(y))$ commute with all operators of the representation T'. Besides, the spectral mapping theorem implies that $\Psi_1(a_1) \in \mathrm{spec}(A_1)$, $\Psi_2(a_2) \in \mathrm{spec}(A_2)$, and $\Psi_1(a_1) \neq 0$, $\Psi_2(a_2) \neq 0$, so the operators are nonzero. Hence the closures of the images of A_1, A_2 are nonzero common invariant subspaces of the operators of the representation T'. But T' is irreducible. Therefore the images of A_1, A_2 are dense in H', which contradicts the fact that $A_1 A_2 = \Psi_1(T'(y))\Psi_2(T'(y)) = \Psi_1\Psi_2\,(T'(y)) = 0$. □

1.2.5. Continuous functions in the quantum disc. Since the Fock representation T_F is faithful, the functional

$$\|f\| \stackrel{\mathrm{def}}{=} \|T_F(f)\|, \qquad f \in \mathrm{Pol}(\mathbb{C})_q,$$

endows $\mathrm{Pol}(\mathbb{C})_q$ with a structure of a normed $*$-algebra. It is shown in Subsection 1.2.3 that its completion

$$\mathrm{Pol}(\mathbb{C})_q \hookrightarrow C(\overline{\mathbb{D}})_q$$

is a q-analog of the embedding $\mathrm{Pol}(\mathbb{C}) \hookrightarrow C(\overline{\mathbb{D}})$. Following [**236, 173**], we consider the C^*-algebra $C(\overline{\mathbb{D}})_q$ as the algebra of continuous functions in the quantum disc.

Now we would like to compare the classical and the quantum disc.

First of all, let us compare the spectrum of $1 - |z|^2 \in C(\overline{\mathbb{D}})$ to the spectrum of $1 - zz^* \in C(\overline{\mathbb{D}})_q$. The first spectrum is the set of values of the continuous function $1 - |z|^2$ in the closed unit disc, i.e., it is just $[0, 1]$. The second spectrum is the closure of the geometric series $q^{2\mathbb{Z}_+}$; this follows from the definition of the Fock representation and the fact that the embedding $C(\overline{\mathbb{D}})_q \hookrightarrow \mathcal{B}(H)$ is isometric. As $q \in (0, 1)$ grows, the geometric series becomes denser and fills the segment $[0, 1]$ under the limit as $q \to 1$, that provides a correspondence between classical and quantum problems on the spectrum of *physical quantity* $y = 1 - zz^*$.

The spectra of z and z^* in either classical or quantum case coincide with $\overline{\mathbb{D}}$. Indeed, $\|T_F(z)\| \leq 1$, and all $\lambda \in \mathbb{D}$ are eigenvalues of the linear operator $T_F(z^*)$.

Now we turn to a noncommutative analog of the fact that the exact sequence

$$0 \to C_0(\mathbb{D}) \to C(\overline{\mathbb{D}}) \xrightarrow{j} C(\mathbb{T}) \to 0 \tag{1.13}$$

does not split; i.e., the circle is not a retract of the disc, which was discussed in Subsection 1.2.1.

Use $y \in C(\overline{\mathbb{D}})_q$ to generate a two-sided ideal and denote by $C_0(\mathbb{D})_q$ its closure.

PROPOSITION 1.14. *An extension by continuity of the Fock representation to $C(\overline{\mathbb{D}})_q$ delivers an isomorphism*

$$C_0(\mathbb{D})_q \cong \mathcal{K}(H).$$

PROOF. It was mentioned in the proof of the irreducibility of the Fock representation that the image of the C^*-algebra $C_0(\mathbb{D})_q$ in $\mathcal{B}(H)$ contains all *matrix units* e_{jk} and, in particular, it contains the orthogonal projection onto the vacuum subspace $\mathbb{C}e_0$. The irreducibility of T_F implies that this algebra contains all linear operators of rank 1. Hence the image of the C^*-algebra $C_0(\mathbb{D})_q$ contains all operators with finite rank and therefore all compact operators in H. On the other hand, $T_F(y)$ is compact. □

We keep the notation T_F for the extension by continuity of the Fock representation to $C(\overline{\mathbb{D}})_q$.

PROPOSITION 1.15. $C(\overline{\mathbb{D}})_q / C_0(\mathbb{D})_q \cong C(\mathbb{T})$.

PROOF. The commutative C^*-algebra $C(\overline{\mathbb{D}})_q / C_0(\mathbb{D})_q$ is naturally isomorphic to $C(\sigma_e(T_F(z)))$. But $\sigma_e(T_F(z)) = \mathbb{T}$. Indeed, $\sigma_e(T_F(z)) \subset \mathbb{T}$ since the element of the Calkin algebra which corresponds to $T_F(z)$ is unitary. It remains to use the fact that $\sigma_e(T_F(z))$ is not empty and is invariant under rotations in $\mathbb{C}$. The invariance follows by observing that the linear operators $e^{i\varphi}T_F(z)$, $\varphi \in \mathbb{R}$, are pairwise unitary equivalent. □

The C^*-subalgebra of linear operators in H generated by the ideal $\mathcal{K}(H)$ of all compact operators and by the isometric operator S of one-sided shift is called the *Toeplitz algebra* $\mathcal{T}$ (see Subsection 1.2.2). Proposition 1.14 delivers the isomorphism of C^*-algebras $T_F : C(\overline{\mathbb{D}})_q \xrightarrow{\approx} \mathcal{T}$ since $T_F(z) - S \in \mathcal{K}$. It follows from Proposition 1.15 that $\mathcal{T}/\mathcal{K}(H) \cong C(\mathbb{T})$. The latter is a well-known fact; see [**334**, p. 68].

We get the exact sequence of C^*-algebras

$$0 \to C_0(\mathbb{D})_q \to C(\overline{\mathbb{D}})_q \xrightarrow{j_q} C(\mathbb{T}) \to 0, \tag{1.14}$$

that is a q-analog of the exact sequence (1.13) and is isomorphic to the exact sequence

$$0 \to \mathcal{K}(H) \to \mathcal{T} \xrightarrow{j} C(\mathbb{T}) \to 0. \tag{1.15}$$

It is not difficult to show [**334**, p. 68] that the exact sequence (1.15) does not split. Indeed, let $i : C(\mathbb{T}) \to \mathcal{T}$ be a splitting morphism of C^*-algebras: $ji = \mathrm{id}$. The image of $C(\mathbb{T})$ in $\mathcal{T}$ consists of normal operators. Hence each operator from $\mathcal{T}$ admits a decomposition into a sum of normal and compact operators. Therefore, the index of each operator from $\mathcal{T}$ equals 0. Now we have a contradiction since the index of S is not 0.

Therefore the exact sequence (1.14) does not split also. This is indeed a q-analog of the fact that the circle is not a retract of the disc.

REMARK 1.16. The homomorphism j_q is a q-analog of the restriction of continuous functions in $\overline{\mathbb{D}}$ to its boundary $\mathbb{T}$. It is clear from the proof of Proposition 1.15 that j_q maps $z \in C(\overline{\mathbb{D}})_q$ to the function $e^{i\varphi}$ on $\mathbb{T} \cong \mathbb{R}/(2\pi\mathbb{Z})$. From $\|j_q\| = 1$ we deduce that

$$|\rho_\varphi(f)| \le \|T_F(f)\|, \qquad f \in \mathrm{Pol}(\mathbb{C})_q.$$

The set of classes of unitary equivalent irreducible representations of the $*$-algebra $\mathrm{Pol}(\mathbb{C})_q$ is exhausted by the representations that appear in this inequality. Therefore $C(\overline{\mathbb{D}})_q$ is a universal enveloping C^*-algebra of the $*$-algebra $\mathrm{Pol}(\mathbb{C})_q$ [**45**, p. 81].

1.2.6. Holomorphic functions in the quantum disc. Let us start with polynomials. Denote by $\mathbb{C}[z]_q$ and $\mathbb{C}[\overline{z}]_q$ the unital subalgebras in $\mathrm{Pol}(\mathbb{C})_q$ generated by z and z^*, respectively. Now $\{z^j\}_{j\in\mathbb{Z}_+}$ constitutes a basis of the vector space $\mathbb{C}[z]_q$ and $\{z^{*j}\}_{j\in\mathbb{Z}_+}$ is a basis of the vector space $\mathbb{C}[\overline{z}]_q$. Indeed, the linear independence of z^j follows from the linear independence of $T_F(z^j)e_0$; also the first claim implies the linear independence of z^{*j}.

Below we assume that the reader is familiar with general notions of polylinear algebra; see, for example, [**202**].

PROPOSITION 1.17. *The linear map*

$$m : \mathbb{C}[z]_q \otimes \mathbb{C}[\overline{z}]_q \to \mathrm{Pol}(\mathbb{C})_q, \qquad m : f_1 \otimes f_2 \mapsto f_1 f_2,$$

is bijective.

PROOF. The required statement means that $\{z^j z^{*k}\}_{j,k\in\mathbb{Z}_+}$ is a basis of the vector space $\mathrm{Pol}(\mathbb{C})_q$. The completeness of this set of vectors follows from the definition of $\mathrm{Pol}(\mathbb{C})_q$, so it remains to check the linear independence.

Let us consider the Fock representation T_F of the $*$-algebra $\mathrm{Pol}(\mathbb{C})_q$ in $\mathcal{H}$. We suppose that the vectors $z^j z^{*k}$ are linearly dependent. Hence for some $j_0, k_0 \in \mathbb{Z}_+$ we get

$$\sum_{j+k\ge j_0+k_0} c_{j,k} T_F(z^j z^{*k}) = 0, \qquad c_{j_0 k_0} \ne 0,$$

where only a finite number of complex coefficients $c_{j,k}$ is nonzero. Then

$$(T_F(z^j z^{*k})e_{k_0}, e_{j_0}) = (T_F(z^{*k})e_{k_0}, T_F(z^{*j})e_{j_0}) = \begin{cases} 0, & k > k_0 \text{ or } j > j_0, \\ \ne 0, & k = k_0 \;\&\; j = j_0, \end{cases}$$

and it leads to $c_{j_0,k_0} = 0$. □

Let $A(\overline{\mathbb{D}})$ be the Banach algebra of all functions that are continuous in $\overline{\mathbb{D}}$ and holomorphic in $\mathbb{D}$. We assume $A(\overline{\mathbb{D}})$ to be isometrically embedded into $C(\overline{\mathbb{D}})$. Now we consider the closure $A(\overline{\mathbb{D}})_q$ of $\mathbb{C}[z]_q$ in $C(\overline{\mathbb{D}})_q$. The pair $(A(\overline{\mathbb{D}})_q, C(\overline{\mathbb{D}})_q)$ is a q-analog of the pair $(A(\overline{\mathbb{D}}), C(\overline{\mathbb{D}}))$.

It is known from standard complex analysis courses that the modulus of a function that is continuous in $\overline{\mathbb{D}}$ and analytic in $\mathbb{D}$ attains its maximum on the boundary $\mathbb{T}$ of the disc $\mathbb{D}$. It means that the restriction of the homomorphism

$$j : C(\overline{\mathbb{D}}) \to C(\mathbb{T}), \qquad j : f \mapsto f|_{\mathbb{T}},$$

to $A(\overline{\mathbb{D}})$ is an isometric linear operator. We intend to prove a similar statement for the quantum disc. For that, the homomorphism $j_q : C(\overline{\mathbb{D}})_q \to C(\mathbb{T})$ introduced in Subsection 1.2.5 will be used.

PROPOSITION 1.18. *The restriction of the homomorphism j_q to the subalgebra $A(\overline{\mathbb{D}})_q$ is an isometric linear operator.*

PROOF. Since $\|j_q\| \le 1$, the restriction of j_q to $A(\overline{\mathbb{D}})_q$ is a contraction. Now it remains to prove that $\|f\| \le \|j_q(f)\|$ for each $f \in \mathbb{C}[z]_q$. For that it is sufficient to use the von Neumann inequality (Proposition 1.23) and the fact that $\|T_F(z)\| \le 1$:

$$\|f\| = \|T_F(f)\| \le \max_{|z|=1} |f(z)| = \max_{\varphi \in \mathbb{R}/(2\pi\mathbb{Z})} \|\rho_\varphi(f)\| = \|j_q(f)\|. \qquad \square$$

REMARK 1.19. It follows from the proof of Proposition 1.23 that the result mentioned above admits a refinement: the restriction of j_q to $A(\overline{\mathbb{D}})_q$ is a *completely* isometric linear operator (the definition can be found, for example, in [**17**, p. 148]).

1.2.7. Appendix on spectral sets. Recall that the spectrum of a bounded linear operator A is the set of $\lambda \in \mathbb{C}$ such that the linear operator $\lambda I - A$ has no bounded inverse. The spectrum is a nonempty compact set [**260**, pp. 213, 214].

Let $\mathfrak{M}$ be a nonempty compact subset of $\mathbb{C}$ with the connected complement. It is said that $\mathfrak{M}$ is a spectral set of a bounded linear operator A in a Hilbert space $\mathcal{H}$ if for each polynomial $f(\lambda) \in \mathbb{C}[\lambda]$ one has

$$\|f(A)\| \le \max_{\lambda \in \mathfrak{M}} |f(\lambda)|. \tag{1.16}$$

Note that the spectrum of a self-adjoint linear operator belongs to the real axis [**260**, p. 217], hence its complement in $\mathbb{C}$ is connected.

PROPOSITION 1.20. *The spectrum $\mathfrak{M}$ of a bounded self-adjoint operator in a Hilbert space is a spectral set of this operator.*

This statement is proved, for example, in [**260**, p. 248], where the authors mention that it ultimately leads to all statements of "the spectral theorem for continuous functions" [**260**, p. 247].

For instance, continuous functions of a bounded self-adjoint operator A in a Hilbert space $\mathcal{H}$ are introduced as follows. Let $\mathfrak{M} \subset \mathbb{R}$ be the spectrum of the bounded self-adjoint operator A. Consider the $*$-algebra $\mathbb{C}[\mathfrak{M}]$ of polynomial functions on $\mathfrak{M}$, i.e., the quotient algebra $\mathbb{C}[\lambda]/J_{\mathfrak{M}}$, where $J_{\mathfrak{M}}$ is the ideal of polynomials that vanish on $\mathfrak{M}$. Proposition 1.20 implies that $f(A) = 0$ for all $f \in J_{\mathfrak{M}}$. Hence the homomorphism $f(\lambda) \mapsto f(A)$ of the $*$-algebra $\mathbb{C}[\mathfrak{M}]$ to the $*$-algebra of bounded linear operators in $\mathcal{H}$ is well defined. Due to (1.16), this homomorphism admits an extension by continuity to the closure of the algebra of polynomial functions in $C(\mathfrak{M})$. The Weierstrass theorem on polynomial approximation [**260**, p. 120] claims that the closure coincides with $C(\mathfrak{M})$. So we obtain a functional calculus for continuous functions, i.e., a homomorphism of the commutative C^*-algebra $C(\mathfrak{M})$ to the C^*-algebra of bounded linear operators such that $f(\lambda)|_{\mathfrak{M}}$ is sent to $f(A)$ for all polynomials $f(\lambda)$.

Here is an important consequence of the spectral theorem for continuous functions: if a continuous function $f(\lambda)$ is well defined on the spectrum of a bounded self-adjoint operator A and does not vanish on at least at one point of the spectrum, then $f(A) \neq 0$.

A linear operator T in a Hilbert space H is called a contraction if $\|T\| \leq 1$. Let $E \supset H$ be a Hilbert space, P the orthogonal projection in E onto the subspace H, and U a unitary operator in E. The latter operator is called a unitary dilation of the contraction T if $T^n = PU^n|_H$ for all $n \in \mathbb{Z}_+$ [**302**]. This implies $T^{*n} = PU^{-n}|_H$.

Similarly, one can introduce the notion of isometric dilation of contraction.

Let us give a well-known B. Szőkefalvi–Nagy result.

PROPOSITION 1.21. *Every contraction in a Hilbert space possesses a unitary dilation.*

A short proof may be found in [**113**, p. 124]. We restrict ourselves to a construction of unitary dilation in the most important case of a contraction T such that

$$\lim_{n\to\infty} \|T^{*\,n} v\| = 0 \tag{1.17}$$

for all $v \in H$. (For instance, the contraction $T_F(z)$ possesses this property.)

Let F be the closure of the image of linear operator $I - TT^*$ and $(I - TT^*)^{\frac{1}{2}}$ the square root of the bounded nonnegative linear operator $I - TT^*$. Consider the Hilbert space $l^2(\mathbb{Z}_+, F)$ of sequences of vectors $\mathbf{v} = (v_0, v_1, v_2, \ldots)$ from F such that $\|\mathbf{v}\|^2 = \sum_{j\in\mathbb{Z}_+} |v_j|^2 < \infty$. Using (1.17), one can show that the linear map

$$v \mapsto ((I - TT^*)^{\frac{1}{2}} v,\ (I - TT^*)^{\frac{1}{2}} T^* v,\ (I - TT^*)^{\frac{1}{2}} T^{*2} v,\ (I - TT^*)^{\frac{1}{2}} T^{*3} v, \ldots)$$

embeds isometrically H to $l^2(\mathbb{Z}_+, F)$. The shift operator

$$(v_0, v_1, v_2, v_3, \ldots) \mapsto (0, v_0, v_1, v_2, v_3, \ldots)$$

in $l^2(\mathbb{Z}_+, F)$ is an isometric dilation of an operator that is unitary equivalent to T. It remains to note that the two-sided shift in $l^2(\mathbb{Z}, F)$ is a unitary dilation of the isometric shift in $l^2(\mathbb{Z}_+, F)$.

REMARK 1.22. The unitary operator U obtained this way is a multiple of the two-sided shift, hence it has an absolutely continuous spectrum. (This is true for each totally nonunitary contraction [**302**].)

PROPOSITION 1.23 (von Neumann inequality). *The closed unit disc $\overline{\mathbb{D}}$ is a spectral set of all contractions in a Hilbert space.*

PROOF. For the special case of a unitary operator, one can prove the required statement similarly to Proposition 1.20. The general case reduces to the special one due to Proposition 1.21. □

1.3. Symmetry

1.3.1. The Hopf algebras $U\mathfrak{sl}_2$ and $U_q\mathfrak{sl}_2$. Recall that a Lie algebra is a generically nonassociative algebra $\mathfrak{g}$ with an anticommutative multiplication $[x,y] = -[y,x]$ satisfying the Jacobi identity $[x,[y,z]] + [z,[x,y]] + [y,[z,x]] = 0$. Every associative algebra A is a Lie algebra with the operation $[x,y] = xy - yx$.

Consider a Lie algebra $\mathfrak{g}$ together with a linear map i to an associative unital algebra $U\mathfrak{g}$ which is a homomorphism of Lie algebras. A pair $(U\mathfrak{g}, i)$ is called a universal enveloping algebra if it possesses the following universality property. Given an associative unital algebra A and a homomorphism of Lie algebras $j : \mathfrak{g} \to A$, there exists a unique homomorphism of associative unital algebras $U\mathfrak{g} \to A$ such that the following diagram is commutative.

$$\begin{array}{ccc} \mathfrak{g} & \xrightarrow{\ i\ } & U\mathfrak{g} \\ & \searrow^{j} & \downarrow \\ & & A \end{array}$$

This definition of a universal enveloping algebra is given, for example, in [**32**, p. 186]. A universal enveloping algebra is essentially unique.

To prove its existence, we choose a basis $\{e_\lambda\}_{\lambda\in\Lambda}$ of the vector space $\mathfrak{g}$. Then $[e_\lambda, e_\mu] = \sum_{\nu\in\Lambda} c^\nu_{\lambda\mu} e_\nu$, where $c^\nu_{\lambda\mu}$ are structure constants of $\mathfrak{g}$. Now it remains to introduce an associative algebra $U\mathfrak{g}$ with the generators $\{e_\lambda\}_{\lambda\in\Lambda}$ and the defining relations $e_\lambda \cdot e_\mu - e_\mu \cdot e_\lambda = \sum_{\nu\in\Lambda} c^\nu_{\lambda\mu} e_\nu$ together with the homomorphism

$$i : \mathfrak{g} \to U\mathfrak{g}, \qquad i : e_\lambda \mapsto e_\lambda.$$

It is easy to show [**32**, p. 186] that i is an embedding, and the pair $(U\mathfrak{g}, i)$ is a universal enveloping algebra.

EXAMPLE 1.24. Consider the complex Lie algebra $\mathfrak{sl}_2$ with generators H, E, F and the defining relations

$$[H,E] = 2E, \qquad [H,F] = -2F, \qquad [E,F] = H.$$

The algebra is isomorphic to the Lie algebra of complex 2×2-matrices with zero trace:

$$H \mapsto \begin{pmatrix} 1 & 0 \\ 0 & -1 \end{pmatrix}, \qquad E \mapsto \begin{pmatrix} 0 & 1 \\ 0 & 0 \end{pmatrix}, \qquad F \mapsto \begin{pmatrix} 0 & 0 \\ 1 & 0 \end{pmatrix}.$$

Let $U\mathfrak{sl}_2$ be the unital associative algebra with the generators H, E, F and the defining relations

$$HE - EH = 2E, \qquad HF - FH = -2F, \qquad EF - FE = H.$$

The homomorphism of Lie algebras

$$i : \mathfrak{sl}_2 \to U\mathfrak{sl}_2, \qquad i : H \mapsto H,\ i : E \mapsto E,\ i : F \mapsto F,$$

possesses the universality property, and the pair $(U\mathfrak{sl}_2, i)$ is a universal enveloping algebra.

In this book we identify a Lie algebra $\mathfrak{g}$ with its image under the embedding in $U\mathfrak{g}$.

There is another, nearly equivalent definition. A universal enveloping algebra is not an arbitrary pair $(U\mathfrak{g}, i)$ with the universality property, but a special distinguished pair whose construction is given in [**168**, p. 167], [**74**, p. 85], [**106**, p. 41], [**46**, p. 20].

Quantum group theory uses the third definition of a universal enveloping algebra $U\mathfrak{g}$ where it appears to be a Hopf algebra. Let us discuss the latter notion in detail.

We start in a roundabout way with motivation. Let G be a complex affine algebraic group and $\mathfrak{g}$ its Lie algebra [**130**]. For example, one can consider the group

$$SL_2 = \left\{ \begin{pmatrix} t_{11} & t_{12} \\ t_{21} & t_{22} \end{pmatrix} \middle| \; t_{11}t_{22} - t_{12}t_{21} = 1 \right\},$$

together with its Lie algebra $\mathfrak{sl}_2$.

The abelian category of rational representations of G possesses, apart from an operation $\oplus$, an operation $\otimes$:

$$(T_1 \otimes T_2)(g) = T_1(g) \otimes T_2(g), \qquad g \in G.$$

In addition, there exists a trivial one-dimensional representation, and one can assign to each representation T the dual representation T^* in the dual vector space

$$T^*(g) = T(g^{-1})^*, \qquad g \in G.$$

It is known [**168**] that the category of $U\mathfrak{g}$-modules inherits these operations. The notion of a Hopf algebra is summoned to reveal algebraic structures in $U\mathfrak{g}$ that are responsible for that.

Now we would like to pass from motivation to definitions. A vector space A is called a coalgebra if there is a linear map $\triangle : A \to A \otimes A$ with the coassociativity property

$$(\triangle \otimes \mathrm{id})\triangle(a) = (\mathrm{id} \otimes \triangle)\triangle(a), \quad a \in A,$$

and a linear map $\varepsilon : A \to \mathbb{C}$ such that

$$(\varepsilon \otimes \mathrm{id})\triangle(a) = (\mathrm{id} \otimes \varepsilon)\triangle(a) = a, \quad a \in A.$$

These requirements are equivalent to the commutativity of the following diagrams.

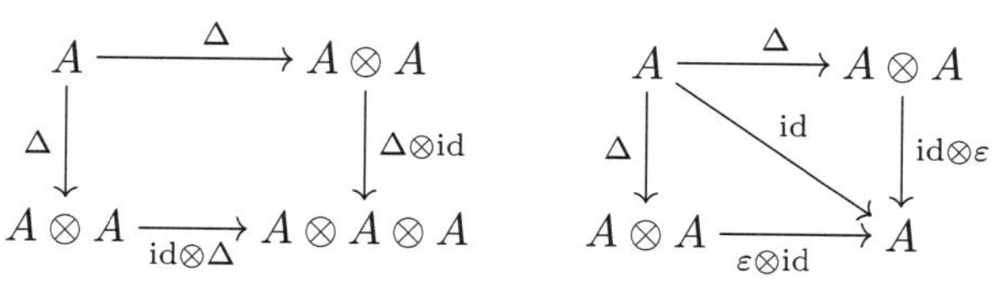

The map $\triangle$ is called a coproduct, while the map ε is called a counit.

REMARK 1.25. Just a few words on etymology. Reversing the arrows, we obtain diagrams that express the associativity of product

$$m : A \otimes A \to A, \qquad m : a_1 \otimes a_2 \mapsto a_1 a_2,$$

and the existence of unit

$$\mathbb{1} : \mathbb{C} \to A, \qquad \mathbb{1} : z \mapsto z \cdot 1,$$

in the algebra A:

$$
\begin{array}{ccc}
A & \xleftarrow{\ m\ } & A\otimes A \\
{\scriptstyle m}\uparrow & & \uparrow{\scriptstyle m\otimes\mathrm{id}} \\
A\otimes A & \xleftarrow[\mathrm{id}\otimes m]{} & A\otimes A\otimes A
\end{array}
\qquad\qquad
\begin{array}{ccc}
A & \xleftarrow{\ m\ } & A\otimes A \\
{\scriptstyle m}\uparrow & \nwarrow^{\mathrm{id}} & \uparrow{\scriptstyle \mathbb{1}\otimes\mathrm{id}} \\
A\otimes A & \xleftarrow[\mathrm{id}\otimes\mathbb{1}]{} & A
\end{array}
$$

A vector space A is called a bialgebra if it is endowed with both structures of unital algebra and coalgebra with counit, where the maps $\triangle$ and ε are homomorphisms of algebras. Here the standard algebra structure in $A\otimes A$ is implicit:

$$(a'\otimes b')\cdot(a''\otimes b'') = a'a''\otimes b'b'', \qquad a', a'', b', b'' \in A.$$

Let us mention that $\triangle(1) = 1\otimes 1$, $\varepsilon(1) = 1$. All the imposed requirements are equivalent to the commutativity of the following diagrams.

$$
\begin{array}{cccc}
 & A\otimes A & \xrightarrow{\triangle\otimes\triangle} & A\otimes A\otimes A\otimes A \\
{\scriptstyle\triangle}\nearrow & & & \\
A & & & \downarrow{\scriptstyle \mathrm{id}\otimes\sigma\otimes\mathrm{id}} \\
{\scriptstyle m}\nwarrow & & & \\
 & A\otimes A & \xleftarrow[m\otimes m]{} & A\otimes A\otimes A\otimes A
\end{array}
$$

$$
\begin{array}{ccc}
A & \xrightarrow{\ \varepsilon\ } & \mathbb{C} \\
{\scriptstyle m}\uparrow & & \uparrow{\scriptstyle m} \\
A\otimes A & \xrightarrow[\varepsilon\otimes\varepsilon]{} & \mathbb{C}\otimes\mathbb{C}
\end{array}
\qquad
\begin{array}{ccc}
A & \xleftarrow{\ \mathbb{1}\ } & \mathbb{C} \\
{\scriptstyle \triangle}\downarrow & & \downarrow{\scriptstyle \triangle} \\
A\otimes A & \xleftarrow[\mathbb{1}\otimes\mathbb{1}]{} & \mathbb{C}\otimes\mathbb{C}
\end{array}
\qquad
A \underset{\mathbb{1}}{\overset{\varepsilon}{\rightleftarrows}} \mathbb{C}
$$

Here σ is the operator of naive permutation of the tensor factors

$$\sigma : A\otimes A \to A\otimes A, \qquad \sigma : a_1\otimes a_2 \mapsto a_2\otimes a_1.$$

A coproduct Δ allows one to define an operation of a tensor product of representations of A ($\equiv$ A-modules) that possesses the usual properties of a tensor product of vector spaces. Namely, let π_1, π_2 be representations of a bialgebra A in vector spaces V_1, V_2, respectively. Then their tensor product $\pi_1\otimes\pi_2$ is the following representation of A in $V_1\otimes V_2$:

$$A \overset{\triangle}{\to} A\otimes A \xrightarrow{\pi_1\boxtimes\pi_2} \mathrm{End}(V_1)\otimes\mathrm{End}(V_2) \hookrightarrow \mathrm{End}(V_1\otimes V_2).$$

The trivial representation of a bialgebra A is the one-dimensional representation defined by the counit $\varepsilon : A \to \mathbb{C} \cong \mathrm{End}(\mathbb{C})$. The corresponding A-module $\mathbb{C}$ is called the trivial A-module.

Given an A-module V, a vector $v \in V$ is called A-invariant if the linear map $\mathbb{C} \to A$ with $1 \mapsto v$ is a morphism of A-modules. Equivalently, the A-invariance of v means that $av = \varepsilon(a)v$ for all $a \in A$.

The properties of coproduct and counit imply that the canonical isomorphisms of vector spaces

$$V_1\otimes(V_2\otimes V_3) \cong (V_1\otimes V_2)\otimes V_3, \qquad V\otimes\mathbb{C} \cong \mathbb{C}\otimes V \cong V,$$

are isomorphisms of A-modules. Once $f_1 : V_1 \to W_1$, $f_1 : V_1 \to W_1$ are morphisms of A-modules, $f_1 \otimes f_2 : V_1 \otimes W_1 \to V_2 \otimes W_2$ is also a morphism of A-modules, i.e., A-modules form a tensor category; see Subsection 1.3.8.

Now we turn from bialgebras to Hopf algebras. Let A be a bialgebra with multiplication $m : A \otimes A \to A$, coproduct $\triangle : A \to A \otimes A$, and counit $\varepsilon : A \to \mathbb{C}$. A linear map $S : A \to A$ is called an antipode if $m(S \otimes \mathrm{id})\triangle(a) = m(\mathrm{id} \otimes S)\triangle(a) = \varepsilon(a)1$ for all $a \in A$, that is equivalent to the commutativity of the following diagram.

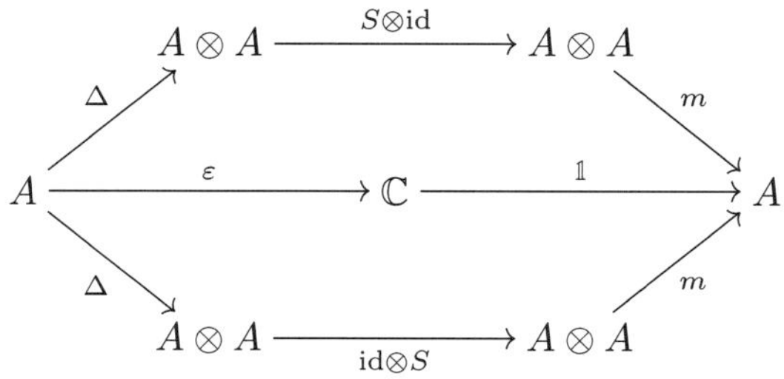

One can check by definitions [**148**, p. 13] that the antipode is unique, it is an antihomomorphism of algebra and an antihomomorphism of coalgebra, i.e., the following diagrams are commutative.

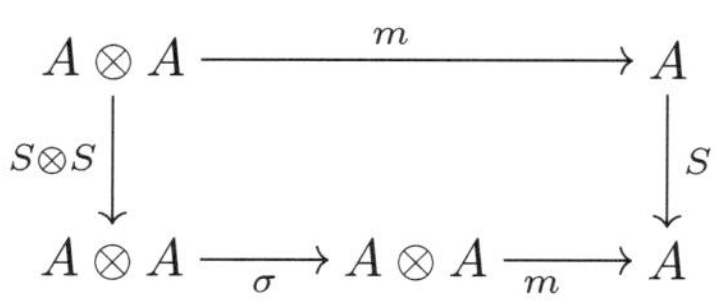

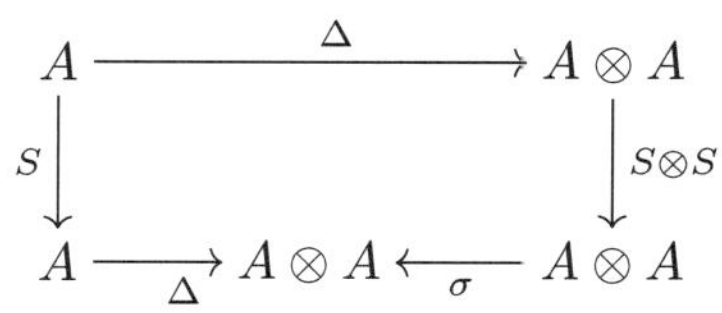

Everywhere further the antipode S is supposed to be invertible.

A bialgebra A with antipode is called a Hopf algebra.

Consider a Hopf algebra A and an A-module V. The dual vector space V^* is endowed with the structure of an A-module

$$(al)(v) = l(S(a)v), \qquad a \in A,\ v \in V,\ l \in V^*.$$

The canonical pairing of vector spaces

$$V^* \otimes V \to \mathbb{C}, \qquad l \otimes v \mapsto l(v),$$

is a morphism of A-modules.

A universal enveloping algebra is a Hopf algebra. Indeed, the homomorphisms of Lie algebras

$$\mathfrak{g} \to U\mathfrak{g} \otimes U\mathfrak{g}, \quad x \mapsto i(x) \otimes 1 + 1 \otimes i(x), \quad \text{and} \quad \mathfrak{g} \to \mathbb{C}, \quad x \mapsto 0,$$

are uniquely extended to homomorphisms of algebras

$$\triangle : U\mathfrak{g} \to U\mathfrak{g} \otimes U\mathfrak{g}, \qquad \varepsilon : U\mathfrak{g} \to \mathbb{C},$$

and the antiautomorphism of $\mathfrak{g}$ which takes x to $-x$ is uniquely extendable to an antiautomorphism S of the algebra $U\mathfrak{g}$.

EXAMPLE 1.26. The coproduct Δ, the counit ε, and the antipode S of the Hopf algebra $U\mathfrak{sl}_2$ are determined as follows:

$$\Delta(H) = H \otimes 1 + 1 \otimes H, \quad \Delta(E) = E \otimes 1 + 1 \otimes E, \quad \Delta(F) = F \otimes 1 + 1 \otimes F,$$
$$\varepsilon(H) = \varepsilon(E) = \varepsilon(F) = 0,$$
$$S(H) = -H, \qquad S(E) = -E, \qquad S(F) = -F.$$

The notions of a tensor product of $U\mathfrak{g}$-modules, the trivial $U\mathfrak{g}$-module and a dual $U\mathfrak{g}$-module agree with the corresponding notions of the Lie group theory [**168**].

In quantum group theory the Hopf algebra $U\mathfrak{g}$ described above is called a *universal enveloping algebra.*

The Hopf algebra $U\mathfrak{sl}_2$ is cocommutative, i.e., $\Delta = \Delta^{\mathrm{cop}}$, where $\Delta^{\mathrm{cop}} = \sigma\Delta$. Besides, $S^2 = \mathrm{id}$. Now we would like to give an example of noncocommutative Hopf algebra with an antipode of infinite order. Recall that $q \in (0,1)$.

EXAMPLE 1.27. ([**139**, Chap. 3], [**161**, Chap. 7]). Consider the algebra $U_q\mathfrak{sl}_2$ with generators K, K^{-1}, E, F and the following defining relations:

$$KK^{-1} = K^{-1}K = 1, \quad K^{\pm1}E = q^{\pm2}EK^{\pm1}, \quad K^{\pm1}F = q^{\mp2}FK^{\pm1}, \tag{1.18}$$

$$EF - FE = \frac{K - K^{-1}}{q - q^{-1}}. \tag{1.19}$$

One can prove the existence of homomorphisms

$$\Delta : U_q\mathfrak{sl}_2 \to U_q\mathfrak{sl}_2 \otimes U_q\mathfrak{sl}_2, \qquad \varepsilon : U_q\mathfrak{sl}_2 \to \mathbb{C},$$

and an antihomomorphism S of the algebra $U_q\mathfrak{sl}_2$ such that

$$\Delta(K^{\pm1}) = K^{\pm1} \otimes K^{\pm1}, \quad \Delta(E) = E \otimes 1 + K \otimes E, \quad \Delta(F) = F \otimes K^{-1} + 1 \otimes F,$$
$$\varepsilon(E) = \varepsilon(F) = 0, \qquad \varepsilon(K^{\pm1}) = 1,$$
$$S(K^{\pm1}) = K^{\mp1}, \quad S(E) = -K^{-1}E, \quad S(F) = -FK.$$

In that way $U_q\mathfrak{sl}_2$ is endowed with a Hopf algebra structure. Its defining relations are connected with the defining relations for $U\mathfrak{sl}_2$ via the substitution

$$q = \exp\left(-\frac{h}{2}\right), \qquad K^{\pm1} = \exp\left(\mp\frac{hH}{2}\right)$$

and a formal passage to the limit as $h \to 0$.

A way that leads to this Hopf algebra is discussed in the Drinfeld article [**79**, p. 807]. This Hopf algebra is the simplest example of a class of Drinfeld–Jimbo quantum universal enveloping algebras [**139**, Chap. 4].

1.3.2. The Hopf $*$-algebras $U\mathfrak{su}_{1,1}$, $U\mathfrak{su}_2$, $U_q\mathfrak{su}_{1,1}$, $U_q\mathfrak{su}_2$. Let $* : \mathfrak{g} \to \mathfrak{g}$ be an antilinear involutive antiautomorphism of a complex Lie algebra $\mathfrak{g}$, i.e.,

$$(\lambda, a + \mu b)^* = \bar{\lambda} a^* + \bar{\mu} b^*, \qquad a^{**} = a, \qquad [a,b]^* = [b^*, a^*],$$

for all $a, b \in \mathfrak{g}$, $\lambda, \mu \in \mathbb{C}$. Such an involution $*$ can be uniquely extended to an antilinear involutive antiautomorphism of the universal enveloping algebra $* : U\mathfrak{g} \to U\mathfrak{g}$.

The set of skew-Hermitian elements $\{a \in \mathfrak{g} | a^* = -a\}$ is a real Lie algebra. Representations of real Lie algebras by skew-Hermitian operators, i.e., $*$-representations of $U\mathfrak{g}$, are crucial in the theory of unitary representations of Lie groups. This allows one to substitute, in a number of cases, the choice of a real form of complex Lie algebra by the choice of an involution. Further, all involutions are supposed to be antilinear antiautomorphisms unless the contrary is stated.

EXAMPLE 1.28. Let us endow the Lie algebra $\mathfrak{sl}_2$ with an involution $\star$ by setting

$$H^\star = H, \qquad E^\star = F, \qquad F^\star = E,$$

then extend this involution to $U\mathfrak{sl}_2$ and use the notation $U\mathfrak{su}_2 = (U\mathfrak{sl}_2, \star)$. Evidently, finite-dimensional $*$-representations of $U\mathfrak{su}_2$ are in one-to-one correspondence with finite-dimensional representations of the real Lie algebra $\mathfrak{su}_2 = \{a \in \mathfrak{sl}_2 | \, a^\star = -a\}$ by skew-Hermitian operators. Note that $\mathfrak{su}_2$ is isomorphic to the Lie algebra of the group of motions of the two-dimensional sphere, cf. [**325**, p. 78], [**322**, p. 109,110].

EXAMPLE 1.29. Now we endow the Lie algebra $\mathfrak{sl}_2$ with an involution $*$ by setting

$$H^* = H, \qquad E^* = -F, \qquad F^* = -E,$$

then extend this involution to $U\mathfrak{sl}_2$ and use the notation $U\mathfrak{su}_{1,1} = (U\mathfrak{sl}_2, *)$. The real Lie algebra $\mathfrak{su}_{1,1} = \{a \in \mathfrak{sl}_2 \,|\, a^* = -a\}$ is interesting for us because it is isomorphic to the Lie algebra of the group of Möbius transformations in the unit disc $\mathbb{D}$, cf. [**124**, p. 45], [**322**, p. 294, 295].

Let us note that the notation introduced is essentially traditional, according to which the universal enveloping algebras $U\mathfrak{su}_{1,1}$, $U\mathfrak{su}_2$ are algebras over $\mathbb{R}$ but not over $\mathbb{C}$.

A Hopf algebra A with an involution $*$ is called a *Hopf $*$-algebra* if $*$ is an automorphism of coalgebra

$$\Delta(a^*) = \Delta(a)^{*\otimes *}, \qquad a \in A.$$

One can easily verify that

$$1^* = 1, \quad *S* = S^{-1}, \quad \varepsilon(a^*) = \overline{\varepsilon(a)}, \quad a \in A.$$

Indeed, let A^{cop} be a Hopf algebra that differs from A by replacing the coproduct by the opposite one. The uniqueness of unit and counit imply the first and the third identities, whereas the second follows from the uniqueness and the invertibility of antipode because $*S*$, S^{-1} are antipodes for the Hopf algebra A^{cop}. A proof of the uniqueness of the antipode can be found in [**161**, p. 65], [**148**, p. 13].

Let π_1, π_2 be $*$-representations of a Hopf $*$-algebra A in pre-Hilbert spaces $\mathcal{H}_1$, $\mathcal{H}_2$. By definition, their tensor product is a $*$-representation of A in the tensor product of the pre-Hilbert spaces. This claim is a counterpart to the fact that the tensor product of unitary representations of a group is again its unitary representation.

The definition of a Hopf algebra A in terms of commutative diagrams imply that once $\dim A < \infty$, the dual vector space A^* is a Hopf algebra with multiplication, unit, coproduct, counit, and antipode being, respectively, dual to coproduct, counit, multiplication, unit, and antipode of the Hopf algebra A.

REMARK 1.30. In the above discussion, the requirement that A is finite dimensional can be weakened. Indeed, let $A^\star$ be the set of linear functionals f on A such that the kernel f contains a two-sided ideal of finite codimension. It is known [**148**, p. 27] that $A^\star$ can be endowed with the structure of Hopf algebra by duality. The latter algebra is called the dual Hopf algebra. If A is a Hopf $*$-algebra, then the equality

$$l^*(a) \stackrel{\text{def}}{=} \overline{l(S(a)^*)}, \qquad l \in A^*, \quad a \in A, \tag{1.20}$$

endows $A^\star$ with the structure of Hopf $*$-algebra.

By definition, $U\mathfrak{su}_{1,1}$ and $U\mathfrak{su}_2$ are Hopf $*$-algebras. Now we introduce their quantum analogs.

There exists a unique involution $*$ on $U_q\mathfrak{sl}_2$ such that

$$(K^{\pm 1})^* = K^{\pm 1}, \qquad E^* = -KF, \qquad F^* = -EK^{-1}.$$

It endows $U_q\mathfrak{sl}_2$ with the structure of Hopf $*$-algebra that will be denoted by $U_q\mathfrak{su}_{1,1}$.

In a similar way, there exists a unique involution $\star$ on $U_q\mathfrak{sl}_2$ such that

$$(K^{\pm 1})^\star = K^{\pm 1}, \qquad E^\star = KF, \qquad F^\star = EK^{-1}. \tag{1.21}$$

It endows $U_q\mathfrak{sl}_2$ with the structure of Hopf $*$-algebra that will be denoted by $U_q\mathfrak{su}_2$.

The Hopf $*$-algebra $U_q\mathfrak{su}_2$ has the two-dimensional $*$-representation π in the Hilbert space $\mathbb{C}^2$:

$$\pi(K^{\pm 1}) = \begin{pmatrix} q^{\pm 1} & 0 \\ 0 & q^{\mp 1} \end{pmatrix}, \quad \pi(E) = \begin{pmatrix} 0 & q^{-\frac{1}{2}} \\ 0 & 0 \end{pmatrix}, \quad \pi(F) = \begin{pmatrix} 0 & 0 \\ q^{\frac{1}{2}} & 0 \end{pmatrix}. \tag{1.22}$$

The corresponding representation of $U_q\mathfrak{sl}_2$ in $\mathbb{C}^2$ is called its vector representation.

1.3.3. The $U_q\mathfrak{sl}_2$-module algebras $\mathbb{C}[z]_q$, $\mathbb{C}[z^*]_q$. Let A be a Hopf algebra. A nonunital algebra F is called an A-module algebra if it is an A-module such that the multiplication

$$m : F \otimes F \to F, \qquad m : f_1 \otimes f_2 \mapsto f_1 f_2,$$

is a morphism of A-modules. This means that $a(f_1 f_2) = \sum_i b_i(f_1)c_i(f_2)$ for $a \in A$ such that $\triangle(a) = \sum_i b_i \otimes c_i$. If F is unital, the definition contains an additional requirement that the linear map

$$\mathbb{1} : \mathbb{C} \to F, \qquad \mathbb{1} : 1 \mapsto 1,$$

is a morphism of A-modules. Equivalently, the unit of F has to be an A-invariant element.

In the special case $A = U_q\mathfrak{sl}_2$, these requirements are just

$$\begin{aligned} K^{\pm 1}(f_1 f_2) &= K^{\pm 1}(f_1) K^{\pm 1}(f_2), & K^{\pm 1}(1) &= 1, \\ E(f_1 f_2) &= E(f_1) f_2 + K(f_1) E(f_2), & E(1) &= 0, \\ F(f_1 f_2) &= K^{-1}(f_1) F(f_2) + f_1 F(f_2), & F(1) &= 0. \end{aligned} \tag{1.23}$$

A homomorphism of A-module algebras is a linear map which is both a homomorphism of algebras and a morphism of A-modules.

Similarly to the notion of an A-module algebra, one can introduce the notions of an A-covariant left module, an A-covariant right module, and an A-covariant bimodule M over an A-module algebra F by requiring the following conditions:

- M is a module over a Hopf algebra A,

- the actions $m_{\rm left} : F \otimes M \to M$, and $m_{\rm right} : M \otimes F \to M$ are morphisms of A-modules,
- the unit of F corresponds to the identity map in M.

A morphism of such modules is a linear map that is both a morphism of F-modules and a morphism of A-modules.

Now we give some examples.

EXAMPLE 1.31. Given a module V over a Hopf algebra A, the algebra $\mathrm{End}(V)$ of all linear operators in V is an A-module algebra. Indeed, for $a \in A$, $\triangle(a) = \sum_i b_i \otimes c_i$ one has

$$(aT)v = \sum_i b_i T(S(c_i)v), \qquad T \in \mathrm{End}(V), \quad v \in V. \tag{1.24}$$

The natural embedding

$$\mathbb{C} \hookrightarrow \mathrm{End}(V), \qquad 1 \mapsto I,$$

is a homomorphism of A-module algebras.

Furthermore, given A-modules V_1 and V_2, the vector space $\mathrm{Hom}(V_1, V_2)$ of linear maps from V_1 to V_2 is endowed with an A-module structure in a way similar to (1.24).

EXAMPLE 1.32. Consider first-order differential operators in the space $\mathbb{C}[t_1^{\pm 1}, t_2^{\pm 1}]$ of Laurent polynomials in two variables

$$H = t_1 \frac{\partial}{\partial t_1} - t_2 \frac{\partial}{\partial t_2}, \qquad E = t_1 \frac{\partial}{\partial t_2}, \qquad F = t_2 \frac{\partial}{\partial t_1}.$$

These operators satisfy the defining relations of the algebra $U\mathfrak{sl}_2$ and are differentiations of the algebra $\mathbb{C}[t_1^{\pm 1}, t_2^{\pm 1}]$; hence they define the structure of an $U\mathfrak{sl}_2$-module algebra. Note that

$$(Ht_1, Ht_2) = (t_1, t_2) \begin{pmatrix} 1 & 0 \\ 0 & -1 \end{pmatrix}, \tag{1.25}$$

$$(Et_1, Et_2) = (t_1, t_2) \begin{pmatrix} 0 & 1 \\ 0 & 0 \end{pmatrix}, \qquad (Ft_1, Ft_2) = (t_1, t_2) \begin{pmatrix} 0 & 0 \\ 1 & 0 \end{pmatrix}. \tag{1.26}$$

EXAMPLE 1.33. Consider the algebra $\mathbb{C}[z]$ of polynomials in one variable and the embedding of algebras

$$\mathbb{C}[z] \hookrightarrow \mathbb{C}[t_1^{\pm 1}, t_2^{\pm 1}], \qquad f(z) \mapsto f\left(\frac{t_1}{t_2}\right).$$

The image $\mathbb{C}[z]$ under this embedding is a submodule of the $U\mathfrak{sl}_2$-module $\mathbb{C}[t_1^{\pm 1}, t_2^{\pm 1}]$. Hence $\mathbb{C}[z]$ inherits the $U\mathfrak{sl}_2$-module algebra structure:

$$H = 2z\frac{d}{dz}, \qquad E = -z^2\frac{d}{dz}, \qquad F = \frac{d}{dz}. \tag{1.27}$$

Evidently,

$$Hz = 2z, \qquad Ez = -z^2, \qquad Fz = 1. \tag{1.28}$$

Let us construct $U_q\mathfrak{sl}_2$-module algebras with an important special case. Consider a vector space M together with its tensor square $M \otimes M$ and a subspace $N \subset M \otimes M$. By the quadratic algebra with space of generators M and space of relations N, we mean the quotient algebra $T(M)/(N)$ of the tensor algebra $T(M)$ by the two-sided ideal generated by N [**161**, p. 118], [**227**, p. 19]. Once A is a

Hopf algebra, M is an A-module, and N is a submodule of $M \otimes M$, the quadratic algebra $T(M)/(N)$ is an A-module algebra.

EXAMPLE 1.34. Consider the algebra $\mathbb{C}[t_1, t_2]_q = \mathbb{C}\langle t_1, t_2\rangle/(t_1t_2 - qt_2t_1)$ with generators t_1, t_2 and the defining relation $t_2t_2 = qt_2t_1$. This algebra is called the algebra of regular functions on the quantum vector space $\mathbb{C}^2$. It is easy to show that the equalities

$$(K^{\pm 1}t_1, K^{\pm 1}t_2) = (t_1, t_2)\begin{pmatrix} q^{\pm 1} & 0 \\ 0 & q^{\mp 1}\end{pmatrix}, \tag{1.29}$$

$$(Et_1, Et_2) = (t_1, t_2)\begin{pmatrix} 0 & q^{-\frac{1}{2}} \\ 0 & 0\end{pmatrix}, \qquad (Ft_1, Ft_2) = (t_1, t_2)\begin{pmatrix} 0 & 0 \\ q^{\frac{1}{2}} & 0\end{pmatrix} \tag{1.30}$$

endow $\mathbb{C}[t_1, t_2]_q$ with a $U_q\mathfrak{sl}_2$-module algebra structure. For that we consider the vector representation of the Hopf algebra $U_q\mathfrak{sl}_2$ in $M = \mathbb{C}^2$ and the standard basis $\{t_1, t_2\}$. The generators of $U_q\mathfrak{sl}_2$ act on t_1, t_2 by formulas (1.29), (1.30). One can easily verify that the vector $t_1 \otimes t_2 - qt_2 \otimes t_1$ is $U_q\mathfrak{sl}_2$-invariant. Assuming $N = \mathbb{C}(t_1 \otimes t_2 - qt_2 \otimes t_1)$, we get a $U_q\mathfrak{sl}_2$-module quadratic algebra that is naturally isomorphic to $\mathbb{C}[t_1, t_2]_q$.

Let us mention that (1.29), (1.30) are q-analogs of (1.25), (1.26).

q-analogs of binomial coefficients were introduced by Carl Gauss [**6**]. Using $\mathbb{C}[t_1, t_2]_q$, one can easily explain an appearance of q-binomial coefficients in the topics involved. The set $\{t_2^{j_2}t_1^{j_1}\}_{j_1, j_2 \in \mathbb{Z}_+}$ is a basis of the vector space $\mathbb{C}[t_1, t_2]_q$. (Indeed, the linear independence follows from the fact that $\mathbb{C}[t_1, t_2]_q$ is a $\mathbb{Z}^2$-graded algebra via

$$\deg(t_1) = (1, 0), \qquad \deg(t_2) = (0, 1),$$

and the bidegrees of the considered monomials are pairwise different. Now it remains to introduce a representation T of $\mathbb{C}[t_1, t_2]_q$ in the vector space of functions on the set $q^{\mathbb{Z}}$ by setting $T(t_1)f(u) = f(qu)$, $T(t_2)f(u) = uf(u)$. One can observe that the operators $T(t_2^{j_2}\, t_1^{j_1})$ are nonzero, so the same holds for these monomials.)

q-analogs $\binom{n}{k}_q$ of binomial coefficients $\binom{n}{k}$ can be introduced as follows:

$$(t_1 + t_2)^n = \sum_{k=0}^{n} \binom{n}{k}_q t_2^k t_1^{n-k}.$$

It is obvious that $\lim_{q \to 1} \binom{n}{k}_q = \binom{n}{k}$. An explicit form of q-binomial coefficients is easily computable (see [**161**, p. 96]):

$$\binom{n}{k}_q = \frac{(q; q)_n}{(q; q)_k (q; q)_{n-k}},$$

where we use the standard notation

$$(a; q)_m = \prod_{j=0}^{m-1} (1 - aq^j).$$

Now turn to a description of a q-analog of the $U\mathfrak{sl}_2$-module algebra $\mathbb{C}[t_1^{\pm 1}, t_2^{\pm 1}]$. The following statement is evident.

LEMMA 1.35. *If a Laurent polynomial in n variables $u_1, u_2, \ldots, u_n$ vanishes on the set $\{ (q^{j_1}, q^{j_2}, \ldots, q^{j_n}) \mid j_1, j_2, \ldots, j_n \in \mathbb{N}\}$, then it is identically zero.*

LEMMA 1.36. *There exists a unique triple of Laurent polynomials $k(u_1,u_2)$, $e(u_1,u_2)$, $f(u_1,u_2) \in \mathbb{C}[u_1^{\pm 1}, u_2^{\pm 1}]$ such that for all $j_1, j_2 \in \mathbb{N}$ one has in $\mathbb{C}[t_1,t_2]_q$*

$$K^{\pm 1}(t_2^{j_2} t_1^{j_1}) = k(q^{\pm j_1}, q^{\pm j_2}) t_2^{j_2} t_1^{j_1}, \tag{1.31}$$

$$E(t_2^{j_2} t_1^{j_1}) = e(q^{j_1}, q^{j_2}) t_2^{j_2+1} t_1^{j_1-1}, \qquad F(t_2^{j_2} t_1^{j_1}) = f(q^{j_1}, q^{j_2}) t_2^{j_2-1} t_1^{j_1+1}. \tag{1.32}$$

PROOF. The existence of such Laurent polynomials is established by a direct computation with summation of geometric series. The uniqueness follows from Lemma 1.35. □

EXAMPLE 1.37. Using $\mathbb{Z}^2$-grading, it is easy to show that the algebra $\mathbb{C}[t_1,t_2]_q$ does not contain zero divisors and its multiplicative subset $(t_1t_2)^{\mathbb{Z}_+}$ is an Ore set (see Subsection 1.3.7). Denote by $\mathbb{C}[t_1^{\pm 1}, t_2^{\pm 1}]_q$ the corresponding localization. This algebra can be defined by generators $t_1^{\pm 1}$, $t_2^{\pm 1}$ and the defining relations

$$t_1 t_2 = q t_2 t_1, \qquad t_1 t_1^{-1} = t_1^{-1} t_1 = 1, \qquad t_2 t_2^{-1} = t_2^{-1} t_2 = 1.$$

As before, one can verify that the set $\{t_2^{j_2} t_1^{j_1}\}_{j_1,j_2 \in \mathbb{Z}}$ is a basis of the vector space $\mathbb{C}[t_1^{\pm 1}, t_2^{\pm 1}]_q$.

Now we show that the $U_q\mathfrak{sl}_2$-module algebra structure is uniquely extendable from $\mathbb{C}[t_1,t_2]_q$ to $\mathbb{C}[t_1^{\pm 1}, t_2^{\pm 1}]_q$. The uniqueness follows from the definitions. It remains to produce an extension.

Using Lemma 1.36, introduce linear operators $K^{\pm 1}$, E, F in $\mathbb{C}[t_1^{\pm 1}, t_2^{\pm 1}]_q$ by defining the actions on $\{t_2^{j_2} t_1^{j_1}\}_{j_1,j_2 \in \mathbb{Z}}$ according to (1.31), (1.32). It is easy to show that these operators satisfy the defining relations (1.18), (1.19) of the algebra $U_q\mathfrak{sl}_2$ and endow $\mathbb{C}[t_1^{\pm 1}, t_2^{\pm 1}]_q$ with a $U_q\mathfrak{sl}_2$-module algebra structure. Here we just verify the relation (1.19).

Since the restrictions of $EF - FE$ and $\frac{K-K^{-1}}{q-q^{-1}}$ to the subspace $\mathbb{C}[t_1,t_2]_q$ coincide, one has

$$e(qu_1, q^{-1}u_2) f(u_1,u_2) - f(q^{-1}u_1, qu_2) e(u_1,u_2) = \frac{k(u_1,u_2) - k(u_1^{\pm 1}, u_2^{\pm 1})}{q - q^{-1}} \tag{1.33}$$

for all $u_1, u_2 \in q^{\mathbb{N}}$. Lemma 1.35 implies that (1.33) holds for all $u_1, u_2 \in q^{\mathbb{Z}}$. Therefore the considered operators in $\mathbb{C}[t_1^{\pm 1}, t_2^{\pm 1}]_q$ are equal.

It is clear that the $U_q\mathfrak{sl}_2$-module algebra $\mathbb{C}[t_1^{\pm 1}, t_2^{\pm 1}]_q$ is a q-analog of the $U\mathfrak{sl}_2$-module algebra $\mathbb{C}[t_1^{\pm 1}, t_2^{\pm 1}]$.

EXAMPLE 1.38. The homomorphism

$$\mathbb{C}[z] \to \mathbb{C}[t_1^{\pm 1}, t_2^{\pm 1}]_q, \qquad f(z) \mapsto f(t_2^{-1} t_1),$$

is injective, the image of this homomorphism is a $U_q\mathfrak{sl}_2$-submodule since $\mathbb{C}[t_1^{\pm 1}, t_2^{\pm 1}]_q$ does not have zero divisors, and $\mathbb{C}[z]$ inherits the $U_q\mathfrak{sl}_2$-module algebra structure:

$$K^{\pm 1} z = q^{\pm 2} z, \qquad Ez = -q^{1/2} z^2, \qquad Fz = q^{1/2}. \tag{1.34}$$

We denote this $U_q\mathfrak{sl}_2$-module algebra by $\mathbb{C}[z]_q$. The equalities (1.34) are q-analogs of (1.28). From (1.34), (1.23), one gets $K^{\pm 1} f(z) = f(q^{\pm 1} z)$,

$$Ef(z) = -q^{1/2} z^2 \frac{f(z) - f(q^2 z)}{z - q^2 z}, \qquad Ff(z) = q^{1/2} \frac{f(q^{-2} z) - f(z)}{q^{-2} z - z}. \tag{1.35}$$

The latter equalities are q-analogs of (1.27).

Up to here, we have dealt with holomorphic Laurent polynomials in $\mathbb{C}^2$ and their quantum analogs. Now let us turn to antiholomorphic Laurent polynomials. Let $\mathbb{C}[(t_1^*)^{\pm 1}, (t_2^*)^{\pm 1}]_q$ be the algebra with generators t_1^*, $(t_1^*)^{-1}$, t_2^*, $(t_2^*)^{-1}$ and the defining relations

$$t_2^* t_1^* = q t_1^* t_2^*, \qquad (t_1^*)^{-1} t_1^* = t_1^* (t_1^*)^{-1} = 1, \qquad (t_2^*)^{-1} t_2^* = t_2^* (t_2^*)^{-1} = 1.$$

Consider the antilinear antiisomorphism

$$* : \mathbb{C}[(t_1^*)^{\pm 1}, (t_2^*)^{\pm 1}]_q \to \mathbb{C}[t_1^{\pm 1}, t_2^{\pm 1}]_q, \qquad * : t_1^* \mapsto t_1, \quad * : t_2^* \mapsto t_2.$$

The inverse antilinear anti-isomorphism will be denoted by the same symbol $*$. Then $\mathbb{C}[(t_1^*)^{\pm 1}, (t_2^*)^{\pm 1}]_q$ is endowed with the $U_q\mathfrak{sl}_2$-module algebra structure by transferring it from $\mathbb{C}[t_1^{\pm 1}, t_2^{\pm 1}]_q$ as follows:

$$af \stackrel{\text{def}}{=} (S(a)^* f^*)^*, \qquad a \in U_q\mathfrak{sl}_2, \quad f \in \mathbb{C}[(t_1^*)^{\pm 1}, (t_2^*)^{\pm 1}]_q. \tag{1.36}$$

(Relation (1.36) endows $\mathbb{C}[(t_1^*)^{\pm 1}, (t_2^*)^{\pm 1}]_q$ with a $U_q\mathfrak{sl}_2$-module structure since $*S$ is an antiautomorphism of the coalgebra $U_q\mathfrak{sl}_2$. The equalities (1.23) follow from the similar relations in $\mathbb{C}[t_1^{\pm 1}, t_2^{\pm 1}]_q$ and from the fact that $*S$ is an antiautomorphism of the coalgebra $U_q\mathfrak{sl}_2$.)

Just as before, using the embedding

$$\mathbb{C}[z^*] \hookrightarrow \mathbb{C}[(t_1^*)^{\pm 1}, (t_2^*)^{\pm 1}]_q, \qquad f(z^*) \mapsto f(t_1^*(t_2^*)^{-1}),$$

one can introduce a $U_q\mathfrak{sl}_2$-module algebra such that

$$K^{\pm 1} z^* = q^{\mp 2} z^*, \qquad E z^* = q^{-3/2}, \qquad F z^* = -q^{5/2} z^{*2}. \tag{1.37}$$

This algebra will be denoted by $\mathbb{C}[\bar{z}]_q$.

1.3.4. The universal R-matrix. A $U_q\mathfrak{sl}_2$-module V is called weight if

$$V = \bigoplus_{\lambda \in \mathbb{R}} V_\lambda, \qquad V_\lambda = \{v \in V \mid K^{\pm 1} v = q^{\pm \lambda} v\}.$$

Further on all $U_q\mathfrak{sl}_2$-modules are supposed to be weight.[5]

EXAMPLE 1.39. $\mathbb{C}[z]_q$ is an infinite-dimensional weight $U_q\mathfrak{sl}_2$-module. The simplest nonweight $U_q\mathfrak{sl}_2$-module is one dimensional: $K^{\pm 1} = -1$, $E = F = 0$.

In a weight $U_q\mathfrak{sl}_2$-module V one can introduce the linear operator

$$H : V \to V, \qquad H|_{V_\lambda} = \lambda.$$

Obviously, $K^{\pm 1} v = q^{\pm H} v$ for all $v \in V$.

The full subcategory of weight $U_q\mathfrak{sl}_2$-modules is closed under the passage to submodules,[6] quotient modules, and tensor products; i.e., it is an abelian tensor category. Each $U_q\mathfrak{sl}_2$-module possesses the largest weight submodule that allows one to introduce the notion of a dual weight $U_q\mathfrak{sl}_2$-module. Indeed,

$$V^* \stackrel{\text{def}}{=} \bigoplus_{\lambda \in \mathbb{R}} (V^*)_\lambda, \qquad (V^*)_\lambda = (V_{-\lambda})^*,$$

for $V = \bigoplus_{\lambda \in \mathbb{R}} V_\lambda$.

Let B be a subalgebra of an algebra A. An A-module V is called locally B-finite dimensional if $\dim(Bv) < \infty$ for all $v \in V$.[7]

---

[5] Finite-dimensional weight $U_q\mathfrak{sl}_2$-modules are also called $U_q\mathfrak{sl}_2$-modules of type 1 [**139**].

[6] See, for example, [**151**] Proposition 1.5.

[7] If $B = A$, then the module V is called just locally finite dimensional.

Fix the notation $U_q\mathfrak{b}^+$ and $U_q\mathfrak{b}^-$ for the Hopf subalgebras in $U_q\mathfrak{sl}_2$ generated by $\{K^{\pm 1}, E\}$ and $\{K^{\pm 1}, F\}$, respectively. Denote by $\mathcal{C}^+$ the full subcategory of weight locally $U_q\mathfrak{b}^+$-finite-dimensional $U_q\mathfrak{sl}_2$-modules and by $\mathcal{C}^-$ the full subcategory of weight locally $U_q\mathfrak{b}^-$-finite-dimensional $U_q\mathfrak{sl}_2$-modules.

Given any $U_q\mathfrak{sl}_2$-modules V', V'' and their tensor products $V' \otimes V''$, $V'' \otimes V'$, the naive permutation of tensor factors

$$\sigma_{V',V''} : V' \otimes V'' \to V'' \otimes V', \quad \sigma_{V',V''} : v' \otimes v'' \mapsto v'' \otimes v',$$

is not in general a morphism of $U_q\mathfrak{sl}_2$-modules. V. Drinfeld introduced the notion of a universal R-matrix and then used it to define linear operators that substitute permutations $\sigma_{V',V''}$ in the theory of $U_q\mathfrak{sl}_2$-modules. Here is a description of these linear operators.

Set

$$\exp_q(t) = \sum_{i=0}^{\infty} \left(\prod_{j=1}^{i} \frac{1-q}{1-q^j} \right) t^i.$$

Given weight $U_q\mathfrak{sl}_2$-modules V', V'', we introduce a linear operator in $V' \otimes V''$ by setting

$$R_{V',V''}(v' \otimes v'') = \exp_{q^2}((q^{-1} - q)E \otimes F)q^{-\frac{H\otimes H}{2}}(v' \otimes v''). \tag{1.38}$$

This operator is well defined once either $V' \in \mathcal{C}^+$ or $V'' \in \mathcal{C}^-$.

The following propositions show that the linear operators

$$\check{R}_{V',V''} = \sigma_{V',V''}\, R_{V',V''}$$

are q-analogs of $\sigma_{V',V''}$.

PROPOSITION 1.40. (1) *Let V', V'' be weight $U_q\mathfrak{sl}_2$-modules such that either $V' \in \mathcal{C}^+$ or $V'' \in \mathcal{C}^-$. The linear operator*

$$\check{R}_{V',V''} : V' \otimes V'' \to V'' \otimes V'$$

is an invertible morphism of $U_q\mathfrak{sl}_2$-modules.

(2) *Let $f' : V' \to W'$, $f'' : V'' \to W''$ be morphisms of weight $U_q\mathfrak{sl}_2$-modules, and either $V', W' \in \mathcal{C}^+$ or $V'', W'' \in \mathcal{C}^-$. Then one has*

$$(f'' \otimes f') \cdot \check{R}_{V',V''} = \check{R}_{W',W''} \cdot (f' \otimes f'').$$

PROPOSITION 1.41. (1) *Let V, V', V'' be weight $U_q\mathfrak{sl}_2$-modules such that either $V', V'' \in \mathcal{C}^+$ or $V \in \mathcal{C}^-$. Then one has*

$$\check{R}_{V'\otimes V'',V} = (\check{R}_{V',V} \otimes \mathrm{id}_{V''})(\mathrm{id}_{V'} \otimes \check{R}_{V'',V}).$$

(2) *Let V, V', V'' be weight $U_q\mathfrak{sl}_2$-modules and either $V \in \mathcal{C}^+$ or $V', V'' \in \mathcal{C}^-$. Then we have*

$$\check{R}_{V,V'\otimes V''} = (\mathrm{id}_{V'} \otimes \check{R}_{V,V''})(\check{R}_{V,V'} \otimes \mathrm{id}_{V''}).$$

(3) *For each weight $U_q\mathfrak{sl}_2$-module V one has*

$$\check{R}_{V,\mathbb{C}} = \check{R}_{\mathbb{C},V} = \mathrm{id}_V .$$

The above properties of $\check{R}_{V',V''}$ imply that $\mathcal{C}^-$ and $\mathcal{C}^+$ are braided tensor categories; see Subsection 1.3.8. Normally, Propositions 1.40 and 1.41 are formulated under the assumption of finite dimensionality of permuting weight $U_q\mathfrak{sl}_2$-modules [**62**, p. 329]), but the proofs show that the requirement of finite dimensionality may be weakened.

EXAMPLE 1.42. The definitions imply that the $U_q\mathfrak{sl}_2$-module $\mathbb{C}[z]_q$ is weight and locally $U_q\mathfrak{b}^-$-finite dimensional. Then one has

$$m\check{R}_{\mathbb{C}[z]_q,\mathbb{C}[z]_q}(f_1\otimes f_2)=f_1f_2,\qquad f_1,f_2\in\mathbb{C}[z]_q, \tag{1.39}$$

where $m:\mathbb{C}[z]_q\otimes\mathbb{C}[z]_q\to\mathbb{C}[z]_q$ is the multiplication in $\mathbb{C}[z]_q$. It suffices to prove this equality for $f_1,f_2\in\{1,z\}$. The only nonevident case is $f_1=f_2=z$, where one has

$$\begin{aligned}m\check{R}_{\mathbb{C}[z]_q,\mathbb{C}[z]_q}(z\otimes z)&=m\left((1+(q^{-1}-q)F\otimes E)q^{-\frac{H\otimes H}{2}}(z\otimes z)\right)\\&=m\left(q^{-2}z\otimes z-(q^{-2}-1)(1\otimes z^2)\right)=z^2.\end{aligned}$$

The equality (1.39) means that the $U_q\mathfrak{sl}_2$-module algebra $\mathbb{C}[z]_q$ is commutative in the sense of category theory; see Subsection 1.3.8.

The following remarks do not pretend to be complete or rigorous. They are presented to explain a relation between Propositions 1.40 and 1.41 and the notion of the universal R-matrix introduced by Drinfeld [**79**]. For a good exposition, see [**161**, Chap. 16, 17].

Consider the topological ring of formal power series $\mathbb{C}[[h]]$ with h-adic topology, i.e., the topology whose base of neighborhoods of zero is $\{h^n\mathbb{C}[[h]]\}_{n\in\mathbb{Z}_+}$ [**161**, p. 481]. (Note that $\mathbb{C}[[h]]$ is not a complex topological vector space in the h-adic topology since the bilinear map

$$\mathbb{C}\times V\to V[[h]],\qquad (z,v)\mapsto zv,$$

is not continuous.)

Following the conventions of deformation theory [**92**], we introduce the additive category $\mathcal{A}$ of topologically free $\mathbb{C}[[h]]$-modules. To each vector space V one can assign the topological $\mathbb{C}[[h]]$-module $V[[h]]$ of formal series with coefficients from V and the h-adic topology with the base of neighborhoods of zero $\{h^nV[[h]]\}_{n\in\mathbb{Z}_+}$.

A topological $\mathbb{C}[[h]]$-module is called topologically free if it is isomorphic to $V[[h]]$ for some vector space V. These modules are objects of the category $\mathcal{A}$, and the morphisms of $\mathcal{A}$ are all morphisms of topological $\mathbb{C}[[h]]$-modules. Direct sums and tensor products in $\mathcal{A}$ are determined on objects by setting

$$V'[[h]]\bigoplus V''[[h]]\stackrel{\text{def}}{=}(V'\bigoplus V'')[[h]],\qquad V'[[h]]\hat{\otimes}V''[[h]]\stackrel{\text{def}}{=}(V'\otimes V'')[[h]]$$

[**161**, p. 487] (on morphisms they are determined in a natural way). The "naive" flip of tensors in coefficients of formal series from $(V'\otimes V'')[[h]]$ endows the obtained tensor category with a symmetric tensor category structure; see Subsection 1.3.8.

Following Drinfeld [**81**], we construct a Hopf algebra $U_h\mathfrak{sl}_2$ in $\mathcal{A}$.

Consider the free noncommutative algebra $\mathbb{C}\langle H,E,F\rangle$ with generators H, E, F and the subspace $V\subset\mathbb{C}\langle H,E,F\rangle$ spanned by the set $\{H^iE^jF^k\,|\,i,j,k\in\mathbb{Z}_+\}$.

The topological submodule $V[[h]]\subset\mathbb{C}\langle F,H,E\rangle[[h]]$ is not a subring since we have to change the order of generators in the product $(F^{i'}H^{j'}E^{k'})(F^{i''}H^{j''}E^{k''})$. Transfer H to the left and F to the right using the commutation relations

$$HE-EH=2E,\qquad HF-FH=-2F,\qquad EF-FE=\frac{\operatorname{sh}(hH/2)}{\operatorname{sh}(h/2)},$$

where $\mathrm{sh}(t) = \sum_{n=0}^{\infty} \frac{t^{2n+1}}{(2n+1)!} \in \mathbb{C}[[t]]$ is the Taylor series of hyperbolic sine. Thus we get a multiplication that endows $U_h\mathfrak{sl}_2 \stackrel{\text{def}}{=} V[[h]]$ with the structure of algebra in the tensor category $\mathcal{A}$.

Using the equalities

$$\begin{aligned} \Delta(H) &= H \otimes 1 + 1 \otimes H, & \varepsilon(H) &= 0, & S(H) &= -H, \\ \Delta(E) &= E \otimes 1 + e^{-hH/2} \otimes E, & \varepsilon(E) &= 0, & S(E) &= -e^{hH/2}E, \\ \Delta(F) &= F \otimes e^{hH/2} + 1 \otimes F, & \varepsilon(F) &= 0, & S(F) &= -Fe^{-hH/2}, \end{aligned}$$

one can introduce a coproduct $\Delta : U_h\mathfrak{sl}_2 \to U_h\mathfrak{sl}_2 \hat{\otimes} U_h\mathfrak{sl}_2$, a counit $\varepsilon : U_h\mathfrak{sl}_2 \to \mathbb{C}[[h]]$, and an antipode $S : U_h\mathfrak{sl}_2 \to U_h\mathfrak{sl}_2$ that endow $U_h\mathfrak{sl}_2$ with the structure of Hopf algebra in the symmetric tensor category $\mathcal{A}$.

REMARK 1.43. The Hopf algebra $U_h\mathfrak{sl}_2$ is related to $U_q\mathfrak{sl}_2$ by the following change of variables and the deformation parameter:

$$q = e^{-h/2}, \quad K^{\pm 1} = e^{\mp hH/2}.$$

The element

$$R = \exp_{q^2}((q^{-1} - q)E \otimes F)q^{-\frac{H \otimes H}{2}}$$

of the tensor square $U_h\mathfrak{sl}_2 \hat{\otimes} U_h\mathfrak{sl}_2$ is called a universal R-matrix of the Hopf algebra $U_h\mathfrak{sl}_2$. Applying the standard embeddings

$$i_{1,2}, i_{1,3}, i_{2,3} : U_h\mathfrak{g} \hat{\otimes} U_h\mathfrak{g} \to U_h\mathfrak{g} \hat{\otimes} U_h\mathfrak{g} \hat{\otimes} U_h\mathfrak{g},$$

we obtain formal series $R^{1,2}$, $R^{1,3}$, $R^{2,3}$. In this context, analogs of Propositions 1.40 and 1.41 look like

$$\Delta^{\mathrm{cop}}(\xi) = R \cdot \Delta(\xi) \cdot R^{-1}, \qquad \xi \in U_q\mathfrak{sl}_2, \tag{1.40}$$

$$(\Delta \otimes \mathrm{id})(R) = R^{1,3} \cdot R^{2,3}, \tag{1.41}$$

$$(\mathrm{id} \otimes \Delta)(R) = R^{1,3} \cdot R^{1,2}, \tag{1.42}$$

where Δ^{cop} is the opposite coproduct. V. Drinfeld introduced Hopf algebras with a distinguished element of the tensor product satisfying such properties. They are called *braided Hopf algebras* [**161**, p. 219].

Consider the antiautomorphism τ of the topologically free algebra $U_h\mathfrak{sl}_2$ defined by

$$\tau(H) = H, \qquad \tau(E) = e^{-\frac{hH}{2}}F, \qquad \tau(F) = Ee^{\frac{hH}{2}}.$$

The following frequently used equalities were obtained in [**80**, p. 34, 37]:

$$(S \otimes S)R = R, \qquad (\tau \otimes \tau)R = R^{21}, \tag{1.43}$$

where R^{21} differs from R by the flip of tensor factors.

1.3.5. The $U_q\mathfrak{su}_{1,1}$-module $*$-algebra $\mathrm{Pol}(\mathbb{C})_q$. Subsection 1.3.3 contains the definition of A-module algebra F. If A is a Hopf $*$-algebra and F is a $*$-algebra, one should include in that definition an additional requirement of compatibility of involutions:

$$(af)^* = S(a)^* f^* \tag{1.44}$$

for all $a \in A$, $f \in F$.

EXAMPLE 1.44. Consider a Hopf $*$-algebra $A = (U\mathfrak{g}, *)$ and the corresponding real Lie subalgebra $\{a \in \mathfrak{g} | \, a^* = -a\}$; see Subsection 1.3.2. Assume that

- X is a homogeneous space of a Lie group with this real Lie algebra,
- $a \in \mathfrak{g}$, $a^* = -a$,
- F is the algebra of smooth complex functions on X,
- $f \in F$ is a self-adjoint element.

Then (1.44) becomes $(af)^* = (-a)^* f^* = af$. This means that the real vector field on X assigned to a (viewed as the corresponding first-order differential operator) maps the space of real smooth functions into itself.

We are about to embed the $U_q\mathfrak{sl}_2$-module algebra $\mathbb{C}[z]_q$ of *holomorphic polynomials* to a $U_q\mathfrak{su}_{1,1}$-module $*$-algebra of *all polynomials on a quantum complex plane*. Let us endow the $U_q\mathfrak{sl}_2$-module $\mathbb{C}[z]_q \otimes \mathbb{C}[\bar{z}]_q$ with a $U_q\mathfrak{sl}_2$-module algebra structure.

By obvious reasons, the $U_q\mathfrak{sl}_2$-modules $\mathbb{C}[z]_q$, $\mathbb{C}[\bar{z}]_q$ are weight, the $U_q\mathfrak{sl}_2$-module $\mathbb{C}[z]_q$ is locally $U_q\mathfrak{b}^-$-finite dimensional, and the $U_q\mathfrak{sl}_2$-module $\mathbb{C}[\bar{z}]_q$ is locally $U_q\mathfrak{b}^+$-finite dimensional. Hence there exists a morphism of $U_q\mathfrak{sl}_2$-modules

$$\check{R} = \check{R}_{\mathbb{C}[\bar{z}]_q,\mathbb{C}[z]_q} : \mathbb{C}[\bar{z}]_q \otimes \mathbb{C}[z]_q \to \mathbb{C}[z]_q \otimes \mathbb{C}[\bar{z}]_q.$$

Let us endow $\mathbb{C}[z]_q \otimes \mathbb{C}[\bar{z}]_q$ with an algebra structure by defining a multiplication

$$m : (\mathbb{C}[z]_q \otimes \mathbb{C}[\bar{z}]_q) \otimes (\mathbb{C}[z]_q \otimes \mathbb{C}[\bar{z}]_q) \to \mathbb{C}[z]_q \otimes \mathbb{C}[\bar{z}]_q$$

via the multiplications

$$m^+ : \mathbb{C}[\bar{z}]_q \otimes \mathbb{C}[\bar{z}]_q \to \mathbb{C}[\bar{z}]_q, \qquad m^- : \mathbb{C}[z]_q \otimes \mathbb{C}[z]_q \to \mathbb{C}[z]_q$$

in $\mathbb{C}[\bar{z}]_q$, $\mathbb{C}[z]_q$, respectively. Namely, we set

$$m = (m^- \otimes m^+)(\mathrm{id}_{\mathbb{C}[z]_q} \otimes \check{R} \otimes \mathrm{id}_{\mathbb{C}[\bar{z}]_q}).$$

Equivalently,

$$\begin{array}{c}
(\mathbb{C}[z]_q \otimes \mathbb{C}[\bar{z}]_q) \otimes (\mathbb{C}[z]_q \otimes \mathbb{C}[\bar{z}]_q) \\
\Big\downarrow \cong \\
\mathbb{C}[z]_q \otimes (\mathbb{C}[\bar{z}]_q \otimes \mathbb{C}[z]_q) \otimes \mathbb{C}[\bar{z}]_q \\
\Big\downarrow \mathrm{id}\otimes\check{R}\otimes\mathrm{id} \\
\mathbb{C}[z]_q \otimes (\mathbb{C}[z]_q \otimes \mathbb{C}[\bar{z}]_q) \otimes \mathbb{C}[\bar{z}]_q \\
\Big\downarrow \cong \\
(\mathbb{C}[z]_q \otimes \mathbb{C}[z]_q) \otimes (\mathbb{C}[\bar{z}]_q \otimes \mathbb{C}[\bar{z}]_q) \\
\Big\downarrow m^-\otimes m^+ \\
\mathbb{C}[z]_q \otimes \mathbb{C}[\bar{z}]_q.
\end{array}$$

Propositions 1.40 and 1.41 imply that

$$\begin{aligned}
(m^- \otimes \mathrm{id})\check{R}_{\mathbb{C}[\bar{z}]_q,\mathbb{C}[z]_q\otimes\mathbb{C}[z]_q} &= \check{R}_{\mathbb{C}[\bar{z}]_q,\mathbb{C}[z]_q}(\mathrm{id} \otimes m^-), \\
(\mathrm{id} \otimes m^+)\check{R}_{\mathbb{C}[\bar{z}]_q\otimes\mathbb{C}[\bar{z}]_q,\mathbb{C}[z]_q} &= \check{R}_{\mathbb{C}[\bar{z}]_q,\mathbb{C}[z]_q}(m^+ \otimes \mathrm{id}), \\
\check{R}_{\mathbb{C}[\bar{z}]_q,\mathbb{C}[z]_q\otimes\mathbb{C}[z]_q} &= (\mathrm{id}_{\mathbb{C}[z]_q} \otimes \check{R}_{\mathbb{C}[\bar{z}]_q,\,\mathbb{C}[z]_q}) \cdot (\check{R}_{\mathbb{C}[\bar{z}]_q,\,\mathbb{C}[z]_q} \otimes \mathrm{id}_{\mathbb{C}[z]_q}), \\
\check{R}_{\mathbb{C}[\bar{z}]_q\otimes\mathbb{C}[\bar{z}]_q,\,\mathbb{C}[z]_q} &= (\check{R}_{\mathbb{C}[\bar{z}]_q,\,\mathbb{C}[z]_q} \otimes \mathrm{id}_{\mathbb{C}[\bar{z}]_q}) \cdot (\mathrm{id}_{\mathbb{C}[\bar{z}]_q} \otimes \check{R}_{\mathbb{C}[\bar{z}]_q,\,\mathbb{C}[z]_q}).
\end{aligned}$$

One can prove the associativity of m using these relations and the associativity of $m^{\pm}$ in $\mathbb{C}[\bar{z}]_q$, $\mathbb{C}[z]_q$. (This proof is well known in the theory of braided tensor categories; see [**225**, p. 439].) The remaining parts in the statements of the following proposition can be deduced from Proposition 1.40.

PROPOSITION 1.45. *The multiplication m endows $\mathbb{C}[z]_q \otimes \mathbb{C}[\bar{z}]_q$ with a structure of $U_q\mathfrak{sl}_2$-module algebra with unit $1 \otimes 1$.*

Now (1.38), (1.34), (1.37) lead to the following:

$$\begin{aligned}\check{R}_{\mathbb{C}[\bar{z}]_q, \mathbb{C}[z]_q}(z^* \otimes z) &= (1 + (q^{-1} - q)F \otimes E)q^{-\frac{H \otimes H}{2}}(z \otimes z^*)\\ &= q^2 z \otimes z^* + 1 - q^2.\end{aligned}$$

Hence,

$$(1 \otimes z^*)(z \otimes 1) = q^2(z \otimes 1)(1 \otimes z^*) + 1 - q^2. \tag{1.45}$$

Therefore, the map

$$z \mapsto z \otimes 1, \qquad z^* \mapsto 1 \otimes z^*,$$

can be extended in a unique way to an isomorphism of algebras $\mathrm{Pol}(\mathbb{C})_q$ and $(\mathbb{C}[z]_q \otimes \mathbb{C}[\bar{z}]_q, m)$. With the involution

$$* : z \otimes 1 \mapsto 1 \otimes z^*, \qquad * : 1 \otimes z^* \mapsto z \otimes 1,$$

on $(\mathbb{C}[z]_q \otimes \mathbb{C}[\bar{z}]_q)$, we obtain an isomorphism of $*$-algebras. Now (1.45) can be rewritten in a more compact form: $z^*z = q^2 zz^* + 1 - q^2$.

We identify $\mathrm{Pol}(\mathbb{C})_q$ with the $*$-algebra constructed in this subsection.

Now we verify the compatibility of involutions (1.44). Consider the representations of $U_q\mathfrak{sl}_2$ in $\mathrm{Pol}(\mathbb{C})_q$ defined as follows:

$$\pi(a)f = af, \qquad a \in U_q\mathfrak{sl}_2, \quad f \in \mathrm{Pol}(\mathbb{C})_q,$$

and

$$\pi'(a)f = (\pi(S(a)^*)f^*)^*, \qquad a \in U_q\mathfrak{sl}_2, \quad f \in \mathrm{Pol}(\mathbb{C})_q.$$

Here we use the fact that $*S$ is an antilinear automorphism of $U_q\mathfrak{sl}_2$. The representation π', just as π, endows $\mathrm{Pol}(\mathbb{C})_q$ with a $U_q\mathfrak{sl}_2$-module algebra structure, since also $*S$ is an antiautomorphism of the coalgebra $U_q\mathfrak{sl}_2$ (cf. with similar reasoning in Subsection 1.3.3). Then the equalities of the operators

$$\pi(K^{\pm 1}) = \pi'(K^{\pm 1}), \qquad \pi(E) = \pi'(E), \qquad \pi(F) = \pi'(F)$$

follow from their coincidence on the three-dimensional subspace spanned by 1, z, z^*. The last claim is an easy consequence of the definitions.

In Subsection 1.2.3 $\mathrm{Pol}(\mathbb{C})_q$ was chosen without serious motivation, as a kind of guess. In this subsection we describe the way to the $*$-algebra $\mathrm{Pol}(\mathbb{C})_q$ that not just yields this algebra, but also endows it with a $U_q\mathfrak{su}_{1,1}$-module algebra structure.

1.3.6. The quantum group SL_2. Recall that $q \in (0,1)$. However, the results of this subsection hold for all q that are not roots of unity [**174**, Chap. 3, 4], [**161**, Chap. 6, 7].

Let $n \in \mathbb{Z}_+$ and let $\bar{\omega}$ be a fundamental weight of the Lie algebra $\mathfrak{sl}_2$, that is, $\bar{\omega}(H) = 1$. Denote by $L(n\bar{\omega})$ the $U_q\mathfrak{sl}_2$-module with a generator $v(n\bar{\omega})$ and the defining relations

$$Ev(n\bar{\omega}) = 0, \qquad K^{\pm 1}v(n\bar{\omega}) = q^{\pm n}v(n\bar{\omega}), \qquad F^{n+1}v(n\bar{\omega}) = 0.$$

One can observe that (1.18), (1.19) imply the following statement.

LEMMA 1.46. *The element*

$$C_q = FE + \frac{1}{(q^{-1}-q)^2}\left(q^{-1}K^{-1} + qK - (q^{-1}+q)\right) \tag{1.46}$$

belongs to the center of $U_q\mathfrak{sl}_2$.

Evidently,

$$C_q = EF + \frac{1}{(q^{-1}-q)^2}\left(q^{-1}K + qK^{-1} - (q^{-1}+q)\right). \tag{1.47}$$

Lemma 1.46 and the formula

$$C_q\, v(n\bar{\omega}) = \frac{(q^{-\frac{n}{2}} - q^{\frac{n}{2}})(q^{-(\frac{n}{2}+1)} - q^{\frac{n}{2}+1})}{(q^{-1}-q)^2} v(n\bar{\omega})$$

lead to

COROLLARY 1.47.

$$C_q|_{L(n\bar{\omega})} = \frac{(q^{-\frac{n}{2}} - q^{\frac{n}{2}})(q^{-(\frac{n}{2}+1)} - q^{\frac{n}{2}+1})}{(q^{-1}-q)^2}.$$

This is a q-analog of the well-known formula that shows how the Casimir element $C = FE + \frac{H(H+2)}{4}$ of $\mathfrak{sl}_2$ acts in finite-dimensional $\mathfrak{sl}_2$-modules [**348**, p. 149].

LEMMA 1.48. (1) *$L(n\bar{\omega})$ is a simple weight $U_q\mathfrak{sl}_2$-module of dimension $n+1$.*
(2) *Each finite-dimensional simple weight $U_q\mathfrak{sl}_2$-module is isomorphic to $L(n\bar{\omega})$ for some n.*
(3) *In the category of $U_q\mathfrak{sl}_2$-modules*

$$L(n_1\bar{\omega}) \otimes L(n_2\bar{\omega}) \approx \bigoplus_{n=0}^{\min(n_1,n_2)} L((n_1+n_2-2n)\bar{\omega}), \qquad n_1, n_2 \in \mathbb{Z}_+. \tag{1.48}$$

The proof is similar to the proof in the classical case $q = 1$; see [**174**, p. 61–62, 72].

COROLLARY 1.49. *Let L' and L'' be finite-dimensional pairwise nonisomorphic simple weight $U_q\mathfrak{sl}_2$-modules. Then $C_q|_{L'} \neq C_q|_{L''}$.*

Using Lemma 1.48, one can show [**174**, p. 65] that each finite-dimensional weight $U_q\mathfrak{sl}_2$-module is semisimple, i.e., each of its submodules admits an invariant complement. Using semisimplicity, one can deduce the following result (see the proof of Proposition 2.34):

LEMMA 1.50. *Let V be a weight $U_q\mathfrak{sl}_2$-module such that* $\dim(U_q\mathfrak{sl}_2 \cdot v) < \infty$ *for all $v \in V$. Then it is isomorphic to a direct sum of simple weight finite-dimensional $U_q\mathfrak{sl}_2$-modules.*

More material on quantum universal enveloping algebras and their representations can be found in Section 2.1.

Given a representation $A \to \mathrm{End}(V)$ of an algebra A, its matrix elements are elements of the image of the dual linear operator $\mathrm{End}(V)^* \to A^*$. A direct sum and a tensor product of finite-dimensional weight $U_q\mathfrak{sl}_2$-modules are both finite-dimensional weight $U_q\mathfrak{sl}_2$-modules. Thus matrix elements of such $U_q\mathfrak{sl}_2$-modules form a Hopf subalgebra $\mathbb{C}[SL_2]_q \subset (U_q\mathfrak{sl}_2)^\star$; see Remark 1.30. It is called an algebra of regular functions on the quantum group SL_2 [**199**, p. 96].

As $L(n\bar{\omega})$ is simple and finite dimensional, the Burnside theorem [**65**, p. 177], [**348**, p. 100] allows one to prove that the natural homomorphism of coalgebras $\mathrm{End}(L(n\bar{\omega}))^* \to \mathbb{C}[SL_2]_q$ is injective. Further we identify the vector space $\mathrm{End}(L(n\bar{\omega}))^*$ with its image under the embedding to $\mathbb{C}[SL_2]_q$. Now we formulate an analog of the Peter–Weyl decomposition.

PROPOSITION 1.51. $\mathbb{C}[SL_2]_q \cong \bigoplus_{n=0}^{\infty} \mathrm{End}(L(n\bar{\omega}))^*$.

PROOF. Lemmas 1.48 and 1.50 imply that $\mathbb{C}[SL_2]_q$ is a sum of its subspaces $\mathrm{End}(L(n\bar{\omega}))^*$. Now we should prove their linear independence. But they are eigensubspaces of the linear operator

$$f(\xi) \mapsto f(\xi\, C_q), \qquad \xi \in U_q\mathfrak{sl}_2,$$

in the vector space $\mathbb{C}[SL_2]_q$ with pairwise different eigenvalues. The latter claim can be proved by Corollary 1.49. □

REMARK 1.52. Let $U_q\mathfrak{sl}_2^{\mathrm{op}}$ be the Hopf algebra that differs from $U_q\mathfrak{sl}_2$ by replacing the multiplication with the opposite one. We endow $\mathbb{C}[SL_2]_q$ with a $U_q\mathfrak{sl}_2$-bimodule structure by setting actions of $U_q\mathfrak{sl}_2$ and $U_q\mathfrak{sl}_2^{\mathrm{op}}$, respectively, as follows:

$$R_{\mathrm{reg}}(\xi) : f(\eta) \mapsto f(\eta\xi), \qquad L_{\mathrm{reg}}(\xi) : f(\eta) \mapsto f(\xi\eta),$$

where $f \in \mathbb{C}[SL_2]_q$, $\xi, \eta \in U_q\mathfrak{sl}_2$. It follows from the definitions that $\mathbb{C}[SL_2]_q$ is a $U_q\mathfrak{sl}_2^{\mathrm{op}} \otimes U_q\mathfrak{sl}_2$-module algebra. Proposition 1.51 provides a decomposition of $\mathbb{C}[SL_2]_q$ into a sum of simple pairwise nonisomorphic $U_q\mathfrak{sl}_2^{\mathrm{op}} \otimes U_q\mathfrak{sl}_2$-modules. (In the sequel we prefer representation R_{reg} to representation L_{reg} and use the brief notation $\xi f = R_{\mathrm{reg}}(\xi) f$.)

There exists another more visual description of the Hopf algebra $\mathbb{C}[SL_2]_q$ in terms of generators and relations, which is frequently used. Let $\{e_1, e_2\}$ be the standard basis of the Hermitian space $\mathbb{C}^2$. Consider the representation π of $U_q\mathfrak{sl}_2$ in $\mathbb{C}^2$ defined by (1.22) and its matrix elements $t_{ij}(\xi) \stackrel{\mathrm{def}}{=} (\pi(\xi)e_j, e_i)$. It is easy to prove the following statements; see [**321**].

PROPOSITION 1.53. *The set* $\{t_{11}, t_{12}, t_{21}, t_{22}\}$ *generates the algebra* $\mathbb{C}[SL_2]_q$, *and one has*

$$t_{11}t_{12} = qt_{12}t_{11}, \quad t_{11}t_{21} = qt_{21}t_{11}, \quad t_{12}t_{22} = qt_{22}t_{12}, \quad t_{21}t_{22} = q\,t_{22}t_{21},$$
$$t_{12}t_{21} = t_{21}t_{12}, \quad t_{11}t_{22} - t_{22}t_{11} = (q - q^{-1})\,t_{12}t_{21},$$

$$t_{11}t_{12} - qt_{12}t_{21} = 1. \tag{1.49}$$

This set of relations between $\{t_{ij}\}_{i,j=1,2}$ *is defining.*[8] *An action of a coproduct* Δ, *a counit* ε, *and an antipode* S *is determined by the following:*

$$\Delta(t_{ij}) = t_{i1} \otimes t_{1j} + t_{i2} \otimes t_{2j}, \quad \varepsilon(t_{ij}) = \delta_{i,j}, \quad i, j = 1, 2,$$
$$\begin{pmatrix} S(t_{11}) & S(t_{12}) \\ S(t_{21}) & S(t_{22}) \end{pmatrix} = \begin{pmatrix} t_{22} & -q^{-1}t_{12} \\ -qt_{21} & t_{11} \end{pmatrix}.$$

[8]In other words, the corresponding set of elements of the free algebra $\mathbb{C}\langle t_{11}, t_{12}, t_{21}, t_{22}\rangle$ generates the kernel of its natural homomorphism to $\mathbb{C}[SL_2]_q$.

Let us explain the first claim. Due to (1.48), the simple $U_q\mathfrak{sl}_2$-module $L(n\bar{\omega})$ admits an embedding to $L(\bar{\omega})^{\otimes n}$. Thus Proposition 1.51 implies that the subspace $\mathrm{End}(L(\bar{\omega}))^*$ generates the algebra $\mathbb{C}[SL_2]_q$. But this subspace is the linear span of the set $\{t_{11}, t_{12}, t_{21}, t_{22}\}$.

COROLLARY 1.54. (1) *The equalities*

$$\pi_+(t_{11})e_j = \sqrt{q^{-2(j+1)}-1}\,e_{j+1}, \qquad \pi_+(t_{12})e_j = q^{-j}e_j,$$

$$\pi_+(t_{21})e_j = -q^{-(j+1)}e_j, \qquad \pi_+(t_{22})e_j = \begin{cases} -\sqrt{q^{-2j}-1}\,e_{j-1}, & j \neq 0, \\ 0, & j = 0 \end{cases}$$

determine an irreducible infinite-dimensional representation π_+ *of* $\mathbb{C}[SL_2]_q$ *in a vector space with the basis* $\{e_j\}_{j\in\mathbb{Z}_+}$.

(2) *The equalities*

$$\rho(t_{11})f(z) = zf(z), \quad \rho(t_{12})f(z) = \rho(t_{21})f(z) = 0, \quad \rho(t_{22})f(z) = z^{-1}f(z)$$

determine an infinite-dimensional representation ρ *of* $\mathbb{C}[SL_2]_q$ *in the vector space* $\mathbb{C}[z^{\pm 1}]$ *of Laurent polynomials in variable* z.

Using the representation $\pi_+ \otimes \rho$, one can easily check the linear independence of the elements

$$\{t_{11}^a t_{12}^b t_{21}^c, \quad t_{12}^b t_{21}^c t_{22}^d \mid a, b, c, d \in \mathbb{Z}_+\} \tag{1.50}$$

of $\mathbb{C}[SL_2]_q$. Therefore we have

COROLLARY 1.55. *The set* (1.50) *is a basis of the vector space* $\mathbb{C}[SL_2]_q$.

REMARK 1.56. One can easily show that $\mathbb{C}[SL_2]_q$ is a Noetherian domain [**49**, p. 18] using the description discussed above in terms of generators and relations.

Consider a coalgebra A with coproduct $\triangle_A$ and counit ε_A. A vector space F is called a (right) A-comodule if there exists a linear map $\triangle_F : F \to F \otimes A$ such that

$$(\triangle_F \otimes \mathrm{id}_A)\cdot\triangle_F = (\mathrm{id}_F \otimes \triangle_A)\cdot\triangle_F, \qquad (\mathrm{id}_F \otimes \varepsilon_A)\cdot\triangle_F = \mathrm{id}_F.$$

To each $U_q\mathfrak{sl}_2$-module F one can assign a linear map

$$F \to \mathrm{Hom}(U_q\mathfrak{sl}_2, F)$$

that sends $f \in F$ to the linear map

$$U_q\mathfrak{sl}_2 \to F, \qquad a \mapsto af.$$

Given a weight locally finite-dimensional $U_q\mathfrak{sl}_2$-module F, the image of this map belongs to the subspace $F \otimes \mathbb{C}[SL_2]_q \hookrightarrow \mathrm{Hom}(U_q\mathfrak{sl}_2, F)$, and the corresponding linear map $\triangle_F : F \to F \otimes \mathbb{C}[SL_2]_q$ endows F with a $\mathbb{C}[SL_2]_q$-comodule structure. The linear map is a homomorphism of unital algebras in case F is a unital $U_q\mathfrak{sl}_2$-module algebra with weight and locally finite action of $U_q\mathfrak{sl}_2$ in F. Thus F is endowed with a $\mathbb{C}[SL_2]_q$-comodule algebra structure.

EXAMPLE 1.57. For obvious reasons, $\mathbb{C}[t_1, t_2]_q$ is a $\mathbb{C}[SL_2]_q$-comodule algebra:

$$\triangle : \mathbb{C}[t_1, t_2]_q \to \mathbb{C}[t_1, t_2]_q \otimes \mathbb{C}[SL_2]_q, \qquad \triangle : t_j \mapsto t_1 \otimes t_{1j} + t_2 \otimes t_{2j}.$$

It is important to note that the $U_q\mathfrak{sl}_2$-modules $\mathbb{C}[t_1^{\pm 1}, t_2^{\pm 1}]_q$, $\mathbb{C}[z]_q$, which are crucial in the subject considered, are not locally finite dimensional. Hence in the sequel we prefer to use the language of $U_q\mathfrak{sl}_2$-module algebras even in the cases when an action of $U_q\mathfrak{sl}_2$ in F is weight and locally finite dimensional; cf. [**161**, Chap. 4].

1.3.7. Appendix on rings and modules of fractions. Let S be a multiplicative subset of a ring R. It means that $S \cdot S \subset S$ and $1 \in S$. A ring R_S together with a homomorphism $\varphi : R \to R_S$ is called a ring of fractions R with respect to S if the following requirement of universality holds.

Given any homomorphism of rings $\psi : R \to Q$, the invertibility of all $\psi(s)$, $s \in S$, implies the existence and uniqueness of a homomorphism $\psi_S : R_S \to Q$ such that $\psi = \psi_S \circ \varphi$.

Given two rings of fractions $\varphi' : R \to R'_S$, $\varphi'' : R \to R''_S$, there exists a unique isomorphism of rings $f : R'_S \to R''_S$ such that $\varphi'' = f\,\varphi'$. In other words, the ring of fractions is essentially unique.

In proving its existence, one can assume that R is defined by generators and defining relations. To each $s \in S$ we assign an additional generator $\overline{s}$ together with two additional relations $s\overline{s} = 1$, $\overline{s}s = 1$. Let R_S be the ring defined by expanded sets of generators and relations. The natural embedding of sets of generators leads to a homomorphism $\varphi : R \to R_S$. Obviously, it is a localization of R with respect to S.

Let R be a graded ring and let S consist of homogeneous elements. Then by a *ring of fractions* one means a graded ring R_S together with a homomorphism of graded rings $\varphi : R \to R_S$ if a requirement of universality similar to the above holds. One can prove the existence and the uniqueness of a ring of fractions in this case just as before.

Given a right R-module M, then $M_S = M \otimes_R R_S$ is a right R_S-module. It is called a module of fractions.

In the definitions above a multiplicative set S can be arbitrary, and the ring of fractions R_S may possess some hardly predictable properties. Let us introduce additional definitions and some related restrictions on S [**210**, p. 299].

A ring of fractions R_S together with a homomorphism $\varphi : R \to R_S$ is called a classical *right* ring of fractions if

(1) $\varphi(s)$ is an invertible element of R_S for all $s \in S$,
(2) every element of R_S has the form $\varphi(r)\varphi(s)^{-1}$, $r \in R$, $s \in S$,
(3) $\operatorname{Ker}\varphi = \{r \in R \mid rs = 0 \text{ for some } s \in S\}$.

The following result is well known.

Proposition 1.58. [**43**, p. 110], [**210**, p. 300] *A ring R_S is a classical right ring of fractions if and only if S has the following properties:*

(1) *for all $s \in S$, $r \in R$*

$$rS \cap sR \neq 0; \tag{1.51}$$

(2) *if $s'r = 0$ for some $s' \in S$, $r \in R$, then $rs = 0$ for some $s \in S$.*

In the same way one can introduce classical left rings of fractions, and an analogous statement holds for them. Instead of (1.51) we have the requirement

$$Sr \cap Rs \neq \varnothing \quad \text{for all} \quad s \in S,\ r \in R. \tag{1.52}$$

In most applications (1.52), (1.51) (which are called the Ore properties) hold and S does not contain zero divisors.

1.3.8. Appendix on algebras and modules in tensor categories. Many of the properties of the universal R-matrix listed in Subsection 1.3.4 admit an interpretation within category theory [**150**].

For example, it is explained in the concluding remarks of this subsection that the category $\mathcal{C}^-$ introduced in Subsection 1.3.4 is an abelian braided tensor category. Recall that a $U_q\mathfrak{sl}_2$-module belongs to this category if and only if it is weight and locally $U_q\mathfrak{b}^-$-finite dimensional.

An algebra in the category $\mathcal{C}^-$ is a unital $U_q\mathfrak{sl}_2$-module algebra which belongs to $\mathcal{C}^-$. A left module over an algebra F in the category $\mathcal{C}^-$ is a $U_q\mathfrak{sl}_2$-covariant left F-module which belongs to $\mathcal{C}^-$. In a similar way one can introduce the notions of a right module and bimodule over an algebra F in the category $\mathcal{C}^-$.

An algebra F in the category $\mathcal{C}^-$ is called commutative in this category if $m = m\,\check{R}_{F,F}$, where $m : F \otimes F \to F$ is a multiplication in F.

Consider a bimodule E over a commutative algebra F in the category $\mathcal{C}^-$:

$$m_{\text{left}} : F \otimes E \to E, \qquad m_{\text{right}} : E \otimes F \to E.$$

As in [**221**], we call E symmetric if

$$m_{\text{right}} = m_{\text{left}}\,\check{R}_{E,F}.$$

EXAMPLE 1.59. An algebra F in the category $\mathcal{C}^-$ is a bimodule over any subalgebra F' of F. Given a commutative algebra F in $\mathcal{C}^-$, this bimodule is symmetric.

PROPOSITION 1.60. [**246**, Remark 1.1] *Let E_1, E_2 be symmetric bimodules over an algebra F in the category $\mathcal{C}^-$. Then the bimodule $E_1 \otimes_F E_2$ is also symmetric.*

PROPOSITION 1.61. [**246**, Remark 1.2] *Let E be a left module over a commutative algebra F in $\mathcal{C}^-$ (with $m_{\text{left}} : F \otimes E \to E$). The morphism*

$$m_{\text{right}} = m_{\text{left}}\check{R}_{E,F}$$

endows E with a structure of symmetric bimodule over F in $\mathcal{C}^-$.

This means that each left module E over a commutative algebra F in $\mathcal{C}^-$ is a symmetric bimodule over F in $\mathcal{C}^-$.

A bimodule E over a commutative algebra F in the category $\mathcal{C}^-$ is called local if

$$m_{\text{right}} = m_{\text{left}}\,\check{R}_{E,F}, \qquad m_{\text{left}} = m_{\text{right}}\,\check{R}_{F,E}$$

(this counterpart for the term "dyslectic", introduced by Pareigis in [**246**], is used in works on conformal quantum field theory).

PROPOSITION 1.62. [**246**, Pr. 2.4, Th. 2.5] *Let E_1, E_2 be local bimodules over a commutative algebra F in the category $\mathcal{C}^-$, j the canonical epimorphism $E_1 \otimes E_2 \to E_1 \otimes_F E_2$. Then*

1. *$E_1 \otimes_F E_2$ is a local bimodule over F;*
2. *there exists a unique morphism*

$$\bar{R}_{E_1,E_2} : E_1 \otimes_F E_2 \to E_2 \otimes_F E_1,$$

in the category $\mathcal{C}^-$ such that $\bar{R}_{E_1,E_2}\, j = j\,\check{R}_{E_1,E_2}$.

This means that the category of local bimodules over a commutative algebra in the category $\mathcal{C}^-$ is an abelian braided tensor category (see definitions in the concluding part of this subsection).

Let us discuss some notions that are not usually mentioned in books on category theory, for example in [**50**].

We start with tensor categories. The category $\mathcal{C}_A$ of modules over a Hopf algebra A serves as a principal example.

Consider a category $\mathcal{C}$ with the distinguished object I. To define a functor $\otimes : \mathcal{C} \times \mathcal{C} \to \mathcal{C}$, one should introduce tensor products of objects and tensor products of morphisms so that

$$\mathrm{id}_{U\otimes V} = \mathrm{id}_U \otimes \mathrm{id}_V, \qquad (f'' \otimes g'')(f' \otimes g') = f''f' \otimes g''g',$$

where U, V are objects of the category $\mathcal{C}$ and $f' \in \mathrm{Hom}(U, U')$, $f'' \in \mathrm{Hom}(U', U'')$, $g' \in \mathrm{Hom}(V, V')$, $g'' \in \mathrm{Hom}(V', V'')$ are morphisms. Details can be found in [**161**, p. 350] and [**159**, p. 22].

The structure of the tensor category in $\mathcal{C}$ is determined by families of isomorphisms

$$\alpha_{X,Y,Z} : X \otimes (Y \otimes Z) \overset{\approx}{\to} (X \otimes Y) \otimes Z, \tag{1.53}$$

$$\lambda_X : I \otimes X \overset{\approx}{\to} X, \qquad \rho_X : X \otimes I \overset{\approx}{\to} X, \tag{1.54}$$

subject to the natural requirements that hold for the category of modules over a Hopf algebra [**161**, pp. 351–357]. In the latter special case the trivial A-module $\mathbb{C}$ serves as a distinguished element, and $\alpha_{X,Y,Z}$, λ_X, ρ_X are the canonical isomorphisms of vector spaces. This allows one to use the equality sign instead of $\approx$ with no risk of gaps in arguments. In the general case one usually replaces the above tensor category by an equivalent tensor category in which (1.53), (1.54) are equalities [**161**, p. 361], i.e., by a *strict* tensor category.

An algebra in a tensor category $\mathcal{C}$ is an object F together with a pair of morphisms

$$m : F \otimes F \to F, \qquad \mathbb{1} : I \to F,$$

subject to the following requirements [**225**, p.438]:

$$m(m \otimes \mathrm{id}_F) = m(\mathrm{id}_F \otimes m), \qquad m(\mathbb{1} \otimes F) = m(F \otimes \mathbb{1}) = \mathrm{id}_F.$$

A left module over an algebra F in $\mathcal{C}$ is an object E and a morphism $m_{\mathrm{left}} : F \otimes E \to E$ so that

$$m_{\mathrm{left}}(m \otimes \mathrm{id}_E) = m_{\mathrm{left}}(\mathrm{id}_F \otimes m_{\mathrm{left}}), \qquad m_{\mathrm{left}}(\mathbb{1} \otimes \mathrm{id}_{\mathrm{E}}) = \mathrm{id}_{\mathrm{E}}.$$

EXAMPLE 1.63. Consider the tensor category $\mathcal{C}_A$ of modules over a Hopf algebra A. Each unital A-module algebra F is an algebra in $\mathcal{C}_A$, and each A-covariant left F-module is a left module over F in $\mathcal{C}_A$.

In a similar way one can introduce the notions of a right module and a bimodule in a tensor category. In the case of a bimodule, it is required that left and right actions

$$m_{\mathrm{left}} : F \otimes E \to E, \qquad m_{\mathrm{right}} : E \otimes F \to E,$$

are permutable as follows:

$$m_{\mathrm{right}}(\mathrm{id}_F \otimes m_{\mathrm{left}}) = m_{\mathrm{left}}(\mathrm{id}_F \otimes m_{\mathrm{right}}).$$

A strict tensor category $\mathcal{C}$ is called a *braided tensor category* if to each pair V', V'' of objects of $\mathcal{C}$ one assigns an isomorphism $\check{R}_{V',V''} : V' \otimes V'' \to V'' \otimes V'$ which possesses the properties listed in Propositions 1.40 and 1.41. The isomorphism $\check{R}_{V',V''}$ is called a *braiding*. For nonstrict tensor categories the isomorphisms $\alpha_{X,Y,Z}$ appear in the definition; see [**161**, p. 395].

A braided tensor category is called *symmetric* if $\check{R}_{V'',V'}\check{R}_{V',V''} = \mathrm{id}_{V'\otimes V''}$ [**161**, p. 400].

An algebra F in a braided tensor category is called *commutative* if $m = m\check{R}_{F,F}$, where $m : F \otimes F \to F$ is a multiplication in F. Modules over commutative algebras in braided abelian tensor categories are used in algebraic K-theory [**221**] and naturally arise in conformal quantum field theory [**169, 97**].[9]

1.4. An invariant integral

1.4.1. Invariant integrals and $*$-representations. Consider an A-module algebra F. A linear functional $\nu : F \to \mathbb{C}$ is called an A-invariant integral if ν is a morphism of A-modules; that is

$$\nu(af) = \varepsilon(a)\nu(f), \qquad a \in A, \quad f \in F,$$

where ε is the counit of the Hopf algebra A. This could be rephrased by saying that an invariant integral is an A-invariant element of the A-module F^*.

Consider a module V over a Hopf algebra A, the A-module algebra $\operatorname{End} V$ of all linear operators in V, and an A-module subalgebra

$$\operatorname{End}_f V = \{T \in \operatorname{End} V |\ \dim(TV) < \infty\}.$$

Note that the algebra $\operatorname{End}_f V$ has no unit if $\dim V = \infty$. Our immediate goal is to produce a nonzero A-invariant integral on $\operatorname{End}_f V$ under the assumption that the square of the antipode S of the Hopf algebra A is an inner automorphism; that is,

$$S^2(a) = gag^{-1}, \qquad a \in A, \tag{1.55}$$

for some invertible $g \in A$. Let π be the representation of A corresponding to the A-module V. Then the representation $\pi \circ S^2$ is equivalent to π because

$$\pi(S^2(a)) = \pi(g)\pi(a)\pi(g)^{-1}, \qquad a \in A.$$

Thus the linear map

$$V \to V^{**}, \qquad v \mapsto \pi(g)v,$$

is a morphism of A-modules, and the sequence of morphisms of A-modules

$$\operatorname{End}_f V \overset{\cong}{\to} V \otimes V^* \hookrightarrow V^{**} \otimes V^* \to \mathbb{C}$$

leads to the following well-known result.

PROPOSITION 1.64. *The linear functional*

$$\nu : \operatorname{End}_f(V) \to \mathbb{C}, \qquad \nu : T \mapsto \operatorname{tr}(T\pi(g)),$$

is an A-invariant integral.

[9] Abelian tensor categories are defined in an obvious way [**112**]: the functor $\otimes$ has to be biadditive [**50**, p. 106, 98].

If g from (1.55) satisfies the additional conditions

$$\triangle(g) = g \otimes g, \qquad \varepsilon(g) = 1, \tag{1.56}$$

the value $\mathrm{tr}_q(T) = \mathrm{tr}(T\pi(g))$ is called a q-trace of a linear operator T in the space V. Due to (1.56),

$$\mathrm{tr}_q(T \otimes U) = \mathrm{tr}_q T \cdot \mathrm{tr}_q U.$$

In the case of finite-dimensional representation π one can consider the q-trace of the identity operator, referred to as a q-dimension of the representation space:

$$\dim_q V \stackrel{\mathrm{def}}{=} \mathrm{tr}_q \mathbf{1} = \mathrm{tr}\, \pi(g).$$

We are about to produce an analog for the *integration by parts* formula.

PROPOSITION 1.65. *Let ν be an invariant integral on an A-module algebra F. One has*

$$\nu((af_1) \cdot f_2) = \nu(f_1 \cdot S(a)f_2), \qquad f_1, f_2 \in F, \quad a \in A. \tag{1.57}$$

PROOF. Let us demonstrate that for A-modules U, V and an arbitrary morphism of A-modules $l : V \otimes U \to \mathbb{C}$ one has

$$l(av \otimes u) = l(v \otimes S(a)u), \qquad u \in U,\ v \in V,\ a \in A. \tag{1.58}$$

Once this is done, it then suffices to set $U = V = F$ and $l(f_1 \otimes f_2) = \nu(f_1 f_2)$.

Obviously, (1.58) is valid in the special case of *natural pairing* $l_{\mathrm{nat}} : U^* \otimes U \to \mathbb{C}$. The general case can be reduced to this special case via an application of the morphism of A-modules $V \to U^*$ determined by l:

$$V \cong V \otimes \mathbb{C} \to V \otimes U \otimes U^* \stackrel{l \otimes \mathrm{id}}{\longrightarrow} \mathbb{C} \otimes U^* \cong U^*. \qquad \square$$

Now we are about to describe one of methods to produce $*$-representations of Hopf $*$-algebras. Recall the terminology. A form $v' \times v'' \mapsto (v', v'')$ on a vector space V is called a *sesquilinear form* if it is linear in v' and antilinear in v''. It is said to be *Hermitian* if $(v'', v') = \overline{(v', v'')}$ for all $v', v'' \in V$. A Hermitian form is called nonnegative if $(v, v) \geq 0$ for all $v \in V$, and positive if $(v, v) > 0$ for all nonzero $v \in V$.

Consider a vector space V endowed with a Hermitian form $(\cdot, \cdot)$. A representation π of a $*$-algebra A in V is called a *$*$-representation* if

$$(\pi(a)v', v'') = (v', \pi(a^*)v''), \quad a \in A, \quad v', v'' \in V.$$

In terms of A-modules this means that

$$(av', v'') = (v', a^*v''), \quad a \in A, \quad v', v'' \in V. \tag{1.59}$$

PROPOSITION 1.66. *Consider a Hopf $*$-algebra A, an A-module $*$-algebra F, and an invariant integral $\nu : F \to \mathbb{C}$. Suppose that ν is real in the sense that*

$$\nu(f^*) = \overline{\nu(f)}, \quad f \in F. \tag{1.60}$$

Then the representation of A in the vector space F with the Hermitian form

$$(f_1, f_2) = \nu(f_2^* f_1), \quad f_1, f_2 \in F, \tag{1.61}$$

is a $*$-representation.

PROOF. We intend to introduce the notion of an *A-invariant sesquilinear form*, to prove that (1.59) is equivalent to A-invariance, and finally to establish the A-invariance of the Hermitian form (1.61).

It is known [**211**, p. 388] that each vector space V admits the associated antispace $\overline{V}$ which differs from V just in a specific action of $\mathbb{C}$:

$$\lambda : v \mapsto \bar{\lambda} v, \qquad \lambda \in \mathbb{C}, \qquad v \in V.$$

The A-module structure transfers from V to $\overline{V}$ as follows:

$$a : v \mapsto S(a)^* v, \qquad a \in A, \quad v \in V,$$

and so the vector space of sesquilinear forms $V \times V \to \mathbb{C}$ becomes naturally isomorphic to $(\overline{V} \otimes V)^*$. We endow the vector space of sesquilinear forms on V with a structure of A-module via this isomorphism.

Consider an A-invariant sesquilinear form $(\cdot,\cdot)$ on V. It determines the linear map

$$\overline{V} \hookrightarrow V^*, \quad v \mapsto (\cdot, v),$$

which is a morphism of A-modules. Equivalently,

$$(v', S(a)^* v'') = (S(a)v', v''), \quad a \in A, \quad v', v'' \in V.$$

Since the antipode S is invertible, one has

$$(v', a^* v'') = (av', v''), \quad a \in A, \quad v', v'' \in V.$$

Thus the invariance of the sesquilinear form implies (1.59).

Conversely, suppose that the sesquilinear form $(\cdot,\cdot)$ on V satisfies (1.59). Let $\Delta(a) = \sum_i b_i \otimes c_i$, then one can deduce from (1.59) that

$$\Delta(a)(v', v'') = \sum_i (c_i v', S(b_i)^* v'') = \left(v', \sum_i c_i^* S(b_i)^* v'' \right)$$
$$= \left(v', \left(\sum_i S(b_i) c_i \right)^* v'' \right) = (v', \overline{\varepsilon(a)} v'') = \varepsilon(a)(v', v'').$$

It is exactly the A-invariance condition for the sesquilinear form $(\cdot,\cdot)$.

It remains to prove the A-invariance of the Hermitian form (1.61). As above, we suppose $\Delta(a) = \sum_i b_i \otimes c_i$. Using $(*\, S)^2 = \mathrm{id}$ and (1.44), we obtain

$$\sum_i (c_i f_1, S(b_i)^* f_2) = \sum_i \nu(b_i(f_2^*) \cdot c_i(f_1)), \qquad f_1, f_2 \in F.$$

On the other hand, F is an A-module algebra and ν is an A-invariant integral. Hence,

$$\sum_i \nu(b_i(f_2^*) \cdot c_i(f_1)) = \nu(a(f_2^* f_1)) = \varepsilon(a)\nu(f_2^* f_1) = \varepsilon(a) \cdot (f_1, f_2). \qquad \square$$

1.4.2. Does there exist a nonzero nonnegative $U_q\mathfrak{sl}_2$-invariant integral? We start with some standard definitions.

Vector subspaces $\{F^{(j)}A\}_{j \in \mathbb{Z}}$ of an algebra A determine a filtration of this algebra if

(1) $F^{(j)}A \subset F^{(k)}A$, for $j < k$,
(2) $1 \in F^{(0)}A$,
(3) $F^{(j)}A \cdot F^{(k)}A \subset F^{(j+k)}A$, for $j, k \in \mathbb{Z}$.

An algebra together with a distinguished filtration is called a *filtered algebra.*

EXAMPLE 1.67. Each two-sided ideal I of A determines a filtration

$$F^{(j)}A = \begin{cases} A, & j \in \mathbb{Z}_+, \\ I^{-j}, & j \in -\mathbb{N}. \end{cases}$$

This filtration is called *I-adic.*

The filtrations considered in the book are assumed to possess the following additional properties: $\bigcup_{j\in\mathbb{Z}} F^{(j)}A = A$, $\bigcap_{j\in\mathbb{Z}} F^{(j)}A = 0$. The degree of a nonzero element $f \in A$ of such algebra is defined by $\deg(f) = \min\left\{j \in \mathbb{Z} | \, f \in F^{(j)}A\right\}$.

EXAMPLE 1.68. If $A = \bigoplus_{j\in\mathbb{Z}} A_j$ is a graded algebra, then the sequence

$$F^{(j)}A = \bigoplus_{k\leq j} A_k, \qquad j \in \mathbb{Z},$$

is a filtration of A.

The vector space $\mathrm{gr}(A) = \bigoplus_{j\in\mathbb{Z}}(F^{(j)}A/F^{(j-1)}A)$ is endowed with a structure of algebra as follows:

$$\left[x + F^{(j-1)}A\right] \cdot \left[y + F^{(k-1)}A\right] = xy + F^{(j+k-1)}A,$$

where $x \in F^{(j)}A/F^{(j-1)}A$, $y \in F^{(k)}A/F^{(k-1)}A$. The algebra $\mathrm{gr}(A)$ is $\mathbb{Z}$-graded with homogeneous components $\mathrm{gr}(A)_j = F^{(j)}A/F^{(j-1)}A$. It is called the associated graded algebra. The functors related to the passage from filtered algebras to associated graded algebras are described in [**241**, pp. 278–292].

EXAMPLE 1.69. Consider the algebra $\mathrm{Pol}(\mathbb{C})_q$ together with the filtration determined by

$$\deg(z) = \deg(z^*) = 1.$$

Here the associated graded algebra appears to be the noncommutative algebra $\mathbb{C}[z, z^*]_{q^2}$ whose list of generators z, z^* is reproduced literally and the defining relation is $z^*z = q^2zz^*$.

The passage from A to $\mathrm{gr}(A)$ is a frequently used trick. For example, it was explained in Subsection 1.3.3 that the set $\{z^j z^{*k} | \, j, k \in \mathbb{Z}_+\}$ is a basis of the vector space $\mathbb{C}[z, z^*]_{q^2}$. Thus one has

PROPOSITION 1.70. *The set* $\left\{z^j z^{*k} | \, j, k \in \mathbb{Z}_+\right\}$ *is a basis of the vector space* $\mathrm{Pol}(\mathbb{C})_q$.

Similar arguments can be used to prove that the set $\left\{z^{*j} z^k | \, j, k \in \mathbb{Z}_+\right\}$ is a basis of the vector space $\mathrm{Pol}(\mathbb{C})_q$.

The passage to $\mathrm{gr}(\mathrm{Pol}(\mathbb{C})_q)$ makes it easy to show that the algebra $\mathrm{Pol}(\mathbb{C})_q$ has no zero divisors. In fact, the following simple statement is valid [**49**, p. 109].

PROPOSITION 1.71. *If the associated graded algebra* $\mathrm{gr}(A)$ *is a domain, then* A *is also a domain.*

A linear functional ν on a $*$-algebra A is called nonnegative if $\nu(f^*f) \geq 0$ for all $f \in A$. A nonnegative linear functional ν is said to be positive if $\nu(f^*f) \neq 0$ for $f \neq 0$.

A nonnegative linear functional is certainly real (see (1.60)) because the nonnegative matrix $(\nu(f_i^* f_j))_{i,j=1,2}$ is Hermitian.

To answer the question formulated in the title of this section, the central element (1.47) of $U_q\mathfrak{sl}_2$ will be used, together with the relation

$$C_q f = 0, \qquad f \in \mathbb{C}[z]_q, \tag{1.62}$$

whose proof readily reduces to the obvious special cases $f = 1$ and $f = z$.

PROPOSITION 1.72. *There exists no nonzero nonnegative $U_q\mathfrak{sl}_2$-invariant integral on* $\mathrm{Pol}(\mathbb{C})_q$.

PROOF. Let ν be a nonzero real $U_q\mathfrak{sl}_2$-invariant integral on $\mathrm{Pol}(\mathbb{C})_q$.

The set $\{z^{*m}z^n\}_{m,n\in\mathbb{Z}_+}$ is a basis of the vector space $\mathrm{Pol}(\mathbb{C})_q$. It follows from $\nu(K^{\pm 1}f) = \nu(f)$ that $\nu(z^{*m}z^n) = 0$ for $m \neq n$. The relations $\nu(Ff) = 0$, $Fz = q^{\frac{1}{2}}$ imply $\nu(1) = 0$.

Thus, $\nu(z^{*n}z^n) \neq 0$ for some $n \in \mathbb{N}$. On the other hand, z^{n+1} differs from $E^n z$ only by a scalar nonzero factor. Hence $\nu((E^{n-1}z)^*(E^{n-1}z)) \neq 0$. Now nonnegativity of ν implies $\nu((E^{n-1}z)^*(E^{n-1}z)) > 0$.

To obtain a desired contradiction, it suffices to prove the presence of negative numbers in the sequence $\{\nu((E^n z)^*(E^n z))\}_{n\in\mathbb{Z}_+}$. Due to (1.44), (1.57), this is equivalent to the presence of negative numbers in the sequence $\{\nu(z^*(F^n E^n z))\}_{n\in\mathbb{Z}_+}$.

The signs of these numbers interchange since

$$F^n E^n z = -F^{n-1}\frac{1}{(q^{-1}-q)^2}\left(q^{-1}K^{-1} + qK - (q^{-1}+q)\right)E^{n-1}z.$$

The latter relation follows from (1.46), together with

$$F^n E^n z = F^{n-1}C_q E^{n-1}z - F^{n-1}\frac{1}{(q^{-1}-q)^2}\left(q^{-1}K^{-1} + qK - (q^{-1}+q)\right)E^{n-1}z$$

and (1.62). □

This result was just the one to be expected. In fact, the $*$-algebra $\mathrm{Pol}(\mathbb{C})_q$ is closely related to the quantum analog of the unit disc $\mathbb{D} = \{z \in \mathbb{C} \mid |z| < 1\}$, as it was demonstrated in Section 1.2. In the classical case $q = 1$ an invariant integral on $\mathbb{D}$ is well known [**124**, p. 46]:

$$\nu(f) = \int_{\mathbb{D}} f \cdot (1 - |z|^2)^{-2} d\,\mathrm{Re}\, z\, d\,\mathrm{Im}\, z,$$

so that $\nu(f^* f) = +\infty$ for every nonzero polynomial f.

To avoid divergent integrals, it suffices to replace the polynomial algebra by the algebra $\mathcal{D}(\mathbb{D})$ of smooth functions with compact support in $\mathbb{D}$. The Hilbert space $L^2(d\nu)$ used in harmonic analysis on $\mathbb{D}$ does not become out of context, as it is a completion of $\mathcal{D}(\mathbb{D})$ with respect to the norm $\|f\| = \nu(f^* f)^{\frac{1}{2}}$.

We will need a $U_q\mathfrak{su}_{1,1}$-module $*$-algebra $\mathcal{D}(\mathbb{D})_q$, which is a q-analog of the algebra $\mathcal{D}(\mathbb{D})$, to be introduced in the next subsection. It will be shown that a positive $U_q\mathfrak{sl}_2$-invariant integral on $\mathcal{D}(\mathbb{D})_q$ exists and is unique up to a scalar factor.

1.4.3. Distributions and finite functions. We are about to introduce a space $\mathcal{D}(\mathbb{D})'_q$, which is a q-analog of the space $\mathcal{D}(\mathbb{D})'$ of distributions on the unit disc. To do this, we endow the vector space $\mathrm{Pol}(\mathbb{C})_q$ with a topology and then pass to the associated completion.

Consider the faithful $*$-representation T_F of the $*$-algebra $\mathrm{Pol}(\mathbb{C})_q$ in the pre-Hilbert space $\mathcal{H}$ with the distinguished orthonormal basis $\{e_n\}_{n\in\mathbb{Z}_+}$ as in Subsection 1.2.3. Endow $\mathrm{Pol}(\mathbb{C})_q$ with the weakest topology $\mathcal{T}$ which makes continuous all the linear functionals

$$l'_{j,k}(f) \stackrel{\mathrm{def}}{=} (T_F(f)e_j, e_k), \qquad j,k\in\mathbb{Z}_+.$$

It is essential to observe that these are just the matrix elements of the Fock representation in the distinguished basis.

A different description of the topology $\mathcal{T}$ is as follows. It was shown in Subsection 1.2.3 that each element $f\in\mathrm{Pol}(\mathbb{C})_q$ is a finite sum of the form

$$f=\sum_{m\in\mathbb{N}} z^m\psi_m(y)+\psi_0(y)+\sum_{m\in\mathbb{N}}\psi_{-m}(y)z^{*m}, \tag{1.63}$$

and this expansion is unique. Here $y=1-zz^*$, and ψ_m are polynomials in a single variable.

PROPOSITION 1.73. *$\mathcal{T}$ is the weakest topology which makes continuous all the linear functionals*

$$l''_{m,n}(f) \stackrel{\mathrm{def}}{=} \psi_m\left(q^{2n}\right), \qquad m\in\mathbb{Z}, \quad n\in\mathbb{Z}_+, \tag{1.64}$$

i.e., just the values of the coefficients $\psi_m(y)$ of the expansion (1.63) *at points of the geometric series $q^{2\mathbb{Z}_+}$.*

PROOF. Let $v_m=T_F(z^m)e_0$ and $l_{j,k}(f)$ be the matrix elements of the operators $T_F(f)$ in the basis $\{v_m\}_{m\in\mathbb{Z}_+}$. It follows from the definitions that $\mathcal{T}$ is the weakest topology which makes continuous all the linear functionals $l_{j,k}$, $j,k\in\mathbb{Z}_+$. It remains to note that $l_{j,k}$ differs from the linear functional $l''_{k-j,\min(j,k)}$ only by a nonzero scalar factor. □

Let us endow the space of all functions on the set $q^{2\mathbb{Z}_+}$ with the topology of pointwise convergence. It is complete in this topology [**84**, p. 50], and the restrictions of polynomials form a dense linear subvariety. Thus the space of all functions on $q^{2\mathbb{Z}_+}$ with the topology of pointwise convergence is the completion of the space of polynomials with the induced topology [**84**, p. 51].

Similarly, consider the vector space $\mathcal{D}(\mathbb{D})'_q$ of *formal series* of the form (1.63) with coefficients from the space of functions on $q^{2\mathbb{Z}_+}$. We endow $\mathcal{D}(\mathbb{D})'_q$ with the topology of pointwise convergence of the coefficients ψ_j on $q^{2\mathbb{Z}_+}$. Obviously, $\mathcal{D}(\mathbb{D})'_q$ is a complete topological vector space, and $\mathrm{Pol}(\mathbb{C})_q$ is naturally isomorphic to its dense linear subvariety. Hence $\mathcal{D}(\mathbb{D})'_q$ is a completion of the topological vector space $\mathrm{Pol}(\mathbb{C})_q$.

The involution $*$ admits an extension by continuity to $\mathcal{D}(\mathbb{D})'_q$.

Let $\overline{\mathcal{H}}$ be the direct product of one-dimensional vector spaces $\mathcal{H}_j=\mathbb{C}e_j$ with the product topology. Evidently, $\mathcal{H}\subset\overline{\mathcal{H}}$ is a dense linear subvariety.

The linear maps from $\mathcal{H}$ to $\overline{\mathcal{H}}$ have the form

$$Av=\sum_{k,j=0}^{\infty} a_{kj}(v,e_j)e_k, \qquad v\in\mathcal{H}, \tag{1.65}$$

where $a_{kj} \in \mathbb{C}$. We call them generalized linear operators in $\mathcal{H}$ and endow the vector space of such operators with the topology of coefficientwise convergence of formal series (1.65). Here is a consequence of Proposition 1.73.

COROLLARY 1.74. (1) *The Fock representation of* $\mathrm{Pol}(\mathbb{C})_q$ *admits an extension by continuity to a linear map from* $\mathcal{D}(\mathbb{D})'_q$ *to the space of generalized linear operators.*

(2) *This linear map is an isomorphism of topological vector spaces.*

We keep the notation T_F for the above extension to the vector space $\mathcal{D}(\mathbb{D})'_q$.

Recall that $\mathrm{Pol}(\mathbb{C})_q$ is a $U_q\mathfrak{su}_{1,1}$-module algebra, hence a $U_q\mathfrak{sl}_2$-covariant $\mathrm{Pol}(\mathbb{C})_q$-bimodule. It can be demonstrated that the operators of representations of $\mathrm{Pol}(\mathbb{C})_q$ and $U_q\mathfrak{sl}_2$ admit extensions by continuity from $\mathrm{Pol}(\mathbb{C})_q$ to the completion $\mathcal{D}(\mathbb{D})'_q$, thus endowing it with a structure of $U_q\mathfrak{sl}_2$-covariant $\mathrm{Pol}(\mathbb{C})_q$-bimodule. A proof will be presented later on in an essentially wider generality; see Section 2.4.

EXAMPLE 1.75. An obvious induction argument allows one to prove the relations $K^{\pm 1}\psi(y) = \psi(y)$,

$$E\psi(y) = -q^{\frac{1}{2}} zy \frac{\psi(y) - \psi(q^2 y)}{y - q^2 y}, \qquad F\psi(y) = -q^{\frac{5}{2}} y \frac{\psi(y) - \psi(q^2 y)}{y - q^2 y} z^* \tag{1.66}$$

for $\psi(y) = y^n$. Thus they also hold for all polynomials $\psi(y)$, hence for all functions $\psi(y)$ on $q^{2\mathbb{Z}_+}$.

A linear operator A in $\mathcal{H}$ is called finite if $Ae_j = 0$ for all but finitely many $j \in \mathbb{Z}_+$. A good example is given by the one-dimensional orthogonal projection P_0 onto the vacuum subspace $\mathcal{H}_0 = \mathbb{C}e_0$. The notion of an adjoint operator to a finite linear operator is introduced in a standard way.

Proposition 1.73 implies

PROPOSITION 1.76. *For* $f \in \mathcal{D}(\mathbb{D})'_q$, *the following claims are equivalent:*

(1) $T_F(f)$ *is a finite linear operator in* $\mathcal{H}$;
(2) *Both the number of nonzero terms of the formal series* (1.65) *and the supports of* $\psi_j(y)$ *in* (1.63) *are finite;*
(3) $z^{*N} f = f z^N = 0$ *for some* $N \in \mathbb{Z}_+$.

The elements $f \in \mathcal{D}(\mathbb{D})'_q$ which satisfy the assumptions of Proposition 1.76 will be called *finite functions on the quantum disc.* They form a dense linear subvariety in $\mathcal{D}(\mathbb{D})'_q$, which will be denoted by $\mathcal{D}(\mathbb{D})_q$.

It follows from the definitions that the map T_F is an isomorphism of the vector space $\mathcal{D}(\mathbb{D})_q$ and the vector space $\mathrm{End}_f(\mathcal{H})$ of finite linear operators in $\mathcal{H}$. This allows one to endow $\mathcal{D}(\mathbb{D})_q$ with a structure of $*$-algebra. The multiplication in $\mathcal{D}(\mathbb{D})_q$ can be obtained in a different way, namely by continuity from the multiplication in $\mathrm{Pol}(\mathbb{C})_q$. One can prove this way that $\mathcal{D}(\mathbb{D})_q$ is a $U_q\mathfrak{su}_{1,1}$-module algebra. A more general approach is expounded in Section 2.4.

EXAMPLE 1.77. There exists a unique element $f_0 \in \mathcal{D}(\mathbb{D})_q$ such that $T_F f_0 = P_0$. Evidently,

$$z^* f_0 = f_0 z = 0, \tag{1.67}$$

$$f_0^2 = f_0, \qquad f_0^* = f_0. \tag{1.68}$$

The coefficients in the expansion (1.63) for f_0 are

$$\psi_0(y)=\begin{cases}1 & \text{for } y=1,\\ 0 & \text{for } y\neq 1,\end{cases} \qquad \psi_j(y)=0, \qquad \text{for } j\neq 0.$$

Thus

$$K^{\pm 1}f_0=f_0, \qquad Ff_0=-\frac{q^{1/2}}{q^{-2}-1}f_0z^*, \qquad Ef_0=-\frac{q^{1/2}}{1-q^2}zf_0. \tag{1.69}$$

It follows from the definitions of the Fock representation T_F and f_0 that

$$\mathcal{D}(\mathbb{D})_q=\mathbb{C}[z]_q\cdot f_0\cdot\mathbb{C}[\bar{z}]_q. \tag{1.70}$$

REMARK 1.78. The ideas that led us to the $U_q\mathfrak{su}_{1,1}$-module algebra $\mathcal{D}(\mathbb{D})_q$ are transparent and substantiated well enough. In the case of a general quantum bounded symmetric domain, the algebra of finite functions will be produced in a different way, more formally, and the element f_0 will be crucial in that construction. Shortly after that, the equivalence of both approaches will become clear.

1.4.4. An invariant integral on $\mathcal{D}(\mathbb{D})_q$. We start with the following auxiliary statement.

LEMMA 1.79. *Consider an A-module algebra F, an A-covaricnt left F-module V, together with the corresponding representation π of F in V*

$$\pi(f)v=fv, \qquad f\in F, \quad v\in V.$$

The homomorphism of algebras $\pi: F\to \mathrm{End}(V)$ is also a morphism of A-modules.

PROOF. Let $a\in A$, and

$$(\mathrm{id}\otimes\Delta)\Delta(a)=(\Delta\otimes\mathrm{id})\Delta(a)=\sum_j b_j\otimes c_j\otimes d_j.$$

One has to verify that $\pi(af)=a\pi(f)$ for all $f\in F$, that is, the relation

$$(af)v=\sum_j(b_jf)(c_jS(d_j)v), \qquad f\in F, \quad v\in V.$$

However $\sum_j b_j\otimes c_jS(d_j)=a\otimes 1$ by the definition of the antipode S. □

It follows from the results of the previous subsection that $\widetilde{\mathcal{H}}=\mathbb{C}[z]_q f_0$ is both a $U_q\mathfrak{b}^+$-covariant $\mathcal{D}(\mathbb{D})_q$-module as well as a $U_q\mathfrak{b}^+$-covariant $\mathrm{Pol}(\mathbb{C})_q$-module. Obviously,

$$\widetilde{\mathcal{H}}=\{f\in\mathcal{D}(\mathbb{D})_q|\ fz=0\}=\{f\in\mathcal{D}(\mathbb{D})_q|\ T_F(f)e_j=0 \text{ for all } j\neq 0\}.$$

The linear map

$$\widetilde{\mathcal{H}}\to\mathcal{H}, \qquad \psi(z)f_0\mapsto T_F(\psi(z))e_0, \tag{1.71}$$

is an isomorphism of $\mathrm{Pol}(\mathbb{C})_q$-modules.

We apply (1.71) to transfer the structure of $U_q\mathfrak{b}^+$-module from $\widetilde{\mathcal{H}}$ to $\mathcal{H}$, and introduce the notation Γ for the representation of the algebra $U_q\mathfrak{b}^+$ in the pre-Hilbert space $\mathcal{H}$ produced this way.

PROPOSITION 1.80. *The linear functional*

$$\int\limits_{\mathbb{D}_q} f d\nu \overset{\text{def}}{=} (1-q^2)\mathrm{tr}_q T_F(f) = (1-q^2)\mathrm{tr}(T_F(f)\Gamma(K^{-1})), \qquad f \in \mathcal{D}(\mathbb{D})_q, \tag{1.72}$$

is a positive $U_q\mathfrak{sl}_2$-invariant integral.

PROOF. For $f \in \mathcal{D}(\mathbb{D})_q$, the operators $T_F(f)$ are finite in $\mathcal{H}$. Thus (1.72) determines a nonnegative linear functional ν on $\mathcal{D}(\mathbb{D})_q$:

$$\mathrm{tr}_q T_F(f^*f) = \mathrm{tr}(T_F(f)\Gamma(K^{-1})T_F(f)^*) \geq 0.$$

Positivity follows from the fact that T_F is a faithful representation of $\mathcal{D}(\mathbb{D})_q$; see Subsection 1.4.3.

By Lemma 1.79, the homomorphism of algebras $T_F : \mathcal{D}(\mathbb{D})_q \to \mathrm{End}_f(\mathcal{H})$ is a morphism of $U_q\mathfrak{b}^+$-modules. Thus the $U_q\mathfrak{b}^+$-invariance of ν follows from Proposition 1.64 and the relation

$$S^2(a) = K^{-1}aK, \qquad a \in U_q\mathfrak{sl}_2.$$

Now the $U_q\mathfrak{b}^+$-invariance and (1.44) imply the $U_q\mathfrak{b}^-$-invariance, because ν is real:

$$\nu(f^*) = \overline{\nu(f)}, \qquad f \in \mathcal{D}(\mathbb{D})_q.$$

It remains to observe that the Hopf subalgebras $U_q\mathfrak{b}^\pm$ generate $U_q\mathfrak{sl}_2$. □

REMARK 1.81. Recall the Johnson notation

$$\int_0^1 f(t)d_{q^2}t = (1-q^2)\sum_{j=0}^{\infty} f(q^{2j})q^{2j}.$$

If $f = \sum_{m\in\mathbb{N}} z^m\psi_m(y) + \psi_0(y) + \sum\limits_{m\in\mathbb{N}} \psi_{-m}(y)z^{*m} \in \mathcal{D}(\mathbb{D})_q$, then

$$\int\limits_{\mathbb{D}_q} f d\nu = \int_0^1 \psi_0(y)y^{-2}d_{q^2}y.$$

A formal passage to a limit as $q \to 1$ yields $y = 1 - |z|^2$,

$$d\nu = \frac{d\,\mathrm{Re}\,z\,d\,\mathrm{Im}\,z}{\pi(1-|z|^2)^2}.$$

PROPOSITION 1.82. $U_q\mathfrak{sl}_2 \cdot f_0 = \mathcal{D}(\mathbb{D})_q$.

PROOF. The $*$-algebra $\mathcal{D}(\mathbb{D})_q$ is a $U_q\mathfrak{su}_{1,1}$-module algebra, and the relations (1.34), (1.69), (1.37) hold. Thus,

$$E(z^j f_0) = a_j z^{j+1} f_0, \tag{1.73}$$

$$\begin{cases} F(f_0 z^{*k}) = b_k f_0 z^{*(k+1)}, \\ F(z^{j+1} f_0 z^{*k}) = c_{jk}\, z^j f_0 z^{*k} + d_{jk} z^{j+1} f_0 z^{*(k+1)} \end{cases} \tag{1.74}$$

for some nonzero scalar factors a_j, b_k, c_{jk}, d_{jk}.

Consider the subspaces

$$L_m = \left\{ f = \sum_{j=0}^{m}\sum_{i=0}^{\infty} a_{ij} z^i f_0 z^{*j} \;\middle|\; a_{ij} \in \mathbb{C} \right\} \subset \mathcal{D}(\mathbb{D})_q, \qquad m \in \mathbb{Z}_+.$$

Due to (1.70), it suffices to prove that $U_q\mathfrak{sl}_2 \cdot f_0 \supset L_m$ for all $m \in \mathbb{Z}_+$. We proceed by induction in m. First, (1.73) implies that $L_0 = U_q\mathfrak{b}_+ \cdot f_0$ because $a_j \neq 0$ for all $j \in \mathbb{Z}_+$. Second, (1.74) implies that $L_{m+1} \subset L_m + F \cdot L_m$ because $b_k \neq 0$, $c_{jk} \neq 0$, $d_{jk} \neq 0$ for all $j, k \in \mathbb{Z}_+$. □

COROLLARY 1.83. *A $U_q\mathfrak{sl}_2$-invariant integral on $\mathcal{D}(\mathbb{D})_q$ is unique up to a scalar factor.*

We have proved the existence and uniqueness of a $U_q\mathfrak{sl}_2$-invariant integral; its explicit form (1.72) is also found.

Proposition 1.82 can be refined as follows.

PROPOSITION 1.84. *The element f_0 generates the $U_q\mathfrak{sl}_2$-module $\mathcal{D}(\mathbb{D})_q$, and $K^{\pm 1} f_0 = f_0$ is the defining relation.*

PROOF. Consider the $U_q\mathfrak{sl}_2$-module V_0 with a single generator v_0 and the defining relation $K^{\pm 1} v_0 = v_0$. One has to prove the bijectivity of the morphism of $U_q\mathfrak{sl}_2$-modules $j : V_0 \to \mathcal{D}(\mathbb{D})_q$ which takes v_0 to f_0.

Consider the subspaces $V_0^{(i,k)} \subset V_0$, $\mathcal{D}(\mathbb{D})_q^{(i,k)} \subset \mathcal{D}(\mathbb{D})_q$ which are the linear spans of the sets

$$\left\{ F^{k_1} E^{i_1} v_0 \middle| \ k_1 \leq k \quad \text{and} \quad i_1 \leq i \right\}, \qquad \left\{ z^{i_1} f_0 z^{*k_1} \middle| \ k_1 \leq k \quad \text{and} \quad i_1 \leq i \right\},$$

respectively. The same argument as that in the proof of Proposition 1.82 allows one to obtain the inclusion $j\, V_0^{(i,k)} \supset \mathcal{D}(\mathbb{D})_q^{(i,k)}$ for all $i, k \in \mathbb{Z}_+$. In fact,

$$\mathcal{D}(\mathbb{D})_q^{(i,k)} \subset \mathcal{D}(\mathbb{D})_q^{(i-1,k-1)} + F\mathcal{D}(\mathbb{D})_q^{(i,k-1)}, \qquad i, k \neq 0,$$
$$\mathcal{D}(\mathbb{D})_q^{(0,j)} \subset F\mathcal{D}(\mathbb{D})_q^{(0,j-1)}, \qquad \mathcal{D}(\mathbb{D})_q^{(j,0)} \subset E\mathcal{D}(\mathbb{D})_q^{(j-1,0)}, \qquad j \neq 0;$$

see (1.73), (1.74).

On the other hand, $\dim V_0^{(i,k)} \leq \dim \left(\mathcal{D}(\mathbb{D})_q^{(i,k)} \right) < \infty$. Thus j maps $V_0^{(i,k)}$ onto $\mathcal{D}(\mathbb{D})_q^{(i,k)}$ bijectively. □

1.5. Differential calculi

1.5.1. Differential forms with polynomial coefficients. We reproduce here some definitions from [**174**]. The algebras to be considered are assumed unital unless the contrary is stated explicitly.

Consider a graded algebra $\Omega = \bigoplus_{i \in \mathbb{Z}_+} \Omega_i$ together with a linear map $d : \Omega \to \Omega$ of degree one. The pair (Ω, d) is called a differential graded algebra if $d^2 = 0$ and

$$d(\omega' \cdot \omega'') = d\omega' \cdot \omega'' + (-1)^n \omega' \cdot d\omega'', \qquad \omega' \in \Omega_n,\ \omega'' \in \Omega. \tag{1.75}$$

A homomorphism of differential graded algebras is defined in an obvious way. To denote the multiplication in Ω, $\cdot$ is sometimes replaced by $\wedge$. Evidently, $d1 = 0$.

A differential calculus over an algebra F is a differential graded algebra (Ω, d) such that $\Omega_0 = F$ and Ω is generated by $F \oplus dF$. Consider a Hopf algebra A and an A-module algebra F. A differential calculus (Ω, d) is said to be covariant if Ω is an A-module algebra and d is an endomorphism of the A-module Ω.

Let $A = U_q\mathfrak{sl}_2$ and $(\Lambda(\mathbb{C})_q, d)$ be a covariant differential calculus over $\mathbb{C}[z]_q$. It follows from (1.34) that

$$Fdz = 0, \qquad K^{\pm 1}dz = q^{\pm 2}dz, \tag{1.76}$$

$$Edz = -q^{1/2}(dz \cdot z + zdz). \tag{1.77}$$

Recall that $\mathbb{C}[z]_q$ is a commutative algebra in the braided tensor category $\mathcal{C}^-$; see Subsection 1.3.8. Suppose that $\Lambda(\mathbb{C})_q = \bigoplus_{i\in\mathbb{Z}_+} \Lambda^i(\mathbb{C})_q$ is a supercommutative algebra in this category, i.e.,

$$m\check{R}_{\Lambda(\mathbb{C})_q,\Lambda(\mathbb{C})_q}(\omega_1 \otimes \omega_2) = (-1)^{ln}\omega_1 \cdot \omega_2$$

for each $\omega_1 \in \Lambda^l(\mathbb{C})_q$, $\omega_2 \in \Lambda^n(\mathbb{C})_q$; cf. (1.39). Now (1.38) and (1.76) imply the relation

$$zdz = q^{-2}dz \cdot z. \tag{1.78}$$

An application of d yields

$$dz \wedge dz = 0. \tag{1.79}$$

This makes plausible the fact that the "natural" covariant differential calculus over $\mathbb{C}[z]_q$ is unique: the graded algebra $\Lambda(\mathbb{C})_q$ is generated by the elements z, dz with

$$\deg(z) = 0, \qquad \deg(dz) = 1,$$

the defining relations are (1.78)–(1.79), and the structure of $U_q\mathfrak{sl}_2$-module algebra is given by (1.76)–(1.77), together with the action in $\Omega_0 = \mathbb{C}[z]_q$; see (1.34).

Of course, one needs to prove the existence of a covariant differential calculus determined by (1.78)–(1.79), (1.76)–(1.77). The proof is postponed until Subsections 2.6.1 and 2.6.2, where it will be given in much wider generality. Note that a differential calculus on the complex quantum plane was found in [**271**, p. 1158].

Let us pass to $*$-algebras. A differential calculus over a $*$-algebra F is called a $*$-calculus [**174**, p. 462] if the antilinear involutive map $* : \Omega \to \Omega$ coincides with the involution $*$ on $\Omega_0 = F$ and is such that for all $\omega_1 \in \Omega_l$, $\omega_2 \in \Omega_n$, $\omega \in \Omega$ one has

$$(\omega_1 \wedge \omega_2)^* = (-1)^{ln}\omega_2^* \wedge \omega_1^*, \qquad (d\omega)^* = d\omega^*.$$

If A is a Hopf $*$-algebra and F is a $*$-algebra, then the covariant differential $*$-calculus over F should be also subject to the compatibility condition for involutions (1.44): $(a\omega)^* = S(a)^*\omega^*$.

Consider a special case $A = U_q\mathfrak{su}_{1,1}$, $F = \mathrm{Pol}(\mathbb{C})_q$. Now (1.78)–(1.79) imply the relations

$$z^*dz^* = q^2dz^*z^*, \qquad dz^* \wedge dz^* = 0.$$

Suppose that the $\mathrm{Pol}(\mathbb{C})_q$-subbimodules generated by dz, dz^* have zero intersection. An application of the differential to (1.8) yields

$$z^*dz = q^2dzz^*, \qquad dz^*z = q^2zdz^*,$$

$$dz^* \wedge dz = -q^2dz \wedge dz^*. \tag{1.80}$$

Now (1.37) implies

$$K^{\pm 1}dz^* = q^{\mp 2}dz^*, \qquad Ez^* = 0, \qquad Fz^* = -q^{5/2}(dz^*z^* + z^*dz^*). \tag{1.81}$$

With a complete list of relations at hand, we now intend to produce differential calculi over $\mathrm{Pol}(\mathbb{C})_q$. We postpone some verifications which demonstrate that our definitions are correct until Subsection 2.6.3. Therein those verifications will be done in a much wider generality.

Consider a bigraded algebra

$$\Omega(\mathbb{C})_q = \bigoplus_{i,j\in\{0,1\}} \Omega^{i,j}(\mathbb{C})_q$$

given by its generators z, z^*, dz, dz^* of bidegrees

$$\deg(z) = \deg(z^*) = (0,0), \qquad \deg(dz) = (1,0), \qquad \deg(dz^*) = (0,1)$$

and the defining relations (1.8), (1.78)–(1.80):

$$z^*z = q^2zz^* + 1 - q^2,$$

$$zdz = q^{-2}dzz, \qquad z^*dz = q^2dzz^*, \qquad dz^*z = q^2zdz^*, \qquad dz^*z^* = q^{-2}z^*dz^*,$$

$$dz \wedge dz = 0, \qquad dz^* \wedge dz^* = 0, \qquad dz^* \wedge dz = -q^2dz \wedge dz^*.$$

We endow it with grading $\Omega^k(\mathbb{C})_q = \bigoplus_{i+j=k} \Omega^{i,j}(\mathbb{C})_q$.

Each $\mathrm{Pol}(\mathbb{C})_q$-bimodule $\Omega^{i,j}(\mathbb{C})_q$ is both a free left and a free right $\mathrm{Pol}(\mathbb{C})_q$-module of rank 1. In particular, every $\omega \in \Omega(\mathbb{C})_q$ admits a unique decomposition

$$\omega = f_{0,0} + dz \cdot f_{1,0} + f_{0,1} \cdot dz^* + dz \cdot f_{1,1} \cdot dz^*, \tag{1.82}$$

with $f_{i,j} \in \mathrm{Pol}(\mathbb{C})_q$.

The structure of $U_q\mathfrak{sl}_2$-module algebra in $\Omega(\mathbb{C})_q$ is introduced by (1.37), (1.34), (1.76), (1.77), (1.81), and the differential d is determined by its values on the generators ($d : z \mapsto dz$, $d : z^* \mapsto dz^*$), together with the requirement $d^2 = 0$. An involution $*$ and a grading are defined in an obvious way.

The pair $(\Omega(\mathbb{C})_q, d)$ is a covariant $*$-calculus over the $U_q\mathfrak{su}_{1,1}$-module algebra $\mathrm{Pol}(\mathbb{C})_q$.

The differential d splits uniquely as a sum $d = \partial + \bar\partial$ of the linear maps

$$\partial : \Omega^{i,j}(\mathbb{C})_q \to \Omega^{i+1,j}(\mathbb{C})_q, \qquad \bar\partial : \Omega^{i,j}(\mathbb{C})_q \to \Omega^{i,j+1}(\mathbb{C})_q$$

such that

$$\partial^2 = 0, \qquad \bar\partial^2 = 0, \qquad \partial\bar\partial + \bar\partial\partial = 0, \tag{1.83}$$

which is due to $d^2 = 0$.

Setting

$$\Omega^{0,*}(\mathbb{C})_q = \Omega^{0,0}(\mathbb{C})_q \oplus \Omega^{0,1}(\mathbb{C})_q, \qquad \Omega^{*,0}(\mathbb{C})_q = \Omega^{0,0}(\mathbb{C})_q \oplus \Omega^{1,0}(\mathbb{C})_q,$$

we get another two covariant differential calculi

$$(\Omega^{0,*}(\mathbb{C})_q, \bar\partial), \qquad (\Omega^{*,0}(\mathbb{C})_q, \partial).$$

Certainly, these are not $*$-calculi and are nonisomorphic to each other, as $\mathrm{Ker}\,\bar\partial \neq \mathrm{Ker}\,\partial$. Note that $\bar\partial$ is significant in complex analysis [**127**].

1.5.2. Differential forms with finite coefficients. Our first topic here is a q-analog of the vector space of differential forms with distribution coefficients.

Consider a vector space $\Omega(\mathbb{D})'_q$ formed by the elements ω which admit decompositions similar to (1.82):

$$\omega = f_{0,0} + dz \cdot f_{1,0} + f_{0,1} \cdot dz^* + dz \cdot f_{1,1} \cdot dz^*, \qquad f_{i,j} \in \mathcal{D}(\mathbb{D})'_q$$

(see Subsection 1.4.3). Let us endow $\Omega(\mathbb{D})'_q$ with a topology using the isomorphism of vector spaces under which $\omega \mapsto (f_{0,0}, f_{1,0}, f_{0,1}, f_{1,1})$ and the topology in $\mathcal{D}(\mathbb{D})'_q$.

The vector space $\Omega(\mathbb{C})_q$ introduced in Subsection 1.5.1 is a dense linear subvariety in $\Omega(\mathbb{D})'_q$. In other words, $\Omega(\mathbb{D})'_q$ is a completion of $\Omega(\mathbb{C})_q$ with respect to the corresponding topology. The involution $*$, the action of $U_q\mathfrak{sl}_2$, and the structure of the $\Omega(\mathbb{C})_q$-bimodule are transferred by continuity from $\Omega(\mathbb{C})_q$ to $\Omega(\mathbb{D})'_q$, just as in Subsection 1.4.3.

In a similar way, ∂, $\bar{\partial}$, d admit extensions by continuity from $\Omega(\mathbb{C})_q$ to $\Omega(\mathbb{D})'_q$:

$$\partial\psi(y) = -dz\frac{\psi(y)-\psi(q^2 y)}{y-q^2y}z^*, \qquad \bar{\partial}\psi(y) = -z\frac{\psi(y)-\psi(q^2y)}{y-q^2y}dz^*, \tag{1.84}$$

where $y = 1 - zz^*$ and ψ is a function on $q^{2\mathbb{Z}_+}$. (It suffices to obtain (1.84) in the special case $\psi(y) = y^n$, which is easily accessible using (1.10).)

In particular, for f_0 as in (1.67) and (1.68) one has

$$df_0 = -\frac{1}{1-q^2}dz f_0 z^* - \frac{1}{1-q^2}z f_0 dz^*.$$

Certainly $\Omega(\mathbb{D})'_q$ is a $U_q\mathfrak{sl}_2$-covariant $\Omega(\mathbb{C})_q$-bimodule, d is an endomorphism of the $U_q\mathfrak{sl}_2$-module $\Omega(\mathbb{D})'_q$, and the relations (1.75) and (1.83) hold.

Partial derivatives of a distribution $f \in \Omega(\mathbb{D})'_q$ are introduced by

$$\partial f = dz\frac{\partial f}{\partial z}, \qquad \bar{\partial} f = dz^*\frac{\partial f}{\partial z^*}.$$

The dense linear subvariety

$$\Omega(\mathbb{D})_q = \{f_{0,0} + dz\cdot f_{1,0} + f_{0,1}\cdot dz^* + dz\cdot f_{1,1}\cdot dz^* \in \Omega(\mathbb{D})'_q | \, f_{i,j} \in \mathcal{D}(\mathbb{D})_q\}$$

will be called the space of differential forms with finite coefficients in the quantum disc. Due to Proposition 1.76, $\omega \in \Omega(\mathbb{D})_q$ if and only if $z^{*N}\cdot\omega = \omega\cdot z^N = 0$ for some $N \in \mathbb{N}$.

The multiplication of differential forms is transferred by continuity from $\Omega(\mathbb{C})_q$ to $\Omega(\mathbb{D})_q$. Now $f_0 z = z^* f_0 = 0$ implies

$$df_0\, z = -f_0 dz = -dz f_0, \qquad z^* df_0 = -dz^* f_0 = -f_0 dz^*.$$

This can be readily used to prove that the algebra $\Omega(\mathbb{D})_q$ is generated by its subspace $\mathcal{D}(\mathbb{D})_q \oplus d\mathcal{D}(\mathbb{D})_q$. The pair $(\Omega(\mathbb{D})_q, d)$ is a covariant $*$-calculus over the $U_q\mathfrak{su}_{1,1}$-module algebra $\mathcal{D}(\mathbb{D})_q$ of finite functions in the quantum disc.

Since $\Omega(\mathbb{D})'_q$ is a $U_q\mathfrak{sl}_2$-covariant $\Omega(\mathbb{C})_q$-bimodule, one can use a continuity argument to endow $\Omega(\mathbb{D})'_q$ with a structure of $U_q\mathfrak{sl}_2$-covariant $\Omega(\mathbb{C})_q \oplus \Omega(\mathbb{D})_q$-bimodule.

Now we introduce a differential form, which is a q-analog of the fundamental 2-form associated to the invariant Kähler metric on the disc [**177**, p. 140], [**108**, p. 120].

PROPOSITION 1.85. *The differential form*

$$\omega_0 = \frac{1}{2i}y^{-2}dz\wedge dz^* \in \Omega(\mathbb{D})'_q \tag{1.85}$$

is $U_q\mathfrak{sl}_2$-invariant, $\partial\omega_0 = \bar{\partial}\omega_0 = 0$, and

$$\omega\wedge\omega_0 = \omega_0\wedge\omega \tag{1.86}$$

for all $\omega \in \Omega(\mathbb{C})_q \oplus \Omega(\mathbb{D})_q$.

PROOF. The only thing that does not look obvious is the $U_q\mathfrak{sl}_2$-invariance of the $(1,1)$-form ω_0. Of course, $K^{\pm 1}\omega_0 = \omega_0$. It suffices to prove just a single one of the relations $E\omega_0 = 0$, $F\omega_0 = 0$. However, it follows from (1.66) that $E(y^{-2}) = q^{\frac{1}{2}}(1+q^2)y^{-2}z$. Moreover,

$$E(dz \wedge dz^*) = d(Ez) \wedge dz^* = d(-q^{\frac{1}{2}}z^2) \wedge dz^* = -q^{\frac{1}{2}}(1+q^2)zdz \wedge dz^*.$$

Hence, $E(2i\omega_0) = q^{\frac{1}{2}}(1+q^2)y^{-2}zdz \wedge dz^* - q^{\frac{1}{2}}(1+q^2)y^{-2}zdz \wedge dz^* = 0$. □

REMARK 1.86. It follows from (1.84) that the differential form ω_0, up to a constant factor, coincides with $\partial\bar{\partial}\log y$ (cf. [**108**, p. 122]); i.e., the function $\varphi = \log y$ is a q-analog of the Kähler potential.

The linear functional

$$\int_{\mathbb{D}_q} f d\mu \stackrel{\text{def}}{=} \int_{\mathbb{D}_q} f \cdot y^2 d\nu, \qquad f \in \mathcal{D}(\mathbb{D})_q, \tag{1.87}$$

is a q-analog of the integral with respect to the normed Lebesgue measure on $\mathbb{D}$. Recall the decomposition

$$\Omega(\mathbb{D})_q = \Omega^{0,0}(\mathbb{D})_q \oplus \Omega^{1,0}(\mathbb{D})_q \oplus \Omega^{0,1}(\mathbb{D})_q \oplus \Omega^{1,1}(\mathbb{D})_q.$$

PROPOSITION 1.87. *The linear functional*

$$\int_{\mathbb{D}_q} f dz \wedge dz^* \stackrel{\text{def}}{=} -2\pi i \int_{\mathbb{D}_q} f d\mu, \qquad f \in \mathcal{D}(\mathbb{D})_q, \tag{1.88}$$

on $\Omega^{1,1}(\mathbb{D})_q$ *is* $U_q\mathfrak{sl}_2$*-invariant.*

PROOF. Proposition 1.85 implies that the map

$$\mathcal{D}(\mathbb{D})_q \to \Omega^{1,1}(\mathbb{D})_q, \qquad f \mapsto f\omega_0,$$

is an isomorphism of $U_q\mathfrak{sl}_2$-modules. It remains to apply Proposition 1.80 on $U_q\mathfrak{sl}_2$-invariance of the integral ν in $\mathcal{D}(\mathbb{D})_q$. □

Note that the linear functional (1.87) admits an extension by continuity to $\mathrm{Pol}(\mathbb{C})_q$.

COROLLARY 1.88. *The linear functional* (1.88) *on* $\Omega^{1,1}(\mathbb{C})_q$ *is* $U_q\mathfrak{sl}_2$*-invariant.*

Consider the two-sided ideal $J \subset \mathrm{Pol}(\mathbb{C})_q$ generated by $1 - zz^*$, together with the commutative quotient algebra $\mathrm{Pol}(\mathbb{C})_q/J$. The latter will be identified to the algebra of Laurent polynomials, and the image $f|_{\mathbb{T}}$ of $f \in \mathrm{Pol}(\mathbb{C})_q$ under the canonical homomorphism will be called its restriction onto the unit circle $\mathbb{T}$.

Define an integral on $\Omega^{1,0}(\mathbb{C})_q$ by

$$\int_{\mathbb{T}} dzf = 2\pi i \int_{\mathbb{T}} (z \cdot f)|_{\mathbb{T}} d\nu, \qquad f \in \mathrm{Pol}(\mathbb{C})_q,$$

where $\int_{\mathbb{T}} \psi d\nu \stackrel{\text{def}}{=} \int_0^{2\pi} \psi(e^{i\theta}) \frac{d\theta}{2\pi}$. The following claim is a q-analog of the Green formula.

PROPOSITION 1.89. *For all* $\psi \in \Omega^{1,0}(\mathbb{C})_q$ *one has*

$$\int_{\mathbb{D}_q} \bar{\partial}\psi = \int_{\mathbb{T}} \psi. \tag{1.89}$$

PROOF. Let $\psi \in \Omega^{1,0}(\mathbb{C})_q$,

$$\psi = dz\left(\sum_{m>0} z^m \psi_m(y) + \psi_0(y) + \sum_{m>0} \psi_{-m}(y) z^{*m}\right).$$

It suffices to consider the case $\psi = dz\psi_{-1}(y)z^*$ because the terms $dz\psi_j(y)z^{*j}$ with $j \neq -1$ do not contribute to (1.89).

The relation (1.84) implies

$$\overline{\partial}(dz\psi_{-1}(y)z^*) = dz\left(\psi_{-1}(y) - q^{-2}\frac{\psi_{-1}(q^{-2}y) - \psi_{-1}(y)}{q^{-2}y - y}(1-y)\right)dz^*.$$

Set $g(y) = \psi_{-1}(y) - q^{-2}\frac{\psi_{-1}(q^{-2}y)-\psi_{-1}(y)}{q^{-2}y-y}(1-y)$. If $\psi_{-1}(0) = 0$, then one can easily deduce that $\sum_{n\in\mathbb{Z}_+} g(q^{2n})q^{2n} = 0$. So, in this case $\int_{\mathbb{D}_q}\overline{\partial}\psi = \int_{\mathbb{T}}\psi = 0$, and Proposition 1.89 is proved for (1,0)-forms from a linear subvariety of codimension 1. It remains to make sure it is valid for $\psi = dzz^*$:

$$\int_{\mathbb{D}_q} \bar{\partial}(dz\, z^*) = -\int_{\mathbb{D}_q} dz \wedge dz^* = 2\pi i, \qquad \int_{\mathbb{T}} dz\, z^* = 2\pi i. \qquad \square$$

The linear functionals in the relation proved above admit extensions by continuity to $\Omega^{1,0}(\mathbb{C})_q \oplus \Omega^{1,0}(\mathbb{D})_q$.

COROLLARY 1.90. *$\int_{\mathbb{D}_q}\overline{\partial}\psi = 0$ for all $\psi \in \Omega^{1,0}(\mathbb{D})_q$.*

1.5.3. Invariant Hermitian metrics and the invariant Laplacian. The differential operator

$$\square = -\left(1-|z|^2\right)^2 \frac{\partial}{\partial z}\frac{\partial}{\partial \overline{z}} \tag{1.90}$$

in the unit disc is called the invariant Laplacian. Here we follow the standards of complex analysis [**29**, p. 159] in choosing the sign in (1.90).

Recall the proof of $SU_{1,1}$-invariance for $\square$. The vector spaces of smooth finite functions and smooth $(0,1)$-forms with finite coefficients are endowed with the $SU_{1,1}$-invariant positive Hermitian forms

$$(f_1, f_2) = \int_{\mathbb{D}} \overline{f}_2 f_1 \frac{d\operatorname{Re} z \wedge d\operatorname{Im} z}{(1-|z|^2)^2}, \qquad (f_1 d\overline{z}, f_2 d\overline{z}) = \int_{\mathbb{D}} \overline{f}_2 f_1 d\operatorname{Re} z \wedge d\operatorname{Im} z.$$

The $SU_{1,1}$-invariance of the differential operator $\overline{\partial}$ implies the $SU_{1,1}$-invariance for the formal dual differential operator $\overline{\partial}^* : fd\overline{z} \mapsto -\left(1-|z|^2\right)^2 \frac{\partial f}{\partial z}$. Therefore, the operator $\square = \overline{\partial}^*\overline{\partial}$ is also $SU_{1,1}$-invariant. A similar argument is to be used in the case of a quantum disc.

In Subsection 1.4.1 we considered invariant sesquilinear forms. Now we are about to introduce more general notions via replacing the complex field by an A-module algebra F. In doing this, we follow an unpublished work by D. Shklyarov on q-analogs of Kähler manifolds.

Consider a $*$-algebra F. If V is a left F-module, then the structure of right F-module on the antispace $\overline{V}$ (see Subsection 1.4.1) is introduced as follows:

$$vf \stackrel{\text{def}}{=} f^*v, \qquad f \in F, \quad v \in V.$$

Likewise, if V is a right F-module, then $\overline{V}$ is endowed with the structure of left F-module:

$$fv \stackrel{\text{def}}{=} vf^*, \qquad f \in F,\ v \in V.$$

Consider an F-bimodule V. A sesquilinear map $\mathbf{h} : V \times V \to F$ is called a *Hermitian metric* if it is real

$$\mathbf{h}(v_2, v_1) = \mathbf{h}(v_1, v_2)^*, \qquad v_1, v_2 \in V,$$

and the associated linear map

$$\mathbf{h} : \overline{V} \otimes_F V \to F \tag{1.91}$$

is a morphism of F-bimodules.

A Hermitian metric is said to be *invariant* if V is an A-covariant F-bimodule and the map (1.91) is a morphism of A-modules.

A Hermitian metric is called nonnegative if $\mathbf{h}(v,v) \geq 0$ for all $v \in V$, and positive if, additionally, $\mathbf{h}(v,v) \neq 0$ for $v \neq 0$.

Let us present examples of Hermitian metrics. Obviously, $\Omega^{0,0}(\mathbb{D})_q$, $\Omega^{1,0}(\mathbb{D})_q$, $\Omega^{0,1}(\mathbb{D})_q$, $\Omega^{1,1}(\mathbb{D})_q$ are both free right and free left $\mathcal{D}(\mathbb{D})_q$-modules generated by 1, dz, dz^*, $dz \wedge dz^*$, respectively.

PROPOSITION 1.91. *Consider the Hopf $*$-algebra $U_q\mathfrak{su}_{1,1}$, the $U_q\mathfrak{su}_{1,1}$-module $*$-algebra $\mathcal{D}(\mathbb{D})_q$, and the $U_q\mathfrak{su}_{1,1}$-covariant $\mathcal{D}(\mathbb{D})_q$-bimodule $\Omega^{0,1}(\mathbb{D})_q$. The following sesquilinear maps are positive invariant Hermitian metrics:*

$$\mathbf{h}_{0,0}(f_1, f_2) = f_2^* f_1, \tag{1.92}$$

$$\mathbf{h}_{0,1}(dz^* f_1, dz^* f_2) = f_2^*(1 - zz^*)^2 f_1, \tag{1.93}$$

where $f_1, f_2 \in \mathcal{D}(\mathbb{D})_q$.

PROOF. The invariance of the Hermitian metrics (1.92), (1.93) follows from Proposition 1.85 and the covariance of the $*$-calculus $(\Omega(\mathbb{D})_q, d)$. The positivity is due to the fact that the linear operator

$$T_F(1 - zz^*) : e_n \mapsto q^{2n} e_n, \qquad n \in \mathbb{Z}_+,$$

is positive. The latter property holds because the Fock representation provides an isomorphism of the $*$-algebra $\mathcal{D}(\mathbb{D})_q$ and the $*$-algebra of finite linear operators in $\mathcal{H}$. □

COROLLARY 1.92. *The Hermitian forms*

$$(f_1, f_2) = \int_{\mathbb{D}_q} f_2^* f_1 d\nu, \qquad (dz^* f_1, dz^* f_2) = \int_{\mathbb{D}_q} f_2^*(1 - zz^*)^2 f_1 d\nu,$$

in $\mathcal{D}(\mathbb{D})_q$, $\Omega^{0,1}(\mathbb{D})_q$, respectively, are invariant and positive.

REMARK 1.93. As a consequence of the definitions, one has

$$(f_1 dz^*, f_2 dz^*) = \int_{\mathbb{D}_q} f_2^* f_1 (1 - zz^*)^2 d\nu, \qquad f_1, f_2 \in \mathcal{D}(\mathbb{D})_q. \tag{1.94}$$

Consider the linear map $\bar{\partial}$ from the pre-Hilbert space $\mathcal{D}(\mathbb{D})_q$ to the pre-Hilbert space $\Omega^{0,1}(\mathbb{D})_q$.

PROPOSITION 1.94. *The linear map*

$$\bar{\partial}^* : \Omega^{0,1}(\mathbb{D})'_q \to \mathcal{D}(\mathbb{D})'_q, \qquad \bar{\partial}^* : dz^* f \mapsto -q^2(1-zz^*)^2 \frac{\partial f}{\partial z},$$

is dual to $\bar{\partial}$, *that is, for all* $f_1, f_2 \in \mathcal{D}(\mathbb{D})_q$ *one has*

$$(\bar{\partial} f_1, dz^* f_2) = \left(f_1, -q^2(1-zz^*)^2 \frac{\partial f_2}{\partial z} \right).$$

PROOF. An application of the relation $\bar{\partial}(f_2^*) = (\partial f_2)^*$ and Corollary 1.90 yields

$$0 = \int\limits_{\mathbb{D}_q} \bar{\partial}(f_2^* dz\, f_1) = \int\limits_{\mathbb{D}_q} \bar{\partial}(f_2^*) \wedge dz\, f_1 - \int\limits_{\mathbb{D}_q} f_2^* dz \wedge \bar{\partial} f_1$$

$$= \int\limits_{\mathbb{D}_q} (\partial f_2)^* \wedge dz\, f_1 - \int\limits_{\mathbb{D}_q} f_2^* dz \wedge \bar{\partial} f_1 = \int\limits_{\mathbb{D}_q} \left(\frac{\partial f_2}{\partial z} \right)^* dz^* \wedge dz\, f_1 - \int\limits_{\mathbb{D}_q} f_2^* dz \wedge \bar{\partial} f_1$$

for all $f_1, f_2 \in \mathcal{D}(\mathbb{D})_q$. Thus,

$$\int\limits_{\mathbb{D}_q} (dz^* f_2)^* \wedge \bar{\partial} f_1 = \int\limits_{\mathbb{D}_q} \left(\frac{\partial f_2}{\partial z} \right)^* dz^* \wedge dz\, f_1.$$

Hence,

$$\int\limits_{\mathbb{D}_q} f_2^* dz \wedge dz^* \frac{\partial f_1}{\partial z^*} = -q^2 \int\limits_{\mathbb{D}_q} \left(\frac{\partial f_2}{\partial z} \right)^* dz \wedge dz^* f_1.$$

It remains to use (1.85), (1.86), and (1.88). □

COROLLARY 1.95. *The linear operator*

$$\Box_q = \bar{\partial}^* \bar{\partial} = -q^2(1-zz^*)^2 \frac{\partial}{\partial z} \frac{\partial}{\partial z^*}$$

is an endomorphism of the $U_q\mathfrak{sl}_2$*-module* $\mathcal{D}(\mathbb{D})_q$.

PROOF. The invariance of the Hermitian form $(\Box_q f_1, f_2)$ in $\mathcal{D}(\mathbb{D})_q$ follows from Proposition 1.94, Corollary 1.92, and the fact that $\bar{\partial}$ is a morphism of $U_q\mathfrak{sl}_2$-modules. □

The linear map $\Box_q$ extends by continuity to an endomorphism of the $U_q\mathfrak{sl}_2$-module $\mathcal{D}(\mathbb{D})'_q$. We call this extended map the *invariant Laplacian in the quantum disc*. Introduce a completion $L^2(d\nu)_q$ of the vector space $\mathcal{D}(\mathbb{D})_q$ with respect to the norm $\|f\| = \sqrt{\int_{\mathbb{D}} f^* f d\nu}$. Proposition 1.94 implies that the invariant Laplacian is a symmetric linear operator in the Hilbert space $L^2(d\nu)_q$, hence admits a closure [**82**, pp. 355, 356]. We keep the notation $\Box_q$ for this completion. It will be demonstrated in Subsection 1.5.5 that the operator $\Box_q$ is self-adjoint and bounded.

1.5.4. The radial part of invariant Laplacian. Consider the one-parameter group of linear operators

$$\sum_{j\in\mathbb{Z}_+} z^j\psi_j(y)+\sum_{k\in\mathbb{N}}\psi_k(y)z^{*k}\mapsto\sum_{j\in\mathbb{Z}_+} e^{ij\varphi}z^j\psi_j(y)+\sum_{k\in\mathbb{N}} e^{-ik\varphi}\psi_k(y)z^{*k},\qquad \varphi\in\mathbb{R},$$

in $\mathcal{D}(\mathbb{D})'_q$ (see (1.63)) together with the subspace of its invariants. The restriction of the linear operator $\square_q$ to the above subspace of invariants will be denoted by $\square_q^{(0)}$ and called the radial part of $\square_q$. Let

$$\mathcal{D}_q\psi(x)=\frac{\psi(q^{-1}x)-\psi(qx)}{q^{-1}x-qx}.$$

We endow the vector space of functions on $q^{-2\mathbb{Z}_+}$ with the topology of pointwise convergence and consider the continuous linear operator $\mathcal{D}_q x(1-q^{-1}x)\mathcal{D}_q$ in this space.

PROPOSITION 1.96. *For each function $\psi(x)$ on $q^{-2\mathbb{Z}_+}$ one has*

$$\square_q^{(0)}\psi(x)=\mathcal{D}_q x(1-q^{-1}x)\mathcal{D}_q\psi(x), \tag{1.95}$$

with $x=y^{-1}$ in the left-hand side of (1.95).

PROOF. It suffices to consider the special case of $\psi(x)$ with finite support. Let

$$\int_1^\infty \psi(t)d_{q^{-2}}t\overset{\text{def}}{=}(q^{-2}-1)\sum_{m=0}^\infty \psi(q^{-2m})q^{-2m}.$$

Obviously, $\|\psi(x)\|^2=q^2\int_1^\infty|\psi(t)|^2d_{q^{-2}}t$. It suffices to prove that

$$\|\bar\partial\psi(x)\|^2=q^2\int_1^\infty(\mathcal{D}_q t(1-q^{-1}t)\mathcal{D}_q\psi(t))\overline{\psi(t)}d_{q^{-2}}t.$$

Now (1.84) implies that $\bar\partial\psi(x)=-z\frac{\psi(x)-\psi(q^{-2}x)}{x^{-1}-q^2x^{-1}}dz^*$ for each $\psi(x)$ with finite support in $q^{-2\mathbb{Z}_+}$.

An application of (1.94) yields

$$\begin{aligned}\left\|\bar\partial\psi(x)\right\|^2&=(1-q^2)^{-1}\sum_{m=0}^\infty\left|\psi(q^{-2m})-\psi(q^{-(2m+2)})\right|^2(1-q^{2m+2})q^{-2m}\\&=q^{-2}(q^{-2}-1)^{-2}\int_1^\infty\left|\psi(t)-\psi(q^{-2}t)\right|^2(1-q^2t^{-1})d_{q^{-2}}t\\&=-\int_1^\infty\left|\frac{\psi(t)-\psi(q^{-2}t)}{t-q^{-2}t}\right|^2\cdot t(1-q^{-2}t)d_{q^{-2}}t.\end{aligned}$$

To complete the proof, one should use a q-analog of "integration by parts" (1.121). □

Below we use the following basis in the space of functions with finite support on $q^{-2\mathbb{Z}_+}$:

$$f_j(x)=\begin{cases}1, & x=q^{-2j},\\ 0, & x\neq q^{-2j},\end{cases}\qquad j\in\mathbb{Z}_+.$$

Consider the central element C_q of $U_q\mathfrak{sl}_2$ introduced in Subsection 1.3.6.

COROLLARY 1.97. *For all* $f \in \mathcal{D}(\mathbb{D})_q$ *one has* $C_q f = -q\Box_q f$.

PROOF. The linear maps to be compared commute with the action of $U_q\mathfrak{sl}_2$ in $\mathcal{D}(\mathbb{D})_q$. Set $x = y^{-1}$. Due to Proposition 1.82, it suffices to prove the relation $C_q f_0 = -q\Box_q f_0$. An application of (1.66) yields

$$C_q f_0 = -\frac{1}{q^{-1}-q} f_0 + \frac{q^2}{q^{-1}-q} f_1,$$

and then by Proposition 1.96 one has

$$\Box_q f_0 = \mathcal{D}_q x(1-q^{-1}x)\mathcal{D}_q f_0 = \frac{f_0}{1-q^2} - \frac{q^2 f_1}{1-q^2}. \qquad \Box$$

COROLLARY 1.98. *For each function* $\psi(t)$ *on* $q^{-2\mathbb{Z}_+}$, *one has*

$$C_q\psi(x) = -q\mathcal{D}_q x(1-q^{-1}x)\mathcal{D}_q\psi(x)$$

with $x = y^{-1}$.

Let $L^2(1,\infty)_q$ be the Hilbert space of functions $\psi(x)$ on $q^{-2\mathbb{Z}_+}$ with $\|\psi\| = \sqrt{\int_1^\infty |\psi(x)|^2 d_{q^{-2}}x} < \infty$. We are about to prove a lemma that will readily imply that the operator

$$L = \mathcal{D}_q x(1-q^{-1}x)\mathcal{D}_q \tag{1.96}$$

in $L^2(1,\infty)_q$ is well defined, bounded, and self-adjoint.

LEMMA 1.99. *The vectors*

$$e_j = (q^{-2}-1)^{-\frac{1}{2}} q^j f_j, \qquad j \in \mathbb{Z}_+,$$

form an orthonormal basis in $L^2(1,\infty)_q$ *and*

$$Le_j = \begin{cases} b_{j-1}e_{j-1} + a_j e_j + b_j e_{j+1}, & j \in \mathbb{N}, \\ a_j e_j + b_j e_{j+1}, & j = 0, \end{cases}$$

where

$$a_j = \frac{1+q^2-2q^{2(j+1)}}{(1-q^2)^2}, \qquad b_j = -q\frac{1-q^{2(j+1)}}{(1-q^2)^2}.$$

PROOF. The first claim is obvious, and the second follows from the relation

$$Lf_j = -\frac{1-q^{2j}}{(1-q^2)^2} f_{j-1} + \frac{1+q^2-2q^{2(j+1)}}{(1-q^2)^2} f_j - q^2\frac{1-q^{2(j+1)}}{(1-q^2)^2} f_{j+1}, \tag{1.97}$$

which is easily deducible from (1.96). (It is implicit that $f_{-1} = 0$ in (1.97).) $\Box$

Obviously, the linear span of $\{L^k f_0 |\ k \in \mathbb{Z}_+\}$ is dense in $L^2(1,\infty)_q$. Hence the bounded self-adjoint linear map L is unitarily equivalent to the multiplication operator $f(\lambda) \mapsto \lambda f(\lambda)$ in the Hilbert space $L^2(d\mu)$, where $d\mu(\lambda)$ is a Borel measure with compact support on the real axis (see [**3**, p. 279]).

Let us fix this measure, together with the operator $U : L^2(1,\infty)_q \to L^2(d\mu)$ which provides the above unitary equivalence and has the property $Uf_0 = q^{-2}-1$.

We intend to find the measure $d\mu$ and an explicit form for the unitary U.

We assume the reader to be familiar with the basic notation of the theory of q-special functions; see Subsection 1.5.7. Let $d\sigma(\rho)$ be an absolutely continuous measure on $\left[0, \frac{\pi}{2\log(q^{-1})}\right]$ determined by

$$\frac{d\sigma(\rho)}{d\rho} = \frac{1}{\pi}\frac{\log(q^{-1})}{1-q^2}\frac{1}{c(-\frac{1}{2}+i\rho)c(-\frac{1}{2}-i\rho)}, \tag{1.98}$$

where

$$c(l) = \frac{(q^{2(l+1)};q^2)^2_\infty}{(q^{2(l+1)};q^2)_\infty(q^2;q^2)_\infty} = \frac{\Gamma_{q^2}(2l+1)}{(\Gamma_{q^2}(l+1))^2}, \tag{1.99}$$

where Γ_q is the q-Gamma function.

Note that $c(l)$ is a q-analog of Harish-Chandra's c-function; see Subsection 1.7.4.

PROPOSITION 1.100. (1) *Consider the function*

$$\Phi_l(x) = {}_3\phi_2\left[\begin{matrix} x, q^{-2l}, q^{2(l+1)} \\ q^2, 0 \end{matrix}; q^2, q^2\right], \qquad l \in \mathbb{C},$$

on $q^{-2\mathbb{Z}_+}$. *It takes the value* 1 *at* $x=1$ *and is such that*

$$\mathcal{D}_q x(1-q^{-1}x)\mathcal{D}_q\Phi_l(x) = \lambda(l)\Phi_l(x), \tag{1.100}$$

where

$$\lambda(l) = \frac{(1-q^{-2l})(1-q^{2(l+1)})}{(1-q^2)^2}. \tag{1.101}$$

(2) *The function* $\Phi_{-\frac{1}{2}+i\rho}(x)$ *on* $q^{-2\mathbb{Z}_+}$ *is real for real* ρ. *The function* $\int_1^\infty f(x)\Phi_{-\frac{1}{2}+i\rho}(x)d_{q^{-2}}x$ *belongs to* $L^2(d\sigma)$ *for* $f(x)$ *with finite support in* $q^{-2\mathbb{Z}_+}$. *The linear map*

$$f(x) \mapsto \int_1^\infty f(x)\Phi_{-\frac{1}{2}+i\rho}(x)d_{q^{-2}}x$$

admits an extension by continuity to a unitary map from $L^2(1,\infty)_q$ *to* $L^2(d\sigma)$.

PROOF. The vectors $P_j = Ue_j \in L^2(d\mu)$, $j \in \mathbb{Z}_+$, are orthogonal polynomials in λ:

$$\int P_i(\lambda)P_j(\lambda)d\mu(\lambda) = \delta_{ij}, \qquad i,j \in \mathbb{Z}_+.$$

It follows from the definition of U and Lemma 1.99 that

$$\lambda P_j(\lambda) = -q\frac{1-q^{2(j+1)}}{(1-q^2)^2}P_{j+1}(\lambda) + \left(\frac{1+q^2}{(1-q^2)^2} - \frac{2q}{(1-q^2)^2}q^{2j+1}\right)P_j(\lambda) - q\frac{1-q^{2j}}{(1-q^2)^2}P_{j-1}(\lambda), \qquad j \in \mathbb{Z}_+, \tag{1.102}$$

$$P_0(\lambda) = Ue_0 = (q^{-2}-1)^{-\frac{1}{2}}Uf_0 = (q^{-2}-1)^{\frac{1}{2}}. \tag{1.103}$$

It is implicit that $P_{-1}=0$ in (1.102):

$$\lambda P_0(\lambda) = -\frac{q}{1-q^2}P_1(\lambda) + \frac{1}{1-q^2}P_0(\lambda). \tag{1.104}$$

The recurrence (1.102), together with the initial data (1.103) and (1.104), determine the orthogonal polynomials $P_j(\lambda)$, $j \in \mathbb{Z}_+$, uniquely. Let us compare them to the Al Salam–Chihara orthogonal polynomials $Q_j(x; q, q|q^2)$ (see (1.125)). One can deduce from (1.126) the following recurrence relations for the "normed" polynomials $\widetilde{Q}_j \stackrel{\text{def}}{=} \frac{1}{(q^2;q^2)_j} Q_j$:

$$x\widetilde{Q}_j = \frac{1}{2}\left(1 - q^{2(j+1)}\right)\widetilde{Q}_{j+1} + q^{2j+1}\widetilde{Q}_j + \frac{1}{2}\left(1 - q^{2j}\right)\widetilde{Q}_{j-1}, \tag{1.105}$$

where it is implicit that $\widetilde{Q}_{-1} = 0$. The explicit formula (1.125) implies

$$Q_j(x; q, q|q^2) = \frac{(q^2;q^2)_j}{q^j}\,{}_3\phi_2\left[\begin{matrix} q^{-2j}, qt, qt^{-1} \\ q^2, 0 \end{matrix}; q^2, q^2\right],$$

with $x = \frac{1}{2}(t + t^{-1})$, whence one obtains the "initial data"

$$\widetilde{Q}_0(x) = 1. \tag{1.106}$$

Now a comparison of (1.102) and (1.103) with (1.105) and (1.106) allows one to observe that the polynomials P_j and $\widetilde{Q}_j$ relate to each other by the change of variable

$$\lambda = \frac{1+q^2}{(1-q^2)^2} - \frac{2q}{(1-q^2)^2}x. \tag{1.107}$$

Since $x = \frac{1}{2}(t + t^{-1})$, one has $\lambda = \frac{(1-qt)(1-qt^{-1})}{(1-q^2)^2}$,

$$P_j(\lambda) = (q^{-2} - 1)^{\frac{1}{2}} q^{-j}\,{}_3\phi_2\left[\begin{matrix} q^{-2j}, qt, qt^{-1} \\ q^2, 0 \end{matrix}; q^2, q^2\right],$$

$$Uf_j = (q^{-2} - 1)\cdot q^{-2j}\,{}_3\phi_2\left[\begin{matrix} q^{-2j}, qt, qt^{-1} \\ q^2, 0 \end{matrix}; q^2, q^2\right].$$

The substitution $t = q^{-2i\rho}$ yields $Uf_j = \int_1^\infty f_j(x)\Phi_{-\frac{1}{2}+i\rho}(x)d_{q^{-2}}x$. Hence for each function $f(x)$ on $q^{-2\mathbb{Z}_+}$ with finite support, one has

$$Uf = \int_1^\infty f(x)\Phi_{-\frac{1}{2}+i\rho}(x)d_{q^{-2}}x. \tag{1.108}$$

Now we finish the proof of Proposition 1.100. Obviously,

$$\Phi_l(1) = 1, \qquad \operatorname{Im}\Phi_{-\frac{1}{2}+i\rho}(q^{-2j}) = 0, \qquad \rho \in \mathbb{R},\ j \in \mathbb{Z}_+.$$

The relation (1.100) is due to (1.108) and

$$U\mathcal{D}_q x(1 - q^{-1}x)\mathcal{D}_q f = \lambda Uf.$$

The rest of the claims are easily deducible from (1.108) by the change of variable (1.107) and the orthogonality relations (1.126) for the Al Salam–Chihara polynomials. □

REMARK 1.101. Since $\lim_{q\to 1}\Gamma_q(x) = \Gamma(x)$ for $-x \notin \mathbb{Z}_+$, the passage to a limit as $q \to 1$ yields

$$\frac{d\sigma(\rho)}{d\rho} = \frac{1}{2\pi}\frac{\Gamma(\frac{1}{2}+i\rho)^2}{\Gamma(2i\rho)^2}\,\frac{\Gamma(\frac{1}{2}-i\rho)}{\Gamma(-2i\rho)}.$$

One can apply the well-known properties of the Gamma function

$$\Gamma(x)\Gamma(1-x)=\frac{\pi}{\sin\pi x},\qquad \Gamma\left(\frac{1}{2}+x\right)\Gamma\left(\frac{1}{2}-x\right)=\frac{\pi}{\cos\pi x},$$

to obtain the relation

$$\frac{d\sigma(\rho)}{d\rho}=\frac{1}{2\pi}\left(\frac{\pi}{\mathrm{ch}(\pi\rho)}\right)^2\frac{2\rho\,\mathrm{sh}(2\pi\rho)}{\pi}=2\rho\,\mathrm{th}(\pi\rho),$$

which agrees with the well-known Meller–Fock formula [**4**, p. 232].

REMARK 1.102. The above simple proof of Proposition 1.100 has been communicated by E. Koelink. The initial proof (see [**278**]) is based on known results of the spectral theory of self-adjoint operators in a Hilbert space [**82**, Ch. 10, Section 6]. The method used in that proof is very widely used and completely standard.

1.5.5. Boundedness and invertibility of the invariant Laplacian. Proposition 1.100 implies that the radial part $\square_q^{(0)}$ of the invariant Laplacian is unitary equivalent to the operator of multiplication by

$$\lambda\left(-\frac{1}{2}+i\rho\right)=\frac{1-2q\cos(2\log(q^{-1})\rho)+q^2}{(1-q^2)^2} \tag{1.109}$$

in the Hilbert space $L^2(d\sigma(\rho))$ of functions on the segment $\left[0,\frac{\pi}{2\log(q^{-1})}\right]$. The function (1.109) increases monotonically at this segment from $(1+q)^{-2}$ up to $(1-q)^{-2}$. Hence the operator $\square_q^{(0)}$ is bounded and invertible:

$$(1+q)^{-2}\le\square_q^{(0)}\le(1-q)^{-2}.$$

We intend to introduce a series of $*$-representations of $U_q\mathfrak{su}_{1,1}$, to be used for producing similar estimates for $\square_q$ in the whole space. Furthermore these estimates will imply boundedness and invertibility of the invariant Laplacian.

Consider the vector space $\mathcal{E}$ with a basis $\{e_k\}_{k\in\mathbb{Z}}$, and a complex number l such that

$$-\frac{\pi}{2\log(q^{-1})}\le\mathrm{Im}\,l\le\frac{\pi}{2\log(q^{-1})}.$$

It is easy to demonstrate that the equalities

$$\pi_l(E)e_k=q^k\frac{q^{-(k-l)}-q^{k-l}}{q^{-1}-q}e_{k+1},\qquad \pi_l(F)e_k=-q^{-(k-1)}\frac{q^{-(k-l)}-q^{k+l}}{q^{-1}-q}e_{k-1},$$
$$\pi_l(K^{\pm1})e_k=q^{\pm2k}e_k,\qquad k\in\mathbb{Z},$$

determine a representation π_l of $U_q\mathfrak{sl}_2$; moreover,

$$\pi_l(C_q)=\frac{(q^{-l}-q^l)(q^{-(l+1)}-q^{l+1})}{(q^{-1}-q)^2}=-q\frac{(1-q^{-2l})(1-q^{2(l+1)})}{(1-q^2)^2}. \tag{1.110}$$

This family of representations (up to unitary equivalence) of $U_q\mathfrak{sl}_2$ is called the principal spherical series.

The representations π_l admit natural realizations in the spaces of homogeneous distributions on the quantum cone; see Subsection 1.6.3.

Let $V^{(l)}$ be a $U_q\mathfrak{sl}_2$-module corresponding to π_l. Now the definitions imply

LEMMA 1.103. *If $l \neq \mathbb{Z}$, then the $U_q\mathfrak{sl}_2$-module $V^{(l)}$ is simple and can be described in terms of its generator e_0 and the defining relations*

$$K^{\pm 1}e_0 = e_0, \qquad C_q e_0 = \frac{(q^{-l}-q^l)(q^{-(l+1)}-q^{l+1})}{(q^{-1}-q)^2}e_0.$$

Let $\mathcal{J}_l$ be the smallest left ideal containing

$$K^{\pm 1}-1, \qquad C_q - \frac{(q^{-l}-q^l)(q^{-(l+1)}-q^{l+1})}{(q^{-1}-q)^2}.$$

The latter claim of the lemma readily implies that the morphism of $U_q\mathfrak{sl}_2$-modules

$$U_q\mathfrak{sl}_2/\mathcal{J}_l \to V^{(l)}, \qquad 1 \mapsto e_0,$$

is bijective.

REMARK 1.104. Note that one may view the edges of the stripe $-\frac{\pi}{2\log(q^{-1})} \leq \operatorname{Im} l \leq \frac{\pi}{2\log(q^{-1})}$ glued together since

$$\pi_{(s-i\frac{\pi}{2\log(q^{-1})})} = \pi_{(s+i\frac{\pi}{2\log(q^{-1})})}, \qquad s \in \mathbb{R}.$$

Lemma 1.103 shows that the representations π_l and π_{-1-l} are equivalent if $l \notin \mathbb{Z}$. It is easy to show that this description of pairs of equivalent representations of the spherical principal series is exhaustive.

Let $\operatorname{Re} l = -\frac{1}{2}$. Let us endow the vector space $\mathcal{E} = V^{(l)}$ with a structure of pre-Hilbert space in such a way that the basis $\{e_k\}_{k\in\mathbb{Z}}$ is orthonormal. It follows from the definitions that π_l is a $*$-representation of the Hopf $*$-algebra $U_q\mathfrak{su}_{1,1}$. These representations constitute the spherical principal unitary series.

PROPOSITION 1.105. *For all nonzero $f \in \mathcal{D}(\mathbb{D})_q$, one has*

$$(1+q)^{-2} \leq \frac{(\overline{\partial} f, \overline{\partial} f)}{(f,f)} \leq (1-q)^{-2}.$$

PROOF. Consider the measure $d\sigma(\rho)$ on $\left[0, \frac{\pi}{2\log(q^{-1})}\right]$ introduced in the previous subsection. The tensor product $V = \mathcal{E} \otimes L^2(d\sigma(\rho))$ of pre-Hilbert spaces is a $U_q\mathfrak{sl}_2$-module:

$$\xi : v \otimes f \mapsto \pi_{-\frac{1}{2}+i\rho}(\xi)v \otimes f, \qquad \xi \in U_q\mathfrak{sl}_2,\ v \in \mathcal{E},\ f \in L^2(d\sigma).$$

In view of Corollary 1.97, it suffices to produce an isometric linear map $\mathcal{I} : \mathcal{D}(\mathbb{D})_q \to V$ such that $\mathcal{I}C_q f = C_q\mathcal{I}f$ for all $f \in \mathcal{D}(\mathbb{D})_q$.

By Proposition 1.84, the $U_q\mathfrak{sl}_2$-module $\mathcal{D}(\mathbb{D})_q$ is generated by f_0 and the relations $K^{\pm 1}f_0 = f_0$ are defining. Hence there exists a unique morphism of $U_q\mathfrak{sl}_2$-modules

$$\mathcal{I} : \mathcal{D}(\mathbb{D})_q \to V, \qquad \mathcal{I} : f_0 \mapsto e_0 \otimes (q^{-2}-1).$$

It remains to prove this is an isometry, or equivalently, that the invariant Hermitian form $\langle f_1, f_2\rangle \stackrel{\text{def}}{=} (\mathcal{I}f_1, \mathcal{I}f_2) - \int_{\mathbb{D}_q} f_2^* f_1 d\nu$ vanishes on $\mathcal{D}(\mathbb{D})_q$.

Recall the notation $l = -\frac{1}{2} + i\rho$, $x = (1-zz^*)^{-1}$. The relation $\mathcal{I}C_q f = C_q\mathcal{I}f$, $f \in \mathcal{D}(\mathbb{D})_q$, together with the results of the previous subsection imply that for each element of the form $\psi(x)$ in $\mathcal{D}(\mathbb{D})_q$, one has

$$\mathcal{I}\psi = e_0 \otimes \left(\int_1^\infty \psi(x)\Phi_{-\frac{1}{2}+i\rho}(x) d_{q^{-2}}x \right),$$

and for each pair of elements of the form $\psi_1(x)$, $\psi_2(x)$ in $\mathcal{D}(\mathbb{D})_q$, one has

$$\langle \psi_1, \psi_2 \rangle = 0.$$

The latter relation and the orthogonality of the weight subspaces with respect to the invariant Hermitian form $\langle \cdot, \cdot \rangle$ imply that every element of the form $\psi(x)$ belongs to the kernel of $\langle \cdot, \cdot \rangle$. In particular, this kernel contains the generator f_0 of the $U_q\mathfrak{sl}_2$-module $\mathcal{D}(\mathbb{D})_q$. □

COROLLARY 1.106. $(1+q)^{-2} \leq \Box_q \leq (1-q)^{-2}$.

We are about to produce a presentation of $L^2(d\nu)_q$ as a direct integral of Hilbert spaces [**48**, pp. 450–451]). Let $\overline{V^{(l)}}$ be the completion of the pre-Hilbert space V^l and $\overline{V}$ the completion of the pre-Hilbert space V; that is

$$\overline{V} = \bigoplus \int_0^{\frac{\pi}{2\log(q^{-1})}} \overline{V^{(-1/2+i\rho)}}\, d\sigma(\rho).$$

PROPOSITION 1.107. *The isometry $\mathcal{I}$ admits an extension by continuity to a unitary operator $\overline{\mathcal{I}}$ which maps $L^2(d\nu)_q$ onto $\overline{V}$.*

PROOF. It suffices to show that the linear subvariety $\mathcal{I}(\mathcal{D}(\mathbb{D})_q)$ is dense in the Hilbert space $\overline{V}$. Note that the closure L of this linear subvariety is invariant with respect to the operator of multiplication by the continuous function (1.109). The latter function is real on $\left[0, \frac{\pi}{2\log(q^{-1})}\right]$ and separates points. Thus L is a common invariant subspace for the operators of multiplication by functions from $L^\infty(d\sigma)$. Hence it suffices to prove that the morphisms of $U_q\mathfrak{sl}_2$-modules

$$\mathcal{I}_l : \mathcal{D}(\mathbb{D})_q \to V^{(l)}, \qquad \mathcal{I}_l : f_0 \mapsto e_0,$$

are onto maps for $\operatorname{Re} l = -\frac{1}{2}$; see [**48**, pp. 452–453]. It remains to use the simplicity of $U_q\mathfrak{sl}_2$-modules $V^{(l)}$ for the nonintegral l; see Lemma 1.103. □

1.5.6. A passage from functions to sections of linear bundles. In the classical case $q = 1$, the spaces of differentiable sections of holomorphic vector bundles are essentially involved in producing the Dolbeault complexes [**335**, p. 83]. Here are examples of their q-analogs.

Let $\lambda \in \mathbb{R}$ and $\mathbb{C}[z]_{q,\lambda}$ be a bimodule over $\mathbb{C}[z]_q$ with a generator v_λ and the defining relation

$$zv_\lambda = q^{-\lambda} v_\lambda z. \tag{1.111}$$

PROPOSITION 1.108. *There exists a unique structure of $U_q\mathfrak{sl}_2$-module on $\mathbb{C}[z]_{q,\lambda}$ such that*

$$K^{\pm 1} v_\lambda = q^{\pm\lambda} v_\lambda, \qquad Ev_\lambda = -q^{\frac{1}{2}} \frac{1-q^{2\lambda}}{1-q^2} zv_\lambda, \qquad Fv_\lambda = 0, \tag{1.112}$$

and $\mathbb{C}[z]_{q,\lambda}$ is a $U_q\mathfrak{sl}_2$-covariant $\mathbb{C}[z]_q$-bimodule.

PROOF. The uniqueness is obvious. In proving the existence, one may stick to the special case $\lambda \in \mathbb{Z}$ because the coefficients in (1.111) and (1.112) are Laurent polynomials in the variable $u = q^\lambda$ (a similar argument has been used in Subsection 1.3.3).

Consider the $U_q\mathfrak{sl}_2$-module algebra $\mathbb{C}[t_1^{\pm 1}, t_2^{\pm 1}]_q$. The embedding

$$\mathbb{C}[z]_q \hookrightarrow \mathbb{C}[t_1^{\pm 1}, t_2^{\pm 1}]_q, \qquad z \mapsto t_2^{-1} t_1,$$

endows $\mathbb{C}[t_1^{\pm 1}, t_2^{\pm 1}]_q$ with a structure of a $U_q\mathfrak{sl}_2$-covariant $\mathbb{C}[z]_q$-bimodule.

The relation $(t_2^{-1}t_1)t_2^{-\lambda} = q^{-\lambda} t_2^{-\lambda}(t_2^{-1}t_1)$ allows one to introduce a homomorphism of $\mathbb{C}[z]_q$-bimodules

$$\mathbb{C}[z]_{q,\lambda} \to \mathbb{C}[t_1^{\pm 1}, t_2^{\pm 1}]_q, \qquad v_\lambda \mapsto t_2^{-\lambda}.$$

It is injective because $\{t_1^j t_2^k\}_{j,k \in \mathbb{Z}}$ is a basis of the vector space $\mathbb{C}[t_1^{\pm 1}, t_2^{\pm 1}]_q$. It remains to note that

$$K^{\pm 1} t_2^{-\lambda} = q^{\pm \lambda} t_2^{-\lambda}, \quad E t_2^{-\lambda} = -q^{\frac{1}{2}} \frac{1 - q^{2\lambda}}{1 - q^2} (t_2^{-1} t_1) t_2^{-\lambda}, \quad F t_2^{-\lambda} = 0. \qquad \square$$

Consider the $U_q\mathfrak{sl}_2$-module $\mathrm{Pol}(\mathbb{C})_{q,\lambda} \overset{\mathrm{def}}{=} \mathbb{C}[z]_{q,\lambda} \otimes \mathbb{C}[\bar{z}]_q$. Let us endow it with a structure of $U_q\mathfrak{sl}_2$-covariant $\mathrm{Pol}(\mathbb{C})_q$-bimodule using the universal R-matrix the way it was done in Subsection 1.3.5. The $\mathrm{Pol}(\mathbb{C})_q$-bimodule obtained from this procedure is determined by a generator v_λ and the defining relations

$$zv_\lambda = q^{-\lambda} v_\lambda z, \qquad z^* v_\lambda = q^\lambda v_\lambda z^*, \tag{1.113}$$

and the action of $U_q\mathfrak{sl}_2$ is described as above by (1.112).

An extension by continuity leads to a $U_q\mathfrak{sl}_2$-covariant $\mathrm{Pol}(\mathbb{C})_q \oplus \mathcal{D}(\mathbb{D})_q$-bimodule, which can be described in terms of generator v_λ, relations (1.113), and the additional relation $\psi(y)v_\lambda = v_\lambda \psi(y)$, where ψ is a function with finite support on $q^{2\mathbb{Z}_+}$.

Introduce the notation $\mathcal{D}(\mathbb{D})_{q,\lambda} \overset{\mathrm{def}}{=} \mathcal{D}(\mathbb{D})_q v_\lambda = v_\lambda \mathcal{D}(\mathbb{D})_q$ for the above $U_q\mathfrak{sl}_2$-covariant $\mathcal{D}(\mathbb{D})_q$-bimodule.

Recall that $\mathcal{D}(\mathbb{D})'_q$ is a $\mathcal{D}(\mathbb{D})_q$-bimodule and $y^\lambda \in \mathcal{D}(\mathbb{D})'_q$. We use the notion of Hermitian metric introduced in Subsection 1.5.3.

PROPOSITION 1.109. *The sesquilinear map*

$$\mathbf{h}_\lambda : \mathcal{D}(\mathbb{D})_{q,\lambda} \times \mathcal{D}(\mathbb{D})_{q,\lambda} \to \mathcal{D}(\mathbb{D})_q, \qquad \mathbf{h}_\lambda : v_\lambda f_1 \times v_\lambda f_2 \mapsto f_2^*(1 - zz^*)^\lambda f_1,$$

is an invariant positive Hermitian metric.

A proof is postponed until Subsection 1.7.2.

COROLLARY 1.110. *The Hermitian form*

$$(v_\lambda f_1, v_\lambda f_2) = \int_{\mathbb{D}_q} f_2^*(1 - zz^*)^\lambda f_1 d\nu, \qquad f_1, f_2 \in \mathcal{D}(\mathbb{D})_q,$$

in $\mathcal{D}(\mathbb{D})_{q,\lambda}$ is positive and invariant.

REMARK 1.111. It is easy to prove that

$$(f_1 v_\lambda, f_2 v_\lambda) = \int_{\mathbb{D}_q} f_2^* f_1 (1 - zz^*)^\lambda d\nu, \qquad f_1, f_2 \in \mathcal{D}(\mathbb{D})_q.$$

Let $\lambda > 1$. Consider the positive linear functional $\int_{\mathbb{D}_q} \cdot d\nu_\lambda$ on the $*$-algebra $\mathrm{Pol}(\mathbb{C})_q$ given by

$$\int_{\mathbb{D}_q} f d\nu_\lambda = \frac{1 - q^{2(\lambda - 1)}}{1 - q^2} \int_{\mathbb{D}_q} f \cdot (1 - zz^*)^\lambda d\nu = (1 - q^{2(\lambda-1)}) \operatorname{tr} T_F(f \cdot (1 - zz^*)^{\lambda - 1}).$$

The constant factor at the integral is chosen so that $\int_{\mathbb{D}_q} 1 d\nu_\lambda = 1$.

Let $L^2(d\nu_\lambda)_q$ be the completion of $\mathrm{Pol}(\mathbb{C})_q$ with respect to the norm $\|f\|_\lambda = \left(\int_{\mathbb{D}_q} f^* f d\nu_\lambda\right)^{\frac{1}{2}}$ and $\square_{q,\lambda}$ an invariant Laplacian in $L^2(d\nu_\lambda)_q$ defined the same way as in the above special case $\lambda = 0$. Note that $L^2(d\nu_\lambda)_q$ admits a natural embedding into the space of distributions $\mathcal{D}(\mathbb{D})'_q$.

The expansion in eigenfunctions of the radial part of $\square_{q,\lambda}$ was obtained in a slightly different context in Groenevelt's thesis [**109**].

1.5.7. Appendix on q-special functions. One of the main features of the hypergeometric series $\sum_{k=0}^{\infty} c_k z^k$ is that the ratio $\frac{c_{k+1}}{c_k}$ is a rational function in the variable k [**27**, p. 183]. In the case of basic hypergeometric series these ratios are rational functions in q^k, where $q \in \mathbb{C}$, $q \neq 0$. More precisely, given $a_1, a_2, \ldots, a_r; b_1, b_2, \ldots, b_s \in \mathbb{C}$, the basic hypergeometric series is defined by

$$ {}_r\phi_s \left[\begin{matrix} a_1, a_2, \ldots, a_r \\ b_1, b_2, \ldots, b_s \end{matrix} ; q, z \right] = \sum_{k=0}^{\infty} \frac{(a_1;q)_k (a_2;q)_k \cdots (a_r;q)_k}{(q;q)_k (b_1;q)_k \cdots (b_s;q)_k} \left[(-1)^k q^{\frac{k(k-1)}{2}} \right]^{1+s-r} z^k, $$

where

$$ (a;q)_k = \begin{cases} (1-a)(1-aq)\cdots(1-aq^{k-1}), & k \in \mathbb{N}, \\ 1, & k = 0; \end{cases} $$

see [**99**, p. 23]. This series is also denoted by ${}_r\phi_s(a_1, a_2, \ldots, a_r; b_1, b_2, \ldots, b_s; q, z)$.

We assume in what follows that $q \in (0,1)$, $b_1, b_2, \ldots, b_s \notin q^{-2\mathbb{Z}_+}$.

The series ${}_r\phi_s$ is a polynomial in z (it terminates) if at least one of $a_1, a_2, \ldots, a_r$ belongs to $q^{-2\mathbb{Z}_+}$. Otherwise, one should care about the convergence of this power series. In the most interesting case $r = s+1$ the radius of convergence is 1.[10]

We reproduce some facts from [**99**] about q-special functions. Here, we introduce the notation

$$ (a;q)_\infty = \prod_{n=0}^{\infty} (1 - aq^n). $$

One of the simplest basic hypergeometric series is the q-binomial series

$$ {}_1\phi_0(a; -; q, z) = \sum_{n=0}^{\infty} \frac{(a;q)_n}{(q;q)_n} z^n. $$

It is easy to prove that for $|z| < 1$, one has

$$ \sum_{n=0}^{\infty} \frac{(a;q)_n}{(q;q)_n} z^n = \frac{(az;q)_\infty}{(z;q)_\infty}; \tag{1.114} $$

see [**99**, p. 26]. In particular, for $a = 0$ and $|z| < 1$

$$ e_q(z) \stackrel{\text{def}}{=} \sum_{n=0}^{\infty} \frac{1}{(q;q)_n} z^n = \frac{1}{(z;q)_\infty}. $$

[10]The well-known [**27**, p. 183] generalized hypergeometric series ${}_{s+1}F_s$ can be derived from the basic hypergeometric series ${}_{s+1}\phi_s$ via the substitution $a_j = q^{\alpha_j}$, $b_j = q^{\beta_j}$ and a formal passage to the limit as $q \to 1$.

The series $e_q(z)$ is called *q-exponential*, because

$$\lim_{q\to 1} e_q((1-q)z) = \lim_{q\to 1} \sum_{n=0}^{\infty} \frac{(1-q)^n}{(q;q)_n} z^n = \sum_{n=0}^{\infty} \frac{z^n}{n!} = e^z$$

in the topology of coefficientwise convergence. In [**99**, p. 285] one can find one more q-analog for the exponential series

$$E_q(z) \stackrel{\text{def}}{=} \sum_{n=0}^{\infty} \frac{q^{\frac{n(n-1)}{2}}}{(q;q)_n} z^n = (-z;q)_\infty .$$

Let us recall the Heine q-analogs of the Euler formula

$$(1.115) \qquad {}_2\phi_1(a,b;c;q,z) = \frac{(b;q)_\infty (az;q)_\infty}{(c;q)_\infty (z;q)_\infty} \, {}_2\phi_1\left(c/b, z; az; q, b\right)$$

and of the Gauss formula

$$(1.116) \qquad {}_2\phi_1(a,b;c;q,c/(ab)) = \frac{(c/a;q)_\infty (c/b;q)_\infty}{(c;q)_\infty (c/(ab);q)_\infty},$$

both presented in [**99**, pp. 29, 30]. The first of the above formulas assumes $|b| < 1$, $|z| < 1$, and the second $|b| < 1$, $|c/(ab)| < 1$.

The Jackson integral is defined by

$$\int_0^a f(x) d_q x = a(1-q) \sum_{k=0}^{\infty} f(aq^k) q^k .$$

Similarly to the Euler Beta function

$$B(x,y) = \int_0^1 t^{x-1}(1-t)^{y-1} dt, \qquad \operatorname{Re} x > 0, \quad \operatorname{Re} y > 0,$$

one defines its q-analog

$$B_q(x,y) = \int_0^1 t^{x-1} \frac{(tq;q)_\infty}{(tq^y;q)_\infty} d_q t, \qquad \operatorname{Re} x > 0, \quad y \notin -\mathbb{Z}_+ .$$

A q-analog of Euler's Gamma function $\Gamma(x)$ is given by

$$\Gamma_q(x) = \frac{(q;q)_\infty}{(q^x;q)_\infty} (1-q)^{1-x} .$$

One can prove that $\lim_{q\to 1} \Gamma_q(x) = \Gamma(x)$,

$$B_q(x,y) = \frac{\Gamma_q(x)\Gamma_q(y)}{\Gamma_q(x+y)};$$

see [**99**, p. 39]. Now (1.115) can be rewritten as

$${}_2\phi_1(q^a, q^b; q^c; q, z) = \frac{\Gamma_q(c)}{\Gamma_q(b)\Gamma_q(c-b)} \int_0^1 t^{b-1} \frac{(tzq^a;q)_\infty (tq;q)_\infty}{(tz;q)_\infty (tq^{c-b};q)_\infty} d_q t$$

(see [**99**, p. 40]), which is a q-analog of Euler's integral representation for the hypergeometric function; see [**27**, p. 72].

As a consequence of the definition of $\Gamma_q(x)$, one has

$$\Gamma_q(n+1) = 1(1+q)(1+q+q^2)\cdots(1+q+\cdots+q^{n-1}).$$

Let

$$\binom{a}{b}_q = \frac{\Gamma_q(a+1)}{\Gamma_q(b+1)\Gamma_q(a-b+1)}.$$

For $n, k \in \mathbb{Z}_+$, $n \geq k$, one gets q-binomial coefficients

$$\binom{n}{k}_q = \frac{(q;q)_n}{(q;q)_k(q;q)_{n-k}},$$

which is also known as Gauss polynomials. These are extensively involved in the theory of partitions and in combinatorics [**6**].[11]

Consider the field $\mathbb{Q}(q)$ and the algebra over $\mathbb{Q}(q)$ given by two generators a, b and the commutation relation $ab = qba$. It is easy to prove that

$$(a+b)^n = \sum_{k=0}^{n} \binom{n}{k}_q b^k a^{n-k}.$$

This means that $(a+b)^n$ is a course-of-value "function" for q-binomial coefficients.

Using (1.115), one can easily obtain the Pfaff–Saalschütz summation formula,

$$(1.117) \qquad {}_3\phi_2\left[\begin{matrix} a, b, q^{-n} \\ c, abc^{-1}q^{1-n} \end{matrix}; q, q\right] = \frac{\left(\frac{c}{a};q\right)_n \left(\frac{c}{b};q\right)_n}{(c;q)_n \left(\frac{c}{ab};q\right)_n}, \qquad n \in \mathbb{Z}_+;$$

see [**99**, p. 286]. Moreover, [**99**] contains a collection of summation formulas similar to (1.117) and transformation formulas for basic hypergeometric series similar to (1.115). For example, [**99**, pp. 45 and 290–291]:

$$(1.118) \qquad {}_2\phi_1\left[\begin{matrix} q^{-n}, b \\ c \end{matrix}; q, z\right] = \frac{\left(\frac{c}{b};q\right)_n}{(c;q)_n} {}_3\phi_2\left[\begin{matrix} q^{-n}, b, \frac{bzq^{-n}}{c} \\ \frac{bq^{1-n}}{c}, 0 \end{matrix}; q, q\right],$$

$$(1.119) \qquad {}_3\phi_2\left[\begin{matrix} a, b, c \\ d, e \end{matrix}; q, \frac{de}{abc}\right] = \frac{\left(\frac{e}{a};q\right)_\infty \left(\frac{de}{bc};q\right)_\infty}{(e;q)_\infty \left(\frac{de}{abc};q\right)_\infty} {}_3\phi_2\left[\begin{matrix} a, \frac{d}{b}, \frac{d}{c} \\ d, \frac{de}{bc} \end{matrix}; q, \frac{e}{a}\right]$$
$$= \frac{(b;q)_\infty \left(\frac{de}{ab};q\right)_\infty \left(\frac{de}{bc};q\right)_\infty}{(d;q)_\infty (e;q)_\infty \left(\frac{de}{abc};q\right)_\infty} {}_3\phi_2\left[\begin{matrix} \frac{d}{b}, \frac{e}{b}, \frac{de}{abc} \\ \frac{de}{ab}, \frac{de}{bc} \end{matrix}; q, b\right],$$

$$(1.120) \qquad {}_3\phi_2\left[\begin{matrix} q^{-n}, b, c \\ d, e \end{matrix}; q, q\right] = \frac{\left(\frac{de}{bc};q\right)_n}{(e;q)_n} \left(\frac{bc}{d}\right)^n {}_3\phi_2\left[\begin{matrix} q^{-n}, \frac{d}{b}, \frac{d}{c} \\ d, \frac{de}{bc} \end{matrix}; q, q\right]$$
$$= \frac{\left(\frac{e}{c};q\right)_n}{(e;q)_n} c^n {}_3\phi_2\left[\begin{matrix} q^{-n}, c, \frac{d}{b} \\ d, \frac{cq^{1-n}}{e} \end{matrix}; q, \frac{bq}{e}\right],$$

$${}_3\phi_2\left[\begin{matrix} q^{-n}, b, c \\ d, e \end{matrix}; q, \frac{deq^n}{bc}\right] = \frac{\left(\frac{e}{c};q\right)_n}{(e;q)_n} {}_3\phi_2\left[\begin{matrix} q^{-n}, c, \frac{d}{b} \\ d, \frac{cq^{1-n}}{e} \end{matrix}; q, q\right].$$

Introduce q-analogs for differentiation operators

$$D_q^- f(z) = \frac{f(z) - f(qz)}{z - qz}, \qquad D_q^+ f(z) = \frac{f(q^{-1}z) - f(z)}{z - qz}.$$

---

[11] For example, if $\mathbb{F}$ is a finite field with q elements, then the vector space $\mathbb{F}^n$ consists of q^n elements, and the number of k-dimensional subspaces is $\binom{n}{k}_q$; see [**6**, p. 218].

One has the following q-analog for integration by parts:

$$\int_0^a (D_q^- u)(x)v(x)d_qx = u(a)v(q^{-1}a) - u(0)v(0) - \int_0^a u(x)(D_q^+ v)(x)d_qx, \tag{1.121}$$

where the functions u, v are continuous on the compact set $\mathfrak{M}_a = \{0\} \cup (aq^{-1})q^{\mathbb{Z}_+}$ and $a \in \mathbb{C}$.

It follows from the definitions that the formal series $e_q((1-q)z)$ is a solution of the q-difference equation $D_q^- y = y$, with value 1 at $z = 0$.

The formal series ${}_2\phi_1(a, b; q, z)$ satisfies the q-difference equation

$$z(c - abqz)(D_q^-)^2 y(z) + \left(\frac{1-c}{1-q} + \frac{(1-a)(1-b) - (1-abq)}{1-q} z\right) D_q^- y(z) - \frac{(1-a)(1-b)}{(1-q)^2} y(z) = 0; \tag{1.122}$$

see [**99**, p. 44]. After replacing a, b, c by q^a, q^b, q^c, a formal passage to a limit as $q \to 1$ leads to the differential equation

$$z(1-z)\frac{d^2}{dz^2}y(z) + (c - (a+b+1)z)\frac{dy(z)}{dz} - aby(z) = 0.$$

Chapters 4–5 of [**93**] are devoted to solving q-difference equations using basic hypergeometric series and q-analogs of the Sturm–Liouville problem.

Let us turn to q-hypergeometric polynomials in one variable.

In the mid-1980s R. Askey and J. Wilson introduced a four-parameter family of polynomials with many remarkable properties [**20**]; those were called later the Askey–Wilson polynomials. They are defined using the algebra of "invariant" Laurent polynomials

$$\mathbb{C}[t, t^{-1}]^{\mathrm{inv}} = \{f(t) \in \mathbb{C}[t, t^{-1}] \,|\, f(t^{-1}) = f(t)\}$$

and the canonical isomorphism

$$\mathbb{C}[x] \xrightarrow{\approx} \mathbb{C}[t, t^{-1}]^{\mathrm{inv}}, \qquad f(x) \mapsto f\left(\frac{t + t^{-1}}{2}\right). \tag{1.123}$$

Namely,

$$p_n(x; a, b, c, d|q) = \frac{(ab; q)_n (ac; q)_n (ad; q)_n}{a^n} {}_4\phi_3 \left[\begin{matrix} q^{-n}, abcdq^{n-1}, at, at^{-1} \\ ab, ac, ad \end{matrix} ; q, q \right]; \tag{1.124}$$

see [**99**, p. 212]. The Askey–Wilson polynomials are solutions of the second-order q-difference equation

$$D_q\Big(\sqrt{1-x^2}\, w\left(x; aq^{\frac{1}{2}}, bq^{\frac{1}{2}}, cq^{\frac{1}{2}}, dq^{\frac{1}{2}} \middle| q\right) D_q p_n\Big) + \lambda_n \sqrt{1-x^2}\, w(x; a, b, c, d|q) p_n = 0;$$

see [**99**, p. 225]. Here $\lambda_n = 4q^{-n+1}(1-q^n)(1-abcdq^{n-1})(1-q)^{-2}$, and $w(x; a, b, c, d|q)$ is given by an explicit formula in [**99**, p. 180], while D_q is determined by the canonical isomorphism (1.123):

$$D_q : \mathbb{C}[t, t^{-1}]^{\mathrm{inv}} \to \mathbb{C}[t, t^{-1}]^{\mathrm{inv}}, \qquad D_q : f(t) \mapsto \frac{f(q^{\frac{1}{2}}t) - f(q^{-\frac{1}{2}}t)}{q^{\frac{1}{2}}t - q^{-\frac{1}{2}}t}.$$

Note that the passage from polynomials in x to invariant Laurent polynomials in t is a step towards the theory of symmetric orthogonal polynomials in several variables produced by I. Macdonald [**224**, Chapter VI] and T. Koornwinder [**192**]. The polynomials introduced in [**192**] are generalizations of the Askey–Wilson polynomials.

We follow [**178**] in reproducing the known results about two families of orthogonal polynomials, which are special or limit cases of the Askey–Wilson polynomials.

The simplest representatives of the first family are the Al Salam–Chihara polynomials, and in the second family these are the little q-Jacobi polynomials. The first family is related to representation theory of the Hopf $*$-algebra $U_q\mathfrak{su}_{1,1}$, whereas the second family is related to representation theory of Hopf $*$-algebra $U_q\mathfrak{su}_2$.

In what follows we use the notation

$$t = e^{i\theta}, \qquad x = \cos\theta,$$
$$(a_1, a_2, \ldots, a_m; q)_k = (a_1; q)_k (a_2; q)_k \cdots (a_m; q)_k, \qquad k \in \mathbb{N} \cup \{\infty\}.$$

By definition [**20**, p. 25], the Al Salam–Chihara polynomials are degree n polynomials in the variable $x = \cos\theta$, which depend on the parameters a, b, q,

$$Q_n(x; a, b|q) = \frac{(ab; q)_n}{a^n} \, {}_3\phi_2 \left[\begin{matrix} q^{-n}, ae^{i\theta}, ae^{-i\theta} \\ ab, 0 \end{matrix} ; q, q \right]. \tag{1.125}$$

If a and b are either real or complex conjugate numbers and $|a| < 1$, $|b| < 1$, then the polynomials $Q_n(x; a, b|q)$ satisfy the following orthogonality relations:

$$\frac{1}{2\pi} \int_{-1}^{1} Q_m(x; a, b|q) Q_n(x; a, b|q) \frac{w(x; a, b|q)dx}{\sqrt{1 - x^2}} = \frac{\delta_{mn}}{(q^{n+1}, abq^n; q)_\infty}, \tag{1.126}$$

where

$$w(x; a, b|q) = \left| \frac{(e^{2i\theta}; q)_\infty}{(ae^{i\theta}, be^{i\theta}; q)_\infty} \right|^2 = \frac{h(x, 1)h(x, -1)h(x, q^{\frac{1}{2}})h(x, -q^{\frac{1}{2}})}{h(x, a)h(x, b)},$$

$$h(x, \alpha) = (\alpha e^{i\theta}, \alpha e^{-i\theta}; q)_\infty = \prod_{k=0}^{\infty} \left(1 - 2\alpha x q^k + \alpha^2 q^{2k}\right).$$

There are recurrence relations for these polynomials:

$$2xQ_n(x) = Q_{n+1}(x) + (a + b)q^n Q_n(x) + (1 - q^n)(1 - abq^{n-1})Q_{n-1}(x).$$

Note that (1.125) are solutions of the q-difference equation

$$(1 - q)^2 D_q \left[\widetilde{w} \left(x; aq^{\frac{1}{2}}, bq^{\frac{1}{2}} \middle| q \right) D_q y(x) \right] + 4q^{-n+1}(1 - q^n)\widetilde{w}(x; a, b|q)y(x) = 0,$$

where $\widetilde{w}(x; a, b|q) = w(x; a, b|q)(1 - x^2)^{-\frac{1}{2}}$. The generating function for the Al Salam–Chihara polynomials has the form

$$\frac{(at, bt; q)_\infty}{(e^{i\theta}t, e^{-i\theta}t; q)_\infty} = \sum_{n=0}^{\infty} \frac{Q_n(x; a, b|q)}{(q; q)_n} t^n, \qquad x = \cos\theta.$$

The Al Salam–Chihara polynomials were studied by many authors. We mention the works [**19**], [**185**]. The latter work contains a q-analog of Graf's addition formula for the Bessel functions [**28**, p. 53].

The Al Salam–Chihara polynomials are special cases of the continuous dual q-Hahn polynomials

$$p_n(x; a, b, c|q) = \frac{(ab, ac; q)_n}{a^n} \, {}_3\phi_2 \left[\begin{matrix} q^{-n}, ae^{i\theta}, ae^{-i\theta} \\ ab, ac \end{matrix} ; q, q \right],$$

where $x = \cos\theta$; see [**110**]. Note that to obtain the Al Salam–Chihara polynomials, it suffices to set $c = 0$. If either all the parameters a, b, c are real or one of them is real and the other two are complex conjugate and $|a| < 1$, $|b| < 1$, $|c| < 1$, then one has the orthogonality relations which generalize (1.126):

$$\frac{1}{2\pi}\int_{-1}^{1} p_m(x;a,b,c|q)p_n(x;a,b,c|q)\frac{w(x;a,b,c|q)dx}{\sqrt{1-x^2}} = h_n\delta_{mn},$$

where

$$w(x;a,b,c|q) = \left|\frac{(e^{2i\theta};q)_\infty}{(ae^{i\theta},be^{i\theta},ce^{i\theta};q)_\infty}\right|^2 = \frac{h(x,1)h(x,-1)h(x,q^{\frac12})h(x,-q^{\frac12})}{h(x,a)h(x,b)h(x,c)},$$

$$h_n = \frac{1}{(q^{n+1},abq^n,acq^n,bcq^n;q)_\infty}.$$

There exist recurrence relations

$$2x\widetilde{p}_n(x) = A_n\widetilde{p}_{n+1}(x) + [a + a^{-1} - (A_n + C_n)]\widetilde{p}_n(x) + C_n\widetilde{p}_{n-1}(x),$$

where

$$\widetilde{p}_n(x) = \frac{a^n p_n(x;a,b,c|q)}{(ab,ac;q)_n} = {}_3\phi_2\left[\begin{matrix} q^{-n}, ae^{i\theta}, ae^{-i\theta} \\ ab, ac \end{matrix}; q, q\right],$$

$$A_n = a^{-1}(1-abq^n)(1-acq^n), \qquad C_n = a(1-q^n)(1-bcq^{n-1}).$$

The polynomials p_n are solutions of the following q-difference equation:

$$(1.127)\quad (1-q)^2 D_q\left[\widetilde{w}\left(x; aq^{\frac12}, bq^{\frac12}, cq^{\frac12}\middle| q\right) D_q y(x)\right] + 4q^{-n+1}(1-q^n)\widetilde{w}(x;a,b,c|q)y(x) = 0,$$

where $\widetilde{w}(x;a,b,c|q) = w(x;a,b,c|q)(1-x^2)^{-\frac12}$. One also has course-of-value functions

$$(1.128)\quad \frac{(ct;q)_\infty}{(e^{i\theta}t;q)_\infty}\,{}_2\phi_1\left[\begin{matrix} ae^{i\theta}, be^{i\theta} \\ ab \end{matrix}; q, e^{-i\theta}t\right] = \sum_{n=0}^{\infty}\frac{p_n(x;a,b,c|q)}{(ab,q;q)_n}t^n, \qquad x = \cos\theta,$$

$$(1.129)\quad \frac{(bt;q)_\infty}{(e^{i\theta}t;q)_\infty}\,{}_2\phi_1\left[\begin{matrix} ae^{i\theta}, ce^{i\theta} \\ ac \end{matrix}; q, e^{-i\theta}t\right] = \sum_{n=0}^{\infty}\frac{p_n(x;a,b,c|q)}{(ac,q;q)_n}t^n, \qquad x = \cos\theta,$$

$$(1.130)\quad \frac{(at;q)_\infty}{(e^{i\theta}t;q)_\infty}\,{}_2\phi_1\left[\begin{matrix} be^{i\theta}, ce^{i\theta} \\ bc \end{matrix}; q, e^{-i\theta}t\right] = \sum_{n=0}^{\infty}\frac{p_n(x;a,b,c|q)}{(bc,q;q)_n}t^n, \qquad x = \cos\theta.$$

There are extensions of the above results to the Askey–Wilson polynomial; see (1.124). The orthogonality relations are valid under more general assumptions than those discussed above. In the general case the measure includes both absolutely continuous and discrete parts [**99**, p. 216].

Let us turn to the second family of q-hypergeometric orthogonal polynomials we are interested in.

The little q-Jacobi polynomials were introduced by W. Hahn in [**111**, p. 29]:

$$p_n(x;a,b|q) = {}_2\phi_1\left[\begin{matrix} q^{-n}, abq^{n+1} \\ aq \end{matrix}; q, qx\right].$$

The corresponding orthogonality relations have the form

$$(1.131)\quad \sum_{k=0}^{\infty}\frac{(bq;q)_k}{(q;q)_k}(aq)^k p_m(q^k;a,b|q)p_n(q^k;a,b|q)$$
$$=\frac{(abq^2;q)_\infty}{(aq;q)_\infty}\frac{(1-abq)(aq)^n}{(1-abq^{2n+1})}\frac{(q,bq;q)_n}{(aq,abq;q)_n}\delta_{mn},$$

where $0<a<q^{-1}$, $b<q^{-1}$, and the recurrence relations are

$$-xp_n(x;a,b|q)=A_np_{n+1}(x;a,b|q)-(A_n+C_n)p_n(x;a,b|q)+C_np_{n-1}(x;a,b|q),$$

where

$$A_n=q^n\frac{(1-aq^{n+1})(1-abq^{n+1})}{(1-abq^{2n+1})(1-abq^{2n+2})},\qquad C_n=aq^n\frac{(1-q^n)(1-bq^n)}{(1-abq^{2n})(1-abq^{2n+1})};$$

see [**7**, p. 110]. The little q-Jacobi polynomials are solutions of a q-difference equation derived from (1.122) and are equivalent to the equation

$$(1.132)\quad q^{-n}(1-q^n)(1-abq^{n+1})x\,y(x)$$
$$=B(x)y(qx)-[B(x)+D(x)]y(x)+D(x)y(q^{-1}x),$$

where

$$B(x)=a(bqx-1),\qquad D(x)=x-1.$$

The generating function has the form

$$(1.133)\quad {}_0\phi_1\left[\begin{matrix}-\\ aq\end{matrix};q,aqxt\right]{}_2\phi_1\left[\begin{matrix}x^{-1},0\\ bq\end{matrix};q,xt\right]$$
$$=\sum_{n=0}^{\infty}\frac{(-1)^nq^{\frac{n(n-1)}{2}}}{(bq,q;q)_n}p_n(x;a,b|q)t^n.$$

If $d\mu$ is concentrated on a finite set which consists of $N+1$ points, then the vector space $L^2(d\mu)$ is $N+1$-dimensional. In this case an orthogonal basis can be produced by applying the orthogonalization procedure to $1,x,\dots,x^N$. An important example of such orthogonal polynomials are the q-Hahn polynomials [**111**]. These are defined by

$$Q_n(q^{-x};\alpha,\beta,N|q)={}_3\phi_2\left[\begin{matrix}q^{-n},\alpha\beta q^{n+1},q^{-x}\\ \alpha q,q^{-N}\end{matrix};q,q\right],\qquad n=0,1,\dots,N,$$

and are polynomials in q^{-x}; see [**99**, p. 204]. The little q-Jacobi polynomials can be derived from the q-Hahn polynomials by the passage to a limit as follows:

$$\lim_{N\to\infty}Q_n(q^{x-N};\alpha,\beta,N|q)=p_n(q^x;\alpha,\beta|q).$$

The orthogonality relations

$$(1.134)\quad \sum_{x=0}^{N}\frac{(\alpha q,q^{-N};q)_x}{(q,\beta^{-1}q^{-N};q)_x}(\alpha\beta q)^{-x}Q_m(q^{-x};\alpha,\beta,N|q)Q_n(q^{-x};\alpha,\beta,N|q)$$
$$=\frac{(\alpha\beta q^2;q)_N}{(\beta q;q)_N(\alpha q)^N}\frac{(q,\alpha\beta q^{N+2},\beta q;q)_n}{(\alpha q,\alpha\beta q,q^{-N};q)_n}\frac{(1-\alpha\beta q)(-\alpha q)^n}{(1-\alpha\beta q^{2n+1})}q^{\frac{n(n-1)}{2}-Nn}\delta_{mn}$$

are valid for either $0 < \alpha < q^{-1}$ and $0 < \beta < q^{-1}$, or $\alpha > q^{-N}$ and $\beta > q^{-N}$. The recurrence relations have the form

$$-(1-q^{-x})Q_n(q^{-x}) = A_n Q_{n+1}(q^{-x}) - (A_n + C_n)Q_n(q^{-x}) + C_n Q_{n-1}(q^{-x}),$$

where

$$\begin{aligned} Q_n(q^{-x}) &= Q_n(q^{-x};\alpha,\beta,N|q), \\ A_n &= \frac{(1-q^{n-N})(1-\alpha q^{n+1})(1-\alpha\beta q^{n+1})}{(1-\alpha\beta q^{2n+1})(1-\alpha\beta q^{2n+2})}, \\ C_n &= -\frac{\alpha q^{n-N}(1-q^n)(1-\alpha\beta q^{n+N+1})(1-\beta q^n)}{(1-\alpha\beta q^{2n})(1-\alpha\beta q^{2n+1})}. \end{aligned}$$

The q-Hahn polynomials are solutions of the q-difference equation

$$\begin{aligned} (1.135)\quad & q^{-n}(1-q^n)(1-\alpha\beta q^{n+1})y(x) \\ &= B(x)y(x+1) - [B(x)+D(x)]y(x) + D(x)y(x-1), \end{aligned}$$

where $y(x) = Q_n(q^{-x};\alpha,\beta,N|q)$,

$$B(x) = (1-q^{x-N})(1-\alpha q^{x+1}), \qquad D(x) = \alpha q(1-q^x)(\beta - q^{x-N-1}).$$

For all $x = 0,1,2,\ldots,N$, one has

$$\begin{aligned} (1.136)\quad & {}_1\phi_1\left[\begin{matrix} q^{-x} \\ \alpha q \end{matrix};q,\alpha qt\right] {}_2\phi_1\left[\begin{matrix} q^{x-N},0 \\ \beta q \end{matrix};q,q^{-x}t\right] \\ &= \sum_{n=0}^{N} \frac{(q^{-N};q)_n}{(\beta q,q;q)_n} Q_n(q^{-x};\alpha,\beta,N|q)t^n. \end{aligned}$$

The above results on q-Hahn polynomials are transferred to the case of q-Racah polynomials [**99**, p. 202], i.e., those whose images under the morphism of algebras

$$\mathbb{C}[x] \to \mathbb{C}[u,u^{-1}], \qquad f(x) \mapsto f(u^{-1} + \gamma\delta q u),$$

are just

$${}_4\phi_3\left[\begin{matrix} q^{-n},\alpha\beta q^{n+1},u^{-1},\gamma\delta q u \\ \alpha q,\beta\delta q,\gamma q \end{matrix};q,q\right], \qquad n = 0,1,2,\ldots,N.$$

The q-Racah polynomials are related to the Askey–Wilson polynomials (1.124) by the substitution

$$\alpha = abq^{-1}, \quad \beta = cdq^{-1}, \quad \gamma = adq^{-1}, \quad \delta = ad^{-1}, \quad u = a^{-1}e^{-i\theta}.$$

1.6. Integral representations

1.6.1. Algebras of kernels of integral operators. In this section we consider q-analogs of continuous linear operators, which are important tools in harmonic analysis or in complex analysis. The Schwartz kernel theorem [**128**, p. 156] leads to the fact that such operators are integral, and their kernels are distributions. Our typical problem will be as follows: given a q-analog of a continuous linear operator, find a q-analog of the kernel of this integral operator.

In this subsection auxiliary notions are introduced. First of all, we show that an A-invariant integral can be used to construct morphisms of A-modules.

Consider A-module algebras F_1, F_2. To a linear functional $\nu : F_2 \to \mathbb{C}$ and an element $\mathcal{K} \in F_1 \otimes F_2$, we assign the linear integral operator

$$K : F_2 \to F_1, \qquad K : f \mapsto (\mathrm{id} \otimes \nu)(\mathcal{K}(1 \otimes f)).$$

The element $\mathcal{K}$ is called the kernel of this integral operator, and it is standard to use the brief notation

$$Kf = \int \mathcal{K}(1 \otimes f)\, d\nu = \int \mathcal{K} f d\nu.$$

PROPOSITION 1.112. *Consider A-module algebras F_1, F_2 and an A-invariant integral $\nu : F_2 \to \mathbb{C}$. Suppose that the bilinear form*

$$f' \times f'' \mapsto \nu(f' \cdot f''), \qquad f', f'' \in F_2, \tag{1.137}$$

is nondegenerate on F_2. The integral operator with kernel $\mathcal{K}$ is a morphism of A-modules if and only if this kernel is A-invariant.

PROOF. 1. If the kernel $\mathcal{K}$ is invariant, then the linear map

$$F_2 \to F_1 \otimes F_2, \qquad f \mapsto \mathcal{K}(1 \otimes f),$$

which is the integral operator with kernel $\mathcal{K}$, is a morphism of A-modules.

2. Now suppose that the integral operator with kernel $\mathcal{K}$ is a morphism of A-modules. Then (1.57) and the nondegeneracy of the bilinear form (1.137) imply that this kernel satisfies the following system of equations:

$$(a \otimes 1)\, \mathcal{K} = (1 \otimes S^{-1}(a))\, \mathcal{K}, \qquad a \in A, \tag{1.138}$$

which leads to its invariance, since $(\mathrm{id} \otimes S^{-1})\Delta^{\mathrm{cop}} = \mathrm{id}$. □

One of the principal problems of noncommutative harmonic analysis is the study of invariant kernels and the corresponding integral operators [**101**]. The following proposition will be used extensively to construct A-invariant kernels

PROPOSITION 1.113. *Let F_1^{op} be an algebra derived from F_1 by replacement the multiplication by the opposite one. A-invariant kernels form a subalgebra in $F_1^{\mathrm{op}} \otimes F_2$.*

PROOF. It suffices to show that the product of A-invariant elements in the algebra of kernels is an A-invariant kernel. To the kernels

$$\mathcal{K}' = \sum_i a_i' \otimes b_i', \qquad \mathcal{K}'' = \sum_j a_j'' \otimes b_j'',$$

we assign the following linear operators in $F_1 \otimes F_2$:

$$\mathcal{K}' : f_1 \otimes f_2 \mapsto \sum_i f_1 a_i' \otimes b_i' f_2, \quad \mathcal{K}'' : f_1 \otimes f_2 \mapsto \sum_j f_1 a_j'' \otimes b_j'' f_2.$$

The A-invariance of $\mathcal{K}'$, $\mathcal{K}''$ implies that the operators $\mathcal{K}'$, $\mathcal{K}''$ are endomorphisms of the A-module $F_1 \otimes F_2$. Hence the same statement holds for their product $\mathcal{K}'\mathcal{K}''$ which maps $f_1 \otimes f_2$ to $\sum_i \sum_j f_1 a_j'' a_i' \otimes b_i' b_j'' f_2$. Applying this endomorphism to the A-invariant element $1 \otimes 1$, we get an A-invariant one which is just the product $\mathcal{K}''\mathcal{K}'$ in the algebra $F_1^{\mathrm{op}} \otimes F_2$. □

In the remainder of the subsection we consider a Hopf $*$-algebra A, A-module $*$-algebras F_1, F_2, a real invariant integral $\nu : F_2 \to \mathbb{C}$; also, the bilinear form

$$F_2 \times F_2 \to \mathbb{C}, \qquad f' \times f'' \mapsto \nu(f'\, f''),$$

is supposed to be nondegenerate.

It is easy to prove the uniqueness of an involution $*$ in $F_1^{\mathrm{op}} \otimes F_2$ such that for all $a \in A$, $f \in F_2$, $\mathcal{K} \in F_1^{\mathrm{op}} \otimes F_2$ one has

$$\int \mathcal{K}^*(1 \otimes f) d\nu = \left(\int \mathcal{K}(1 \otimes f^*) d\nu \right)^* . \tag{1.139}$$

In the classical case $q = 1$, the equality (1.139) means that the integral operator with real kernel is real.

An involution satisfying (1.139) admits a restriction to the subalgebra of invariant kernels; this can be easily deduced from Proposition 1.112. Indeed, if $\mathcal{K} \in F_1^{\mathrm{op}} \otimes F_2$ is an invariant kernel, then for all $a \in A$, $f \in F_2$ one has

$$\int \mathcal{K}^*(1 \otimes af) d\nu = \left(\int \mathcal{K}(1 \otimes (af)^*) d\nu \right)^* = \left(\int \mathcal{K}(1 \otimes S(a)^* f^*) d\nu \right)^*$$
$$= \left(S(a)^* \int \mathcal{K}(1 \otimes f^*) d\nu \right)^* = a \left(\int \mathcal{K}(1 \otimes f^*) d\nu \right)^* = a \int \mathcal{K}^*(1 \otimes f) d\nu.$$

Thus if the integral operator with kernel $\mathcal{K}$ is a morphism of A-modules, then the integral operator with kernel $\mathcal{K}^*$ possesses the same property.

One can find an involution that satisfies (1.139) using an explicit form of an invariant integral. For example, if an algebra F_2 possesses a finite-dimensional faithful $*$-representation T, and for some invertible self-adjoint element $Q \in F_2$ one has

$$\nu(f) = \mathrm{tr}(T(f)Q), \qquad f \in F_2,$$

then $\mathcal{K}^* = (1 \otimes Q)^{-1}\, \mathcal{K}^{*\otimes *}\, (1 \otimes Q)$ is the required involution by definitions.

Note that the latter formula is true under very mild assumptions since the finite dimensionality of the representation T is used in the proof only to permute factors under the trace sign.

1.6.2. The Green function for the operator $\square_q$. Let us start from the classical case $q = 1$. For a wide class of functions f, a solution of the Poisson equation $\square u = f$ in the unit disc $\mathbb{D}$ has the form

$$u(z) = \int_{\mathbb{D}} G(z, \zeta)\, f(\zeta)\, d\nu(\zeta),$$

where the Green function G equals

$$G(z, \zeta) = -\ln(|z - \zeta|^2) + \ln(|1 - z\overline{\zeta}|^2). \tag{1.140}$$

The equality (1.140) can be proved by the reflection method: the first summand is a contribution of the main source, while the second summand is a contribution of a source obtained by the inversion of the first with respect to the unit circle; see [**329**, p. 429].

Quantum analogs of integral operators and their kernels are needed to turn to the quantum disc.

Consider the algebra $\mathrm{Pol}(\mathbb{C})_q^{\mathrm{op}}$ derived from $\mathrm{Pol}(\mathbb{C})_q$ by replacing the multiplication with the opposite one. In this algebra one has $zz^* = q^2 z^* z + 1 - q^2$.

Starting from $\mathrm{Pol}(\mathbb{C})_q^{\mathrm{op}}$, we can repeat all the arguments used in Subsection 1.4.3 to construct the algebra $\mathcal{D}(\mathbb{D})_q$ of finite functions in the quantum disc and the $\mathcal{D}(\mathbb{D})_q$-bimodule $\mathcal{D}(\mathbb{D})'_q$ of distributions in the quantum disc. As a result we obtain an algebra $\mathcal{D}(\mathbb{D})_q^{\mathrm{op}}$ and a $\mathcal{D}(\mathbb{D})_q^{\mathrm{op}}$-bimodule $\mathcal{D}(\mathbb{D})_q'^{\mathrm{op}}$.

Introduce the abbreviated notation $\mathrm{Pol}(\mathbb{C}\times\mathbb{C})_q$ for the algebra $\mathrm{Pol}(\mathbb{C})_q^{\mathrm{op}}\otimes\mathrm{Pol}(\mathbb{C})_q$ and z, z^*, ζ, ζ^* for its generators $z\otimes 1$, $z^*\otimes 1$, $1\otimes z$, $1\otimes z^*$. The elements

$$(1-z^*z)\otimes 1, \qquad 1\otimes(1-zz^*)$$

will be denoted by y, η, respectively.

Again, by reproducing the arguments used to construct $\mathcal{D}(\mathbb{D})_q$ and $\mathcal{D}(\mathbb{D})'_q$ and applying them to $\mathrm{Pol}(\mathbb{C}\times\mathbb{C})_q$, we obtain the algebra $\mathcal{D}(\mathbb{D}\times\mathbb{D})_q$ of finite functions on the Cartesian product of quantum discs and a $\mathcal{D}(\mathbb{D}\times\mathbb{D})_q$-bimodule $\mathcal{D}(\mathbb{D}\times\mathbb{D})'_q$ of distributions.

The elements of $\mathcal{D}(\mathbb{D}\times\mathbb{D})'_q$ will be called generalized kernels, because one can use them to introduce integral operators as in the previous subsection. Namely, to a generalized kernel $\mathcal{K}$ one assigns the linear map from $\mathcal{D}(\mathbb{D})_q$ to $\mathcal{D}(\mathbb{D})_q'^{\mathrm{op}}$ defined as follows

$$Kf = \int_{\mathbb{D}_q}\mathcal{K}\cdot f\,d\nu = (\mathrm{id}\otimes\nu)(K(1\otimes f)).$$

REMARK 1.114. Using the opposite multiplication in the first tensor factor in $\mathrm{Pol}(\mathbb{C})_q^{\mathrm{op}}\otimes\mathrm{Pol}(\mathbb{C})_q$, we follow the conventions of the previous subsection. It allows us to introduce an algebra of $U_q\mathfrak{sl}_2$-invariant kernels; see Proposition 1.113.

The next lemma follows from definitions.

LEMMA 1.115. *For any set* $\{\psi_{ij}(y,\eta)\}_{i,j\in\mathbb{Z}_+}$ *of functions on* $q^{2\mathbb{Z}_+}\times q^{2\mathbb{Z}_+}$*, the series* $\sum_{i,j\in\mathbb{Z}_+} z^{*i}\zeta^i\,\psi_{ij}(y,\eta)\,z^j\zeta^{*j}$ *converges in* $\mathcal{D}(\mathbb{D}\times\mathbb{D})'_q$.

EXAMPLE 1.116. For all $m\geq 0$, there exists the generalized kernel

$$G_m = (1-\zeta\zeta^*)^m\cdot(q^{-2(m-1)}z^*\zeta;q^2)_m^{-1}\cdot(q^{-2m}z\zeta^*;q^2)_m^{-1}\cdot(1-z^*z)^m, \tag{1.141}$$

where $(t;q^2)_m=\prod_{j=0}^{m-1}(1-tq^{2j})$ and

$$\frac{1}{(t;q^2)_m} = \sum_{n=0}^{\infty}\frac{(q^{2m};q^2)_n}{(q^2;q^2)_n}t^n.$$

The latter equality follows from (1.114).

The following statement will be proved in Subsection 1.7.3.

LEMMA 1.117. *The generalized kernel* G_m *is a* $U_q\mathfrak{sl}_2$*-invariant element of the* $U_q\mathfrak{sl}_2$*-module* $\mathcal{D}(\mathbb{D}\times\mathbb{D})'_q$.

To formulate the main result of this subsection, we turn to the classical case $q=1$ and use the decomposition of the Green function (1.140):

$$-\ln\frac{|z-\zeta|^2}{|1-z\overline{\zeta}|^2} = -\ln\left(1-\frac{(1-|z|^2)(1-|\zeta|^2)}{|1-z\overline{\zeta}|^2}\right) = \sum_{m=1}^{\infty}\frac{1}{m}\left(\frac{(1-|z|^2)(1-|\zeta|^2)}{|1-z\overline{\zeta}|^2}\right)^m. \tag{1.142}$$

Passing to a formal limit in (1.141), we get

$$\lim_{q\to 1} G_m = \left(\frac{(1-|z|^2)(1-|\zeta|^2)}{|1-z\overline{\zeta}|^2}\right)^m.$$

The following result is a q-analog of the decomposition (1.142). Due to Corollary 1.106, the invariant Laplacian $\square_q$ in $L^2(d\nu)_q$ is bounded and possesses a bounded inverse.

PROPOSITION 1.118. *The restriction of* $\square_q^{-1}$ *from* $L^2(d\nu)_q$ *to the dense linear subvariety* $\mathcal{D}(\mathbb{D})_q$ *is an integral operator with kernel*

$$G_q = \sum_{m=1}^{\infty} \frac{q^{-2}-1}{q^{-2m}-1} G_m,$$

i.e., $\square_q^{-1} f = \int_{\mathbb{D}_q} G_q(z,\zeta) f(\zeta) d\nu$ *for all* $f \in \mathcal{D}(\mathbb{D})_q$.

PROOF. Both linear operators under comparison are morphisms of $U_q\mathfrak{sl}_2$-modules. Moreover, $U_q\mathfrak{sl}_2 \cdot f_0 = \mathcal{D}(\mathbb{D})_q$, due to Proposition 1.82. Hence it is sufficient to prove the formulas

$$\int_{\mathbb{D}_q} G_q\, f_0\, d\nu = (1-q^2) \sum_{m=1}^{\infty} \frac{q^{-2}-1}{q^{-2m}-1} y^m, \tag{1.143}$$

$$\square_q^{-1} f_0 \;=\; (1-q^2) \sum_{m=1}^{\infty} \frac{q^{-2}-1}{q^{-2m}-1} y^m, \tag{1.144}$$

which imply the equality

$$\square_q^{-1} f_0 = \int_{\mathbb{D}_q} G_q\, f_0\, d\nu.$$

The validity of (1.143) follows from the definition of G_q and the equalities

$$z^* \cdot f_0 = 0, \qquad \psi(y) \cdot f_0 = \psi(1), \qquad \int_{\mathbb{D}_q} z^k\, f_0\, d\nu = (1-q^2)\delta_{k,0}.$$

It remains to verify (1.144). Let $x = y^{-1}$. The invertibility of $\square_q$ implies that there exists a unique element $\psi \in L^2(d\nu)_q$ such that $\square_q \psi \;=\; f_0$; that is, there exists a unique function $\psi(x)$ on $q^{-2\mathbb{Z}_+}$ such that

$$x(1-q^{-1}x)\mathcal{D}_q \psi(x) = q^{-1} - q, \qquad \sum_{j=0}^{\infty} |\psi(q^{-2j})|^2 \cdot q^{-2j} < \infty \tag{1.145}$$

(see Proposition 1.96). By (1.145), one has

$$\begin{aligned}
\psi(x) &= \psi(q^{-2}x) + (q^{-2}-1)^2 \frac{q^4 x^{-1}}{1-q^2 x^{-1}},\\
\psi(x) &= (q^{-2}-1)^2 q^2 \sum_{j=1}^{\infty} \frac{q^{2j} x^{-1}}{1-q^{2j}x^{-1}} \;=\; (q^{-2}-1)^2 q^2 \sum_{m=1}^{\infty} \frac{q^{2m}}{1-q^{2m}} x^{-m}
\end{aligned}$$

for all $x \in q^{-2\mathbb{Z}_+}$. Hence $\psi(x) = (1-q^2)\sum_{m=1}^{\infty} \frac{q^{-2}-1}{q^{-2m}-1} x^{-m}$. $\square$

1.6.3. Quantum cones. Recall that the algebra $\mathbb{C}[SL_2]_q$ of regular functions on the quantum group SL_2 is generated by the matrix elements t_{ij} of the vector representation of $U_q\mathfrak{sl}_2$; see (1.22). In Subsection 1.3.6 the algebra $\mathbb{C}[SL_2]_q$ is endowed with the structure of a $U_q\mathfrak{sl}_2$-module algebra. It follows from the definitions that

$$\begin{pmatrix} K^{\pm 1}t_{11} & K^{\pm 1}t_{12} \\ K^{\pm 1}t_{21} & K^{\pm 1}t_{22} \end{pmatrix} = \begin{pmatrix} t_{11} & t_{12} \\ t_{21} & t_{22} \end{pmatrix} \cdot \begin{pmatrix} q^{\pm 1} & 0 \\ 0 & q^{\mp 1} \end{pmatrix} = \begin{pmatrix} q^{\pm 1}t_{11} & q^{\mp 1}t_{12} \\ q^{\pm 1}t_{21} & q^{\mp 1}t_{22} \end{pmatrix}, \tag{1.146}$$

$$\begin{pmatrix} Et_{11} & Et_{12} \\ Et_{21} & Et_{22} \end{pmatrix} = \begin{pmatrix} t_{11} & t_{12} \\ t_{21} & t_{22} \end{pmatrix} \cdot \begin{pmatrix} 0 & q^{-\frac{1}{2}} \\ 0 & 0 \end{pmatrix} = \begin{pmatrix} 0 & q^{-\frac{1}{2}}t_{11} \\ 0 & q^{-\frac{1}{2}}t_{21} \end{pmatrix},$$

$$\begin{pmatrix} Ft_{11} & Ft_{12} \\ Ft_{21} & Ft_{22} \end{pmatrix} = \begin{pmatrix} t_{11} & t_{12} \\ t_{21} & t_{22} \end{pmatrix} \cdot \begin{pmatrix} 0 & 0 \\ q^{\frac{1}{2}} & 0 \end{pmatrix} = \begin{pmatrix} q^{\frac{1}{2}}t_{12} & 0 \\ q^{\frac{1}{2}}t_{22} & 0 \end{pmatrix}. \tag{1.147}$$

We introduce quantum analogs of spaces of functions on the cone

$$\widetilde{\Xi} = \{(t_{11}, t_{12}) \in \mathbb{C}^2 \quad | \quad |t_{11}| = |t_{12}|\}.$$

The algebra $\mathbb{C}[SL_2]_q$ can be defined by generators t_{ij} and relations that are all homogeneous except (1.49); see Proposition 1.53. Let us make the following change of variables in the defining relations:

$$t_{ij} = q^{-N} \cdot t_{ij}^{(N)}, \qquad N \in \mathbb{N}.$$

A formal passage to a limit as $N \to +\infty$ does not affect the homogeneous relations, but transforms (1.49) to $t_{11}t_{22} - qt_{12}t_{21} = 0$.

Denote by $\mathbb{C}[\widetilde{\Xi}]_q$ the algebra defined by the new set of relations among generators t_{ij}.

REMARK 1.119. Set $\deg(t_{ij}) = 1$. This endows the algebra $\mathbb{C}[SL_2]_q$ with a filtration, and the algebra $\mathbb{C}[\widetilde{\Xi}]_q$ with a grading. The passage from $\mathbb{C}[SL_2]_q$ to $\mathbb{C}[\widetilde{\Xi}]_q$ as above is just a "utilitarian" description of the standard passage from a filtered algebra to the associated graded algebra.

Introduce the involution $*$ in $\mathbb{C}[\widetilde{\Xi}]_q$ by setting

$$t_{11}^* = -t_{22}, \qquad t_{12}^* = -qt_{21}.$$

By the definitions, the $*$-algebra $\mathbb{C}[\widetilde{\Xi}]_q$ with the $U_q\mathfrak{su}_{1,1}$-action given by (1.146)–(1.147) is a $U_q\mathfrak{su}_{1,1}$-module algebra.

Let $x = -qt_{12}t_{21}$. We are about to prove that each element of $\mathbb{C}[\widetilde{\Xi}]_q$ admits a unique decomposition

$$f = \sum_{ad=bc=0} t_{11}^a t_{12}^b \psi_{abcd}(x) t_{21}^c t_{22}^d, \tag{1.148}$$

where $a, b, c, d \in \mathbb{Z}_+$ and ψ_{abcd} are polynomials in one variable. The existence of such decomposition is evident.

Consider a pre-Hilbert space with an orthonormal basis $\{e_j\}_{j \in \mathbb{Z}_+}$ and the $*$-representation π of the algebra $\mathbb{C}[\widetilde{\Xi}]_q$ given by

$$\pi(t_{11})e_j = q^{-(j+1)}e_{j+1}, \qquad \pi(t_{12})e_j = q^{-j}e_j,$$

$$\pi(t_{21})e_j = -q^{-(j+1)}e_j, \qquad \pi(t_{22})e_j = -q^{-j}e_{j-1}.$$

One can use it to verify that

$$\{t_{11}^a t_{12}^b t_{21}^c,\ t_{12}^b t_{21}^c t_{22}^d \mid a,b,c,d \in \mathbb{Z}_+\}$$

are linearly independent in $\mathbb{C}[\widetilde{\Xi}]_q$, as is done in Subsection 1.3.6. This implies the uniqueness of the decomposition (1.148).

One can easily show that $x^{\mathbb{Z}_+}$ is an Ore set and the corresponding localization $\mathbb{C}[\widetilde{\Xi}]_{q,x}$ of the algebra $\mathbb{C}[\widetilde{\Xi}]_q$ inherits the structure of $U_q\mathfrak{su}_{1,1}$-module algebra. The elements of $\mathbb{C}[\widetilde{\Xi}]_{q,x}$ admit a decomposition (1.148), where ψ_{abcd} are Laurent polynomials in one variable.

Consider the $U_q\mathfrak{su}_{1,1}$-module subalgebras $\mathbb{C}[\Xi]_q \subset \mathbb{C}[\widetilde{\Xi}]_q$ and $\mathbb{C}[\Xi]_{q,x} \subset \mathbb{C}[\widetilde{\Xi}]_{q,x}$ of invariant elements under the one-parameter group of automorphisms

$$t_{1j} \mapsto e^{i\varphi}t_{1j}, \quad t_{2j} \mapsto e^{-i\varphi}t_{2j}, \qquad \varphi \in \mathbb{R}/(2\pi\mathbb{Z}),\ j=1,2.$$

These are quantum analogs of algebras of functions on the two-dimensional cone. Their elements possess the decomposition (1.148) with nonzero summands only if $a+b \neq c+d$.

Let us endow the algebra $\mathbb{C}[\Xi]_{q,x}$ with the following grading:

$$\deg(x^{\pm 1}) = \pm 1, \qquad \deg(t_{1i}t_{2j}) = 1, \quad i,j=1,2.$$

It follows from (1.189)–(1.190) that the homogeneous components are $U_q\mathfrak{sl}_2$-submodules of the $U_q\mathfrak{sl}_2$-module $\mathbb{C}[\Xi]_{q,x}$. We show that they are isomorphic to $U_q\mathfrak{sl}_2$-modules of principal spherical series. First of all, note that t_{12}, t_{21}, t_{11}, t_{22} are invertible elements of the algebra $\mathbb{C}[\widetilde{\Xi}]_{q,x}$. Therefore $z = t_{12}^{-1}t_{11} \in \mathbb{C}[\Xi]_{q,x}$ is well defined and invertible.

PROPOSITION 1.120. *Let $l \in \mathbb{Z}$. The linear span $M^{(l)}$ of $\{z^k x^l \mid k \in \mathbb{Z}\}$ is a $U_q\mathfrak{sl}_2$-module which is isomorphic to the $U_q\mathfrak{sl}_2$-module $V^{(l)}$ of principal spherical series.*

PROOF. One can easily show that for all l and all Laurent polynomials $f(z)$

$$K^{\pm 1}\left(f(z)x^l\right) = f(q^{\pm 2}z)x^l,$$

$$E\left(f(z)x^l\right) = -q^{\frac{1}{2}}z^2\frac{f(z)-f(q^2z)}{z-q^2z}x^l + q^{-\frac{3}{2}}\frac{q^{-2l}-1}{q^{-2}-1}zf(q^2z)x^l,$$

$$F\left(f(z)x^l\right) = q^{\frac{1}{2}}\frac{f(q^{-2}z)-f(z)}{q^{-2}z-z}x^l + q^{\frac{5}{2}}\frac{1-q^{2l}}{1-q^2}z^{-1}f(z)x^l.$$

These equalities allow us to find an explicit form of the morphism of $U_q\mathfrak{sl}_2$-modules

$$j^{(l)} : V^{(l)} \to M^{(l)}, \qquad j^{(l)} : e_0 \mapsto x^l.$$

Namely,

$$j^{(l)}e_k = (-q^{-(l+\frac{1}{2})})^k z^k x^l, \qquad k \in \mathbb{Z}.$$

It remains to use the bijectivity of $j^{(l)}$. □

One can easily remove the requirement $l \in \mathbb{Z}$ by considering distributions on quantum cones. Let us outline the required constructions. They are based on the fact that $q^{-2\mathbb{Z}_+}$ is the set of eigenvalues of the operator $\pi(x)$.

Let $\mathcal{D}(\widetilde{\Xi})'_q$ be the vector space of formal series (1.148) with coefficients in the space of functions on $q^{-2\mathbb{Z}_+}$. We endow $\mathcal{D}(\widetilde{\Xi})'_q$ with the weakest topology which makes all the linear functionals $\mathcal{L}_{a,b,c,d,n}$,

$$\mathcal{L}_{a,b,c,d,n}(f) = \psi_{abcd}(q^{-2n}), \qquad ad = bc = 0,\ n \in \mathbb{Z}_+, \tag{1.149}$$

continuous. By continuity, $\mathcal{D}(\widetilde{\Xi})'_q$ is endowed with a structure of $U_q\mathfrak{sl}_2$-covariant $\mathbb{C}[\widetilde{\Xi}]_q$-bimodule.

One can introduce the notion of a *complex degree of homogeneity for distributions on the quantum cone.* Let $l \in \mathbb{C}$. A distribution $f \in \mathcal{D}(\widetilde{\Xi})'_q$ is called homogeneous of degree l if the coefficients $\psi_{abcd}(x)$ of its decomposition (1.148) satisfy the following requirements:

$$\psi_{abcd}(q^2x) = q^{2(l-\frac{a+b+c+d}{2})}\psi(x), \qquad x \in q^{-2\mathbb{Z}_+}.$$

Equivalently, let α be an automorphism of $\mathbb{C}[\widetilde{\Xi}]_q$ defined by $\alpha(t_{ij}) = qt_{ij}$, where $i, j = 1, 2$, and $\overline{\alpha}$ is its extension by continuity to $\mathcal{D}(\widetilde{\Xi})'_q$. The element $\psi \in \mathcal{D}(\widetilde{\Xi})'_q$ is homogeneous of degree l if $\overline{\alpha}(\psi) = q^{2l}\psi$.

It follows from (1.189)–(1.190) that homogeneous distributions of fixed degree form a $U_q\mathfrak{sl}_2$-submodule. It is important to note that the fact $l \in \mathbb{Z}$ is not much in use in the definition of $M^{(l)}$ and the proof of Proposition 1.120.

Using Proposition 1.120 and (1.110), one can easily prove that

$$C_qv = -q\frac{(1-q^{-2l})(1-q^{2(l+1)})}{(1-q^2)^2}v, \qquad v \in M^{(l)}, \tag{1.150}$$

for all $l \in \mathbb{C}$.

PROPOSITION 1.121. *The linear functional on the space of homogeneous distributions of degree* -1 *given by*

$$\int_{\widetilde{\Xi}_q} f\,d\eta \stackrel{\text{def}}{=} \psi_{0000}(1)$$

is $U_q\mathfrak{sl}_2$*-invariant.*

PROOF. One can easily check that

$$\int_{\widetilde{\Xi}_q} (K^{\pm 1} - 1)x^{-1}\,d\eta = 0,$$

$$\int_{\widetilde{\Xi}_q} E(t_{12}x^{-2}t_{22})\,d\eta = \int_{\widetilde{\Xi}_q} F(t_{11}x^{-2}t_{21})\,d\eta = 0.$$

This implies the $U_q\mathfrak{sl}_2$-invariance of the linear functional considered above. □

An element $f \in \mathcal{D}(\widetilde{\Xi})'_q$ is called a *finite function on the quantum cone* if $\mathrm{card}\{(a,b,c,d,n) \mid \mathcal{L}_{a,b,c,d,n}(f) \neq 0\} < \infty$. Denote by $\mathcal{D}(\widetilde{\Xi})_q$ the vector space of finite functions on the quantum cone. It is a $U_q\mathfrak{sl}_2$-covariant $\mathbb{C}[\widetilde{\Xi}]_q$-bimodule and is endowed with a structure of $U_q\mathfrak{su}_{1,1}$-module $*$-algebra by continuity.

1.6.4. Eigenfunctions of the operator $\Box_q$. In the classical case $q = 1$, the function

$$u(z) = \int_{\partial\mathbb{D}} \left(\frac{1 - |z|^2}{(1 - z\overline{\zeta})(1 - \overline{z}\zeta)} \right)^{l+1} f(\zeta) d\nu, \qquad d\nu = \frac{d\theta}{2\pi},$$

is an eigenfunction of the invariant Laplacian $\Box$ under certain wide assumptions on f. Namely,

$$\Box u = -\left((l + \frac{1}{2})^2 - \frac{1}{4} \right) u;$$

see [**124**, p. 51].

To obtain a quantum analog of this result, we introduce quantum analogs of powers of the Poisson kernel and spaces of functions on the unit circle $\partial\mathbb{D}$. The embedding of algebras

$$\mathbb{C}[z, z^{-1}] \hookrightarrow \mathbb{C}[\widetilde{\Xi}]_{q,x}, \qquad z \mapsto t_{12}^{-1} t_{11},$$

allows us to endow the algebra of Laurent polynomials in variable z with a structure of $U_q\mathfrak{su}_{1,1}$-module algebra with involution $z^* = z^{-1}$. Denote by $\mathbb{C}[\partial\mathbb{D}]_q = (\mathbb{C}[z, z^{-1}], *)$ the $U_q\mathfrak{su}_{1,1}$-module algebra obtained in this way. It is a quantum analog of the algebra of regular functions on the unit circle $\partial\mathbb{D}$.

We endow $\mathbb{C}[\partial\mathbb{D}]_q$ with the topology of coefficientwise convergency, that is, the weakest topology which makes continuous all the linear functionals

$$l_j : \sum_{n\in\mathbb{Z}} c_n z^n \mapsto c_j, \qquad j \in \mathbb{Z}. \tag{1.151}$$

The $U_q\mathfrak{sl}_2$-action and the involution $*$ admit an extension by continuity from $\mathbb{C}[\partial\mathbb{D}]_q$ to a completion which is a topological vector space of formal series $\sum_{n\in\mathbb{Z}} c_n z^n$ with complex coefficients and the topology of coefficientwise convergency. Denote by $\mathbb{C}[[\partial\mathbb{D}]]_q$ this $U_q\mathfrak{sl}_2$-covariant $\mathbb{C}[\partial\mathbb{D}]_q$-bimodule.

Finally, endow $\mathrm{Pol}(\mathbb{C})_q^{\mathrm{op}} \otimes \mathbb{C}[\partial\mathbb{D}]_q$ with the weakest topology which makes all the linear functionals $l''_{m,n} \otimes l_j$ continuous; see (1.64), (1.151). The completion $\mathcal{D}(\mathbb{D} \times \partial\mathbb{D})'_q$ may be identified with the vector space of formal series $\sum_{n\in\mathbb{Z}} c_n z^n$, $c_n \in \mathcal{D}(\mathbb{D})'_q$.

Let us introduce q-analogs of powers P^γ of the Poison kernel $P = \frac{1-|z|^2}{|1-z\overline{\zeta}|^2}$ using the notation of Subsection 1.6.2. In the quantum case $q \in (0, 1)$, we replace P^γ by the element

$$P_\gamma = (q^2 z^* \zeta; q^2)_{-\gamma}\, (z\zeta^*; q^2)_{-\gamma} (1 - z^* z)^\gamma\,, \tag{1.152}$$

where the kernels $(z\zeta^*; q^2)_{-\gamma}, (q^2 z^* \zeta; q^2)_{-\gamma} \in \mathcal{D}(\mathbb{D} \times \partial\mathbb{D})'_q$ are given by

$$(z\zeta^*; q^2)_{-\gamma} = \sum_{n=0}^{\infty} \frac{(q^{2\gamma}; q^2)_n}{(q^2; q^2)_n} (q^{-2\gamma} z\zeta^*)^n = \frac{(z\zeta^*; q^2)_\infty}{(q^{-2\gamma} z\zeta^*; q^2)_\infty}, \tag{1.153}$$

$$(q^2 z^* \zeta; q^2)_{-\gamma} = \sum_{n=0}^{\infty} \frac{(q^{2\gamma}; q^2)_n}{(q^2; q^2)_n} (q^{2-2\gamma} z^* \zeta)^n = \frac{(q^2 z^* \zeta; q^2)_\infty}{(q^{-2\gamma+2} z^* \zeta; q^2)_\infty} \tag{1.154}$$

and are q-analogs of $(1 - z\overline{\zeta})^{-\gamma}$ and $(1 - \overline{z}\zeta)^{-\gamma}$, respectively.

The main result of this subsection is the following.

PROPOSITION 1.122. *For all $l \in \mathbb{C}$, $f \in \mathbb{C}[\partial\mathbb{D}]_q$, the element*

$$u = \int_0^{2\pi} P_{l+1}(z, e^{i\theta}) f(e^{i\theta}) \frac{d\theta}{2\pi} \tag{1.155}$$

belongs to $\mathcal{D}(\mathbb{D})'_q$, and

$$\Box_q u = \alpha(l) u,$$

where $\alpha(l) = (1 - q^{-2l})(1 - q^{2l+2})/(1 - q^2)^2$.

PROOF. By Corollary 1.97, it suffices to prove that $C_q u = -q\alpha(l)u$.

Using (1.150), one can easily show that in $\mathcal{D}(\widetilde{\Xi})'_q$ one has $C_q\left(f(z)x^l\right) = -q\alpha(l)\left(f(z)x^l\right)$ for each Laurent polynomial $f(z)$ and all $l \in \mathbb{C}$.

The $U_q\mathfrak{sl}_2$-modules $M^{(l)}$ of homogeneous elements of degree l on a quantum analog of the cone Ξ are introduced in Subsection 1.6.3. To complete the proof, it is sufficient to produce a morphism of the $U_q\mathfrak{sl}_2$-modules $M^{(l)} \to \mathcal{D}(\mathbb{D})'_q$ that maps $f(z)x^l$ to u.

Let us endow the vector space $\text{Pol}(\mathbb{C})_q^{\text{op}} \otimes \mathbb{C}[\Xi]_q$ with the weakest topology which makes continuous all the linear functionals $l''_{m,n} \otimes \mathcal{L}_{a,b,c,d,n}$; see (1.64), (1.149). Following the notation of Subsection 1.6.2, we set $\xi = 1 \otimes x$.

The $U_q\mathfrak{sl}_2$-invariance of the element

$$\mathcal{K}_l = \xi^l \sum_{j=0}^{\infty} \frac{(q^{-2l}; q^2)_j}{(q^2; q^2)_j} (q^{2(l+1)} z^* \zeta)^j \cdot \sum_{m=0}^{\infty} \frac{(q^{-2l}; q^2)_m}{(q^2; q^2)_m} (q^{2l} z \zeta^*)^m (1 - z^* z)^{-l}$$

of the completion of the space $\text{Pol}(\mathbb{C})_q^{\text{op}} \otimes \mathbb{C}[\Xi]_q$ will be proved in Subsection 1.7.3; see Remark 1.173. Subsection 1.6.3 contains the proof of the $U_q\mathfrak{sl}_2$-invariance of the linear functional

$$\eta : M^{(-1)} \to \mathbb{C}, \qquad \eta : \left(\sum_{m=-\infty}^{\infty} a_m z^m\right) x^{-1} \mapsto a_0.$$

The second tensor factor of the product $\mathcal{K}_{-l-1}(1 \otimes f(z)x^l)$ belongs to $M^{(-1)}$. Thus the integral operator

$$M^{(l)} \to \mathcal{D}(\mathbb{D})'_q, \qquad f \mapsto (\text{id} \otimes \eta)\left(\mathcal{K}_{-l-1}(1 \otimes f(z)x^l)\right),$$

is well defined. The results listed above, together with (1.138), (1.58), imply that this integral operator is a morphism of $U_q\mathfrak{sl}_2$-modules:

$$(\text{id} \otimes \eta)((a \otimes 1)\mathcal{K}_{-l-1}(1 \otimes f(z)x^l)) = (\text{id} \otimes \eta)((1 \otimes S^{-1}(a))\mathcal{K}_{-l-1}(1 \otimes f(z)x^l)),$$
$$((1 \otimes S^{-1}(a))\mathcal{K}_{-l-1}(1 \otimes f(z)x^l)) = (\text{id} \otimes \eta)(\mathcal{K}_{-l-1}(1 \otimes a(f(z)x^l))).$$

It remains to observe that $(\text{id} \otimes \eta)\left(\mathcal{K}_{-l-1}(1 \otimes fx^l)\right) = u$. □

REMARK 1.123. Assuming $\gamma = 1$ in (1.152), we get a q-analog of the Poisson kernel $P = \frac{1-|z|^2}{|1-z\bar{\zeta}|^2}$. For $\gamma = 1$ the integral operator (1.155) maps $f = \sum_{n=-\infty}^{\infty} c_n e^{in\theta}$ to $u = \sum_{n=1}^{\infty} c_n z^n + c_0 + \sum_{n=0}^{\infty} c_{-n} z^{*n}$ and admits an extension by continuity to an isometric linear operator from the space $C(\partial\mathbb{D})$ of continuous functions on the circle to the space $C(\mathbb{D})_q$ of continuous functions in the quantum disc. (The last statement can be proved by means of unitary dilation of the contraction $T_F(z)$; see Subsection 1.2.7.) It is clear that $\Box_q u = 0$, $u|_{\partial\mathbb{D}} = f$.

1.6.5. Harmonic analysis. Let us start with well-known results related to the classical case $q = 1$ [**124**, pp. 48, 49]. A non-Euclidean Fourier transform of a function u in the disc $\mathbb{D}$ is given by

$$\widetilde{u}(e^{i\theta}, \rho) = \int_{\mathbb{D}} P^{\frac{1}{2}-i\rho}(z, e^{i\theta}) u(z) d\nu,$$

where P is the Poisson kernel. The inverse transform has the form

$$u(z) = \int_0^\infty \left(\int_0^{2\pi} P^{\frac{1}{2}+i\rho}(z, e^{i\theta}) \widetilde{u}(e^{i\theta}, \rho) \frac{d\theta}{2\pi} \right) d\sigma_{\mathrm{classic}}(\rho), \tag{1.156}$$

where $d\sigma_{\mathrm{classic}}(\rho) = 2\rho\, \mathrm{th}(\pi\rho)\, d\rho$ (an inessential difference of this Plancherel measure from the one presented in the literature is explained by our choice of an invariant measure $d\nu = \frac{d\mathrm{Re}z\; d\mathrm{Im}z}{\pi(1-|z|)^2}$ in $\mathbb{D}$).

The non-Euclidean Fourier transform is a unitary linear operator from $L^2(d\nu)$ to $L^2(d\sigma) \otimes L^2(\frac{d\theta}{2\pi})$.

Note (see Subsection 1.6.4) that (1.156) yields an expansion of $u(z)$ in the eigenfunctions $u_\rho(z) = \int_0^{2\pi} P^{\frac{1}{2}+i\rho}(z, e^{i\theta}) \widetilde{u}(e^{i\theta}, \rho) \frac{d\theta}{2\pi}$ of $\square$. We will obtain a q-analog of this expansion. Let

$$P^t_{l+1} = (q^2 z^* \zeta; q^2)_{-l-1} (z\zeta^*; q^2)_{-l-1} (1 - \zeta\zeta^*)^{1+l};$$

see (1.153), (1.154).

PROPOSITION 1.124. *Let $d\sigma$ be the Borel measure on $\left[0, \frac{\pi}{2\log(q^{-1})}\right]$ given by* (1.98). *The integral operators*

$$u(z) \mapsto \int_{\mathbb{D}_q} P^t_{\frac{1}{2}-i\rho}(e^{i\theta}, \zeta) u(\zeta) d\nu,$$

$$\widetilde{u}(e^{i\theta}, \rho) \mapsto \int_0^{\pi/h} \left(\int_0^{2\pi} P_{\frac{1}{2}+i\rho}(z, e^{i\theta}) \widetilde{u}(e^{i\theta}, \rho) \frac{d\theta}{2\pi} \right) d\sigma(\rho) \tag{1.157}$$

can be extended by continuity from dense linear subvarieties

$$\mathcal{D}(\mathbb{D})_q \subset L^2(d\nu)_q, \quad \mathbb{C}[\partial\mathbb{D}]_q \otimes C^\infty \left[0, \frac{\pi}{2\log(q^{-1})}\right] \subset L^2\left(\frac{d\theta}{2\pi}\right) \otimes L^2(d\sigma),$$

respectively, to mutually inverse unitary operators.

PROOF. Let $-\frac{\pi}{2\log(q^{-1})} \leq \mathrm{Im}\, l \leq \frac{\pi}{2\log(q^{-1})}$. Consider the embedding of vector spaces

$$\mathbb{C}[\partial\mathbb{D}] \to \mathcal{D}(\Xi)'_q, \qquad f(z) \mapsto f(z) x^l.$$

We want to endow $\mathbb{C}[\partial\mathbb{D}]_q$ with a structure of a $U_q\mathfrak{sl}_2$-module in a such way that this embedding will become a morphism of $U_q\mathfrak{sl}_2$-modules. The resulting $U_q\mathfrak{sl}_2$-module will be denoted by $\mathbb{C}[\partial\mathbb{D}]_q^{(l)}$.

Proposition 1.120 claims that $V^{(-1/2+i\rho)} \cong \mathbb{C}[\partial\mathbb{D}]_q^{(-1/2+i\rho)}$. This realization of the principal unitary series of representations of the $*$-algebra $U_q\mathfrak{su}_{1,1}$ in the space of functions on the circle is well known in the classical case $q = 1$. It allows us to replace the unitary operator

$$\overline{\mathcal{I}} : L^2(d\nu)_q \to \bigoplus \int_0^{\frac{\pi}{2\log(q^{-1})}} \overline{V^{(-1/2+i\rho)}}\, d\sigma(\rho), \tag{1.158}$$

introduced in Subsection 1.5.5, by the unitary operator

$$\overline{\mathcal{J}} : L^2(d\nu)_q \to \bigoplus \int_0^{\frac{\pi}{2\log(q^{-1})}} L_2\left(\frac{d\theta}{2\pi}\right) d\sigma(\rho). \tag{1.159}$$

Let us prove that $\overline{\mathcal{J}}\, u = \int_{\mathbb{D}_q} P^t_{\frac{1}{2}-i\rho}(z,\zeta)u(\zeta)d\nu$ for all $u \in \mathcal{D}(\mathbb{D})_q$. This statement is a consequence of two lemmas. Introduce the linear map

$$j_\rho : \mathcal{D}(\mathbb{D})_q \to \mathbb{C}[\partial\mathbb{D}]_q^{(-1/2+i\rho)}, \qquad j_\rho : f \mapsto \int_{\mathbb{D}_q} P^t_{\frac{1}{2}-i\rho}(z,\zeta)f(\zeta)d\nu.$$

LEMMA 1.125. $j_\rho f_0 = 1 - q^2$.

PROOF OF THE LEMMA. It follows from the definitions that

$$P^t_l(z,\zeta) = \sum_{j>0} \zeta^j \cdot \psi_j + \psi_0 + \sum_{j>0} \psi_{-j} \cdot \zeta^*$$

and $\psi_0 = 1$. It remains to observe that only the summand $\psi_0(z,\xi)$ contributes to the integral $j_\rho f_0$, and $\int_{\mathbb{D}_q} f_0\, d\nu = 1 - q^2$. □

LEMMA 1.126. *The linear map* $j_\rho : \mathcal{D}(\mathbb{D})_q \to \mathbb{C}[\partial\mathbb{D}]_q^{(-\frac{1}{2}+i\rho)}$ *is a morphism of* $U_q\mathfrak{sl}_2$*-modules.*

PROOF OF THE LEMMA. Consider the integral operator

$$i_\rho : \mathbb{C}[\partial\mathbb{D}]_q^{(-\frac{1}{2}+i\rho)} \to \mathcal{D}(\mathbb{D})'_q, \qquad i_\rho : f \mapsto \int_0^{2\pi} P_{\frac{1}{2}+i\rho}(z, e^{i\theta})f(e^{i\theta})\frac{d\theta}{2\pi}.$$

The proof of Proposition 1.122 shows that i_ρ is a morphism of $U_q\mathfrak{sl}_2$-modules.

We endow the $U_q\mathfrak{su}_{1,1}$-modules $\mathcal{D}(\mathbb{D})_q$ and $\mathbb{C}[\partial\mathbb{D}]_q^{(-\frac{1}{2}+i\rho)}$ with invariant scalar products

$$\mathcal{D}(\mathbb{D})_q \times \mathcal{D}(\mathbb{D})_q \to \mathbb{C}, \qquad f_1 \times f_2 \mapsto \int_{\mathbb{D}_q} f_2^* f_1 d\nu,$$

$$\mathbb{C}[\partial\mathbb{D}]_q^{(l)} \times \mathbb{C}[\partial\mathbb{D}]_q^{(l)} \to \mathbb{C}, \qquad f_1 \times f_2 \mapsto \int_{\partial\mathbb{D}} f_2^* f_1 \frac{d\theta}{2\pi}.$$

The integral operator with kernel $K = \sum_i k_i'' \otimes k_i'$ is adjoint to the integral operator with kernel $K^t = \sum_i k_i^{'*} \otimes k_i^{''*}$. Hence $j_\rho = i_\rho^*$, and j_ρ is a morphism of $U_q\mathfrak{sl}_2$-modules since i_ρ possesses the same property. □

Let us complete the proof of Proposition 1.124. As one can observe from the proof of Lemma 1.126, the linear operator adjoint to $\mathcal{J}$ is the integral operator (1.157). To prove Proposition 1.124, it remains to use the unitarity of $\mathcal{J}$; see Subsection 1.5.5. □

It follows from the results of Subsection 1.6.4 that Proposition 1.124 yields an expansion of an element u in the eigenfunctions

$$u_\rho = \int_0^{2\pi} P_{\frac{1}{2}+i\rho}(z, e^{i\theta})\widetilde{u}(e^{i\theta}, \rho)\frac{d\theta}{2\pi}$$

of $\Box_q$, where $\widetilde{u} = \int_{\mathbb{D}_q} P^t_{\frac{1}{2}-i\rho}(e^{i\theta}, \zeta)u(\zeta)d\nu$ is a q-analog of the non-Euclidean Fourier transform.

1.6.6. The Toeplitz operators in Bergman spaces. Let $\lambda > 1$. It is explained in Subsection 1.5.6 (see also (1.112), (1.35)) that the equalities

$$Ef(z) = -q^{1/2}z^2\frac{f(z) - f(q^2 z)}{z - q^2 z} - q^{1/2}\frac{1 - q^{2\lambda}}{1 - q^2}zf(q^2 z),$$
$$Ff(z) = q^{1/2-\lambda}\frac{f(q^{-2} z) - f(z)}{q^{-2} z - z}, \qquad K^{\pm 1}f(z) = q^{\pm\lambda} f(q^{\pm 1}z),$$

endow the algebra of polynomials $\mathbb{C}[z]$ with the structure of a unitarizable $U_q\mathfrak{su}_{1,1}$-module. The invariant scalar product

$$(f_1, f_2)_\lambda = \frac{1 - q^{2(\lambda-1)}}{1 - q^2}\int_{\mathbb{D}_q} f_2^* f_1(1 - zz^*)^\lambda d\nu = \int_{\mathbb{D}_q} f_2^* f_1 d\nu_\lambda$$

is defined by means of the embedding into the Hilbert space $L^2(d\nu_\lambda)_q$ introduced in Subsection 1.5.6. It is a q-analog of the spherical holomorphic discrete series.

The closure of $\mathbb{C}[z]$ in $L^2(d\nu_\lambda)_q$ is a q-analog of weighted Bergman space [**122**, p. 2] and is denoted by $L^2_a(d\nu_\lambda)_q$.

LEMMA 1.127. *The monomials $\{z^n\}_{n\in\mathbb{Z}_+}$ are pairwise orthogonal in $L^2_a(d\nu_\lambda)_q$, and*

$$\|z^n\|^2_\lambda = \frac{(q^2; q^2)_n}{(q^{2\lambda}; q^2)_n}. \tag{1.160}$$

PROOF. The pairwise orthogonality of the monomials z^n is evident. Using the properties of the q-Beta and q-Gamma functions in Subsection 1.5.7, we get

$$\frac{1 - q^2}{1 - q^{2(\lambda-1)}}\|z^n\|^2_\lambda = \int_0^1 (q^2 y; q^2)_n \cdot y^{\lambda-2}d_{q^2}y = B_{q^2}(\lambda - 1, n + 1)$$
$$= \frac{\Gamma_{q^2}(\lambda - 1)\cdot\Gamma_{q^2}(n + 1)}{\Gamma_{q^2}(n + \lambda)} = \frac{1 - q^2}{1 - q^{2(\lambda-1)}}\frac{(q^2; q^2)_n}{(q^{2\lambda}; q^2)_n}. \qquad \Box$$

The equality (1.160) was obtained by Klimek and Lesniewski in [**173**], and its classical limit as $q \to 1$ is well known [**122**, p. 4].

Let P_λ be the orthogonal projection in the Hilbert space $L^2(d\nu_\lambda)_q$ to the subspace $L^2_a(d\nu_\lambda)_q$. We assign to each formal series $\mathcal{K}$ in variable $z \otimes \zeta^*$ the integral operator from $\mathcal{D}(\mathbb{D})_q$ to $\mathcal{D}(\mathbb{D})'_q$ given by $Kf(z) = \int_{\mathbb{D}_q}\mathcal{K}(z, \zeta^*)f(\zeta)d\nu_\lambda(\zeta)$.

Lemma 1.127 and the q-binomial formula (1.114) imply that

$$P_\lambda f = \int_{\mathbb{D}_q}\mathcal{K}_\lambda(z, \zeta^*)f(\zeta)d\nu_\lambda(\zeta), \qquad f \in \mathcal{D}(\mathbb{D})_q,$$

where

$$\mathcal{K}_\lambda = \sum_{m=0}^{\infty} \frac{(q^{2\lambda};q^2)_m}{(q^2;q^2)_m} (z\otimes\zeta^*)^m = (q^{2\lambda}\,z\otimes\zeta^*;q^2)_\infty\,(z\otimes\zeta^*;q^2)_\infty^{-1}$$

($\mathcal{K}_\lambda$ is a q-analog of the weighted Bergman kernel $(1-z\overline{\zeta})^{-\lambda}$; see [**122**, p. 5])).

An element $f\in\mathcal{D}(\mathbb{D})'_q$ will be called a *bounded function in the quantum disc* if the generalized linear operator[12] $T_F(f)$ is bounded. We put $\|f\|_\infty = \|T_F(f)\|$. Bounded functions in the quantum disc form a von Neumann algebra $L^\infty(\mathbb{D})_q$, which is naturally isomorphic to the algebra of all bounded linear operators in the representation space of T_F; see Subsection 1.4.3.

Let us turn to Toeplitz operators. It follows from the definition of Hilbert–Schmidt operators of class S_2 (see Subsection 1.6.10) that

$$L^2(d\nu_\lambda)_q = \left\{\psi\in\mathcal{D}(\mathbb{D})'_q \;\middle|\; T_F(\psi)T_F\left(y^{\frac{\lambda-1}{2}}\right)\in S_2\right\}$$

and

$$\|\psi\|_\lambda^2 = (1-q^{2(\lambda-1)})\left\|T_F(\psi)T_F\left(y^{\frac{\lambda-1}{2}}\right)\right\|_2.$$

Consider a bounded function $f\in L^\infty(\mathbb{D})_q$ in the quantum disc. Thus,

$$\|f\psi\|_\lambda \le \|f\|_\infty\,\|\psi\|_\lambda, \qquad \psi\in\mathcal{D}(\mathbb{D})_q,$$

since S_2 is a two-sided ideal in the algebra of bounded linear operators. Hence the linear operator $\psi\mapsto P_\lambda(f\psi)$ admits an extension by continuity to a bounded operator in $L^2(d\nu_\lambda)_q$ with norm not bigger than $\|f\|_\infty$. The restriction of this linear operator to the subspace $L^2_a(d\nu_\lambda)_q$ is called a Toeplitz operator with (contravariant) symbol f and is denoted by $\mathcal{T}_f^{(\lambda)}$ or $\hat f$. Evidently, $\|\hat f\|\le\|f\|_\infty$ and $\hat f^* = \widehat{f^*}$.

EXAMPLE 1.128. We show that for all $k,n\in\mathbb{Z}_+$

$$\widehat{y^k}\,z^n = q^{2kn}\frac{(q^{2(\lambda-1)};q^2)_k}{(q^{2(\lambda+n)};q^2)_k}\,z^n.$$

Indeed, for each function $f(y)\in L^\infty(\mathbb{D})_q$,

$$\hat f\,z^n = \frac{(f(y)z^n,z^n)_\lambda}{(z^n,z^n)_\lambda}\,z^n = \frac{\int_0^1 f(q^{2n}y)\,(q^2y;q^2)_n\,y^{\lambda-2}d_{q^2}y}{\int_0^1 (q^2y;q^2)_n\,y^{\lambda-2}d_{q^2}y}\,z^n.$$

Applying the same argument as in the proof of Lemma 1.127, we obtain

$$\widehat{y^k}\,z^n = q^{2kn}\frac{B_{q^2}(\lambda+k-1,n+1)}{B_{q^2}(\lambda-1,n+1)}\,z^n = q^{2kn}\frac{\Gamma_{q^2}(\lambda+k-1)\Gamma_{q^2}(\lambda+n)}{\Gamma_{q^2}(\lambda+k+n)\Gamma_{q^2}(\lambda-1)}\,z^n.$$

Now it remains to use the definition of the q-Gamma function

$$\frac{\Gamma_{q^2}(\lambda+k-1)\Gamma_{q^2}(\lambda+n)}{\Gamma_{q^2}(\lambda+k+n)\Gamma_{q^2}(\lambda-1)} = \frac{(q^{2(\lambda+k+n)};q^2)_\infty\,(q^{2(\lambda-1)};q^2)_\infty}{(q^{2(\lambda+k-1)};q^2)_\infty\,(q^{2(\lambda+n)};q^2)_\infty} = \frac{(q^{2(\lambda-1)};q^2)_k}{(q^{2(\lambda+n)};q^2)_k}.$$

---

[12] See (1.65).

EXAMPLE 1.129. Let us show that

$$\widehat{z}^*\widehat{z} = q^2\widehat{z}\widehat{z}^* + 1 - q^2 + q^{2(\lambda-1)}\frac{1-q^2}{1-q^{2(\lambda-1)}}(1-\widehat{z}\widehat{z}^*)(1-\widehat{z}^*\widehat{z}). \tag{1.161}$$

Indeed,

$$\widehat{z}\,z^n = z^{n+1}, \qquad \widehat{z}^*\,z^n = \begin{cases}\frac{1-q^{2n}}{1-q^{2(\lambda-1)+2n}}z^{n-1}, & n \neq 0,\\ 0, & n = 0.\end{cases} \tag{1.162}$$

Therefore,

$$(1-\widehat{z}\widehat{z}^*)^{-1}\,z^n = \frac{q^{-2n}-q^{2(\lambda-1)}}{1-q^{2(\lambda-1)}}z^n, \quad (1-\widehat{z}^*\widehat{z})^{-1}\,z^n = \frac{q^{-2n-2}-q^{2(\lambda-1)}}{1-q^{2(\lambda-1)}}z^n.$$

Hence,

$$(1-\widehat{z}\widehat{z}^*)^{-1} = q^2(1-\widehat{z}^*\widehat{z})^{-1} - q^{2(\lambda-1)}\frac{1-q^2}{1-q^{2(\lambda-1)}},$$

that is equivalent to (1.161). The commutation relation (1.161) was obtained in [**173**].

EXAMPLE 1.130. Let $f \in \mathrm{Pol}(\mathbb{C})_q$. Given $f = \sum_{j,k\in\mathbb{Z}_+} a_{j,k}z^{*j}z^k$, one has $\widehat{f} = \sum_{j,k\in\mathbb{Z}_+} a_{j,k}\widehat{z}^{*j}\widehat{z}^k$.

EXAMPLE 1.131.

$$\widehat{f}_0 : z^j \mapsto \begin{cases}1-q^{2(\lambda-1)}, & j=0,\\ 0, & j\neq 0.\end{cases} \tag{1.163}$$

Hankel operators with bounded symbols are introduced in an evident way:

$$\mathcal{H}_f^{(\lambda)} : L_a^2(d\nu_\lambda)_q \to L^2(d\nu_\lambda)_q \ominus L_a^2(d\nu_\lambda)_q, \qquad \mathcal{H}_f^{(\lambda)} : \psi \to (I - P_\lambda)(f\psi).$$

Let us mention that there is a vast literature devoted to the study of Toeplitz and Hankel operators in weighted Bergman spaces in the classical case $q = 1$.

1.6.7. Covariant symbols of linear operators. Let us start with some non-rigorous heuristic considerations. Let A be a bounded linear operator in $L_a^2(d\nu_\lambda)_q$ and

$$A = \sum_{i,j=0}^{\infty} a_{ij}\widehat{z}^i\widehat{z}^{*j} = \sum_{i,j=0}^{\infty} b_{ij}\widehat{z}^{*j}\widehat{z}^i.$$

We would like to call the functions $\sum_{i,j=0}^{\infty} a_{ij}z^iz^{*j}$, $\sum_{i,j=0}^{\infty} b_{ij}z^{*j}z^i$ in the quantum disc the covariant and contravariant symbol of A, respectively. The map that assigns to the contravariant symbols of a linear operator its covariant symbol is the Berezin transform.

Unfortunately, even in the classical case $q = 1$, these constructions can hardly be made rigorous, thus forcing us to use less naive definitions of covariant and contravariant symbols. Their relationship with the above heuristic argument is described by Proposition 1.146 and Example 1.130.

Consider the graded $U_q\mathfrak{sl}_2$-module $\mathbb{C}[z]_{q,\lambda}$ introduced in Subsection 1.5.6, together with the dual graded $U_q\mathfrak{sl}_2$-module $\mathbb{C}[z]_{q,\lambda}^*$ and the $U_q\mathfrak{sl}_2$-module algebra $\mathrm{End}(\mathbb{C}[z]_{q,\lambda})$ of all linear operators in $\mathbb{C}[z]_{q,\lambda}$. Evidently, $\mathbb{C}[z]_{q,\lambda} \otimes \mathbb{C}[z]_{q,\lambda}^* \hookrightarrow \mathrm{End}(\mathbb{C}[z]_{q,\lambda})$. The subalgebra

$$\underline{\mathrm{End}}(\mathbb{C}[z]_{q,\lambda}) \stackrel{\mathrm{def}}{=} \mathbb{C}[z]_{q,\lambda} \otimes \mathbb{C}[z]_{q,\lambda}^*$$

will be called the algebra of finite linear operators in $\mathbb{C}[z]_{q,\lambda}$.

LEMMA 1.132. *For $f \in \mathcal{D}(\mathbb{D})_q$, the integral $I_{m,j} = \int_{\mathbb{D}_q} z^{*m} f z^n \, d\nu_\lambda$ vanishes for almost all $m, n \in \mathbb{Z}_+$ (i.e., for all but finitely many of those).*

PROOF. For each $f \in \mathcal{D}(\mathbb{D})_q$ the equality $z^{*N} f = f z^N = 0$ holds for all big enough $N \in \mathbb{N}$. Thus $I_{mj} = 0$ for $\max(m,n) \geq N$. □

The lemma yields that a Toeplitz operator with finite symbol is a finite linear operator in $\mathbb{C}[z]_{q,\lambda}$.

PROPOSITION 1.133. *The linear operator*

$$\mathcal{T}^{(\lambda)} : \mathcal{D}(\mathbb{D})_q \to \underline{\mathrm{End}}(\mathbb{C}[z]_{q,\lambda}), \qquad \mathcal{T}^{(\lambda)} : f \mapsto \widehat{f},$$

is a morphism of $U_q\mathfrak{sl}_2$-modules.

PROOF. It is shown in Subsection 1.5.6 that $\mathbb{C}[z]_{q,\lambda}$ is a $U_q\mathfrak{sl}_2$-covariant $\mathbb{C}[z]_q$-module and the Hermitian form $(\cdot,\cdot)_\lambda$ is invariant. Therefore, the linear map

$$\mathcal{D}(\mathbb{D})_q \otimes \mathbb{C}[z]_{q,\lambda} \to \mathbb{C}[z]_{q,\lambda}, \qquad f \otimes \psi \mapsto P_{q,\lambda}(f\psi),$$

is a morphism of $U_q\mathfrak{sl}_2$-modules. Hence the corresponding element of the tensor product

$$\mathbb{C}[z]_{q,\lambda} \otimes (\mathcal{D}(\mathbb{D})_q \otimes \mathbb{C}[z]_{q,\lambda})^* \cong \underline{\mathrm{End}}(\mathbb{C}[z]_{q,\lambda}) \otimes \mathcal{D}(\mathbb{D})_q^*$$

is $U_q\mathfrak{sl}_2$-invariant. □

We start the analysis of the dual to $\mathcal{T}^{(\lambda)}$ with a description of dual vector spaces.

Let $\mathbb{C}[[z]]_{q,\lambda}$ be the vector space of formal power series in z with the topology of termwise convergence and the $U_q\mathfrak{sl}_2$-action extended by continuity from $\mathbb{C}[z]_{q,\lambda}$. Linear operators from $\mathbb{C}[z]_{q,\lambda}$ to $\mathbb{C}[[z]]_{q,\lambda}$ will be called generalized linear operators in $\mathbb{C}[z]_{q,\lambda}$ and the notation

$$\overline{\mathrm{End}}(\mathbb{C}[z]_{q,\lambda}) = \mathrm{Hom}(\mathbb{C}[z]_{q,\lambda}, \mathbb{C}[[z]]_{q,\lambda})$$

will be used. Obviously, $\overline{\mathrm{End}}(\mathbb{C}[z]_{q,\lambda})$ is a $U_q\mathfrak{sl}_2$-covariant $\underline{\mathrm{End}}(\mathbb{C}[z]_{q,\lambda})$-bimodule.

By tradition, we introduce an invariant integral tr_q on the $U_q\mathfrak{sl}_2$-module algebra $\underline{\mathrm{End}}(\mathbb{C}[z]_{q,\lambda})$ using the relation

$$S^2(\xi) = K^{-1}\xi K, \qquad \xi \in U_q\mathfrak{sl}_2;$$

see Subsections 1.4.1 and 1.4.4. Once $A : z^j \mapsto \sum_{m\in\mathbb{Z}_+} a_{mj} z^m$ is an element of $\underline{\mathrm{End}}(\mathbb{C}[z]_{q,\lambda})$, we have $\mathrm{tr}_q(A) = \sum_{k\in\mathbb{Z}_+} a_{kk} q^{-2k}$.

Let us prove a lemma which allows one to identify $\overline{\mathrm{End}}(\mathbb{C}[z]_{q,\lambda})$ with the $U_q\mathfrak{sl}_2$-module dual to $\underline{\mathrm{End}}(\mathbb{C}[z]_{q,\lambda})$ and $\mathcal{D}(\mathbb{D})'_q$ with the $U_q\mathfrak{sl}_2$-module dual to $\mathcal{D}(\mathbb{D})_q$.

LEMMA 1.134. (1) *The linear map $i_\lambda : \overline{\mathrm{End}}(\mathbb{C}[z]_{q,\lambda}) \to \underline{\mathrm{End}}(\mathbb{C}[z]_{q,\lambda})^*$ given by*

$$i_\lambda(A)(X) = \frac{1-q^2}{1-q^{2(\lambda-1)}} \cdot \mathrm{tr}_q(A\,X), \qquad X \in \underline{\mathrm{End}}(\mathbb{C}[z]_{q,\lambda}),$$

is an isomorphism of $U_q\mathfrak{sl}_2$-modules.

(2) *The linear map* $i : \mathcal{D}(\mathbb{D})'_q \to (\mathcal{D}(\mathbb{D})_q)^*$ *given by*

$$i(f)(\psi) = \int_{\mathbb{D}_q} f\psi \, d\nu, \qquad \psi \in \mathcal{D}(\mathbb{D})_q,$$

is an isomorphism of $U_q\mathfrak{sl}_2$*-modules.*

PROOF. The linear functional

$$\int A \, d\mu_\lambda = \frac{1-q^2}{1-q^{2(\lambda-1)}} \cdot \mathrm{tr}_q A \tag{1.164}$$

is an invariant integral on $\underline{\mathrm{End}}(\mathbb{C}[z]_{q,\lambda})$, and the linear functional $\int_{\mathbb{D}_q} f \, d\nu$ is an invariant integral on $\mathcal{D}(\mathbb{D})_q$. Now we should recall the definition of dual $U_q\mathfrak{sl}_2$-module and apply integration by parts (1.65). □

REMARK 1.135. The scalar factor in (1.164) is chosen in a such way that

$$\int_{\mathbb{D}_q} f_0 \, d\nu = \int \widehat{f}_0 \, \mu_\lambda;$$

see Example 1.131.

Using the last lemma, we introduce a dual linear map

$$(\mathcal{T}^{(\lambda)})^* : \overline{\mathrm{End}}(\mathbb{C}[z]_{q,\lambda}) \to \mathcal{D}(\mathbb{D})'_q,$$

which is a morphism of $U_q\mathfrak{sl}_2$-modules.

The distribution $(\mathcal{T}^{(\lambda)})^* A$ is called the covariant symbol of generalized linear operator $A \in \overline{\mathrm{End}}(\mathbb{C}[z]_{q,\lambda})$. Berezin used this notion in a different context, so we should agree on the definitions. First of all, let us obtain an integral representation of matrix elements of Toeplitz operators.

LEMMA 1.136. *Let* $\widehat{f} : \mathbb{C}[z]_{q,\lambda} \to \mathbb{C}[z]_{q,\lambda}$ *be the Toeplitz operator with symbol* $f \in \mathcal{D}(\mathbb{D})_q$, *and let* $\{\widehat{f}_{ml}\}$ *be its matrix elements in the basis* $\{z^m\}_{m\in\mathbb{Z}_+}$. *Then*

$$\widehat{f}_{ml} = \frac{1-q^{2(\lambda-1)}}{1-q^2} \int_{\mathbb{D}_q} P_{z,m,l} \, f(z) \, d\nu(z), \tag{1.165}$$

where

$$P_{z,m,l} = \frac{(q^{2\lambda};q^2)_m}{(q^2;q^2)_m} q^{2l} z^l (1-zz^*)^\lambda z^{*m}. \tag{1.166}$$

PROOF. Since

$$(z^m, z^l)_\lambda = \frac{(q^2;q^2)_m}{(q^{2\lambda};q^2)_m} \delta_{ml}, \qquad m, l \in \mathbb{Z}_+,$$

one has

$$\widehat{f}_{ml} = \frac{(f\,z^l, z^m)_\lambda}{(z^m, z^m)_\lambda} = \frac{(q^{2\lambda};q^2)_m}{(q^2;q^2)_m} \int_{\mathbb{D}_q} z^{*m} \, f \, z^l \, d\nu_\lambda$$

$$= \frac{1-q^{2(\lambda-1)}}{1-q^2} \cdot \frac{(q^{2\lambda};q^2)_m}{(q^2;q^2)_m} \int_{\mathbb{D}_q} z^{*m} \, f \, z^l (1-zz^*)^\lambda d\nu.$$

An explicit form of the invariant integral shows that

$$\int\limits_{\mathbb{D}_q} f_1 f_2(1-zz^*)d\nu(z) = \int\limits_{\mathbb{D}_q} f_2 f_1(1-zz^*)d\nu(z)$$

for each $f_1 \in \mathcal{D}(\mathbb{D})'_q$, $f_2 \in \mathcal{D}(\mathbb{D})_q$. Thus,

$$\widehat{f}_{ml} = \frac{1-q^{2(\lambda-1)}}{1-q^2} \cdot \frac{(q^{2\lambda};q^2)_m}{(q^2;q^2)_m} \int\limits_{\mathbb{D}_q} (1-zz^*)z^l(1-zz^*)^{\lambda-1}z^{*m}\, f\, d\nu.$$

It remains to use the commutation relation

$$(1-zz^*)z = q^2 z(1-zz^*). \qquad \square$$

The matrix $P_z = (P_{z,m,l})_{m,l\in\mathbb{Z}_+}$ is a q-analog of the matrix of orthogonal projection onto a one-dimensional subspace generated by an eigenvector of the adjoint to the operator of multiplication by z, that is, by a vector of an overfull system used by Berezin [**31**].

PROPOSITION 1.137. *Let $A \in \overline{\mathrm{End}}(\mathbb{C}[z]_{q,\lambda})$ and let $\{a_{ml}\}$ be its matrix elements in the basis $\{z^l\}_{l\in\mathbb{Z}_+}$: $Az^j = \sum_{m=0}^{\infty} a_{ml} z^m$. Then*

$$(\mathcal{T}^{(\lambda)})^*(A) = \mathrm{tr}_q(A\, P_z), \tag{1.167}$$

where $\mathrm{tr}_q(A\, P_z) = \sum_{l,m=0}^{\infty} a_{lm} P_{z,m,l} q^{-2l}$, and the series converges in the topology of $\mathcal{D}(\mathbb{D})'_q$.

PROOF. By (1.165),

$$\int A\left(\mathcal{T}^{(\lambda)}\psi\right) d\mu_\lambda = \frac{1-q^2}{1-q^{2(\lambda-1)}}\, \mathrm{tr}_q\left(A\, \mathcal{T}^{(\lambda)}\psi\right)$$

$$= \sum_{l,m=0}^{\infty} a_{lm} \left(\int\limits_{\mathbb{D}_q} P_{z,m,l}\psi d\nu \right) q^{-2l} = \int\limits_{\mathbb{D}_q} \left(\sum_{l,m=0}^{\infty} a_{lm} P_{z,m,l} q^{-2l} \right) \psi d\nu.$$

$\square$

EXAMPLE 1.138. Consider the operator A such that $Az^j = \begin{cases} 1, & j=0, \\ 0, & j\neq 0 \end{cases}$. Its covariant symbol is $(1-zz^*)^\lambda$.

It is important to mention that in the Berezin paper [**31**] the covariant symbol of a linear operator is introduced by an equality similar to (1.167).

In the remainder of this subsection we describe the main properties of covariant symbols of generalized linear operators.

Recall that the operators $\widehat{z}$, $\widehat{z}^*$ act in the graded vector space $\mathbb{C}[z]_{q,\lambda}$, and $\deg(\widehat{z}) = +1$, $\deg(\widehat{z}^*) = -1$. Therefore, for each numeric matrix $(a_{ij})_{i,j\in\mathbb{Z}_+}$, the series

$$A = \sum_{i,j=0}^{\infty} a_{ij} \widehat{z}^i \widehat{z}^{*j} \tag{1.168}$$

converges in the strong operator topology of the space

$$\overline{\mathrm{End}}(\mathbb{C}[z]_{q,\lambda}) = \mathrm{Hom}_{\mathbb{C}}(\mathbb{C}[z]_{q,\lambda}, \mathbb{C}[[z]]_{q,\lambda}).$$

One can prove in a similar way that for each matrix $(c_{ij})_{i,j\in\mathbb{Z}_+}$ the series

$$f=\sum_{i,j=0}^{\infty} c_{ij} z^i z^{*j} \tag{1.169}$$

converges in the topological vector space $\mathcal{D}(\mathbb{D})'_q$, which is canonically isomorphic to the vector space of generalized linear operators in $\mathcal{H}$.

From (1.162) one can deduce the following statement; see [**173**].

LEMMA 1.139. *If $A=\sum_{i,j=0}^{\infty} a_{ij}\widehat{z}^i\widehat{z}^{*j}$, then $A\,z^n=\sum_{m=0}^{\infty} b_{mn}z^m$, where*

$$b_{mn}=\sum_{j=0}^{\min(m,n)}\frac{(q^{2n};q^{-2})_{n-j}}{(q^{2(\lambda+n-1)};q^{-2})_{n-j}}a_{m-j,n-j}.$$

This lemma implies

PROPOSITION 1.140. *For each $A\in\overline{\mathrm{End}}(\mathbb{C}[z]_{q,\lambda})$ there exists a unique decomposition* (1.168).

One can similarly prove

PROPOSITION 1.141. *For each distribution $f\in\mathcal{D}(\mathbb{D})'_q$, there exists a unique decomposition* (1.169). *The linear map $(c_{ij})_{i,j\in\mathbb{Z}_+}\mapsto\sum_{i,j=0}^{\infty}c_{ij}z^iz^{*j}$ is an isomorphism between the vector space of matrices $(c_{ij})_{i,j\in\mathbb{Z}_+}$ with the topology of termwise convergence and the topological vector space $\mathcal{D}(\mathbb{D})'_q$.*

EXAMPLE 1.142. Consider the Toeplitz operator $\widehat{f}_0\,z^j=\begin{cases}1-q^{2(\lambda-1)}, & j=0,\\ 0, & j\neq 0,\end{cases}$ with symbol f_0; see (1.163). Let us prove that

$$\widehat{f}_0=\left(1-q^{2(\lambda-1)}\right)\sum_{k=0}^{\infty}\frac{(q^{-2\lambda};q^2)_k}{(q^2;q^2)_k}q^{2\lambda k}\widehat{z}^k\widehat{z}^{*k}. \tag{1.170}$$

We turn from the equality of operators to the equality of their matrix elements in the basis $\{z^n\}_{n\in\mathbb{Z}_+}$. Nondiagonal elements are zero. The comparison of diagonal elements yields

$$\sum_{k=0}^{j}\frac{(q^{-2\lambda};q^2)_k}{(q^2;q^2)_k}\cdot\frac{(q^{2j};q^{-2})_k}{(q^{2(\lambda+j-1)};q^{-2})_k}\cdot q^{2\lambda k}=\delta_{j,0}. \tag{1.171}$$

It suffices to consider the case $j>0$. Multiplying (1.171) by $\frac{(q^{2(\lambda+j-1)};q^{-2})_j}{(q^{2j};q^{-2})_j}$, we get

$$\sum_{k=0}^{j}\frac{(q^{-2\lambda};q^2)_k}{(q^2;q^2)_k}\cdot\frac{(q^{2\lambda};q^2)_{j-k}}{(q^2;q^2)_{j-k}}\cdot q^{2\lambda k}=0,$$

that is,

$$\sum_{k+m=j}\frac{(q^{-2\lambda};q^2)_k}{(q^2;q^2)_k}\cdot q^{2\lambda k}\cdot\frac{(q^{2\lambda};q^2)_m}{(q^2;q^2)_m}=0.$$

Now it remains to consider the q-binomial series (see (1.114))

$$a(t) = \sum_{k=0}^{\infty} \frac{(q^{-2\lambda};q^2)_k}{(q^2;q^2)_k} \cdot q^{2\lambda k} \cdot t^k = \frac{(t;q^2)_\infty}{(q^{2\lambda}t;q^2)_\infty},$$

$$b(t) = \sum_{m=0}^{\infty} \frac{(q^{2\lambda};q^2)_m}{(q^2;q^2)_m} \cdot t^m = \frac{(q^{2\lambda}t;q^2)_\infty}{(t;q^2)_\infty},$$

and observe that $a(t)\cdot b(t) = 1$.

EXAMPLE 1.143. One can similarly prove that in $\mathcal{D}(\mathbb{D})'_q$ one has

$$f_0 = \sum_{j=0}^{\infty} \frac{(-1)^j q^{j(j-1)}}{(q^2;q^2)_j} z^j z^{*j}.$$

EXAMPLE 1.144. Recall that $y = 1 - zz^*$ and $y^\lambda \in \mathcal{D}(\mathbb{D})'_q$ for all λ. Then we get the decomposition

$$(1 - zz^*)^\lambda = \sum_{k=0}^{\infty} \frac{(q^{-2\lambda};q^2)_k}{(q^2;q^2)_k} q^{2\lambda k} z^k z^{*k} \tag{1.172}$$

in $\mathcal{D}(\mathbb{D})'_q$. Consider the operators of a Fock representation and equate their matrix elements in the basis $\{T_F(z^m)e_0\}_{m\in\mathbb{Z}_+}$ (it is sufficient to consider only diagonal matrix elements) to get

$$\sum_{k=0}^{j} \frac{(q^{-2\lambda};q^2)_k}{(q^2;q^2)_k} \cdot (q^{2j};q^{-2})_k \cdot q^{2\lambda k} = q^{2\lambda j}, \qquad j \in \mathbb{Z}_+,$$

i.e.,

$$\sum_{k+m=j} \frac{(q^{-2\lambda};q^2)_k}{(q^2;q^2)_k} \cdot q^{2\lambda k} \cdot \frac{1}{(q^2;q^2)_m} = \frac{q^{2\lambda j}}{(q^2;q^2)_j}, \qquad j \in \mathbb{Z}_+.$$

Now it remains to consider the course-of-value functions

$$a(t) = \frac{(t;q^2)_\infty}{(q^{2\lambda}t;q^2)_\infty}, \qquad b(t) = \frac{1}{(t;q^2)_\infty}, \qquad c(t) = \frac{1}{(q^{2\lambda}t;q^2)_\infty},$$

and apply the q-binomial decomposition (1.114) to both sides of the evident equality $a(t)b(t) = c(t)$.

Now we show that the element $\widehat{f_0}$ generates $\overline{\mathrm{End}}(\mathbb{C}[z]_{q,\lambda})$ as a topological $U_q\mathfrak{sl}_2$-module.

LEMMA 1.145. $U_q\mathfrak{sl}_2\,\widehat{f_0} = \underline{\mathrm{End}}(\mathbb{C}[z]_{q,\lambda})$.

PROOF. Using (1.162) and (1.163), it is not hard to observe that

$$\widehat{z}^i \widehat{f_0} \widehat{z}^{*j} : z^m \mapsto \left(1 - q^{2(\lambda-1)}\right) \frac{(q^2;q^2)_j}{(q^{2\lambda};q^2)_j} \delta_{mj} z^i, \qquad i,j,m \in \mathbb{Z}_+.$$

Thus the linear operators $\{\widehat{z}^i \widehat{f_0} \widehat{z}^{*j}\}_{i,j\in\mathbb{Z}_+}$ generate the vector space $\underline{\mathrm{End}}(\mathbb{C}[z]_{q,\lambda})$. It remains to verify that they all belong to $U_q\mathfrak{sl}_2\widehat{f_0}$. For that, it is enough to replace the generators z, z^*, f_0 by $\widehat{z}$, $\widehat{z}^*$, $\widehat{f_0}$ and to reproduce the proof of Proposition 1.82. Indeed, formulas which described the action of $K^{\pm 1}$, E, F on $\widehat{f_0}$ differ from (1.69) only by numerical factors:

$$K^{\pm 1}\widehat{f_0} = \widehat{f_0}, \quad E\widehat{f_0} = c'\widehat{z}\cdot\widehat{f_0}, \quad F\widehat{f_0} = c''\widehat{f_0}\cdot\widehat{z}^*, \qquad c', c'' \neq 0.$$

These relations follow from Proposition 1.133, the equalities (1.69), and $\mathbb{C}\widehat{zf_0} = \mathbb{C}\widehat{z}\widehat{f}_0$, $\mathbb{C}\widehat{f_0 z^*} = \mathbb{C}\widehat{f}_0\widehat{z}^*$, that can be easily deduced from

$$\operatorname{Im}\widehat{zf_0} = \mathbb{C}z, \quad \operatorname{Im}\widehat{f_0 z^*} = \mathbb{C}\cdot 1, \quad \operatorname{Ker}\widehat{f}_0 = \operatorname{Ker}\widehat{zf_0} = (\mathbb{C}\cdot 1)^{\perp}. \qquad \square$$

PROPOSITION 1.146. *For every numerical matrix $(a_{jk})_{j,k\in\mathbb{Z}_+}$ the covariant symbol of the operator $\sum_{j,k=0}^{\infty} a_{jk}\widehat{z}^j\widehat{z}^{*k}$ is $\sum_{j,k=0}^{\infty} a_{jk}z^j z^{*k}$.*

PROOF. We already know the covariant symbol of the Toeplitz operator $\widehat{f}_0$; see Example 1.138. Using (1.170) and (1.172), we get the required statement in that special case.

Now it remains to apply Lemma 1.145 and the fact that $\left(\mathcal{T}^{(\lambda)}\right)^*$ is a morphism of $U_q\mathfrak{sl}_2$-modules. $\square$

COROLLARY 1.147. *The map $\left(\mathcal{T}^{(\lambda)}\right)^* : \overline{\operatorname{End}}(\mathbb{C}[z]_{q,\lambda}) \to \mathcal{D}(\mathbb{D})'_q$, which assigns to an operator its covariant symbol, is bijective.*

1.6.8. The Berezin transform. In the classical case $q = 1$ the Berezin transform is an important tool in the theory of Bergman spaces; see [**122, 349, 15**]. Consider the linear map $\mathcal{B}_\lambda$ from $\mathcal{D}(\mathbb{D})_q$ to $\mathcal{D}(\mathbb{D})'_q$ given by $\mathcal{B}_\lambda = \left(\mathcal{T}^{(\lambda)}\right)^*\mathcal{T}^{(\lambda)}$. It assigns to a finite function f the covariant symbol of the Toeplitz operator $\hat{f}$ with the contravariant symbol f. The operator $\mathcal{B}_\lambda$ is called the Berezin transform; see [**310**, p. 566]. The results from the previous subsection imply

PROPOSITION 1.148. *The Berezin transform is a morphism of $U_q\mathfrak{sl}_2$-modules.*

EXAMPLE 1.149. $\mathcal{B}_\lambda f_0 = (1-q^{2(\lambda-1)})(1-zz^*)^\lambda$; see Examples 1.131 and 1.138.

The book [**122**] is devoted to the theory of Bergman spaces in the unit disc, and the Berezin transform is defined therein on page 29 as

$$\mathcal{B}_\lambda f(z) = (\lambda-1)\int_{\mathbb{D}}\left(\frac{(1-|z|^2)(1-|\zeta|^2)}{|1-z\bar{\zeta}|^2}\right)^\lambda f(\zeta)\,d\nu(\zeta) \tag{1.173}$$

(the authors use the parameter $\alpha = \lambda - 2$ instead of the Wallach parameter λ). We are about to show that in the case of quantum disc the Berezin transform is also an integral operator, and we will find its kernel. In this way we obtain a q-analog of (1.173).

PROPOSITION 1.150. *For each $f \in \mathcal{D}(\mathbb{D})_q$ one has*

$$(\mathcal{B}_\lambda f)(z) = \int_{\mathbb{D}_q} b_\lambda(z,\zeta)\, f(\zeta) d\nu(\zeta),$$

where $b_\lambda \in \mathcal{D}(\mathbb{D}\times\mathbb{D})'_q$ and

$$b_\lambda(z,\zeta) = \frac{1-q^{2(\lambda-1)}}{1-q^2}(1-\zeta\zeta^*)^\lambda(q^2z^*\zeta;q^2)_{-\lambda}\cdot(z\zeta^*;q^2)_{-\lambda}(1-z^*z)^\lambda.$$

PROOF. Consider the linear integral operator

$$\widetilde{\mathcal{B}}_\lambda : \mathcal{D}(\mathbb{D})_q \to \mathcal{D}(\mathbb{D})'_q, \qquad \widetilde{\mathcal{B}}_\lambda : f \mapsto \int_{\mathbb{D}_q} b_\lambda(z,\zeta) f(\zeta) d\nu(\zeta),$$

with kernel b_λ. This kernel coincides up to a scalar factor with the $U_q\mathfrak{sl}_2$-invariant kernel $k_{22}^{-\lambda} \cdot k_{11}^{-\lambda}$; see Subsection 1.7.3. Thus $\widetilde{\mathcal{B}}_\lambda$ is a morphism of $U_q\mathfrak{sl}_2$-modules. Observe that $\mathcal{B}_\lambda$ possesses the same property and f_0 generates the $U_q\mathfrak{sl}_2$-module $\mathcal{D}(\mathbb{D})_q$. Therefore it is enough to prove the equality $\widetilde{\mathcal{B}}_\lambda f = \mathcal{B}_\lambda f$ only in the special case $f = f_0$. On the other hand, $\widetilde{\mathcal{B}}_\lambda f_0 = (1-q^{2(\lambda-1)})(1-zz^*)^\lambda = \mathcal{B}_\lambda f_0$; see Example 1.149. □

We are about to obtain an expansion of the operator-valued function $\mathcal{B}_\lambda$ in powers of $t = q^{2(\lambda-1)}$. Evidently, $t \to 0$ as $\lambda \to +\infty$. In the classical case $q = 1$ an asymptotic expansion of the Berezin transform as $\lambda \to +\infty$ is well known [**232, 54, 310, 55, 85, 12**].

Let f_n be the element of $\mathcal{D}(\mathbb{D})_q$ given by

$$T_F(f_n)e_j = \begin{cases} e_j, & j = n, \\ 0, & j \neq n. \end{cases}$$

The following lemma can be proved using the spectral decomposition $1 - zz^* = \sum_{n=0}^{\infty} q^{2n} f_n$ of $y = 1 - zz^*$ and (1.166).

LEMMA 1.151. *For all $m, j \in \mathbb{Z}_+$ one has the expansion*

$$P_{z,m,l} = \frac{(q^{2\lambda};q^2)_m}{(q^2;q^2)_m} \sum_{n=0}^{\infty} q^{2(\lambda-1)n} \cdot P_{z,m,l}^{(n)}$$

in $\mathcal{D}(\mathbb{D})_q'$, where $P_{z,m,l}^{(n)} = q^{2(l+n)} z^l \cdot f_n \cdot z^{*m} \in \mathcal{D}(\mathbb{D})_q$.

Let $f, \psi \in \mathcal{D}(\mathbb{D})_q$. Consider the integral $\int_{\mathbb{D}_q} \psi^* \, (\mathcal{B}_\lambda f) \, d\nu$ as a function in variable $t = q^{2(\lambda-1)}$. Proposition 1.137 and Lemma 1.151 yield the analyticity of this function in $(0,1)$. We get

PROPOSITION 1.152. *There exists a unique sequence of morphisms of $U_q\mathfrak{sl}_2$-modules $\mathcal{B}^{(n)} : \mathcal{D}(\mathbb{D})_q \to \mathcal{D}(\mathbb{D})_q'$, $n \in \mathbb{Z}_+$, such that*

$$\mathcal{B}_\lambda f = \sum_{n=0}^{\infty} q^{2(\lambda-1)\,n} \, \mathcal{B}^{(n)} f$$

for all $f \in \mathcal{D}(\mathbb{D})_q$.

Our goal is to prove that the linear operators $\mathcal{B}^{(n)}$ map $\mathcal{D}(\mathbb{D})_q$ into itself and belong to the subalgebra generated by $\Box_q$.

Let

$$p_j(x) = \sum_{k=0}^{j} \frac{(q^{-2j};q^2)_k}{(q^2;q^2)_k^2} q^{2k} \cdot \prod_{i=0}^{k-1} \left(1 + q^{2i}\left((1-q^2)^2 x - 1 - q^2\right) + q^{4i+2}\right). \tag{1.174}$$

LEMMA 1.153. *$p_j(\Box_q) f_0 = q^{2j} \, f_j$ for all $j \in \mathbb{Z}_+$.*

PROOF. We will use results of Subsection 1.5.4. The basic hypergeometric series

$$\Phi_l = {}_3\Phi_2 \left[\begin{matrix} (1-zz^*)^{-1}, q^{-2l}, q^{2(l+1)} \\ q^2, 0 \end{matrix} ; q^2, q^2 \right]$$

converges in $\mathcal{D}(\mathbb{D})'_q$ for all $l \in \mathbb{C}$, and

$$\Box_q \Phi_l = \frac{(1-q^{-2l})(1-q^{2(l+1)})}{(1-q^2)^2}\,\Phi_l.$$

It is sufficient to verify that

$$q^{-2j} \cdot \int\limits_{\mathbb{D}_q} \Phi_l^* \cdot p_j(\Box_q)\, f_0 d\nu = \int\limits_{\mathbb{D}_q} \Phi_l^* \, f_j d\nu \tag{1.175}$$

for all $l \in \mathbb{C}$. Replacing l by $\bar{l}$, we find that (1.175) is equivalent to the equality

$$p_j\left(\frac{(1-q^{-2l})(1-q^{2(l+1)})}{(1-q^2)^2}\right) = {}_3\Phi_2\left[\begin{matrix} q^{-2j}, q^{-2l}, q^{2(l+1)}; q^2; q^2 \\ q^2, 0 \end{matrix}\right].$$

Denoting for brevity $u = q^{2l+2} + q^{-2l}$, let us write it in the following way:

$$p_j\left(\frac{1+q^2-u}{(1-q^2)^2}\right) = {}_3\Phi_2\left[\begin{matrix} q^{-2j}, q^{-2l}, q^{2(l+1)}; q^2; q^2 \\ q^2, 0 \end{matrix}\right].$$

By the definition of ${}_3\Phi_2$ we get

$$\begin{aligned}
&{}_3\Phi_2\left[\begin{matrix} q^{-2j}, q^{-2l}, q^{2(l+1)} & ; q^2, q^2 \\ q^2, 0 & \end{matrix}\right] \\
&= \sum_{k=0}^{j} \frac{(q^{-2j}; q^2)_k}{(q^2; q^2)_k^2} \cdot \prod_{i=0}^{k-1} \left(1 - q^{2i} \cdot q^{2(l+1)})(1 - q^{2i} \cdot q^{-2l})\right) \cdot q^{2k} \\
&= \sum_{k=0}^{j} \frac{(q^{-2j}; q^2)_k}{(q^2; q^2)_k^2} \cdot \prod_{i=0}^{k-1} (1 - q^{2i} \cdot u + q^{4i+2}) \cdot q^{2k}.
\end{aligned}$$

It remains to use the definition of p_j. □

The following statement essentially refines Proposition 1.152.

PROPOSITION 1.154. *For all $f \in \mathcal{D}(\mathbb{D})_q$ and $\lambda > 1$, one has the expansion*

$$\mathcal{B}_\lambda f = (1 - q^{2(\lambda-1)}) \sum_{j \in \mathbb{Z}_+} q^{2(\lambda-1)\,j}\, p_j(\Box_q) f \tag{1.176}$$

in $\mathcal{D}(\mathbb{D})'_q$.

PROOF. In the special case $f = f_0$, the required claim follows from Lemma 1.153 since

$$\mathcal{B}_\lambda f_0 = (1 - q^{2(\lambda-1)}) \sum_{k \in \mathbb{Z}_+} q^{2(\lambda-1)\,k}\, f_k;$$

see Example 1.149. It remains to take into account that f_0 generates the $U_q\mathfrak{sl}_2$-module $\mathcal{D}(\mathbb{D})_q$, and the operators $\mathcal{B}_\lambda$, $\Box_q$ are morphisms of $U_q\mathfrak{sl}_2$-modules. □

COROLLARY 1.155.

$$\mathcal{B}^{(n)} = \begin{cases} I, & n = 0, \\ p_n(\Box_q) - p_{n-1}(\Box_q), & n \in \mathbb{N}. \end{cases}$$

REMARK 1.156. Eric Koelink has drawn our attention to the fact that p_j differ from the orthogonal Al Salam–Chihara polynomials only by affine transformations of variables (1.125).

It follows from Proposition 1.150 that there exist elements $\varphi_{j,k,\lambda} \in \mathrm{Pol}(\mathbb{C})_q$ such that

$$\mathcal{B}_\lambda f = \sum_{j,k \in \mathbb{Z}_+} \left(\int_{\mathbb{D}_q} \varphi_{j,k,\lambda}\, f\, d\nu_\lambda \right) z^j z^{*\,k}, \qquad f \in \mathcal{D}(\mathbb{D})_q.$$

Thus the Berezin transform can be extended by continuity to a linear map from $L^1(d\nu_\lambda)_q$ to $\mathcal{D}(\mathbb{D})'_q$. Let us obtain a decomposition of distribution $\mathcal{B}_\lambda f$ for $f \in \mathrm{Pol}(\mathbb{C})_q$ similar to (1.176).

COROLLARY 1.157. *For all $f \in \mathrm{Pol}(\mathbb{C})_q$ and $\lambda > 1$, one has the expansion*

$$\mathcal{B}_\lambda f = \left(1 - q^{2(\lambda-1)}\right) \sum_{j \in \mathbb{Z}_+} q^{2(\lambda-1)\,j}\, p_j(\Box_q) f \tag{1.177}$$

in $\mathcal{D}(\mathbb{D})'_q$.

PROOF. Recall that the algebra $L^\infty(\mathbb{D})_q$ is naturally isomorphic to the algebra of all bounded linear operators in a Hilbert space. We endow $L^\infty(\mathbb{D})_q$ with the ultraweak topology [**48**, p. 76]. It suffices to show that for a certain sequence of continuous linear maps

$$B^{(n)} : L^\infty(\mathbb{D})_q \to \mathcal{D}(\mathbb{D})'_q, \qquad n \in \mathbb{Z}_+,$$

one has the expansion $\mathcal{B}_\lambda f = \sum_{n=0}^{\infty} q^{2(\lambda-1)\,n}\, B^{(n)} f$ for all $f \in L^\infty(\mathbb{D})_q$.

We apply the same argument as in the proof of Lemma 1.136 and use the fact that $(T_F(1 - zz^*))^{\lambda-1}$ is a trace class operator for $\lambda > 1$ to obtain an integral representation of matrix elements of the Toeplitz operator $\widehat{f}$ with bounded symbol f:

$$\widehat{f}_{mj} = \left(1 - q^{2(\lambda-1)}\right) \frac{(q^{2\lambda}; q^2)_m}{(q^2; q^2)_m} \operatorname{tr}\left(T_F(z^j (1 - zz^*)^{\lambda-1} z^{*m}) T_F(f)\right).$$

Observe that the operator-valued function

$$\left(1 - q^{2(\lambda-1)}\right) \frac{(q^{2\lambda}; q^2)_m}{(q^2; q^2)_m}\, T_F\left(z^j (1 - zz^*)^{\lambda-1} z^{*m}\right)$$

is expanded as a power series in $q^{2(\lambda-1)}$ converging for $\lambda > 1$, and the coefficients of this power series are trace class operators. It remains to use the equality

$$\int_{\mathbb{D}_q} \psi^* \left(\mathcal{B}_\lambda f\right) d\nu = \frac{1 - q^2}{1 - q^{2(\lambda-1)}} \cdot \operatorname{tr}_q(\widehat{f}\, \widehat{\psi}), \qquad f \in L^\infty(\mathbb{D})_q,\ \psi \in \mathcal{D}(\mathbb{D})_q$$

(cf. Lemma 1.134) and take into account that $\widehat{\psi}$ has only finitely many nonzero matrix elements in the basis $\{z^m\}_{m \in \mathbb{Z}_+}$. □

Let us obtain a q-analog of the well-known decomposition of the Berezin transform into an infinite product [**31, 310**]. Recall the notation $y = 1 - zz^*$, $[\lambda]_q = \frac{q^\lambda - q^{-\lambda}}{q - q^{-1}}$.

Using (1.96), it is not hard to show that

$$\Box_q\, y^\lambda = -q^{-1}\, [\lambda]_q [\lambda - 1]_q\, y^\lambda + [\lambda]_q^2\, y^{\lambda+1}.$$

Applying $\Box_q$ to both sides of the equality $\mathcal{B}_\lambda f_0 = (1 - q^{2(\lambda-1)})(1 - zz^*)^\lambda$, we get

$$\left(\Box_q \mathcal{B}_\lambda + q^{-1} [\lambda]_q [\lambda - 1]_q (\mathcal{B}_\lambda - \mathcal{B}_{\lambda+1})\right) f_0 = 0.$$

Thus,

$$\left(1+\frac{q}{[\lambda]_q[\lambda-1]_q}\Box_q\right)\mathcal{B}_\lambda f_0=\mathcal{B}_{\lambda+1}f_0,$$

$$\mathcal{B}_\lambda f_0=\prod_{j=0}^{\infty}\left(1+\frac{q}{[\lambda+j]_q[\lambda+j-1]_q}\Box_q\right)^{-1}f_0 \tag{1.178}$$

(the operators $1+\frac{q}{[\lambda+j]_q[\lambda+j-1]_q}\Box_q$ are invertible in $L^2(d\nu)_q$ for all $\lambda>1$ since $\Box_q=\bar{\partial}^*\bar{\partial}\geq 0$).

As $\Box_q$ and $\mathcal{B}_q$ are morphisms of $U_q\mathfrak{sl}_2$-modules and f_0 generates the $U_q\mathfrak{sl}_2$-module $\mathcal{D}(\mathbb{D})_q$, (1.178) yields the equality

$$\mathcal{B}_\lambda=\prod_{j=0}^{\infty}\left(1+\frac{q}{[\lambda+j]_q[\lambda+j-1]_q}\Box_q\right)^{-1} \tag{1.179}$$

($\mathcal{B}_\lambda:\mathcal{D}(\mathbb{D})_q\to\mathcal{D}(\mathbb{D})'_q$, and the product in the right-hand side of (1.179) converges on each vector from $\mathcal{D}(\mathbb{D})_q$).

1.6.9. The Berezin quantization. Consider a unital algebra A and the $\mathbb{C}[[t]]$-module $A[[t]]$ of formal series with coefficients from A. A $\mathbb{C}[[t]]$-bilinear multiplication

$$*_t: A[[t]]\times A[[t]]\to A[[t]]$$

is called a deformation of the algebra A if $f_1 *_t f_2=f_1f_2+\sum_{k=1}^{\infty}t^k\, m_k(f_1,f_2)$, where m_k are bilinear maps from $A\times A$ to A. The definition includes the requirement of associativity for the multiplication $*_t$ and the existence of a unity, namely, the fact that the natural embedding $A\hookrightarrow A[[t]]$ must be a homomorphism of complex unital algebras: $m_k(1,f)=m_k(f,1)=0$. This allows us to identify A with its image under the embedding to $A[[t]]$.

If A is a $U_q\mathfrak{su}_{1,1}$-module $*$-algebra, then $A[[t]]$ inherits this structure (it is implicit that $t^*=t$). A deformation is called a $U_q\mathfrak{su}_{1,1}$-module if

$$(f_1 *_t f_2)^*=f_2^* *_t f_1^*,\qquad f_1,f_2\in A[[t]],$$

and all the maps

$$A\otimes A\to A,\qquad f_1\otimes f_2\mapsto m_k(f_1,f_2),$$

are morphisms of $U_q\mathfrak{sl}_2$-modules.

We turn to the special case $A=\mathrm{Pol}(\mathbb{C})_q$. A deformation of the algebra $\mathrm{Pol}(\mathbb{C})_q$ is called a deformation with separable variables if $f_1 *_t f=f_1f$ and $f *_t f_2=ff_2$ for all $f_1\in\mathbb{C}[z]_q$, $f_2\in\mathbb{C}[\bar{z}]_q$, $f\in\mathrm{Pol}(\mathbb{C})_q$; see **[157]**.

The commutation relation (1.161) and the substitution $t=q^{2(\lambda-1)}$ assumed everywhere below in this subsection allow us to obtain a real $U_q\mathfrak{sl}_2$-invariant deformation of $\mathrm{Pol}(\mathbb{C})_q$ with separable variables, as one can see from

PROPOSITION 1.158. (1) *There exists a unique deformation of* $\mathrm{Pol}(\mathbb{C})_q$ *with separable variables such that*

$$z^* *_t z=q^2zz^*+1-q^2+\frac{(1-q^2)t}{1-q^2t}(1-z^*z)(1-zz^*). \tag{1.180}$$

(2) *This deformation is* $U_q\mathfrak{su}_{1,1}$*-invariant.*

One can easily prove this proposition starting from the definitions. In the sequel we obtain a more precise result using the Berezin approach.

It is shown in the Shklyarov note from [**317**] that each $U_q\mathfrak{su}_{1,1}$-invariant deformation of $\mathrm{Pol}(\mathbb{C})_q$ with separable variables can be obtained from the deformation described above by the formal substitution $t := \sum_{j=1}^{\infty} a_j t^j$, where $a_j \in \mathbb{R}$. This result agrees with a description of invariant deformations with separable variables for commutative algebras [**235**].

It will be shown that for the deformation described in Proposition 1.158, the bilinear maps m_k are q-bidifferential. Along with the "left" partial derivatives

$$\frac{\partial^{(l)} f}{\partial z} = \frac{\partial f}{\partial z}, \qquad \frac{\partial^{(l)} f}{\partial z^*} = \frac{\partial f}{\partial z^*},$$

introduced in Subsection 1.5.2, we need the "right" partial derivatives defined by

$$\partial f = \frac{\partial^{(r)} f}{\partial z} dz, \qquad \bar{\partial} f = \frac{\partial^{(r)} f}{\partial z^*} dz^*.$$

Consider the algebra of linear operators in the vector space $\mathrm{Pol}(\mathbb{C})_q = \mathbb{C}[z]_q \otimes \mathbb{C}[\bar{z}]_q$ and the algebra of linear operators in $\mathrm{Pol}(\mathbb{C})_q \otimes \mathrm{Pol}(\mathbb{C})_q$. Let i be the isomorphism of vector spaces

$$i : \mathbb{C}[\bar{z}]_q \otimes \mathbb{C}[z]_q \xrightarrow{\approx} \mathrm{Pol}(\mathbb{C})_q, \qquad i : g_1 \otimes g_2 \mapsto g_1 g_2.$$

Consider the homomorphism $A \mapsto \widetilde{A}$ from the first algebra to the second one given by $\widetilde{A} = \mathrm{id}_{\mathbb{C}[z]_q} \otimes (i^{-1} A\, i) \otimes \mathrm{id}_{\mathbb{C}[\bar{z}]_q}$.

EXAMPLE 1.159. It is not hard to show that

$$\Box_q(f(z^*)\, g(z)) = -q^2 \frac{\partial^{(r)} f}{\partial z^*} \left(1 - \left(1 + q^{-2}\right) z^* z + q^{-2} z^{*2} z^2\right) \frac{\partial^{(l)} g}{\partial z}$$

for all $f \in \mathbb{C}[\bar{z}]_q$, $g \in \mathbb{C}[z]_q$. Thus,

$$\widetilde{\Box}_q = -q^{-2} \left(1 - \left(1 + q^{-2}\right) z^* \otimes z + q^{-2} z^{*2} \otimes z^2\right) \frac{\partial^{(r)}}{\partial z^*} \otimes \frac{\partial^{(l)}}{\partial z},$$

where the multiplication implicit here is that of the algebra of kernels $\mathrm{Pol}(\mathbb{C})_q^{\mathrm{op}} \otimes \mathrm{Pol}(\mathbb{C})_q$.

PROPOSITION 1.160. *For all $f_1, f_2 \in \mathrm{Pol}(\mathbb{C})_q$, one has*

$$f_1 *_t f_2 = (1-t) \sum_{j=0}^{\infty} t^j \cdot m(p_j(\widetilde{\Box}_q)(f_1 \otimes f_2)),$$

where p_j are the polynomials (1.174).

PROOF OF PROPOSITIONS 1.158 AND 1.160. Let us write $\mathbb{C}[z]_q$ instead of $\mathbb{C}[z]_{q,\lambda}$ in the cases where a $U_q\mathfrak{sl}_2$-module structure is not essential.

Let $Q : \mathrm{Pol}(\mathbb{C})_q[[t]] \to \mathrm{End}(\mathbb{C}[z]_q)[[t]]$ be continuous in the t-adic topology morphism of $\mathbb{C}[[t]]$-modules such that

$$Q : z^j z^{*k} \mapsto \hat{z}^j \hat{z}^{*k}, \qquad j, k \in \mathbb{Z}_+.$$

This map assigns to a polynomial $f \in \mathrm{Pol}(\mathbb{C})_q$ the linear operator with the covariant symbol f; see Proposition 1.146. The multiplication in $\mathrm{End}(\mathbb{C}[z]_q)[[t]]$ is associative, and, just as in Berezin's paper [**31**], we transfer it to $\mathrm{Pol}(\mathbb{C})_q[[t]]$ using Q.

LEMMA 1.161. *There exists a unique* $\mathbb{C}[[t]]$*-bilinear map* $*_t : \mathrm{Pol}(\mathbb{C})_q[[t]] \times \mathrm{Pol}(\mathbb{C})_q[[t]] \to \mathrm{Pol}(\mathbb{C})_q[[t]]$ *such that*

$$Q(f_1 *_t f_2) = (Qf_1) \cdot (Qf_2) \tag{1.181}$$

for all $f_1, f_2 \in \mathrm{Pol}(\mathbb{C})_q[[t]]$.

PROOF OF THE LEMMA. The uniqueness follows from the injectivity of Q; see Corollary 1.147. Let us produce a multiplication $*_t$ satisfying the requirement (1.181) on the image of the embedding

$$\mathbb{C}[\bar{z}]_q \times \mathbb{C}[z]_q \hookrightarrow \mathrm{Pol}(\mathbb{C})_q[[t]] \times \mathrm{Pol}(\mathbb{C})_q[[t]], \qquad f \times g \mapsto (1 \otimes f) \times (g \otimes 1).$$

To every pair $(f(z^*), g(z)) \in \mathbb{C}[\bar{z}]_q \times \mathbb{C}[z]_q$, we assign the operators $\check{f}$, $\check{g}$ with covariant symbols $f(z^*)$, $g(z)$. These operators and their product are analytic functions in $t = q^{2(\lambda-1)}$ with values in the Banach algebra of bounded linear operators. The covariant symbol of the operator $\check{f} \cdot \check{g}$ is

$$\mathcal{B}_\lambda(f(z^*)g(z)) = (1 - q^{2(\lambda-1)}) \sum_{j \in \mathbb{Z}_+} q^{2(\lambda-1)j}; p_j(\Box_q)(f(z^*)g(z));$$

see Corollary 1.177 and Example 1.130. It remains to replace $q^{2(\lambda-1)}$ with t:

$$f(z^*) *_t g(z) \stackrel{\mathrm{def}}{=} (1-t) \sum_{j \in \mathbb{Z}_+} t^j\, p_j(\Box_q)(f(z^*)g(z)), \quad f(z^*) \in \mathbb{C}[\bar{z}]_q,\ g(z) \in \mathbb{C}[z]_q.$$

Using Proposition 1.146, we extend $*_t$ to

$$\mathrm{Pol}(\mathbb{C})_q \times \mathrm{Pol}(\mathbb{C})_q \subset \mathrm{Pol}(\mathbb{C})_q[[t]] \times \mathrm{Pol}(\mathbb{C})_q[[t]],$$

with (1.181) being still valid:

$$f_1(z)f_2(z^*) *_t g_1(z)g_2(z^*) = (1-t) \sum_{j \in \mathbb{Z}_+} t^j f_1(z) \cdot p_j(\Box_q)(f_2(z^*)g_1(z)) \cdot g_2(z^*),$$

where $f_1(z), g_1(z) \in \mathbb{C}[z]_q$, $f_2(z^*), g_2(z^*) \in \mathbb{C}[\bar{z}]_q$. It remains to extend $*_t$ to $\mathrm{Pol}(\mathbb{C})_q[[t]] \times \mathrm{Pol}(\mathbb{C})_q[[t]]$ by $\mathbb{C}[[t]]$-linearity and continuity in the t-adic topology. The lemma is proved. □

Consider the map $*_t$ defined in the lemma. The manner of its construction and (1.161) yield (1.180). The associativity of $*_t$ and the $U_q\mathfrak{su}_{1,1}$-invariance of this deformation together with the fact that it is a deformation with separable variables are verified from definitions. This implies Propositions 1.158 and 1.160. □

REMARK 1.162. Under the Berezin method of deformation, a function f is assigned to the Toeplitz operator with the covariant symbol f. Under another approach, the so-called Berezin–Toeplitz quantization, a function f is assigned to the Toeplitz operator with the contravariant symbol f; see, for example, [**172, 40, 86**].

1.6.10. Appendix on Schatten–von Neumann classes. Let A be a bounded nonnegative linear operator in a separable Hilbert space H. The sum $\mathrm{tr}(A) = \sum_j (Ae_j, e_j)$ does not depend on the choice of an orthonormal basis $\{e_j\}$ in H, so A is necessarily compact if the trace is finite [**103**, p. 126].

Let $p \geq 1$. A Schatten–von Neumann class S_p is formed by compact operators in H such that $\|A\|_p = (\mathrm{tr}(|A|^p))^{\frac{1}{p}} < \infty$. Here $|A| = \sqrt{A^*A}$. The trace is a noncommutative analog of the integral, and the classes S_p are noncommutative analogs of the spaces L^p. $\mathcal{B}(H)$ is a noncommutative analog of the algebra L^∞.

The classes S_p can be introduced in a different way. Let us define singular numbers of the operator A by the formula $s_j(A) = \inf\{\|A - B\| \mid \operatorname{rk}(B) \leq j-1\}$, [**103**, p. 48]. Here $j \in \mathbb{N}$. One can show that $\|A\|_p = \left(\sum_{j=1}^{\infty} s_j^p(A)\right)^{\frac{1}{p}}$. Thus we obtain a definition of the classes S_p in terms of singular numbers [**103**, p. 121]. These numbers $s_j(A)$ admit another definition; namely, they are eigenvalues of the operator $|A|$ that are taken with multiplicities in a decreasing order [**103**, p. 46].

It is not hard to verify that

$$\|A\|_p^p = \sup_{\{e_j\}} \left\{ \sum_j |(Ae_j, e_j)|^p \right\},$$

where $\{e_j\}$ runs over all orthonormal bases of the Hilbert space H; see [**349**, p. 19]. Thus the functional $\|\cdot\|_p$ is a norm. Another proof can be found in [**103**, p. 107]. The normed spaces S_p are complete [**103**, p. 108].

It follows from the definitions that S_p is a two-sided ideal of the algebra of bounded linear operators in H and $\|A\|_p = \|U_1 A U_2\|_p$ for unitary operators U_1, U_2.

The operators from S_1 are called trace class operators, and the trace is extended by continuity from the two-sided ideal of finite-dimensional linear operators to S_1. The operators from S_2 are called Hilbert–Schmidt operators, and $\|A\|_2 = \sum_{j,k} |(Ae_j, e_k)|^2$ for each orthonormal basis $\{e_n\}$ in H.

Let $p > 1$, $q > 1$, and $\frac{1}{p} + \frac{1}{q} = 1$. Each bounded linear operator $A \in S_p$ determines a continuous linear functional $l_A(X) = \operatorname{tr}(AX)$ on S_q, and its norm is $\|A\|_p$. This is a general form of a continuous linear functional on S_q [**103**, p. 170].

In a similar way, each bounded linear operator A in H determines a continuous linear functional $l_A(X) = \operatorname{tr}(AX)$ on S_1, and its norm is $\|A\|$. This is a general form of a continuous linear functional on S_1 [**103**, p. 166]. It is important to note that S_1 is the space of all ultraweakly continuous linear functionals on $\mathcal{B}(H)$ [**48**, p. 77].

There are well-known interpolation theorems in a scale of Banach spaces S_p [**104**, pp. 140–145]. In particular, S_p can be obtained from $\mathcal{B}(H)$ and S_1 by complex interpolation method:

$$S_p = [\mathcal{B}(H), S_1,]_{p^{-1}}.$$

Let us explain the latter observation. Consider the vector space of bounded operator-valued functions with values in $\mathcal{B}(H)$ which are continuous in the vertical strip $0 \leq \operatorname{Re} z \leq 1$. Let $\mathcal{A}$ be the subspace of operator-valued functions such that they are holomorphic in $0 < \operatorname{Re} z < 1$ and their restrictions to the line $\operatorname{Im} z = 1$ are continuous and bounded in S_1. It is easy to show that $\mathcal{A}$ with the norm

$$\|f\|_{\mathcal{A}} = \max\{\sup_{y \in \mathbb{R}} \|f(iy)\| \, , \sup_{y \in \mathbb{R}} \|f(1+iy)\|_1\}$$

is a Banach space [**203**, p. 288].

It turns out that the values of operator-valued functions from $\mathcal{A}$ at the point p^{-1} are operators from S_p, and

$$\|A\|_p = \inf\{\|f\|_{\mathcal{A}} \mid f \in \mathcal{A} \quad \text{and} \quad f(p^{-1}) = A\}$$

for all $A \in S_p$; see [**349**, p. 31].

This yields the following result [**203**, p. 309]. Let a linear operator T in the Banach space $\mathcal{B}(H)$ map S_1 into itself and $\|T\|_{\mathcal{B}(H)} < \infty$, $\|T\|_1 < \infty$. Then T maps each of the subspaces S_p into itself and the norm $\|T\|_p$ depends on $\theta = p^{-1}$ in a log-convex way.

1.7. On kernels of intertwining integral operators

1.7.1. An embedding $\mathrm{Pol}(\mathbb{C})_q \hookrightarrow \mathbb{C}[w_0 SU_{1,1}]_{q,x}$. The unit circle $\mathbb{D}$ is a $SU_{1,1}$-space, and in the category of $SU_{1,1}$-spaces we have

$$\mathbb{D} \cong U_1 \setminus SU_{1,1}, \tag{1.182}$$

where $U_1 = \{\mathrm{diag}(e^{i\varphi}, e^{-i\varphi}) |\, \varphi \in \mathbb{R}/(2\pi\mathbb{Z})\}$. The isomorphism (1.182) delivers an embedding of the $*$-algebra $\mathrm{Pol}(\mathbb{C})$ of polynomials to the $*$-algebra of rational functions on the real affine algebraic group $SU_{1,1}$ such that

$$z \mapsto t_{11}^{-1} t_{12}. \tag{1.183}$$

Does this embedding have a q-analog? To answer positively, we should slightly alter the question. Let $w_0 = \left(\begin{smallmatrix} 0 & -1 \\ 1 & 0 \end{smallmatrix}\right)$. We replace the group $SU_{1,1}$ by its principal homogeneous space

$$w_0\, SU_{1,1} = \left\{ \begin{pmatrix} t_{11}, & t_{12} \\ t_{21} & t_{22} \end{pmatrix} \in SL_2 \;\middle|\; \bar{t}_{11} = -t_{22},\, \bar{t}_{12} = -t_{21} \right\}, \tag{1.184}$$

and the embedding (1.183) by the embedding

$$z \mapsto t_{12}^{-1} t_{11}. \tag{1.185}$$

Let us introduce a quantum analog of the algebra of regular functions on the $SU_{1,1}$-orbit (1.184). We endow $\mathbb{C}[SL_2]_q$ with the involution

$$t_{11}^* = -t_{22}, \qquad t_{12}^* = -q t_{21} \tag{1.186}$$

and denote by $\mathbb{C}[w_0 SU_{1,1}]_q = (\mathbb{C}[SL_2]_q, *)$ the $*$-algebra produced this way. One of the motives for choosing the involution as above is the fact that under a formal limit as $q \to 1$ we get the system of equations $\bar{t}_{11} = -t_{22}$, $\bar{t}_{12} = -t_{21}$, that separates out the real subset $w_0 SU_{1,1}$ of SL_2. Proposition 1.163 below is a more substantial reason.

Recall that the algebra $\mathbb{C}[SL_2]_q$ of regular functions on the quantum group SL_2 is generated by the matrix elements t_{ij} of the vector representation of $U_q\mathfrak{sl}_2$ (see (1.22)) and the algebra $\mathbb{C}[SL_2]_q$ was endowed with the structure of $U_q\mathfrak{sl}_2$-module algebra in Subsection 1.3.6.

The following proposition can be easily derived from the definitions.

PROPOSITION 1.163. *The $*$-algebra $\mathbb{C}[w_0 SU_{1,1}]_q$ is a $U_q\mathfrak{su}_{1,1}$-module algebra; that is,*

$$(\xi f)^* = (S(\xi))^* f^*, \qquad \xi \in U_q\mathfrak{su}_{1,1},\ f \in \mathbb{C}[w_0 SU_{1,1}]_q. \tag{1.187}$$

As in the classical case $q = 1$ (see (1.185)), the algebra of regular functions is too small for our purposes, so it needs to be extended. Use the fact that the element

$$x = -q t_{12} t_{21}$$

is self-adjoint and quasi-commutes with all the generators t_{ij}:

$$t_{11}\, x = q^2\, x\, t_{11}, \qquad t_{22}\, x = q^{-2}\, x\, t_{22}, \qquad t_{12}\, x = x\, t_{12}, \qquad t_{21}\, x = x\, t_{21}.$$

Thus the multiplicative system $x^{\mathbb{Z}_+}$ satisfies the Ore condition. As was mentioned in Subsection 1.3.6, the algebra $\mathbb{C}[SL_2]_q$ has no zero divisors. Therefore it is embedded into the localization $\mathbb{C}[w_0 SU_{1,1}]_{q,x}$ with respect to the multiplicative system $x^{\mathbb{Z}_+}$. Since x is self-adjoint, one can extend the involution $*$ to $\mathbb{C}[w_0 SU_{1,1}]_{q,x}$ and prove that the natural embedding $\mathbb{C}[w_0 SU_{1,1}]_q \hookrightarrow \mathbb{C}[w_0 SU_{1,1}]_{q,x}$ is a morphism of $*$-algebras.

Obviously, the elements t_{12}, t_{21} are invertible in $\mathbb{C}[w_0 SU_{1,1}]_{q,x}$:

$$t_{12}^{-1} = t_{21}\, x, \qquad t_{21}^{-1} = t_{12}\, x.$$

PROPOSITION 1.164. (1) *There exists a unique extension of the $U_q\mathfrak{su}_{1,1}$-module algebra structure from $\mathbb{C}[w_0 SU_{1,1}]_q$ to $\mathbb{C}[w_0 SU_{1,1}]_{q,x}$.*

(2) *For all $\xi \in U_q\mathfrak{su}_{1,1}$, $f \in \mathbb{C}[w_0 SU_{1,1}]_{q,x}$ there exists a unique Laurent polynomial $p_{f,\xi}$ in one variable with coefficients from $\mathbb{C}[w_0 SU_{1,1}]_{q,x}$ such that*

$$\xi(f \cdot x^n) = p_{f,\xi}(q^n) \cdot x^n \tag{1.188}$$

for all $n \in \mathbb{Z}$.

PROOF. The uniqueness statements are evident. Starting from (1.146)–(1.147), one can easily show that for each polynomial ψ in one variable one has

$$K^{\pm 1}\psi(x) = \psi(x), \tag{1.189}$$

$$E\psi(x) = -q^{1/2} t_{11} \frac{\psi(q^{-2}x) - \psi(x)}{q^{-2}x - x} t_{21} = -q^{1/2} t_{11} t_{21} \frac{\psi(q^{-2}x) - \psi(x)}{q^{-2}x - x},$$

$$F\psi(x) = -q^{3/2} t_{12} \frac{\psi(q^{-2}x) - \psi(x)}{q^{-2}x - x} t_{22} = -q^{3/2} t_{12} t_{22} \frac{\psi(x) - \psi(q^2 x)}{x - q^2 x}. \tag{1.190}$$

These equalities imply that for each $\xi \in U_q\mathfrak{su}_{1,1}$, $f \in \mathbb{C}[w_0 SU_{1,1}]_q$ there exists a Laurent polynomial $p_{f,\xi}$ with coefficients from $\mathbb{C}[w_0 SU_{1,1}]_q$ such that (1.188) holds for all $n \in \mathbb{Z}_+$. The Laurent polynomial $p_{f,\xi}$ is determined uniquely by its values at the points of the geometric series $q^{\mathbb{N}}$. Its uniqueness yields that (1.188) determines a well-defined action of $\xi \in U_q\mathfrak{su}_{1,1}$ in $\mathbb{C}[w_0 SU_{1,1}]_{q,x}$ since

$$p_{f,\xi}(q^n) x = p_{f \cdot x,\xi}(q^{n-1}), \qquad n \in \mathbb{Z}.$$

It is not hard to verify that the map

$$\pi : U_q\mathfrak{su}_{1,1} \to \mathrm{End}(\mathbb{C}[w_0 SU_{1,1}]_{q,x})$$

is linear and is a morphism of algebras, and also to prove that the involutions are compatible in the sense of (1.187). Indeed, the required equalities can be readily reduced to some equalities between Laurent polynomials by substituting involved elements $f \in \mathbb{C}[w_0 SU_{1,1}]_{q,x}$ by the sequences $f_n = f \cdot x^n$, $n \in \mathbb{Z}_+$. It remains to observe that each of the required equalities on Laurent polynomials holds at the points of $\{q^n\}$ for $n \in \mathbb{N}$ large enough.

Finally, we get

$$\xi(f_1 f_2) = \sum_i \xi_i'(f_1)\, \xi_i''(f_2), \qquad f_1, f_2 \in \mathbb{C}[w_0 SU_{1,1}]_{q,x},\ \xi \in U_q\mathfrak{sl}_2, \tag{1.191}$$

where it is implicit that $\Delta\xi = \sum_i \xi_i' \otimes \xi_i''$. As before, one can prove the existence and uniqueness of a Laurent polynomial $p_{f,\xi}$ in two variables such that for all m, n

$$\xi(x^m f x^n) = x^m\, p_{f,\xi}(q^m, q^n)\, x^n.$$

If the numbers $m, n \in \mathbb{Z}_+$ are large enough, then

$$x^{-m}\,\xi(x^m f_1 f_2 x^n)\,x^{-n} = \sum_i (x^{-m}\xi_i'(x^m f_1)), (\xi_i''(f_2 x^n)\,x^{-n}), \tag{1.192}$$

since $\mathbb{C}[SL_2]_q$ is a $U_q\mathfrak{sl}_2$-module algebra. We come to an equality of two Laurent polynomials at all points $\{(q^m, q^n)\}$ with m, n big enough. Thus these Laurent polynomials are equal and (1.192) holds for all $m, n \in \mathbb{Z}$. With $m = n = 0$, we get (1.191). □

PROPOSITION 1.165. *The map $\mathcal{I} : z \mapsto t_{12}^{-1} t_{11}$ can be uniquely extended to an embedding of $U_q\mathfrak{su}_{1,1}$-module algebras*

$$\mathcal{I} : \mathrm{Pol}(\mathbb{C})_q \hookrightarrow \mathbb{C}[w_0 SU_{1,1}]_{q,x}.$$

PROOF. The uniqueness of the morphism of $*$-algebras in question is evident, whereas the existence is obtained from the following equalities in $\mathbb{C}[w_0 SU_{1,1}]_{q,x}$:

$$(t_{12}^{-1}t_{11})^* = q^{-1}t_{22}t_{21}^{-1}, \quad (t_{12}^{-1}t_{11})(t_{22}t_{21}^{-1}) - q^{-2}(t_{22}t_{21}^{-1})(t_{12}^{-1}t_{11}) = q - q^{-1}$$

(the last one is derivable from $t_{11}t_{22} - t_{22}t_{11} = (q - q^{-1})t_{12}t_{21}$).

Let us prove the injectivity of $\mathcal{I}$. Consider a pre-Hilbert space with an orthonormal basis $\{e_j\}_{j\in\mathbb{Z}_+}$ and the representation π_+ of $\mathbb{C}[SL_2]_q$ introduced in Subsection 1.3.6. It is a $*$-representation of the algebra $\mathbb{C}[w_0 SU_{1,1}]_q$. Keep the notation π_+ for its extension to a $*$-representation of the localization $\mathbb{C}[w_0 SU_{1,1}]_{q,x}$.

The definitions yield that the representation $\pi_+ \circ \mathcal{I}$ of $\mathrm{Pol}(\mathbb{C})_q$ is equivalent to its Fock representation. Thus the representation $\pi_+ \circ \mathcal{I}$ is faithful; see Subsection 1.2.3. Therefore $\mathcal{I}$ is injective.

Now we prove that $\mathcal{I}$ is a morphism of $U_q\mathfrak{sl}_2$-modules. Since $\mathrm{Pol}(\mathbb{C})_q$ and $\mathbb{C}[w_0 SU_{1,1}]_{q,x}$ are $U_q\mathfrak{su}_{1,1}$-module algebras, it suffices to verify $\mathcal{I}(\xi f) = \xi\mathcal{I}(f)$ for $\xi \in U_q\mathfrak{su}_{1,1}$, $f \in \mathbb{C}[z]_q$. It remains to use the embedding of $U_q\mathfrak{sl}_2$-module algebras

$$\mathbb{C}[t_1, t_2]_q \to \mathbb{C}[SL_2]_q, \qquad t_1 \mapsto t_{11},\ t_2 \mapsto t_{12},$$

and the embedding of $\mathbb{C}[z]_q$ to a localization of the algebra $\mathbb{C}[t_1, t_2]_q$ described in Subsection 1.3.3. □

The embedding of $U_q\mathfrak{su}_{1,1}$-module algebras $\mathrm{Pol}(\mathbb{C})_q \hookrightarrow \mathbb{C}[w_0 SU_{1,1}]_{q,x}$ described above will be called the canonical embedding.

1.7.2. Invariant integrals. Due to Corollary 1.55, each element $f \in \mathbb{C}[w_0 SU_{1,1}]_q$ can be uniquely represented as a sum

$$f = \sum_{ad=bc=0} t_{11}^a t_{12}^b \psi_{abcd}(x) t_{21}^c t_{22}^d, \tag{1.193}$$

where $a, b, c, d \in \mathbb{Z}_+$, and ψ_{abcd} are polynomials in one variable.

The constructions of Subsection 1.4.3 and the equality

$$\mathcal{I}(y) = x^{-1}$$

suggest that we introduce a vector space $\mathcal{D}(w_0 SU_{1,1})_q'$ of formal series (1.193) with coefficients in the space of functions on $q^{-2\mathbb{Z}_+}$ and endow it with the weakest topology which makes continuous all the linear functionals $\mathcal{L}_{a,b,c,d,n}$:

$$\mathcal{L}_{a,b,c,d,n}(f) = \psi_{abcd}(q^{-2n}), \qquad ad = bc = 0,\ n \in \mathbb{Z}_+. \tag{1.194}$$

Polynomials are dense in the space of functions on $q^{-2\mathbb{Z}_+}$ with the pointwise convergence topology. Thus $\mathbb{C}[w_0SU_{1,1}]_q$ is a dense linear subvariety in $\mathcal{D}(w_0SU_{1,1})'_q$. Obviously, the involution $*$ extends by continuity to $\mathcal{D}(w_0SU_{1,1})'_q$.

Due to (1.189)–(1.190) and

$$\begin{aligned}(t_{11}^a t_{12}^b\,\psi(x)\,t_{21}^c)\,t_{22} &= q^{b+c}t_{11}^{a-1}t_{12}^b\,(1-x)\psi(q^2x)\,t_{21}^c,\\ t_{11}\,(t_{12}^b\psi(x)\,t_{21}^c t_{22}^d) &= q^{b+c}t_{12}^b\,(1-x)\psi(q^2x)\,t_{21}^c t_{22}^{d-1},\end{aligned}$$

the structure of the $U_q\mathfrak{sl}_2$-covariant $\mathbb{C}[w_0SU_{1,1}]_q$-bimodule also admits an extension by continuity from $\mathbb{C}[w_0SU_{1,1}]_q$ to $\mathcal{D}(w_0SU_{1,1})'_q$.

A distribution $f\in\mathcal{D}(w_0SU_{1,1})'_q$ is called finite if

$$\mathrm{card}\{(a,b,c,d,n)\mid \mathcal{L}_{a,b,c,d,n}(f)\neq 0\}<\infty.$$

The vector space $\mathcal{D}(w_0SU_{1,1})_q$ of finite functions is a submodule of the $U_q\mathfrak{sl}_2$-covariant $\mathbb{C}[w_0SU_{1,1}]_q$-bimodule $\mathcal{D}(w_0SU_{1,1})'_q$. Moreover, the structure of $U_q\mathfrak{su}_{1,1}$-module algebra extends to $\mathcal{D}(w_0SU_{1,1})_q$ by continuity.

The embedding of $U_q\mathfrak{su}_{1,1}$-module algebras $\mathcal{I}:\mathrm{Pol}(\mathbb{C})_q\hookrightarrow\mathbb{C}[w_0SU_{1,1}]_{q,x}$, described in the previous subsection, admits an extension by continuity to an embedding of $U_q\mathfrak{su}_{1,1}$-module algebras $\mathcal{I}:\mathcal{D}(\mathbb{D})_q\hookrightarrow\mathcal{D}(w_0SU_{1,1})_q$ and also to an embedding of spaces of distributions $\mathcal{D}(\mathbb{D})'_q\hookrightarrow\mathcal{D}(w_0SU_{1,1})'_q$.

The $U_q\mathfrak{sl}_2$-module algebra $\mathcal{D}(\widetilde{\Xi})_q$ of finite functions on the quantum cone, described in Subsection 1.6.3, appears in the statement of the following proposition.

PROPOSITION 1.166. *The linear functional on $\mathcal{D}(w_0SU_{1,1})_q$ defined by*

$$\int\limits_{(w_0SU_{1,1})_q} f\,d\nu = (1-q^2)\sum_{n=0}^{\infty}\psi_{0000}(q^{-2n})q^{-2n}$$

is positive and $U_q\mathfrak{sl}_2$-invariant. The linear functional on $\mathcal{D}(\widetilde{\Xi})_q$ defined by

$$\int\limits_{\widetilde{\Xi}_q} f\,d\nu = (1-q^2)\sum_{n=-\infty}^{\infty}\psi_{0000}(q^{-2n})q^{-2n}$$

is positive and $U_q\mathfrak{sl}_2$-invariant.

PROOF. Let us restrict ourselves to the first assertion. The second one can be proved in a similar way.

We use the representations π_+,ρ introduced in Subsection 1.3.6. The representation $\pi_+\otimes\rho$ can be extended to a faithful $*$-representation of the algebra $\mathcal{D}(w_0SU_{1,1})_q$. The positivity of the linear functional in question follows from the equality

$$\int\limits_{(w_0SU_{1,1})_q} f\,d\nu = (1-q^2)\,\mathrm{tr}(\pi_+\otimes\rho)(f\cdot x),\qquad f\in\mathcal{D}(w_0SU_{1,1})_q,$$

where the trace is well defined since the matrix of $(\pi_+\otimes\rho)(f\cdot x)$ in the basis $\{e_j\otimes z^k\}_{j\in\mathbb{Z}_+,\,k\in\mathbb{Z}}$ has only finitely many nonzero entries on the principal diagonal.

Let $\psi(x)$ be a function with finite support on the set $q^{-2\mathbb{Z}_+}$. Using (1.146) and (1.147), it is not hard to show that

$$\int\limits_{(w_0SU_{1,1})_q} (K^{\pm 1}-1)\psi(x)\,d\nu = 0,$$

$$\int\limits_{(w_0SU_{1,1})_q} E(\,t_{12}\psi(x)t_{22}\,)\,d\nu \;=\; \int\limits_{(w_0SU_{1,1})_q} F(\,t_{11}\psi(x)t_{21}\,)\,d\nu \;=\; 0.$$

The statement on $U_q\mathfrak{sl}_2$-invariance of the linear functional easily follows from these equalities. □

Let us complete this subsection with a proof of Proposition 1.109.

It is sufficient to verify the invariance of the sesquilinear map

$$\mathbf{h}_\lambda : \mathcal{D}(\mathbb{D})_{q,\lambda}\times\mathcal{D}(\mathbb{D})_{q,\lambda}\to\mathcal{D}(\mathbb{D})_q, \qquad \mathbf{h}_\lambda : v_\lambda f_1 \times v_\lambda f_2 \mapsto f_2^*\,(1-zz^*)^\lambda\,f_1,$$

in the special case $\lambda\in\mathbb{Z}$. By means of the embedding

$$\mathcal{I}_\lambda : \mathcal{D}(\mathbb{D})_{q,\lambda}\hookrightarrow\mathcal{D}(w_0SU_{1,1})_q, \qquad \mathcal{I}_\lambda : v_\lambda f\mapsto t_{12}^{-\lambda}\,\mathcal{I}(f),$$

the proof of invariance reduces to the equality $\mathcal{I}(y^\lambda)=(t_{12}^*)^{-\lambda}t_{12}^\lambda$ since $\mathcal{D}(w_0SU_{1,1})'_q$ is a $U_q\mathfrak{sl}_2$-covariant $\mathcal{D}(w_0SU_{1,1})_q$-bimodule.

Positivity is established easily for $\lambda\in\mathbb{R}$; see the proof of Proposition 1.91.

1.7.3. The proof of Lemma 1.117. Let us introduce algebras of kernels

$$\mathbb{C}[w_0SU_{1,1}\times w_0SU_{1,1}]_q \stackrel{\text{def}}{=} \mathbb{C}[w_0SU_{1,1}]_q^{\text{op}}\otimes\mathbb{C}[w_0SU_{1,1}]_q,$$

$$\mathbb{C}[w_0SU_{1,1}\times\widetilde{\Xi}]_q \stackrel{\text{def}}{=} \mathbb{C}[w_0SU_{1,1}]_q^{\text{op}}\otimes\mathbb{C}[\widetilde{\Xi}]_q, \quad \mathbb{C}[\widetilde{\Xi}\times\widetilde{\Xi}]_q \stackrel{\text{def}}{=} \mathbb{C}[\widetilde{\Xi}]_q^{\text{op}}\otimes\mathbb{C}[\widetilde{\Xi}]_q.$$

Also denote by t_{ij},τ_{ij} their generators $t_{ij}\otimes 1, 1\otimes t_{ij}$, respectively.

From the definitions, one obtains the following result.

LEMMA 1.167. *The kernels*

$$k_{11}=t_{11}\tau_{22}-qt_{12}\tau_{21}, \qquad k_{12}=-q^{-1}t_{11}\tau_{12}+t_{12}\tau_{11},$$
$$k_{21}=t_{21}\tau_{22}-qt_{22}\tau_{21}, \qquad k_{22}=-q^{-1}t_{21}\tau_{12}+t_{22}\tau_{11}$$

are $U_q\mathfrak{sl}_2$-invariant.

PROPOSITION 1.168. *In $\mathbb{C}[w_0SU_{1,1}\times w_0SU_{1,1}]_q$ the following relations hold:*

$$k_{11}k_{12}=q^{-1}k_{12}k_{11}, \qquad k_{21}k_{22}=q^{-1}k_{22}k_{21},$$
$$k_{11}k_{21}=q^{-1}k_{21}k_{11}, \qquad k_{12}k_{22}=q^{-1}k_{22}k_{12},$$
$$k_{12}k_{21}=k_{21}k_{12}, \qquad k_{11}k_{22}-k_{22}k_{11}=(q^{-1}-q)k_{12}k_{21},$$
$$k_{22}k_{11}-qk_{12}k_{21}=1.$$

PROOF. Consider the map

$$\mathbb{C}[SL_2]_q\to\mathbb{C}[SL_2]_q^{\text{op}}\otimes\mathbb{C}[SL_2]_q, \qquad f\mapsto(\text{id}\otimes S)\Delta(f),$$

where Δ is the coproduct and S is the antipode in the Hopf algebra $\mathbb{C}[SL_2]_q$. This map is an antiautomorphism, and $t_{ij}\mapsto k_{ij}$. □

COROLLARY 1.169. *In $\mathbb{C}[w_0 SU_{1,1} \times \widetilde{\Xi}]_q$ and $\mathbb{C}[\widetilde{\Xi} \times \widetilde{\Xi}]_q$, the following relations hold:*

$$k_{11}k_{12} = q^{-1}k_{12}k_{11}, \qquad k_{21}k_{22} = q^{-1}k_{22}k_{21},$$
$$k_{11}k_{21} = q^{-1}k_{21}k_{11}, \qquad k_{12}k_{22} = q^{-1}k_{22}k_{12},$$
$$k_{12}k_{21} = k_{21}k_{12}, \qquad k_{11}k_{22} - k_{22}k_{11} = (q^{-1} - q)k_{12}k_{21},$$
$$k_{22}k_{11} - qk_{12}k_{21} = 0.$$

Introduce the elements $x = t_{12}^* t_{12}$, $\xi = \tau_{12}\tau_{12}^*$ of the algebras of kernels $\mathbb{C}[w_0 SU_{1,1} \times w_0 SU_{1,1}]_q$, $\mathbb{C}[w_0 SU_{1,1} \times \widetilde{\Xi}]_q$, $\mathbb{C}[\widetilde{\Xi} \times \widetilde{\Xi}]_q$ and the localizations $\mathbb{C}[w_0 SU_{1,1} \times w_0 SU_{1,1}]_{q,x\xi}$, $\mathbb{C}[w_0 SU_{1,1} \times \widetilde{\Xi}]_{q,x\xi}$, $\mathbb{C}[\widetilde{\Xi} \times \widetilde{\Xi}]_{q,x\xi}$ of these algebras with respect to multiplicative systems $(x\xi)^{\mathbb{Z}_+}$. Let

$$z = qt_{12}^{-1}t_{11}, \quad z^* = t_{22}t_{21}^{-1}, \quad \zeta = q\tau_{11}\tau_{12}^{-1}, \quad \zeta^* = \tau_{21}^{-1}\tau_{22}.$$

LEMMA 1.170. *In $\mathbb{C}[w_0 SU_{1,1} \times w_0 SU_{1,1}]_{q,x\xi}$, $\mathbb{C}[w_0 SU_{1,1} \times \widetilde{\Xi}]_{q,x\xi}$, $\mathbb{C}[\widetilde{\Xi} \times \widetilde{\Xi}]_{q,x\xi}$, we have*

$$k_{22}^l k_{11}^l = q^{-2l}\xi^l \sum_{j=0}^{\infty} \frac{(q^{-2l};q^2)_j}{(q^2;q^2)_j}(q^{2(l+1)}z^*\zeta)^j \sum_{m=0}^{\infty} \frac{(q^{-2l};q^2)_m}{(q^2;q^2)_m}(q^{2l}z\zeta^*)^m x^l \tag{1.195}$$

for all $l \in \mathbb{Z}$.

PROOF. Using the equalities

$$t_{12}\tau_{21}(z\zeta^*) = q^2(z\zeta^*)t_{12}\tau_{21}, \qquad (\zeta z^*)t_{12}\tau_{21} = q^2 t_{21}\tau_{12}(\zeta z^*),$$

we get

$$k_{11}^l = (-qt_{12}\tau_{21}(1 - q^{-2}z\zeta^*))^l = (-qt_{12}\tau_{21})^l(q^{-2}z\zeta^*;q^{-2})_l,$$
$$k_{22}^l = ((1 - z^*\zeta)(-q^{-1}t_{21}\tau_{12}))^l = (z^*\zeta;q^{-2})_l(-q^{-1}t_{21}\tau_{12})^l.$$

Thus,

$$k_{22}^l k_{11}^l = (z^*\zeta;q^{-2})_l(t_{12}t_{21})^l(\tau_{12}\tau_{21})^l(q^{-2}z\zeta^*;q^{-2})_l = q^{-2l}\xi^l(q^2z^*\zeta;q^2)_l(z\zeta^*;q^2)_l x^l.$$

It remains to use the equality

$$(t;q)_n = \sum_{j=0}^{n} \frac{(q^{-n};q)_j}{(q;q)_j} q^{jn}t^j$$

(see (1.114)) and its consequences

$$(q^2z^*\zeta;q^2)_l = \sum_{n=0}^{\infty} \frac{(q^{-2l};q^2)_n}{(q^2;q^2)_n} q^{2(l+1)n}(z^*\zeta)^n,$$
$$(z\zeta^*;q^2)_l = \sum_{n=0}^{\infty} \frac{(q^{-2l};q^2)_n}{(q^2;q^2)_n} q^{2ln}(z\zeta^*)^n. \qquad \square$$

REMARK 1.171. Let us produce an analytic continuation of the kernel $k_{22}^l k_{11}^l$ with respect to l. To do this we need a topological vector space of generalized kernels. Recall that the vector space $\mathbb{C}[w_0 SU_{1,1}]_q$ is endowed with the weakest topology that makes continuous all the linear functionals $\mathcal{L}_{a,b,c,d,n}$ introduced by (1.194) in Subsection 1.7.2. For the same reasons, we endow the vector space $\mathbb{C}[w_0 SU_{1,1} \times w_0 SU_{1,1}]_q$ with the weakest topology that makes continuous all the linear functionals $\mathcal{L}_{a',b',c',d',n'} \otimes \mathcal{L}_{a'',b'',c'',d'',n''}$, and consider the completion, to be called the space of generalized kernels and to be denoted by $\mathcal{D}(w_0 SU_{1,1} \times w_0 SU_{1,1})_q'$.

As in Subsection 1.7.2, we obtain a description of this completion in terms of formal series and an extension of the $U_q\mathfrak{sl}_2$-action on it by continuity. The algebra of finite kernels

$$\mathcal{D}(w_0SU_{1,1}\times w_0SU_{1,1})_q \stackrel{\text{def}}{=} \mathcal{D}(w_0SU_{1,1})_q^{\text{op}}\otimes\mathcal{D}(w_0SU_{1,1})_q$$

is a dense linear subvariety in $\mathcal{D}(w_0SU_{1,1}\times w_0SU_{1,1})'_q$.

Let us show that $k_{22}^lk_{11}^l$ is a generalized kernel. Indeed, (1.195) yields that

$$\begin{aligned}
k_{22}^lk_{11}^l &= q^{-2l}\sum_{j=0}^{\infty}\sum_{m=0}^{\infty}\frac{(q^{-2l};q^2)_j}{(q^2;q^2)_j}\frac{(q^{-2l};q^2)_m}{(q^2;q^2)_m}q^{2j(l+1)+2ml}z^{*j}z^mx^l\xi^l\zeta^j\zeta^{*m}\\
&= q^{-2l}\sum_{j<m}\frac{(q^{-2l};q^2)_j}{(q^2;q^2)_j}\frac{(q^{-2l};q^2)_m}{(q^2;q^2)_m}q^{2j(l+1)+2ml}(z^{*j}z^j)z^{m-j}x^l\xi^l(\zeta^j\zeta^{*j})\zeta^{*(m-j)}\\
&\qquad + q^{-2l}\sum_{j=0}^{\infty}\frac{(q^{-2l};q^2)_j^2}{(q^2;q^2)_j^2}q^{4jl+2j}(z^{*j}z^j)x^l\xi^l(\zeta^j\zeta^{*j})\\
&+ q^{-2l}\sum_{m<j}\frac{(q^{-2l};q^2)_j}{(q^2;q^2)_j}\frac{(q^{-2l};q^2)_m}{(q^2;q^2)_m}q^{2j(l+1)+2ml}z^{*(j-m)}(z^{*m}z^m)x^l\xi^l\zeta^{j-m}(\zeta^m\zeta^{*m}).
\end{aligned}$$

Let us turn to analytic continuation. Consider a vector-valued function $\mathcal{K}(\lambda)$ of a complex variable λ with values in $\mathcal{D}(w_0SU_{1,1}\times w_0SU_{1,1})'_q$. Such a function is called a polynomial function if for each function $\psi\in\mathcal{D}(w_0SU_{1,1}\times w_0SU_{1,1})'_q$ the integral $\int_{(w_0SU_{1,1})_q}\int_{(w_0SU_{1,1})_q}\mathcal{K}(\lambda)\,\psi\,d\nu\,d\nu$ is a Laurent polynomial in λ.

PROPOSITION 1.172. *There exists a unique polynomial vector-valued function* $\mathcal{K}(\lambda)$ *with values in* $\mathcal{D}(w_0SU_{1,1}\times w_0SU_{1,1})'_q$ *such that* $\mathcal{K}(q^{2l})=k_{22}^lk_{11}^l$ *for all* $l\in\mathbb{Z}_+$.

PROOF. The uniqueness of the rational function $f(\lambda)$ with given values at points of the set $q^{2\mathbb{Z}_+}$ is evident.

Consider a finite function ψ and replace $k_{22}^lk_{11}^l$ in

$$\int_{(w_0SU_{1,1})_q}\int_{(w_0SU_{1,1})_q}k_{22}^lk_{11}^l\psi d\nu d\nu$$

by the right-hand side of (1.195). Only a finite number of summands in (1.195) contribute to this integral. It remains to observe that each of these summands is a Laurent polynomial in q^{2l}. □

Let us keep the notation $k_{22}^lk_{11}^l$ for the generalized kernel described in the previous proposition. Lemma 1.167 and Proposition 1.113 imply that this kernel is $U_q\mathfrak{sl}_2$-invariant for $l\in\mathbb{Z}_+$, hence in a general situation.

Lemma 1.117 now follows from the invariance of $k_{22}^lk_{11}^l$ for $l=-m$ and Lemma 1.170 since

$$\begin{aligned}
q^{-2m}\,k_{22}^{-m}k_{11}^{-m} &= (1-\zeta\zeta^*)^m\sum_{j=0}^{\infty}\frac{(q^{2m};q^2)_j}{(q^2;q^2)_j}(q^{2(-m+1)}z^*\zeta)^j\\
&\qquad\times\sum_{n=0}^{\infty}\frac{(q^{2m};q^2)_n}{(q^2;q^2)_n}(q^{-2m}z\zeta^*)^n(1-z^*z)^m = G_m.
\end{aligned}$$

Let us note that in this proof we implicitly use the canonical embedding of $U_q\mathfrak{sl}_2$-module algebras and the embedding of $U_q\mathfrak{sl}_2$-modules $\mathcal{D}(\mathbb{D}\times\mathbb{D})'_q\to\mathcal{D}(w_0SU_{1,1}\times w_0SU_{1,1})'_q$ produced via an extension by continuity.

REMARK 1.173. One may replace $\mathbb{C}[w_0 SU_{1,1} \times w_0 SU_{1,1}]_q$ by $\mathbb{C}[w_0 SU_{1,1} \times \widetilde{\Xi}]_q$ or by $\mathbb{C}[\widetilde{\Xi} \times \widetilde{\Xi}]_q$ in the previous argument to obtain the claim that for all $l \in \mathbb{C}$, the distributions

$$\mathcal{K}_l = \xi^l \sum_{j=0}^{\infty} \frac{(q^{-2l};q^2)_j}{(q^2;q^2)_j} (q^{2(l+1)} z^* \zeta)^j \sum_{m=0}^{\infty} \frac{(q^{-2l};q^2)_m}{(q^2;q^2)_m} (q^{2l} z \zeta^*)^m x^l$$

on quantum analogs of the Cartesian products $w_0 SU_{1,1} \times \widetilde{\Xi}$ and $\widetilde{\Xi} \times \widetilde{\Xi}$ are $U_q\mathfrak{sl}_2$-invariant.

1.7.4. Standard intertwining operators and the Harish-Chandra c-function. In the classical case $q = 1$, the study of intertwining operators for representations of principal series lead to the Harish-Chandra c-function. Let us describe a q-analog of this approach.

Recall the notation $(a;q^2)_l = (a;q^2)_\infty / (aq^{2l};q^2)_\infty$.

Consider the $U_q\mathfrak{sl}_2$-modules

$$\mathbb{C}[\partial\mathbb{D}]_q^{(l)}, \qquad -\frac{\pi}{2\log(q^{-1})} \le \operatorname{Im} l \le \frac{\pi}{2\log(q^{-1})},$$

of spherical principal series introduced in Subsection 1.6.5.

PROPOSITION 1.174. *For all $l \notin \{-1,-2,-3,\ldots\}$, the linear operator*

$$I_l : f(e^{i\varphi}) \mapsto \int_0^{2\pi} (e^{i(\varphi-\theta)};q^2)_l \, (q^2 e^{i(\theta-\varphi)};q^2)_l \, f(e^{i\theta}) \, \frac{d\theta}{2\pi} \tag{1.196}$$

is a morphism of $U_q\mathfrak{sl}_2$-modules

$$I_l : \mathbb{C}[\partial\mathbb{D}]_q^{(-l-1)} \to \mathbb{C}[\partial\mathbb{D}]_q^{(l)}.$$

PROOF. We may restrict ourselves to the case $\Re l > 0$. The required claim can be proved by reproducing the proof of Proposition 1.122. Indeed, first, (1.153) and (1.154) hold, and the distribution $\mathcal{K}_l$ on the quantum cone $\Xi \times \Xi$ is $U_q\mathfrak{sl}_2$-invariant; see Remark 1.173. Second, the $U_q\mathfrak{sl}_2$-action was transfered to $\mathbb{C}[\partial\mathbb{D}]_q^{(-l-1)}$ from the quantum cone by means of the map $f \mapsto f\,x^{-l-1}$. Third, $\mathcal{K}_l\,(1 \otimes f\,x^{-l-1})$ has the degree of homogeneity -1 along the second tensor factor. Fourth, the linear functional $\eta : \left(\sum_{m=-\infty}^{\infty} a_m z^m\right) x^{-1} \mapsto a_0$ on the subspace of distributions of degree of homogeneity -1 on the quantum cone is $U_q\mathfrak{sl}_2$-invariant. □

By traditions (see, for example, [**144**]) the linear operator I_l in the vector space $\mathbb{C}[\partial\mathbb{D}]$ is called a standard intertwining operator.

Recall that the Harish-Chandra c-function is introduced by (1.99):

$$c(l) = \frac{(q^{2(l+1)};q^2)_\infty^2}{(q^{2(l+1)};q^2)_\infty (q^2;q^2)_\infty} = \frac{\Gamma_{q^2}(2l+1)}{(\Gamma_{q^2}(l+1))^2}.$$

The function $c(l)$ implicitly depends on q and in a formal limit as $q \to 1$ unessentially differs from the standard Harish-Chandra c-function [**124**, p. 55], as it follows from the Legendre doubling formula [**27**, p. 19]

$$\Gamma(2l+1) = 2^{2l}\,\pi^{-1/2}\,\Gamma(l+\frac{1}{2})\,\Gamma(l+1)$$

(its q-analog is given in [**99**, p. 38]).

PROPOSITION 1.175. *For all* $l \notin \{-1,-2,-3,\dots\}$, $n \in \mathbb{Z}$, *one has*

$$I_l\, z^n = c(l)a(l,n)\, z^n,$$

where

$$a(l,n) = q^{-(2l+1)n}\begin{cases}\prod_{j=0}^{n-1}\frac{q^{-(n-l-j-1)}-q^{n-l-j-1}}{q^{-(n+l-j)}-q^{n+l-j}}, & n>0,\\ 1, & n=0,\\ \prod_{j=0}^{n-1}\frac{q^{-(n+l+j+1)}-q^{n+l+j+1}}{q^{-(n-l+j)}-q^{n-l+j}}, & n<0.\end{cases} \tag{1.197}$$

PROOF. It suffices to consider the case $l \notin \mathbb{Z}$. Lemma 1.103 shows that for $l \notin \mathbb{Z}$ there exists a unique isomorphism of the $U_q\mathfrak{sl}_2$-modules $A_l : \mathbb{C}[\partial\mathbb{D}]_q^{(-l-1)} \to \mathbb{C}[\partial\mathbb{D}]_q^{(l)}$ such that $A_l(1) = 1$. Using an explicit form of the isomorphisms of these $U_q\mathfrak{sl}_2$-modules and the modules of spherical principal series, one can easily show that $A_l(z^n) = a(l,n)z^n$ for all $n \in \mathbb{Z}$.

The isomorphism of the $U_q\mathfrak{sl}_2$-modules $\mathbb{C}[\partial\mathbb{D}]_q^{(-l-1)} \to \mathbb{C}[\partial\mathbb{D}]_q^{(l)}$ is unique up to a scalar factor. Thus Proposition 1.174 implies the equality $I_l(z^n) = \mathrm{const}(l)\, a(l,n)\, z^n$. It remains to verify that $\mathrm{const}(l) = c(l)$, that is, to prove that $I_l(1) = c(l)$. But (1.196) leads to

$$I_l(1) = \sum_{k=0}^{\infty}\frac{(q^{-2l};q^2)_k(q^{-2l};q^2)_k}{(q^2;q^2)_k(q^2;q^2)_k}q^{2(2l+1)j} = {}_2\phi_1\left[\begin{matrix}q^{-2l}, q^{-2l}\\ q^2\end{matrix}; q^2, q^{2(2l+1)}\right].$$

By the q-analog (1.116) of the Gauss formula and definition (1.99) of $c(l)$, we get

$$I_l(1) = \frac{(q^{2(l+1)};q^2)_\infty(q^{2(l+1)};q^2)_\infty}{(q^2;q^2)_\infty(q^{2(2l+1)};q^2)_\infty} = c(l). \qquad \square$$

COROLLARY 1.176. *For* $-1 < l < 0$, *the Hermitian form*

$$(f_1, f_2)_l = \int_0^{2\pi}\overline{f_2(e^{i\theta})}(I_{-l-1}f_1)(e^{i\theta})\frac{d\theta}{2\pi} \tag{1.198}$$

in $\mathbb{C}[\partial\mathbb{D}]_q^{(l)}$ *is positive and* $U_q\mathfrak{su}_{1,1}$*-invariant.*

PROOF. The definitions of the $U_q\mathfrak{sl}_2$-modules $\mathbb{C}[\partial\mathbb{D}]_q^{(-l-1)}$, $\mathbb{C}[\partial\mathbb{D}]_q^{(l)}$ and Proposition 1.121 imply the $U_q\mathfrak{su}_{1,1}$-invariance of the sesquilinear form $\mathbb{C}[\partial\mathbb{D}]_q^{(-l-1)} \times \mathbb{C}[\partial\mathbb{D}]_q^{(l)} \to \mathbb{C}$ defined as follows:

$$\langle f_1, f_2\rangle_l = \int_0^{2\pi}\overline{f_2(e^{i\theta})}\, f_1(e^{i\theta})\frac{d\theta}{2\pi}.$$

Equivalently, $\langle \xi f_1, f_2\rangle_l = \langle f_1, \xi^* f_2\rangle_l$ for all $\xi \in U_q\mathfrak{su}_{1,1}$. Thus the Hermitian form (1.198) is $U_q\mathfrak{su}_{1,1}$-invariant. It remains to apply (1.175) and (1.197) to verify its positivity. $\square$

The set of simple unitarizable $U_q\mathfrak{su}_{1,1}$-modules $\mathbb{C}[\partial\mathbb{D}]_q^{(l)}$ with $-\frac{1}{2} < l < 0$ is called a complementary series.

Similarly one can prove the unitarizability of the $U_q\mathfrak{su}_{1,1}$-modules $\mathbb{C}[\partial\mathbb{D}]_q^{(l)}$ for $\mathrm{Im}\, l = \pm\frac{\pi}{2\log(q^{-1})}$. The set of simple unitarizable $U_q\mathfrak{su}_{1,1}$-modules $\mathbb{C}[\partial\mathbb{D}]_q^{(l)}$ with $\mathrm{Im}\, l = \frac{\pi}{2\log(q^{-1})}$ and $\mathrm{Re}\, l > 0$ is called the spherical strange series.

REMARK 1.177. These series, together with the principal unitary spherical series

$$\mathbb{C}[\partial\mathbb{D}]_q^{(l)}, \qquad \operatorname{Re} l = -\frac{1}{2} \quad \text{and} \quad 0 \le \operatorname{Im} l \le \frac{\pi}{2\log(q^{-1})},$$

and the trivial $U_q\mathfrak{su}_{1,1}$-module $\mathbb{C}$ constitute a complete list of simple spherical unitarizable $U_q\mathfrak{su}_{1,1}$-modules; see the Burban and Klymik paper [**53**] where a complete list of simple weight unitarizable $U_q\mathfrak{su}_{1,1}$-modules is given.

Let us show that, just as it is in the classical case $q = 1$, the Harish-Chandra c-function arises in the study of asymptotics for eigenfunctions of the invariant Laplacian $\square_q$ as $|z| \to 1$. This explains the appearance of the c-function in an explicit form for the kernel of integral operator $(\lambda - \square_q)^{-1}$ and therefore, in an explicit form (1.98) of the Plancherel measure [**82**, p. 77].

First, it is shown in Subsection 1.5.4 that the function

$$\Phi_l(x) = {}_3\phi_2\left[\begin{matrix} x, q^{-2l}, q^{2(l+1)} \\ q^2, 0 \end{matrix}; q^2, q^2\right], \qquad x = (1 - zz^*)^{-1},$$

is an eigenfunction of the operator $\square_q^{(0)}$ corresponding to the eigenvalue (1.101). Moreover, $\Phi_l(1) = 1$. Second, it is shown in Subsection 1.6.4 that the distribution $\psi(x) = \int_0^{2\pi} P_{l+1}(z, e^{i\theta}) f(e^{i\theta}) \frac{d\theta}{2\pi}$ is an eigenfunction of the operator $\square_q^{(0)}$ with the same eigenvalue and equals 1 as $x = 1$. Thus,

$$\Phi_l(x) = \int_0^{2\pi} P_{l+1}(z, e^{i\theta}) f(e^{i\theta}) \frac{d\theta}{2\pi}.$$

This is a q-analog of the well-known integral representation of spherical functions [**124**, p. 54].

PROPOSITION 1.178. (1) *For* $\operatorname{Re} l > -\frac{1}{2}$, *one has*

$$\lim_{x \in q^{-2\mathbb{Z}_+},\, x \to \infty} x^{-l} \Phi_l(x) = \frac{\Gamma_{q^2}(2l+1)}{\Gamma_{q^2}^2(l+1)}.$$

(2) *For* $\operatorname{Re} l < -\frac{1}{2}$, *one has*

$$\lim_{x \in q^{-2\mathbb{Z}_+},\, x \to \infty} x^{l+1} \Phi_l(x) = \frac{\Gamma_{q^2}(-2l-1)}{\Gamma_{q^2}^2(-l)}.$$

PROOF. The relation $\Phi_l(x) = \Phi_{-1-l}(x)$ allows us to restrict ourselves to the case $\operatorname{Re} l > -\frac{1}{2}$. In (1.118) replacing q, b, c, z by q^2, q^{-2l}, q^{-2l-2n}, q^{2l+2}, respectively, we get

$$\begin{aligned} \Phi_l(q^{-2n}) &= \frac{(q^{-2l-2n}; q^2)_n}{(q^{-2n}; q^2)_n} \cdot {}_2\phi_1\left[\begin{matrix} q^{-2n}, q^{-2l} \\ q^{-2(l+n)} \end{matrix}; q^2, q^{2(l+1)}\right] \\ &\sim q^{-2nl} \frac{(q^{2(l+1)}; q^2)_\infty}{(q^2; q^2)_\infty} \cdot {}_1\phi_0\left[\begin{matrix} q^{-2l} \\ - \end{matrix}; q^2, q^{2(2l+1)}\right]. \end{aligned}$$

Due to the q-binomial formula (1.114),

$$\Phi_l(q^{-2n}) \sim q^{-2nl} \frac{(q^{2(l+1)}; q^2)_\infty}{(q^2; q^2)_\infty} \cdot \frac{(q^{2(l+1)}; q^2)_\infty}{(q^{2(2l+1)}; q^2)_\infty},$$

and it remains to use the definition of q-Gamma function. □

This proof has been communicated by Leonid Korogodsky.

CHAPTER 2

Basic Quantum Theory of Bounded Symmetric Domains

2.1. Summary on quantum universal enveloping algebras

2.1.1. The Drinfeld–Jimbo algebras. The Cartan classification of simple complex Lie algebras is among the best known results in mathematics of twentieth century [**123, 133**]. Up to isomorphism, such algebras are determined by their Cartan matrices $\mathbf{a}=(a_{ij})_{i,j=1,2,\ldots,l}$, whose complete list is presented in [**44**]. Every matrix $\mathbf{a}$ is indecomposable, integral, and $a_{ii}=2$, $a_{ij}<0$ for $i\neq j$. Furthermore, with some $d_i>0$, $i=1,2,\ldots,l$, the matrix $\mathbf{b}=(d_i a_{ij})_{i,j=1,2,\ldots,l}$ is positive definite.
In what follows, d_i will be assumed coprime positive integers.

The simple complex Lie algebra $\mathfrak{g}$ with Cartan matrix $\mathbf{a}$ admits a description in terms of generators and relations. We give such a description for the universal enveloping algebra $U\mathfrak{g}$ [**139**, p. 51]. It can be determined by the generators $\{H_i,E_i,F_i\}_{i=1,2,\ldots,l}$ and the following defining relations:

$$H_iH_j-H_jH_i=0,\qquad E_iF_j-F_jE_i=\delta_{ij}H_i,$$
$$H_iE_j-E_jH_i=a_{ij}E_j,\qquad H_iF_j-F_jH_i=-a_{ij}F_j$$

for $i,j=1,2,\ldots,l$;

$$\sum_{m=0}^{1-a_{ij}}(-1)^m\binom{1-a_{ij}}{m}E_i^{1-a_{ij}-m}E_jE_i^m=0,$$
$$\sum_{m=0}^{1-a_{ij}}(-1)^m\binom{1-a_{ij}}{m}F_i^{1-a_{ij}-m}F_jF_i^m=0$$

for $i\neq j$.

The universal enveloping algebra $U\mathfrak{g}$ is a Hopf algebra with the comultiplication Δ:

$$\Delta(H_i)=H_i\otimes 1+1\otimes H_i,\qquad \Delta(E_i)=E_i\otimes 1+1\otimes E_i,\qquad \Delta(F_i)=F_i\otimes 1+1\otimes F_i,$$

the counit ε, and the antipode S given by

$$\varepsilon(H_i)=\varepsilon(E_i)=\varepsilon(F_i)=0,$$
$$S(H_i)=-H_i,\qquad S(E_i)=-E_i,\qquad S(F_i)=-F_i,\qquad i=1,2,\ldots,l.$$

The Lie algebra $\mathfrak{g}$ can be identified with the Lie subalgebra generated by $\{H_i,E_i,F_i\}_{i=1,2,\ldots,l}$.

Let us turn to the notion of quantum universal enveloping algebra, in the notation of [**220, 67, 264, 139, 49**]. This Hopf algebra was introduced by Drinfeld and Jimbo in mid-1980s in a much wider generality than it is described below.

Let $\mathbf{a} = (a_{ij})_{i,j=1,2,\dots,l}$ be the Cartan matrix of a simple complex Lie algebra, and let q be a nonzero complex number such that $q^{2d_i} \neq 1$ for all $i = 1, 2, \dots, l$. The algebra $U_q\mathfrak{g}$ is defined by generators $\{K_i, K_i^{-1}, E_i, F_i\}_{i=1,2,\dots,l}$ and relations

$$K_iK_j = K_jK_i, \qquad K_iK_i^{-1} = K_i^{-1}K_i = 1,$$

$$K_iE_j = q_i^{a_{ij}}E_jK_i, \quad K_iF_j = q_i^{-a_{ij}}F_jK_i, \tag{2.1}$$

$$E_iF_j - F_jE_i = \delta_{ij}\frac{K_i - K_i^{-1}}{q_i - q_i^{-1}},$$

$$\sum_{m=0}^{1-a_{ij}}(-1)^m \begin{bmatrix} 1-a_{ij} \\ m \end{bmatrix}_{q_i} E_i^{1-a_{ij}-m}E_jE_i^m = 0,$$

$$\sum_{m=0}^{1-a_{ij}}(-1)^m \begin{bmatrix} 1-a_{ij} \\ m \end{bmatrix}_{q_i} F_i^{1-a_{ij}-m}F_jF_i^m = 0,$$

where $q_i = q^{d_i}$, $1 \le i \le l$,

$$\begin{bmatrix} m \\ n \end{bmatrix}_q = \frac{[m]_q!}{[n]_q![m-n]_q!}, \qquad [n]_q! = [n]_q \cdots [2]_q[1]_q, \qquad [n]_q = \frac{q^n - q^{-n}}{q - q^{-1}}.$$

There exists a unique structure of Hopf algebra on $U_q\mathfrak{g}$ such that for all $i = 1, 2, \dots, l$ one has

$$\Delta(E_i) = E_i \otimes 1 + K_i \otimes E_i, \quad \Delta(F_i) = F_i \otimes K_i^{-1} + 1 \otimes F_i, \quad \Delta(K_i) = K_i \otimes K_i, \tag{2.2}$$

$$\varepsilon(E_i) = \varepsilon(F_i) = 0, \qquad \varepsilon(K_i) = 1,$$

$$S(E_i) = -K_i^{-1}E_i, \qquad S(F_i) = -F_iK_i, \qquad S(K_i) = K_i^{-1}; \tag{2.3}$$

see [**139**, Chapter 4].

In the special case $l = 1$ one obtains this way the Hopf algebra $U_q\mathfrak{sl}_2$, which was used to describe the symmetries of the quantum disc in Chapter 1. As in that chapter, we restrict ourselves to the special case $q \in (0, 1)$. These are just those q which appear in the problems of noncommutative complex analysis in which we are interested.

As in the case $l = 1$, the defining relations for $U\mathfrak{g}$ can be deduced from the Drinfeld–Jimbo relations via the substitution

$$K_i = q_i^{H_i}, \qquad K_i^{-1} = q_i^{-H_i}, \qquad i = 1, 2, \dots, l,$$

and a subsequent formal passage to the limit as $q \to 1$.

2.1.2. Weight $U_q\mathfrak{g}$-modules. We recall here some notions of the theory of Lie algebras [**151**]. Consider a simple complex Lie algebra $\mathfrak{g}$ of rank l given by its generators $\{H_i, E_i, F_i\}_{i=1,2,\dots,l}$ and the well-known defining relations; see the previous subsection.

The linear span $\mathfrak{h}$ of the pairwise commuting elements $\{H_j\}_{j=1,2,\dots,l}$ is called the Cartan subalgebra, the elements of the biorthogonal basis $\{\overline{\omega}_j\}_{j=1,2,\dots,l}$ in $\mathfrak{h}^*$ the fundamental weights, and the Abelian group $P = \sum_{i=1}^{l} \mathbb{Z}\overline{\omega}_i$ the weight lattice.

The real linear span of fundamental weights is denoted by $\mathfrak{h}^*_{\mathbb{R}}$. Given a weight $\lambda \in \mathfrak{h}^*_{\mathbb{R}}$, one has the associated uplet of numbers $(\lambda(H_1), \lambda(H_2), \dots, \lambda(H_l))$ which provides an isomorphism $\mathfrak{h}^*_{\mathbb{R}} \cong \mathbb{R}^l$.

A linear functional λ on $\mathfrak{h}$ is called a weight of a representation π of $\mathfrak{g}$ if there exists a nonzero vector v such that

$$\pi(\xi)v = \lambda(\xi)v, \qquad \xi \in \mathfrak{h}.$$

The common eigenvector v of the linear operators $\pi(\xi)$ involved here is called a weight vector of weight λ.

Nonzero weights of the adjoint representation $\mathrm{ad}_\xi = [\xi, \cdot]$ of $\mathfrak{g}$ are called roots, and the corresponding weight subspaces are called root subspaces. The linear functionals

$$\alpha_j = \sum_{i=1}^{l} a_{ij}\varpi_i \qquad j = 1, 2, \ldots, l, \tag{2.4}$$

are roots because $[H_i, E_j] = a_{ij}E_j$. These are called simple roots. It follows from the invertibility of the Cartan matrix that the simple roots form a basis of the vector space $\mathfrak{h}^*_{\mathbb{R}}$.

All the roots belong to the group $Q = \sum_{i=1}^{l} \mathbb{Z}\alpha_i$, which is called the root lattice. The following notation will be used:

$$P_+ = \sum_{j=1}^{l} \mathbb{Z}_+\varpi_j, \qquad Q_+ = \sum_{j=1}^{l} \mathbb{Z}_+\alpha_j.$$

The vector space $\mathfrak{h}^*_{\mathbb{R}}$ is endowed with a partial order by setting $\lambda' \geq \lambda''$ iff all the coefficients of the vector $\lambda' - \lambda''$ in the basis $\alpha_1, \alpha_2, \ldots, \alpha_l$ are nonnegative.

Let $\mathfrak{n}^+$, $\mathfrak{n}^-$ be the Lie subalgebras generated by $\{E_1, E_2, \ldots, E_l\}$ and $\{F_1, F_2, \ldots, F_l\}$, respectively. The subspaces $\mathfrak{n}^\pm$ are ad_ξ-invariant for all $\xi \in \mathfrak{h}$. The decomposition $\mathfrak{g} = \mathfrak{n}^- \oplus \mathfrak{h} \oplus \mathfrak{n}^+$ leads to splitting the set of roots Φ as a union of the subset Φ^+ of positive roots and the subset Φ^- of negative roots. Obviously, $\Pi = \{\alpha_1, \alpha_2, \ldots, \alpha_l\} \subset \Phi^+$.

One can show that all the root subspaces are one dimensional, and there exists the highest root.

Let us turn to Lie groups. Consider a simple affine algebraic group G whose Lie algebra is $\mathfrak{g}$. The weight lattice of rational representations of G is canonically isomorphic to a subgroup $\mathcal{L}$ that lies between Q and P: $Q \subset \mathcal{L} \subset P$. The pair $(\mathfrak{g}, \mathcal{L})$ determines the group G up to an isomorphism [**130**, p. 312]. Equivalently, G is determined up to an isomorphism by the pair $(\mathbf{a}, \mathcal{L})$, where $\mathbf{a}$ is the Cartan matrix of $\mathfrak{g}$.

Now let us turn to quantum groups. A module V over $U_q\mathfrak{g}$ is called an $\mathcal{L}$-weight module if

$$V = \bigoplus_{\lambda \in \mathcal{L}} V_\lambda, \qquad V_\lambda = \left\{ v \in V \,\middle|\, K_i v = q_i^{\lambda_i} v, \quad i = 1, 2, \ldots, l \right\}, \tag{2.5}$$

where $\lambda = \sum_{j=1}^{l} \lambda_j \varpi_j$. A nonzero V_λ in the above decomposition is called the weight subspace of weight λ. The full subcategory of finite-dimensional $\mathcal{L}$-weight $U_q\mathfrak{g}$-modules (endowed with some additional structures) is treated as a quantum algebraic group G [**49**].

EXAMPLE 2.1. The adjoint representation of the Hopf algebra $U_q\mathfrak{g}$ is given by $\mathrm{ad}(\xi)\eta = \sum_j \xi_j' \eta S(\xi'')$, where $\Delta(\xi) = \sum_j \xi_j' \otimes \xi_j''$. It endows $U_q\mathfrak{g}$ with the structure of Q-weight $U_q\mathfrak{g}$-module: $U_q\mathfrak{g} = \bigoplus_{\lambda \in Q} (U_q\mathfrak{g})_\lambda$.

The assumption $\mathcal{L} \subset P$ may turn out to be too restrictive. Let us weaken it. As in Subsection 1.3.4, the term "weight $U_q\mathfrak{g}$-module" will mean the $U_q\mathfrak{g}$-module which satisfies (2.5) with $\mathcal{L} = \mathfrak{h}_{\mathbb{R}}^* \cong \mathbb{R}^l$.[1] We introduce the linear maps H_i in V by setting $H_i|_{V_\lambda} = \lambda_i$, so that

$$K_i^{\pm 1} v = q_i^{\pm H_i} v, \qquad v \in V.$$

The full subcategory of weight $U_q\mathfrak{g}$-modules is invariant under the passage to submodules,[2] quotients, and tensor products. Every $U_q\mathfrak{g}$-module has the largest weight submodule. This fact allows one to introduce the notion of dual weight $U_q\mathfrak{g}$-module. Namely, if $V = \bigoplus_\lambda V_\lambda$, then $V^* \stackrel{\text{def}}{=} \bigoplus_\lambda (V_\lambda)^*$.

To each element $\lambda = \sum_{i=1}^l n_i \alpha_i$ of the root lattice Q, we associate the element

$$K_\lambda = K_1^{n_1} K_2^{n_2} \cdots K_l^{n_l}$$

of $U_q\mathfrak{g}$. The definition of $U_q\mathfrak{g}$ implies

$$K_\lambda K_j K_\lambda^{-1} = K_j, \qquad K_\lambda E_j K_\lambda^{-1} = q_j^{\lambda(H_j)} E_j, \qquad K_\lambda F_j K_\lambda^{-1} = q_j^{-\lambda(H_j)} F_j.$$

Let us demonstrate that the automorphism S^2 of $U_q\mathfrak{g}$ is inner. Due to (2.3),

$$S^2(K_j^{\pm 1}) = K_j^{\pm 1}, \qquad S^2(E_j) = q_j^{-2} E_j, \qquad S^2(F_j) = q_j^2 F_j, \qquad j = 1, 2, \ldots, l.$$

Hence,

$$S^2(\xi) = K_{-2\rho} \xi K_{-2\rho}^{-1}, \qquad \xi \in U_q\mathfrak{g}, \tag{2.6}$$

where

$$\rho \stackrel{\text{def}}{=} \frac{1}{2} \sum_{\beta \in \Phi^+} \beta = \sum_{i=1}^l \overline{\omega}_i$$

(the latter relation is well known; see [**44**, p. 210]).

Consider a weight $U_q\mathfrak{g}$-module V, together with the corresponding representation π. Proposition 1.64 and (2.6) allow one to conclude that the linear functional

$$\operatorname{tr}_q A \stackrel{\text{def}}{=} \operatorname{tr}(A \pi(K_{-2\rho}))$$

is an invariant integral on the $U_q\mathfrak{g}$-module algebra $\operatorname{End}_f(V)$ of finite rank linear operators in V.

The linear functional tr_q is called a q-trace. Obviously (1.56) holds: $\Delta(K_{-2\rho}) = K_{-2\rho} \otimes K_{-2\rho}$, $\varepsilon(K_{-2\rho}) = 1$.

2.1.3. Monomial bases. Consider the Hopf algebra $U_q\mathfrak{g}$ and the Hopf subalgebras $U_q\mathfrak{h}$, $U_q\mathfrak{n}^+$, $U_q\mathfrak{n}^-$ generated by $\{K_i^{\pm 1}\}_{i=1,2,\ldots,l}$, $\{E_i\}_{i=1,2,\ldots,l}$, $\{F_i\}_{i=1,2,\ldots,l}$, respectively.

We want to construct monomial bases of the vector spaces $U_q\mathfrak{n}^\pm$. This will lead to bases of $U_q\mathfrak{g}$, as the monomials $\{K_1^{j_1} K_2^{j_2} \cdots K_l^{j_l}\}_{j_1, j_2, \ldots, j_l \in \mathbb{Z}}$ form a basis in $U_q\mathfrak{h}$, and the linear maps

$$U_q\mathfrak{n}^- \otimes U_q\mathfrak{h} \otimes U_q\mathfrak{n}^+ \to U_q\mathfrak{g}, \qquad u^- \otimes u^0 \otimes u^+ \mapsto u^- u^0 u^+,$$
$$U_q\mathfrak{n}^+ \otimes U_q\mathfrak{h} \otimes U_q\mathfrak{n}^- \to U_q\mathfrak{g}, \qquad u^+ \otimes u^0 \otimes u^- \mapsto u^+ u^0 u^-$$

are isomorphisms of vector spaces [**139**, p. 66].

[1] Finite-dimensional weight $U_q\mathfrak{g}$-modules are normally called $U_q\mathfrak{g}$-modules of type 1 [**139**].
[2] See, e.g., [**151**], Proposition 1.5.

In the classical case $q = 1$, the Poincaré–Birkhoff–Witt theorem [**46**, p. 28] implies that monomial bases of $U\mathfrak{n}^{\pm}$ are derivable from bases in $\mathfrak{n}^{\pm}$, e.g., the bases of root vectors. In this way we get q-analogs of those bases.

To begin with, we introduce the Weyl group. We endow $\mathfrak{h}^*_{\mathbb{R}}$ with the scalar product

$$(\alpha_i, \alpha_j) = d_i a_{ij}, \qquad i, j = 1, 2, \dots, l. \tag{2.7}$$

It follows from (2.4) that

$$(\overline{\omega}_i, \alpha_j) = d_i \delta_{ij}, \qquad i, j = 1, 2, \dots, l. \tag{2.8}$$

This scalar product provides the isomorphism

$$\mathfrak{h}_{\mathbb{R}} \xrightarrow{\approx} \mathfrak{h}^*_{\mathbb{R}}, \qquad H_i \mapsto \frac{2}{(\alpha_i, \alpha_i)} \alpha_i, \tag{2.9}$$

which allows one to identify $\mathfrak{h}_{\mathbb{R}}$ with $\mathfrak{h}^*_{\mathbb{R}}$.

The linear maps $s_i : \lambda \mapsto \lambda - \lambda(H_i)\alpha_i$ in the above Euclidean space are reflections with respect to the hyperplanes orthogonal to the simple roots α_i. In fact, the relation $\alpha_i(H_i) = 2$ implies that $s_i^2 = \mathrm{id}$, and the relations

$$(s_i(\alpha_j), \alpha_k) = d_j a_{jk} - a_{ij} d_i a_{ik} = d_j a_{jk} - a_{ik} d_j a_{ji} = (\alpha_j, s_i(\alpha_k))$$

already give orthogonality of the operators s_i.

The Weyl group is the group W of invertible linear operators in $\mathfrak{h}^*_{\mathbb{R}}$ generated by the simple reflections $s_1, s_2, \dots, s_l$.

REMARK 2.2. It is easy to show that the finite subset $\Phi \subset E$ is an (abstract) root system. More precisely, it generates $\mathfrak{h}^*_{\mathbb{R}}$ and

(1) $s_\alpha(\beta) \in \Phi$ for all $\alpha, \beta \in \Phi$;
(2) $s_\alpha(\beta) - \beta \in \mathbb{Z}\alpha$ for all $\alpha, \beta \in \Phi$, where $s_\alpha(\lambda) = \lambda - 2\frac{(\lambda,\alpha)}{(\alpha,\alpha)}\alpha$.

A list of defining relations between the generators s_i of W is of the form [**44**, pp. 116, 184]

$$s_i^2 = 1, \quad (s_i s_j)^{m_{ij}} = 1, \quad i \neq j,$$

where

$$m_{ij} = \begin{cases} 2, & \text{if} \quad a_{ij} a_{ji} = 0, \\ 3, & \text{if} \quad a_{ij} a_{ji} = 1, \\ 4, & \text{if} \quad a_{ij} a_{ji} = 2, \\ 6, & \text{if} \quad a_{ij} a_{ji} = 3. \end{cases}$$

Consider the longest element w_0 of W. Choose a reduced expression

$$w_0 = s_{i_1} s_{i_2} s_{i_3} \cdots s_{i_M},$$

i.e., such that M is just the length of w_0. This decomposition determines the linear order on the set of positive roots Φ^+

$$\beta_1 = \alpha_{i_1}, \quad \beta_2 = s_{i_1}(\alpha_{i_2}), \dots, \quad \beta_M = s_{i_1} s_{i_2} \cdots s_{i_{M-1}}(\alpha_{i_M});$$

see [**44**, pp. 196–197]. To obtain q-analogs of the root vectors $E_{\beta_1}, E_{\beta_2}, \dots, E_{\beta_M}$ associated to these roots, we use the automorphisms $T_1, T_2, \dots, T_l$ of $U_q\mathfrak{g}$ introduced by Lusztig [**220**]. Their action on the generators is given by

$$T_i(K_j) = K_j K_i^{-a_{ij}}, \tag{2.10}$$

$$T_i(E_i) = -F_iK_i, \qquad T_i(F_i) = -K_i^{-1}E_i,$$
$$T_i(E_j) = \sum_{r+s=-a_{ij}} \frac{(-1)^s q_i^{-r}}{[s]_{q_i}![r]_{q_i}!} E_i^s E_j E_i^r, \qquad i \neq j,$$
$$T_i(F_j) = \sum_{r+s=-a_{ij}} \frac{(-1)^s q_i^{r}}{[s]_{q_i}![r]_{q_i}!} F_i^r F_j F_i^s, \qquad i \neq j.$$

Consider the adjoint representation of the Hopf algebra $U_q\mathfrak{g}$. By (2.10), T_i permutes the weight subspaces $T_i(U_q\mathfrak{g})_{\boldsymbol{\lambda}} = (U_q\mathfrak{g})_{s_i(\boldsymbol{\lambda})}$, where

$$(U_q\mathfrak{g})_{\boldsymbol{\lambda}} = \left\{ \xi \in U_q\mathfrak{g} \,\middle|\, K_i\xi K_i^{-1} = q_i^{\lambda_i}\xi \right\}, \qquad \boldsymbol{\lambda} = (\lambda_1, \lambda_2, \ldots, \lambda_l) \in \mathbb{Z}^l.$$

Lusztig demonstrated that these automorphisms are subject to the braid relations

$$\underbrace{T_iT_jT_i\cdots}_{m_{ij}} = \underbrace{T_jT_iT_j\cdots}_{m_{ij}}, \qquad i \neq j$$

(the number of factors in each side of this equality is m_{ij}).

Let us define *q-analogs of the root vectors* as follows:

$$E_{\beta_k} = \begin{cases} T_{i_1}T_{i_2}T_{i_3}\cdots T_{i_{k-1}}(E_{i_k}), & k \geq 2, \\ E_{i_1}, & k = 1, \end{cases}$$

$$F_{\beta_k} = \begin{cases} T_{i_1}T_{i_2}T_{i_3}\cdots T_{i_{k-1}}(F_{i_k}), & k \geq 2, \\ F_{i_1}, & k = 1. \end{cases}$$

The following result is again due to Lusztig.

PROPOSITION 2.3. *The monomials*

$$E_{\beta_1}^{j_1}E_{\beta_2}^{j_2}\cdots E_{\beta_M}^{j_M}, \qquad j_1, j_2, \ldots, j_M \in \mathbb{Z}_+,$$

form a basis in $U_q\mathfrak{n}^+$, and the monomials

$$F_{\beta_M}^{j_M}F_{\beta_{M-1}}^{j_{M-1}}\cdots F_{\beta_1}^{j_1}, \qquad j_1, j_2, \ldots, j_M \in \mathbb{Z}_+,$$

form a basis in $U_q\mathfrak{n}^-$.

To conclude, we present commutation relations obtained by Levendorskii and Soibelman [**71**, pp. 67–68].

PROPOSITION 2.4. (1) *For all $i < j$ one has*

$$E_{\beta_i}E_{\beta_j} - q^{(\beta_i,\beta_j)}E_{\beta_j}E_{\beta_i} = \sum_{\mathbf{m}\in\mathbb{Z}_+^M} C'_{\mathbf{m}}(q)\cdot E^{\mathbf{m}},$$
$$F_{\beta_i}F_{\beta_j} - q^{-(\beta_i,\beta_j)}F_{\beta_j}F_{\beta_i} = \sum_{\mathbf{m}\in\mathbb{Z}_+^M} C''_{\mathbf{m}}(q)\cdot F^{\mathbf{m}},$$

where $\mathbf{m} = (m_1, \ldots, m_M)$,

$$E^{\mathbf{m}} = E_{\beta_1}^{m_1}E_{\beta_2}^{m_2}\cdots E_{\beta_M}^{m_M}, \qquad F^{\mathbf{m}} = F_{\beta_M}^{m_M}F_{\beta_{M-1}}^{m_{M-1}}\cdots F_{\beta_1}^{m_1},$$

and the coefficients $C'_{\mathbf{m}}(q)$, $C''_{\mathbf{m}}(q)$ vanish unless $m_1 = m_2 = \cdots = m_i = 0$ and $m_j = m_{j+1} = \cdots = m_M = 0$.

(2) *$C'_{\mathbf{m}}(q), C''_{\mathbf{m}}(q) \in \mathbb{Q}[q, q^{-1}]$, that is $C'_{\mathbf{m}}(q)$ and $C''_{\mathbf{m}}(q)$ are Laurent polynomials with rational coefficients.*

Proposition 2.4 allows one to endow the algebra $U_q\mathfrak{g}$ with a filtration such that in the associated graded algebra one has

$$K_iK_j = K_jK_i, \qquad K_iK_{i-1} = K_{i-1}K_i = 1, \qquad i = 1, 2, \ldots, l,$$
$$E_{\beta_i}E_{\beta_j} = q^{(\beta_i,\beta_j)}E_{\beta_j}E_{\beta_i}, \qquad F_{\beta_i}F_{\beta_j} = q^{-(\beta_i,\beta_j)}F_{\beta_j}F_{\beta_i}, \qquad i < j.$$

COROLLARY 2.5. [**70**] *The algebra $U_q\mathfrak{g}$ is a domain.*

The following claim is a consequence of Proposition 2.3 and of the fact that the generators of the associated graded algebra quasi-commute.

COROLLARY 2.6. *The monomials*

$$E_{\beta_M}^{j_M}E_{\beta_{M-1}}^{j_{M-1}}\cdots E_{\beta_1}^{j_1}, \qquad j_1, j_2, \ldots, j_M \in \mathbb{Z}_+,$$

form a basis in $U_q\mathfrak{n}^+$, and the monomials

$$F_{\beta_1}^{j_1}F_{\beta_2}^{j_2}\cdots F_{\beta_M}^{j_M}, \qquad j_1, j_2, \ldots, j_M \in \mathbb{Z}_+,$$

form a basis in $U_q\mathfrak{n}^-$.

2.1.4. The universal R-matrix. Here we present generalizations of the results of Subsection 1.3.4 using the notation of this subsection.

Consider the Hopf subalgebras $U_q\mathfrak{b}^+$, $U_q\mathfrak{b}^-$ generated by $\{K_i^{\pm 1}, E_i\}_{i=1,2,\ldots,l}$, $\{K_i^{\pm 1}, F_i\}_{i=1,2,\ldots,l}$, respectively. The full subcategory of weight locally $U_q\mathfrak{b}^+$-finite-dimensional $U_q\mathfrak{g}$-modules is to be denoted by $\mathcal{C}^+$, and the full subcategory of weight locally $U_q\mathfrak{b}^-$-finite-dimensional $U_q\mathfrak{g}$-modules by $\mathcal{C}^-$.

PROPOSITION 2.7. [**306**, p. 946], [**139**, p. 114] *There exists a unique bilinear form $(\cdot,\cdot)$: $U_q\mathfrak{b}^+ \times U_q\mathfrak{b}^- \to \mathbb{C}$ such that*

(1) $(\xi'\xi'', \eta) = (\xi'' \otimes \xi', \Delta(\eta))$ *and* $(\xi, \eta'\eta'') = (\Delta(\xi), \eta' \otimes \eta'')$ *for all* $\xi, \xi', \xi'' \in U_q\mathfrak{b}^+$, $\eta, \eta', \eta'' \in U_q\mathfrak{b}^-$,
(2) $(K_\mu, K_\nu) = q^{-(\mu,\nu)}$ *for all* $\mu, \nu \in Q$,
(3) $(E_i, K_j^{\pm 1}) = (K_i^{\pm 1}, F_j) = 0$, $(E_i, F_j) = \delta_{ij}\frac{1}{q_i^{-1} - q_i}$ *for all* $i, j = 1, 2, \ldots, l$.

The first property listed in the formulation means that the bilinear form $(\cdot,\cdot)$ sets a duality between the Hopf algebras $(U_q\mathfrak{b}^-)^{\mathrm{cop}}$ and $U_q\mathfrak{b}^+$. (Recall that the superscript "cop" means that the comultiplication is replaced by its opposite.) This pairing is nondegenerate due to Proposition 2.8.

We recall the standard notation for weight subspaces

$$(U_q\mathfrak{n}^\pm)_\lambda = \left\{\xi \in U_q\mathfrak{n}^\pm \;\middle|\; \mathrm{ad}(K_i)\xi = q_i^{\lambda_i}\xi, \quad i = 1, 2, \ldots, l\right\}.$$

Here we use the conventional notation $\lambda = (\lambda_1, \lambda_2, \ldots, \lambda_l)$. Obviously,

$$U_q\mathfrak{n}^+ = \bigoplus_{\lambda \in Q_+}(U_q\mathfrak{n}^+)_\lambda, \qquad U_q\mathfrak{n}^- = \bigoplus_{\lambda \in -Q_+}(U_q\mathfrak{n}^-)_\lambda.$$

The definition of the bilinear form $(\cdot,\cdot)$ implies that it vanishes on $(U_q\mathfrak{n}^-)_\mu \times (U_q\mathfrak{n}^+)_\lambda$ for $\lambda + \mu \neq 0$. The next claim was proved initially for transcendental q, then for all q which are not roots of unit; see [**139**, p. 105].

PROPOSITION 2.8. *The restriction of the bilinear form $(\cdot,\cdot)$ to $(U_q\mathfrak{b}^-)_{-\lambda} \times (U_q\mathfrak{b}^+)_\lambda$ is nondegenerate for all $\lambda \in Q_+$.*

Let $\{\xi_i\}_{i=1}^{\infty}$ be a basis of weight vectors in $U_q\mathfrak{n}^+$, and let $\{\eta_i\}_{i=1}^{\infty}$ be the biorthogonal basis of weight vectors in $U_q\mathfrak{n}^-$: $(\eta_i, \xi_i) = \delta_{ij}$.

Suppose V', V'' are weight $U_q\mathfrak{g}$-modules and either $V' \in \mathcal{C}^+$ or $V'' \in \mathcal{C}^-$. Then the linear map

$$\Theta_{V',V''} : V' \otimes V'' \to V' \otimes V'', \qquad \Theta_{V',V''} : v' \otimes v'' \mapsto \sum_{i=1}^{\infty} \xi_i v' \otimes \eta_i v'',$$

is well defined. In fact, for all $v' \otimes v'' \in V' \otimes V''$ the series $\sum_{i=1}^{\infty} \xi_i v' \otimes \eta_i v''$ terminates, and its sum does not depend on the choice of a basis of weight vectors $\{\xi_i\}_{i=1}^{\infty}$.

Under the above assumptions, consider the linear map in $V' \otimes V''$ given by

$$R_{V',V''} = \Theta_{V',V''} \cdot q^{-\sum\limits_{i,j=1}^{l} c_{ij} d_i d_j H_i \otimes H_j},$$

where $\mathbf{c} = (c_{ij})_{i,j=1,2,\ldots,l}$ is the inverse matrix for $\mathbf{b} = (d_i a_{ij})_{i,j=1,2,\ldots,l}$.

Setting $\check{R}_{V',V''} = \sigma_{V'V''} R_{V',V''}$, one can prove claims which reproduce Propositions 1.40 and 1.41 literally and can be treated in the sense that $\check{R}_{V'V''}$ is a q-analog of the flip of tensor factors. Here are these claims.

PROPOSITION 2.9. (1) *Consider weight $U_q\mathfrak{g}$-modules V', V'' such that either $V' \in \mathcal{C}^+$ or $V'' \in \mathcal{C}^-$. The linear map*

$$\check{R}_{V',V''} : V' \otimes V'' \to V'' \otimes V'$$

is an invertible morphism of $U_q\mathfrak{g}$-modules.

(2) *If $f' : V' \to W'$, $f'' : V'' \to W''$ are morphisms of $U_q\mathfrak{g}$-modules and either $V', W' \in \mathcal{C}^+$ or $V'', W'' \in \mathcal{C}^-$, then one has*

$$(f'' \otimes f') \cdot \check{R}_{V',V''} = \check{R}_{W',W''} \cdot (f' \otimes f'').$$

PROPOSITION 2.10. (1) *If V, V', V'' are weight $U_q\mathfrak{g}$-modules and either $V', V'' \in \mathcal{C}^+$ or $V \in \mathcal{C}^-$, then one has*

$$\check{R}_{V' \otimes V'',V} = \left(\check{R}_{V',V} \otimes \mathrm{id}_{V''}\right)\left(\mathrm{id}_{V'} \otimes \check{R}_{V'',V}\right).$$

(2) *If V, V', V'' are weight $U_q\mathfrak{g}$-modules and either $V \in \mathcal{C}^+$ or $V', V'' \in \mathcal{C}^-$, then one has*

$$\check{R}_{V,V' \otimes V''} = \left(\mathrm{id}_{V'} \otimes \check{R}_{V,V''}\right)\left(\check{R}_{V,V'} \otimes \mathrm{id}_{V''}\right).$$

(3) *For each weight $U_q\mathfrak{g}$-module V one has*

$$\check{R}_{V,\mathbb{C}} = \check{R}_{\mathbb{C},V} = \mathrm{id}_V .$$

Let us turn to generalizing (1.38) produced before for the universal R-matrix of $U_q\mathfrak{sl}_2$. Choose a reduced expression of the longest element w_0 of the Weyl group W. We intend to use the q-analogs of root vectors E_{β_i}, F_{β_i} corresponding to that reduced expression as in Proposition 2.3. Let

$$q_\beta = q^{\frac{(\beta,\beta)}{2}}, \qquad \beta \in \Phi^+.$$

Consider the product

$$R = \prod_{\beta \in \Phi^+}^{\curvearrowleft} \exp_{q_\beta^2}\left(\left(q_\beta^{-1} - q_\beta\right) E_\beta \otimes F_\beta\right) q^{-\sum\limits_{i,j=1}^{l} c_{ij} d_i d_j H_i \otimes H_j}, \tag{2.11}$$

whose factors go in descending order of indices at the roots β:

$$\exp_{q^2_{\beta_M}}\left(\left(q^{-1}_{\beta_M}-q_{\beta_M}\right)E_{\beta_M}\otimes F_{\beta_M}\right)\exp_{q^2_{\beta_{M-1}}}\left(\left(q^{-1}_{\beta_{M-1}}-q_{\beta_{M-1}}\right)E_{\beta_{M-1}}\otimes F_{\beta_{M-1}}\right)$$
$$\cdots\exp_{q^2_{\beta_1}}\left(\left(q^{-1}_{\beta_1}-q_{\beta_1}\right)E_{\beta_1}\otimes F_{\beta_1}\right)q^{-\sum\limits_{i,j=1}^{l}c_{ij}d_id_jH_i\otimes H_j}.$$

PROPOSITION 2.11. **[215]**, **[171]** *Let V', V'' be weight $U_q\mathfrak{g}$-modules and either $V'\in\mathcal{C}^+$ or $V''\in\mathcal{C}^-$. The linear map in $V'\otimes V''$ given by the right-hand side of* (2.11) *is just $R_{V',V''}$.*

Some important additional properties of the universal R-matrix are presented in the works by Khoroshkin and Tolstoy **[165, 166]**.

REMARK 2.12. One can deduce from

$$\alpha_i(H_j)=a_{ij},\qquad(\alpha_i,\alpha_j)=d_ia_{ij}=d_ja_{ji},$$

and the definition of the matrix $\mathbf{c}=(c_{ij})_{i,j=1,2,\ldots,l}$ that

$$t_0=\sum_{i,j=1}^{l}c_{ij}d_id_jH_i\otimes H_j$$

is the canonical element in $\mathfrak{h}\otimes\mathfrak{h}$ corresponding to the scalar product $(\cdot,\cdot)$ in $\mathfrak{h}^*$:

$$(\lambda,\mu)=(\lambda\otimes\mu)(t_0),\qquad\lambda,\mu\in\mathfrak{h}^*.\tag{2.12}$$

Therefore, for arbitrary weight $U_q\mathfrak{g}$-modules V', V'' and weight vectors $v'\in V'_\lambda$, $v''\in V''_\mu$, one has

$$q^{-t_0}(v'\otimes v'')=q^{-(\lambda,\mu)}v'\otimes v''.\tag{2.13}$$

Given an orthogonal basis $\{I_j\}_{j=1,2,\ldots,l}$ in $\mathfrak{h}\cong\mathfrak{h}^*$, one has

$$t_0=\sum_k\frac{I_k\otimes I_k}{(I_k,I_k)}.\tag{2.14}$$

Arguments similar to those used in Subsection 1.3.4 allow one to pass from the ground field $\mathbb{C}$ and the scalar parameter q to the topological algebra $\mathbb{C}[[h]]$ endowed with the h-adic topology.

The topological Hopf algebra $U_h\mathfrak{g}$ is given by generators $\{H_i,X_i^\pm\}_{i=1,2,\ldots,l}$ and relations

$$H_iH_j-H_jH_i=0,\qquad H_iX_j^\pm-X_j^\pm H_i=\pm a_{ij}X_j^\pm,$$
$$X_i^+X_j^--X_j^-X_i^+=\delta_{ij}\frac{e^{d_ihH_i/2}-e^{-d_ihH_i/2}}{e^{d_ih/2}-e^{-d_ih/2}},$$
$$\sum_{s=0}^{1-a_{ij}}(-1)^s\begin{bmatrix}1-a_{ij}\\ s\end{bmatrix}_{e^{d_ih/2}}X_i^\pm X_j^\pm X_i^\pm=0,\qquad i\neq j,$$
$$\Delta(H_i)=H_i\otimes1+1\otimes H_i,\qquad\Delta(X_i^\pm)=X_i^\pm\otimes e^{d_ihH/4}+e^{-d_ihH/4}\otimes X_i^\pm,$$
$$\varepsilon(H_i)=\varepsilon(X_i^\pm)=0,\qquad S(H_i)=-H_i,$$
$$S(X_i^\pm)=-e^{\pm d_ih/2}X_i^\pm,\qquad i=1,2,\ldots,l.$$

REMARK 2.13. This Hopf algebra over $\mathbb{C}[[h]]$ is derived from the algebra $U_q\mathfrak{g}$ over $\mathbb{C}$ using the following *change of variables and the deformation parameter*:

$$q = e^{-h/2}, \qquad K_i^{\pm 1} = e^{\mp d_i h H_i/2}, \tag{2.15}$$

$$E_i = X_i^+ e^{-d_i h H_i/4}, \qquad F_i = e^{d_i h H_i/4} X_i^-. \tag{2.16}$$

Just as in the case of $U_q\mathfrak{g}$, one defines a pairing $(\cdot,\cdot)$ and a formal series R in the variable h with coefficients from $U\mathfrak{g} \otimes U\mathfrak{g}$, which works here as the universal R-matrix. An application of the standard embeddings $i_{1,2}, i_{1,3}, i_{2,3} : U\mathfrak{g} \otimes U\mathfrak{g} \to U\mathfrak{g} \otimes U\mathfrak{g} \otimes U\mathfrak{g}$ to the coefficients of this series produces the formal series $R^{1,2}$, $R^{1,3}$, $R^{2,3}$. Analogs of Propositions 2.9 and 2.10 are now consequences of the principal properties of the universal R-matrix:

$$\Delta^{\mathrm{cop}}(\xi) = R \cdot \Delta(\xi) \cdot R^{-1}, \quad \xi \in U_q\mathfrak{g}, \tag{2.17}$$

$$(\Delta \otimes \mathrm{id})(R) = R^{1,3} \cdot R^{2,3},$$

$$(\mathrm{id} \otimes \Delta)(R) = R^{1,3} \cdot R^{1,2}, \tag{2.18}$$

where Δ^{cop} is the opposite comultiplication, cf. (1.40)–(1.42). The product (2.11) is called the multiplicative formula for the universal R-matrix [**215, 171**].

Just as in Subsection 1.3.4, one introduces the antiautomorphism τ of the algebra $U_h\mathfrak{g}$,

$$\tau(H_i) = H_i, \qquad \tau(E_i) = e^{-\frac{hd_iH_i}{2}} F_i, \qquad \tau(F_i) = E_i e^{\frac{hd_iH_i}{2}},$$

and then one proves the relations which generalize (1.43),

$$(S \otimes S)R = R, \qquad (\tau \otimes \tau)R = R_{21}, \tag{2.19}$$

where R_{21} differs from R by the permutation of tensor factors; see the work of Drinfeld [**80**, pp. 34, 37].

Due to [**80**, Proposition 4.2], the passage from the original comultiplication to the opposite forces the universal R-matrix to be replaced by its inverse.

2.1.5. Verma modules. Let $\lambda = \sum_{j=1}^{l} \lambda_j \overline{\omega}_j \in \mathfrak{h}_{\mathbb{R}}^*$ and χ_λ be the one dimensional representation of the algebra $U_q\mathfrak{b}^+$ in the vector space $\mathbb{C}$,

$$\chi_\lambda : K_i^{\pm 1} \mapsto q_i^{\pm \lambda_i}, \qquad \chi_\lambda : E_i \mapsto 0, \qquad i = 1, 2, \ldots, l. \tag{2.20}$$

Consider the one-dimensional $U_q\mathfrak{b}^+$-module $\mathbb{C}_\lambda$ corresponding to this representation, and define the Verma module $M(\lambda)$ over $U_q\mathfrak{g}$ the same way as in the classical case $q = 1$: $M(\lambda) = U_q\mathfrak{g} \otimes_{U_q\mathfrak{b}^+} \mathbb{C}_\lambda$.

REMARK 2.14. One may drop the condition that the coordinates λ_j of the weight λ are real, but within the problems we are interested in, this condition always holds.

Let i_λ be the natural embedding of $U_q\mathfrak{b}^+$-modules

$$i_\lambda : \mathbb{C}_\lambda \to M(\lambda), \qquad i_\lambda : z \mapsto 1 \otimes z.$$

For each $U_q\mathfrak{g}$-module M and each morphism of $U_q\mathfrak{b}^+$-modules $f : \mathbb{C}_\lambda \to M$ there exists a unique morphism of $U_q\mathfrak{g}$-modules $\widetilde{f} : M(\lambda) \to M$ such that $f = \widetilde{f} \cdot i_\lambda$. This property of the Verma module $M(\lambda)$ is referred to as its universality.

Let $v(\lambda) = 1 \otimes 1$. The module $M(\lambda)$ admits a description in terms of the generator $v(\lambda)$ and relations

$$E_i v(\lambda) = 0, \qquad K_i^{\pm 1} v(\lambda) = q_i^{\pm \lambda_i} v(\lambda), \qquad i = 1, 2, \dots, l.$$

The weight vectors

$$v_J(\lambda) = F_{\beta_M}^{j_M} F_{\beta_{M-1}}^{j_{M-1}} \cdots F_{\beta_1}^{j_1} v(\lambda), \qquad J = (j_1, j_2, \dots, j_M) \in \mathbb{Z}_+^M,$$

form a basis of the vector space $M(\lambda)$ by Proposition 2.3. Hence $M(\lambda)$ is a weight $U_q\mathfrak{g}$-module, whose set of weights is $\lambda - Q_+$ and the dimensions of weight subspaces are as in the classical case $q = 1$, namely,

$$\dim M(\lambda)_{\lambda - \mu} = \operatorname{card}\left\{ (j_1, j_2, \dots, j_M) \in \mathbb{Z}_+^M \,\middle|\, \sum_{i=1}^{M} j_i \beta_i = \mu \right\}.$$

Recall that the set of weights is partially ordered. As a consequence of the definitions, one has that $M(\lambda)$ is a $U_q\mathfrak{g}$-module with highest weight λ. Another consequence is that the highest weight subspace (containing $\mathbb{C}v(\lambda)$) is one dimensional.

Let us endow $U_q\mathfrak{g}$ with the involution

$$(K_j^{\pm 1})^\star = K_j^{\pm 1}, \qquad E_j^\star = K_j F_j, \qquad F_j^\star = E_j K_j^{-1}, \qquad j = 1, 2, \dots, l, \tag{2.21}$$

which generalizes (1.21). The Hopf $*$-algebra produced this way is a q-analog of the Hopf $*$-algebra $U_q\mathfrak{g}_0 \otimes_{\mathbb{R}} \mathbb{C}$, where $\mathfrak{g}_0$ is the compact real form of the simple complex Lie algebra $\mathfrak{g}$ [**47**, p. 24].

Recall that we did not include the nonnegativity assumption to the definition of Hermitian form, and the $U_q\mathfrak{g}$-invariance of a Hermitian form in $M(\lambda)$ should look as follows:

$$(\xi v', v'') = (v', \xi^\star v''), \qquad v', v'' \in M(\lambda), \qquad \xi \in U_q\mathfrak{g};$$

see Subsections 1.4.1 and 1.5.3.

There exists a unique invariant Hermitian form $(\cdot, \cdot)$ on $M(\lambda)$ such that $(v(\lambda), v(\lambda)) = 1$, which can easily be deduced using the isomorphism of vector spaces $U_q\mathfrak{n}^- \otimes U_q\mathfrak{h} \otimes U_q\mathfrak{n}^+ \approx U_q\mathfrak{g}$ described in Subsection 2.1.3. This isomorphism allows one to introduce the linear map

$$\mathbf{h} : \xi\eta\zeta \mapsto \varepsilon(\xi)\eta\varepsilon(\zeta), \qquad \xi \in U_q\mathfrak{n}^-, \quad \eta \in U_q\mathfrak{h}, \quad \zeta \in U_q\mathfrak{n}^+,$$

from $U_q\mathfrak{g}$ to $U_q\mathfrak{h}$ and to obtain the relation $(\xi v(\lambda), \eta v(\lambda)) = \chi_\lambda(\mathbf{h}(\eta^\star \xi))$; see (2.20).

The weight subspaces $M(\lambda)_\mu$ of the Verma module $M(\lambda)$ are pairwise orthogonal. Every proper submodule of $M(\lambda)$ is weight and is orthogonal to $\mathbb{C}v(\lambda)$, hence it is inside the kernel of the invariant Hermitian form. Therefore this kernel $K(\lambda)$ is the largest nontrivial submodule of $M(\lambda)$, and the quotient module $L(\lambda) = M(\lambda)/K(\lambda)$ is simple and carries a nondegenerate Hermitian form with the property

$$(\xi v', v'') = (v', \xi^\star v''), \qquad v', v'' \in L(\lambda), \quad \xi \in U_q\mathfrak{g}.$$

REMARK 2.15. It is easy to show that $K(\lambda)$ is a unique submodule of the $U_q\mathfrak{g}$-module $M(\lambda)$ whose codimension is finite. In fact, suppose $\dim(M(\lambda)/\mathcal{K}) < \infty$ for some submodule $\mathcal{K} \subset K(\lambda)$. The central characters of the $U_q\mathfrak{g}$-modules $M(\lambda)/\mathcal{K}$ and $L(\lambda)$ coincide. Hence the $U_q\mathfrak{g}$-module $M(\lambda)/\mathcal{K}$ is a multiple of $L(\lambda)$; see Subsections 2.1.6 and 2.1.8. But the multiplicity here is one since $\dim(M(\lambda)/\mathcal{K})_\lambda \leq \dim M(\lambda)_\lambda = 1$. Thus $\mathcal{K} = K(\lambda)$.

Obviously, $L(\lambda) = M(\lambda)$ iff the Gram matrix $\mathbf{g}(\lambda, \mu)$ formed by pairwise scalar products of the vectors

$$\left\{ v_J(\lambda) \,\middle|\, J = (j_1, j_2, \dots, j_M) \in \mathbb{Z}_+^M \quad \text{and} \quad \sum_{i=1}^{M} j_i \beta_i = \mu \right\}$$

is nondegenerate for all $\mu \in Q_+$. One has a q-analog of the well-known result by Shapovalov [**273**].

PROPOSITION 2.16. [**137**, p. 129], [**70**, p. 484] *Up to a nonzero scalar factor, the determinant of the Gram matrix* $\mathbf{g}(\lambda, \mu)$ *has the form*

$$\chi_\lambda \left(\prod_{\beta \in \Phi^+} \prod_{m=1}^{\infty} \left([m]_{q_\beta} \frac{K_\beta \, q^{(\rho - \frac{m}{2}\beta, \beta)} - K_\beta^{-1} q^{-(\rho - \frac{m}{2}\beta, \beta)}}{q^{\frac{(\beta,\beta)}{2}} - q^{-\frac{(\beta,\beta)}{2}}} \right)^{P(\mu - m\beta)} \right),$$

where $q_\beta = q^{\frac{(\beta,\beta)}{2}}$ *and* $[m]_{q_\beta} = \frac{q_\beta^m - q_\beta^{-m}}{q_\beta - q_\beta^{-1}}$.

REMARK 2.17. Given $\lambda = \sum_j \lambda_j \overline{\omega}_j$, $\beta = \sum_i n_i \alpha_i$, it follows from (2.8) that

$$\chi_\lambda \left(K_\beta^{\pm 1} \right) = q^{\pm \sum\limits_i d_i n_i \lambda_i} = q^{\pm \sum\limits_{i,j} (\alpha_i, \overline{\omega}_j) n_i \lambda_j} = q^{\pm(\lambda, \beta)}.$$

This allows one to rewrite the determinant of the Gram matrix as

$$\text{const} \cdot \prod_{\beta \in \Phi^+} \prod_{m=1}^{\infty} \left([m]_{q_\beta} \frac{q^{(\lambda + \rho - \frac{m}{2}\beta, \beta)} - q^{-(\lambda + \rho - \frac{m}{2}\beta, \beta)}}{q^{\frac{(\beta,\beta)}{2}} - q^{-\frac{(\beta,\beta)}{2}}} \right)^{P(\mu - m\beta)},$$

where const $\neq 0$.

COROLLARY 2.18. *The Verma module* $M(\lambda)$ *is simple iff* $\frac{2(\lambda+\rho,\beta)}{(\beta,\beta)} \notin \mathbb{N}$ *for all* $\beta \in \Phi^+$.

2.1.6. Central elements of the algebra $U_q^{\text{ext}}\mathfrak{g}$. In Subsection 2.1.2 to every $\lambda \in Q$ we associated the element

$$K_\lambda = K_1^{a_1} K_2^{a_2} \cdots K_l^{a_l}, \qquad \lambda = \sum_{i=1}^{l} a_i \alpha_i. \tag{2.22}$$

Obviously,

$$K_\lambda K_\mu = K_{\lambda+\mu}, \qquad K_0 = 1. \tag{2.23}$$

Since $(\alpha_i, \alpha_j) = d_i a_{ij}$, one has

$$K_\lambda E_j = q^{(\lambda, \alpha_j)} E_j K_\lambda, \qquad K_\lambda F_j = q^{-(\lambda, \alpha_j)} F_j K_\lambda. \tag{2.24}$$

The elements K_λ corresponding to all $\lambda \in P$ can be helpful in solving some problems of quantum group theory. In this case a_i in (2.22) are rational but not necessarily integers, which requires the Hopf algebra $U_q\mathfrak{g}$ to be extended.

We follow [**174**, p. 281], [**214**, p. 73] in introducing the algebra $U_q^{\text{ext}}\mathfrak{g}$ given by its generators

$$E_i, F_i, \quad i = 1, 2, \dots, l; \qquad K_\lambda, \quad \lambda \in P,$$

and the relations which are derived from those for $U_q\mathfrak{g}$ via replacing (2.1) with (2.23), (2.24). Obviously, $U_q^{\text{ext}}\mathfrak{g}$ is a Hopf algebra:

$$\triangle(K_\lambda) = K_\lambda \otimes K_\lambda, \qquad S(K_\lambda) = K_{-\lambda}, \qquad \varepsilon(K_\lambda) = 1,$$

and $U_q\mathfrak{g}$ is its Hopf subalgebra.

Just as in Subsection 2.1.2, we introduce the notion of weight $U_q^{\text{ext}}\mathfrak{g}$-module V:

$$V = \bigoplus_\mu V_\mu, \qquad V_\mu = \left\{ v \in V \,\middle|\, K_\lambda v = q^{(\lambda,\mu)} v, \quad \lambda \in P \right\}.$$

Obviously, the category of weight $U_q^{\text{ext}}\mathfrak{g}$-modules is canonically isomorphic to the category of weight $U_q\mathfrak{g}$-modules. By some abuse of notation, one may reformulate this by saying that both algebras $U_q^{\text{ext}}\mathfrak{g}$ and $U_q\mathfrak{g}$ correspond to the same quantum group.

Here is a description of the center $Z(U_q^{\text{ext}}\mathfrak{g})$ of $U_q^{\text{ext}}\mathfrak{g}$. Both the linear span $U_q^{\text{ext}}\mathfrak{h}$ of $\{K_\lambda\}_{\lambda\in P}$ and the linear span $U_q^{\text{even}}\mathfrak{h}$ of $\{K_\lambda\}_{\lambda\in 2P}$ are commutative subalgebras. The linear map

$$U_q\mathfrak{n}^- \otimes U_q^{\text{ext}}\mathfrak{h} \otimes U_q\mathfrak{n}^+ \to U_q^{\text{ext}}\mathfrak{g}, \qquad \xi\otimes\eta\otimes\zeta \mapsto \xi\eta\zeta,$$

is one-to-one (cf. the similar result in Subsection 2.1.3). This allows one to consider the linear map

$$\gamma' : U_q^{\text{ext}}\mathfrak{g} \to U_q^{\text{ext}}\mathfrak{h}, \qquad \gamma' : \xi\eta\zeta \mapsto \varepsilon(\xi)\eta\varepsilon(\zeta),$$

where $\xi \in U_q\mathfrak{n}^-$, $\eta \in U_q^{\text{ext}}\mathfrak{h}$, $\zeta \in U_q\mathfrak{n}^+$. Just as in the classical case $q = 1$ [**74**, p. 268], the map γ' is proved to be a morphism of algebras.

Recall the representation-theoretic meaning of this morphism. Let $z \in Z(U_q^{\text{ext}}\mathfrak{g})$. The algebra $U_q^{\text{ext}}\mathfrak{h}$ admits a natural embedding into the algebra of functions on $\mathfrak{h}_\mathbb{R}^*$:

$$K_\mu : \lambda \mapsto q^{(\mu,\lambda)}, \qquad \mu \in P.$$

This embedding takes $\gamma'(z)$ to the function which determines the action of z in the Verma module $M(\lambda)$. Introduce the isomorphism:

$$\tau : U_q^{\text{ext}}\mathfrak{h} \xrightarrow{\approx} U_q^{\text{ext}}\mathfrak{h}, \qquad \tau : K_i^{\pm 1} \mapsto q_i^{\mp 1} K_i^{\pm 1}.$$

The next claim is a q-analog of the well-known Harish-Chandra theorem which describes the center of $U\mathfrak{g}$ [**74**, p. 269]. This claim involves the natural action of W in $U_q^{\text{ext}}\mathfrak{h}$:

$$wK_\lambda \stackrel{\text{def}}{=} K_{w^{-1}\lambda}, \qquad \lambda \in P, \quad w \in W.$$

PROPOSITION 2.19. [**306**, p. 951] *The restriction of the morphism* $\gamma \stackrel{\text{def}}{=} \tau\gamma'$ *to* $Z(U_q^{\text{ext}}\mathfrak{g})$ *is injective, and its image is the subalgebra of all* W*-invariant elements of* $U_q^{\text{even}}\mathfrak{h}$.

The following method of producing central elements of $U_q^{\text{ext}}\mathfrak{g}$ is due to Drinfeld. According to (2.12), the isomorphism

$$\mathfrak{h}^* \xrightarrow{\approx} \mathfrak{h}, \qquad \mu \mapsto \mathrm{id}\otimes\mu(t_0), \tag{2.25}$$

is inverse to (2.9). Let H_μ be its value at μ. Using the notation of Subsection 2.1.2 and (2.9), we obtain

$$H_{\alpha_i} = d_i H_i, \qquad K_{\alpha_i}^{\pm 1} = K_i^{\pm 1} = q_i^{\pm H_i} = q^{\pm H_{\alpha_i}}.$$

Hence,

$$K_{\overline{\omega}_i}^{\pm 1} = q^{\pm H_{\overline{\omega}_i}}, \qquad i = 1, 2, \ldots, l. \tag{2.26}$$

REMARK 2.20. Let V be a finite-dimensional weight $U_q^{\mathrm{ext}}\mathfrak{g}$-module and π_V the corresponding representation of $U_q^{\mathrm{ext}}\mathfrak{g}$. It follows from (2.25), (2.26) that the element $\mathrm{id}\otimes\pi_V(q^{-t_0})$ of $U_q^{\mathrm{ext}}\mathfrak{g}\otimes\mathrm{End}\,V$ is well defined. The associated linear map from V to $U_q^{\mathrm{ext}}\mathfrak{g}\otimes V$ takes the weight vector v_μ of weight μ to $K_{-\mu}\otimes v_\mu$.

Let R be the universal R-matrix (2.11), and R_{21} differs from R by the flip of tensor factors:

$$R_{21}=\overset{\curvearrowleft}{\prod_{\beta\in\Phi^+}}\exp_{q_\beta^2}\left(\left(q_\beta^{-1}-q_\beta\right)F_\beta\otimes E_\beta\right)q^{-t_0}.$$

To every finite-dimensional weight $U_q^{\mathrm{ext}}\mathfrak{g}$-module V, we assign the linear functional

$$\mathrm{tr}_q:\mathrm{End}\,V\to\mathbb{C},\qquad \mathrm{tr}_q:A\mapsto\mathrm{tr}(A\pi_V(K_{-2\rho})),$$

where π_V is the representation of $U_q^{\mathrm{ext}}\mathfrak{g}$ corresponding to V. Just as in Subsection 2.1.2, one proves that this linear functional is $U_q^{\mathrm{ext}}\mathfrak{g}$-invariant.

PROPOSITION 2.21. *The element $C_V=(\mathrm{id}\otimes\mathrm{tr}_q\,\pi_V)(R_{21}R)$ belongs to the center of $U_q^{\mathrm{ext}}\mathfrak{g}$.*

PROOF. This result is known in the context of formal series in h. Namely, it was demonstrated in [**80**, p. 32] that the element $\mathrm{id}\otimes l(R_{21}R)$ is central if the linear functional l on a Hopf algebra is chosen so that $l(\xi\eta)=l(\eta S^2(\xi))$ for all elements ξ, η of this algebra. It remains to set $l(\xi)=\mathrm{tr}_q\,\pi_V(\xi)$. To pass from $q\in(0,1)$ to the formal parameter h in

$$C_V\cdot\xi=\xi\cdot C_V,\qquad \xi\in U_q^{\mathrm{ext}}\mathfrak{g},$$

it suffices to expand both sides of this equality in the Poincaré–Birkhoff–Witt basis of $U_q^{\mathrm{ext}}\mathfrak{g}$ (see Subsection 2.1.3). After substituting $q=e^{-h/2}$, every coefficient of this expansion becomes an analytic function on the half-axis $h>0$ and in a neighborhood of $h=0$ (see Proposition 2.4). It remains to prove that all the derivatives of those functions vanish at $h=0$. This fact follows from the relation $C_V\xi=\xi C_V$ in the context of formal series. □

The following result provides a description of the image of C_V under the homomorphism γ'.

PROPOSITION 2.22. *Let $M(\lambda)$ be the Verma module with highest weight λ and V a finite-dimensional weight $U_q^{\mathrm{ext}}\mathfrak{g}$-module. Then*

$$C_V|_{M(\lambda)}=\sum_{\mu\in P}(\dim V_\mu)q^{-2(\mu,\lambda+\rho)},\tag{2.27}$$

where ρ is half the sum of all positive roots.

PROOF. The relation (2.27) is known in the context of formal series [**80**, p. 39]. The passage from $q\in(0,1)$ to the formal parameter h can be done the same way as in the proof of the previous proposition. □

COROLLARY 2.23. *For each finite-dimensional weight $U_q^{\mathrm{ext}}\mathfrak{g}$-module V one has*

$$\gamma(C_V)=\sum_{\mu\in P}(\dim V_\mu)K_{-2\mu}.\tag{2.28}$$

PROOF. The right-hand side of (2.28) is

$$\sum_{\mu \in P}(\dim V_\mu)q^{-2(\mu\otimes(\lambda+\rho))t_0} = \sum_{\mu \in P}(\dim V_\mu)q^{-2(\lambda+\rho)(H_\mu)}. \qquad \square$$

Consider the finite-dimensional simple weight $U_q\mathfrak{g}$-module $L(\overline{\omega}_i)$ with highest weight $\overline{\omega}_i$, $i = 1, 2, \ldots, l$. We keep the notation $L(\overline{\omega}_i)$ for the corresponding $U_q^{\mathrm{ext}}\mathfrak{g}$-module.

COROLLARY 2.24. *The set* $\{C_{L(\overline{\omega}_i)}\}_{i=1,2,\ldots,l}$ *generates the center* $Z(U_q^{\mathrm{ext}}\mathfrak{g})$ *of* $U_q^{\mathrm{ext}}\mathfrak{g}$.

PROOF. We proceed via the standard argument [**139**, p. 125] which uses Corollary 2.23 and Proposition 2.19. Namely, it suffices to prove that the following subalgebras of $U_q^{\mathrm{ext}}\mathfrak{h}$ coincide:

A_1, the subalgebra generated by $\sum_\mu \dim L(\overline{\omega}_i)K_{-2\mu}$, $i = 1, 2, \ldots, l$;
A_2, the subalgebra generated by $\sum_\mu \dim L(\lambda)_\mu K_{-2\mu}$, $\lambda \in P_+$;
A_3, the subalgebra generated by $\sum_{\mu \in W\lambda} K_{-2\mu}$, $\lambda \in P_+$;
A_4, the algebra of W-invariants of the linear span of $K_{-2\lambda}$, $\lambda \in P_+$;
A_5, $\gamma(Z(U_q^{\mathrm{ext}}\mathfrak{g}))$.

The claim $A_1 = A_2$ is true because the ring of finite-dimensional weight representations of $U_q^{\mathrm{ext}}\mathfrak{g}$ is canonically isomorphic to the ring of finite-dimensional representations of $U\mathfrak{g}$; hence it is generated by the fundamental representations [**47**, p. 175].

To prove the claim $A_2 = A_3$, it suffices to reproduce the argument of [**139**, p. 125]. In the present context it is convenient to proceed by induction in (λ, λ), $\lambda \in P_+$, and to use the fact that for each $\lambda \in P_+$ and each weight $\mu \in P_+$ of $L(\lambda)$, $\mu \neq \lambda$, one has $(\mu, \mu) < (\lambda, \lambda)$ [**175**, p. 234].

The claim $A_3 = A_4$ follows from [**44**, p. 232], and the claim $A_5 = A_6$ is due to Proposition 2.19. $\square$

It is easy to prove that the elements $C_{L(\overline{\omega}_i)}$, $i = 1, 2, \ldots, l$, are algebraically independent, hence the center $Z(U_q^{\mathrm{ext}}\mathfrak{g})$ of $U_q^{\mathrm{ext}}\mathfrak{g}$ is isomorphic to the polynomial algebra in l variables, just as in the classical case $q = 1$.

REMARK 2.25. The center $Z(U_q\mathfrak{g})$ of $U_q\mathfrak{g}$ is a bit more complicated and is the linear span of $C_{L(\lambda)}$, $\lambda \in P_+ \cap \frac{1}{2}Q$; see [**139**, p. 109]. It is easy to prove that for any $\mu_1, \mu_2 \in P_+$ there exists $z \in Z(U_q\mathfrak{g})$ which separates the $U_q\mathfrak{g}$-modules $L(\mu_1), L(\mu_2)$; see [**139**, p. 125–126].

2.1.7. Reducing to the case $q = 1$. Let us sketch a method of reducing problems on $U_q\mathfrak{g}$-modules to corresponding problems on $U\mathfrak{g}$-modules. This method is especially powerful under the additional assumption of q being transcendental [**139**, Chapters 5–7], and the passage to the general case requires some more effort [**139**, Chapter 8].

The equality (2.13) indicates that the formulas related to $U_q\mathfrak{g}$-modules could involve fractional powers of q. Let s be the number of elements in P/Q. As a consequence of (2.8) one has $(\lambda, \mu) \in \frac{1}{s}\mathbb{Z}$ for all $\lambda, \mu \in P$. Consider the field $\mathbb{C}(t^{1/s})$ of rational functions of $t^{1/s}$ and the Hopf algebra $U_t\mathfrak{g}$ over this field given by the same generators and relations as $U_q\mathfrak{g}$, but with the scalar parameter q being replaced by the variable t. This substitution allows one to obtain the elements

E_{β_j}, F_{β_j}, $j = 1, 2, \ldots, M$, of $U_t\mathfrak{g}$, the Verma modules $M(\lambda)$ over this algebra, their simple quotient modules $L(\lambda)$ and the linear maps $\check{R}_{L(\lambda),L(\mu)}$ for $\lambda, \mu \in P$. The assumption $\lambda \in P$ will be implicit within this subsection.

Let $\mathcal{A} = \mathbb{Q}[t^{1/s}, t^{-1/s}]$ be the algebra of Laurent polynomials in $t^{1/s}$ with rational coefficients.

The principal observation here is that a large number of properties of $U_t\mathfrak{g}$-modules admit formulations and can be proved in terms of distinguished $\mathcal{A}$-submodules. Indeed, consider the $\mathcal{A}$-subalgebra $U_{\mathcal{A}}$ in $U_t\mathfrak{g}$ generated by $K_i^{\pm 1}, E_i, F_i, h_i = \frac{K_i - K_i^{-1}}{t_i - t_i^{-1}}$, $i = 1, 2, \ldots, l$, where $t_i = t^{d_i}$. It is known from [**70**] that a list of defining relations between these generators can be produced in two steps. First, one should replace the relation $E_iF_j - F_jE_i = \delta_{ij}\frac{K_i - K_i^{-1}}{t_i - t_i^{-1}}$ in the standard list of relations for $U_t\mathfrak{g}$ by $E_iF_j - F_jE_i = \delta_{ij}h_i$. Second, one adds the relations

$$(t_i - t_i^{-1})h_i = K_i - K_i^{-1}, \qquad i = 1, 2, \ldots, l.$$

Of course, the $\mathcal{A}$-subalgebra $U_{\mathcal{A}}$ inherits the structure of Hopf algebra:

$$\Delta(h_i) = h_i \otimes K_i + K_i^{-1} \otimes h_i, \qquad S(h_i) = -h_i, \qquad \varepsilon(h_i) = 0, \qquad i = 1, 2, \ldots, l.$$

Let v_λ be a highest weight vector of the $U_t\mathfrak{g}$-module $M(\lambda)$. Then the vectors $F_{\beta_\mu}^{j_\mu} F_{\beta_{\mu-1}}^{j_{\mu-1}} \cdots F_1^{j_1} v_\lambda$ generate the $\mathcal{A}$-module $M_{\mathcal{A}}(\lambda) = U_{\mathcal{A}} \cdot v_\lambda$. We denote the image of $M_{\mathcal{A}}(\lambda)$ under the canonical projection $M(\lambda) \to L(\lambda)$ by $L_{\mathcal{A}}(\lambda)$. Obviously, $M_{\mathcal{A}}(\lambda)$ and $L_{\mathcal{A}}(\lambda)$ are $U_{\mathcal{A}}$-modules.

Let us describe two morphisms of algebras. The first of them allows one to transfer the properties of $U\mathfrak{g}$-modules to $U_{\mathcal{A}}$-modules, while the second transfers the properties of $U_{\mathcal{A}}$-modules to $U_q\mathfrak{g}$-modules. Consider the morphism of $\mathbb{Q}$-algebras

$$\mathbb{Q}\left[t^{1/s}, t^{-1/s}\right] \to \mathbb{C}, \qquad t^{\pm 1/s} \mapsto 1.$$

It admits a unique extension to a morphism of $\mathbb{Q}$-algebras $U_{\mathcal{A}} \to U\mathfrak{g}$ with the properties

$$K_i^{\pm 1} \mapsto 1, \quad E_i \mapsto E_i, \quad F_i \mapsto F_i, \quad h_i \mapsto H_i, \qquad i = 1, 2, \ldots, l.$$

This morphism makes it possible to apply the classical theory of $U\mathfrak{g}$-modules in studying $U_{\mathcal{A}}$-modules. Furthermore, consider the morphism of $\mathbb{Q}$-algebras

$$\mathbb{Q}\left[t^{1/s}, t^{-1/s}\right] \to \mathbb{C}, \qquad t^{\pm 1/s} \mapsto q^{\pm 1/s}.$$

There exists a unique extension of this morphism to a morphism of $\mathbb{Q}$-algebras $U_{\mathcal{A}} \to U_q\mathfrak{g}$ such that

$$K_i^{\pm 1} \mapsto K_i^{\pm 1}, \quad E_i \mapsto E_i, \quad F_i \mapsto F_i, \quad h_i \mapsto \frac{K_i - K_i^{-1}}{q_i - q_i^{-1}}, \qquad i = 1, 2, \ldots, l.$$

This morphism is injective for q transcendental; it allows one to transfer the properties of $U_{\mathcal{A}}$-modules to $U_q\mathfrak{g}$-modules.

For example, it is known that

$$S \otimes S(R) = R;$$

see [**80**, p. 34]. With the notation $q = e^{-h/2}$, $\Theta = Rq^{t_0}$, this relation can be rewritten as

$$S \otimes S(\Theta) = q^{-t_0}\Theta q^{t_0}. \tag{2.29}$$

Using transfers from $U_q\mathfrak{g}$ to $U_{\mathcal{A}}\mathfrak{g}$ and from $U_{\mathcal{A}}\mathfrak{g}$ to $U_h\mathfrak{g}$, it is easy to deduce from (2.29) that

$$R^*_{V_1,V_2}(l_1 \otimes l_2) = R_{V_1^*,V_2^*}(l_1 \otimes l_2), \qquad l_1 \in V_1^*, \quad l_2 \in V_2^*,$$

for a wide class of weight $U_q\mathfrak{g}$-modules V_1, V_2.

2.1.8. Locally finite-dimensional $U_q\mathfrak{g}$-modules. The following results are well known [**139**].

PROPOSITION 2.26. (1) *The simple weight $U_q\mathfrak{g}$-modules $L(\lambda)$, $\lambda \in P_+$, are finite dimensional and pairwise nonisomorphic.*

(2) *Every simple weight finite-dimensional $U_q\mathfrak{g}$-module is isomorphic to some $U_q\mathfrak{g}$-module $L(\lambda)$, $\lambda \in P_+$.*

We keep the notation $v(\lambda)$ for the image of the generator of $M(\lambda)$ under the canonical projection $M(\lambda) \to L(\lambda)$.

PROPOSITION 2.27. *Let $\lambda = \sum_{j=1}^{l} \lambda_j \overline{\omega}_j \in P_+$. The element $v(\lambda)$ generates the $U_q\mathfrak{g}$-module $L(\lambda)$, and the following relations are defining:*

$$E_i v(\lambda) = 0, \qquad K_i^{\pm 1} v(\lambda) = q_i^{\pm \lambda_i} v(\lambda), \qquad F_i^{\lambda_i + 1} v(\lambda) = 0, \tag{2.30}$$

where $i = 1, 2, \ldots, l$.

REMARK 2.28. The relations (2.30) provide another equivalent definition of the $U_q\mathfrak{g}$-module $L(\lambda)$ for $\lambda \in P_+$.

PROPOSITION 2.29. (cf. [**139**, pp. 76–77]) Given a nonzero $\xi \in U_q\mathfrak{g}$, there exists $\lambda \in P_+ \bigcap Q$ such that $\xi L(\lambda) \neq 0$.

PROPOSITION 2.30. [**139**, p. 81] *The collection of weights of $L(\lambda)$ and their multiplicities do not change under the passage from $U\mathfrak{g}$ to $U_q\mathfrak{g}$, that is under quantization.*

PROPOSITION 2.31. [**139**, p. 82] *Every weight finite-dimensional $U_q\mathfrak{g}$-module is semisimple; that is, the corresponding representation of $U_q\mathfrak{g}$ is completely reducible.*

A consecutive application of Propositions 2.31, 2.26, 2.30 allows one to obtain

COROLLARY 2.32. *For all $\lambda, \mu \in P_+$ one has*

$$L(\lambda) \otimes L(\mu) \approx \sum_{\nu \in P_+} c^{\nu}_{\lambda\mu} L(\nu),$$

and the multiplicities $c^{\nu}_{\lambda\mu}$ of $L(\lambda)$ in $L(\lambda) \otimes L(\mu)$ are the same as in the classical case $q = 1$.

Let $P^{\nu}_{\lambda\mu}$ be the projection in $L(\lambda) \otimes L(\mu)$ onto the isotypic component which is a multiple of $L(\nu)$, parallel to the sum of all other isotypic components.

PROPOSITION 2.33. [**80**, p. 39] *For all $\lambda, \mu \in P_+$, one has*

$$\check{R}_{L(\mu),L(\lambda)} \check{R}_{L(\lambda),L(\mu)} = \bigoplus_{\nu \in P_+} q^{(\lambda,\lambda+2\rho)+(\mu,\mu+2\rho)-(\nu,\nu+2\rho)} P^{\nu}_{\lambda\mu}. \tag{2.31}$$

Let $\mathcal{C}$ be the full subcategory of all weight locally finite-dimensional $U_q\mathfrak{g}$-modules.

PROPOSITION 2.34. *Every $U_q\mathfrak{g}$-module $V \in \mathcal{C}$ admits a unique decomposition as a direct sum $V = \bigoplus_{\lambda \in P_+} V(\lambda)$ of its submodules $V(\lambda)$ that are multiples of $L(\lambda)$.*

PROOF. Let $\lambda \in P_+$. Consider the vector space

$$X(\lambda) = \left\{ v \in V \,\middle|\, E_j v = 0, \quad K_j^{\pm 1} v = q_j^{\pm \lambda_j} v, \quad j = 1, 2, \ldots, l \right\},$$

together with the $U_q\mathfrak{g}$-module $V(\lambda) = U_q\mathfrak{g} \cdot X(\lambda)$. It suffices to prove that $V(\lambda)$ is the largest among the submodules of V which are multiple to $L(\lambda)$, and that $V = \bigoplus_{\lambda \in P_+} V(\lambda)$. Given a nonzero vector $v_\lambda \in X(\lambda)$, there exists a unique embedding of $U_q\mathfrak{g}$-modules

$$L(\lambda) \hookrightarrow V, \qquad v(\lambda) \mapsto v_\lambda$$

(the uniqueness is obvious, and the existence can be easily proved using the universality property of the Verma module $M(\lambda)$ and Remark 2.15). Thus, every basis of the vector space $X(\lambda)$ determines a decomposition of $V(\lambda)$ as a direct sum of modules multiple to $L(\lambda)$. The decomposition $V = \bigoplus_{\lambda \in P_+} V(\lambda)$ follows from the above properties of finite-dimensional $U_q\mathfrak{g}$-modules. □

COROLLARY 2.35. *Every $U_q\mathfrak{g}$-module in $\mathcal{C}$ is semisimple.*

The category $\mathcal{C}$ is invariant under the passage to submodules, quotient modules, and tensor products.

Consider the functor F from the category of all $U_q\mathfrak{g}$-modules to the category $\mathcal{C}$ which takes every $U_q\mathfrak{g}$-module V to its largest submodule in $\mathcal{C}$. The action on morphisms is defined in an obvious way. The dual object V^* in $\mathcal{C}$ is the result of application of F to the $U_q\mathfrak{g}$-module of all linear functionals on V.

The algebra $U_q\mathfrak{g}$ endowed with the adjoint action is a $U_q\mathfrak{g}$-module algebra, and $F(U_q\mathfrak{g})$ is its proper $U_q\mathfrak{g}$-module subalgebra.

PROPOSITION 2.36. [**148**, Chap. 7] *$U_q\mathfrak{g}$ and $F(U_q\mathfrak{g})$ are both left and right Noetherian algebras.*

Let $\lambda \in P_+$. The results of Subsections 2.1.3 and 2.1.5 and the present subsection allow one to introduce naturally a vector bundle $\mathcal{E}_\lambda$, continuous in $(0, 1]$ and analytic in $(0, 1)$, whose fibers are isomorphic to $L(\lambda)$.

We are about to sketch a construction of $\mathcal{E}_\lambda$. The reader is referred to Subsection 2.1.5 for the definitions of the Hermitian form $(\cdot, \cdot)$ in $L(\lambda)$ and the bases of $M(\lambda)$ associated to a reduced expression of $w_0 \in W$.

For every $q_0 \in (0, 1]$, choose a subset $\{v_j\}_{j=1,2,\ldots,\dim L(\lambda)}$ of elements of a distinguished basis in such a way that the Gram matrix $((v_i, v_j))_{i,j=1,2,\ldots,\dim L(\lambda)}$ is nondegenerate. Due to continuity reasons, it is still nondegenerate for all q close enough to q_0, so we obtain a trivial bundle with the required properties over a neighborhood of q_0 since $\dim L(\lambda)$ does not depend on $q \in (0, 1]$.

The elements of the Gram matrix are continuous in $(0, 1]$ and analytic in $(0, 1)$. Hence, first, the matrices of the operators E_i, F_i, H_i are continuous in $(0, 1]$ and analytic in $(0, 1)$, and, second, the same properties are valid for the transition maps defined on intersections of the neighborhoods. Finally, the vector bundle over $(0, 1]$ so obtained is essentially independent of the choices made in the construction.

An application of $\mathcal{E}_\lambda$ makes it easy to prove that the invariant Hermitian form in $L(\lambda)$ is positive definite. In fact, this Hermitian form is positive definite for

$q = 1$, is analytic in the interval $q \in (0, 1)$, and is nondegenerate for all $q \in (0, 1]$, as one deduces from the simplicity of the $U_q\mathfrak{g}$-module $L(\lambda)$. Thus we obtain

PROPOSITION 2.37. *The Hermitian form $(\cdot, \cdot)$ in $L(\lambda)$ is positive definite.*

The following claim follows from the semisimplicity of $U_q\mathfrak{g}$-modules in $\mathcal{C}$ and Proposition 2.37.

COROLLARY 2.38. *Every $U_q\mathfrak{g}$-module in $\mathcal{C}$ is unitarizable.*

REMARK 2.39. Let $\lambda \in P_+$ and μ be a weight of the $U\mathfrak{g}$-module $L(\lambda)$. Arguments presented in the previous subsection show that there exists a $\mathbb{Q}\left[t^{1/s}, t^{-1/s}\right]$-module which supplies the weight subspaces of $U_q\mathfrak{g}$-modules of weight μ under the specialization $t^{1/s} = q^{1/s}$, $q \in (0, 1]$.

2.2. Summary on algebras of functions on compact quantum groups

2.2.1. Algebraic quantum groups. As it was mentioned in Subsection 2.1.2, a simple complex affine algebraic group G is determined up to an isomorphism by a pair $(\mathbf{a}, \mathcal{L})$, where $\mathbf{a}$ is the Cartan matrix of its Lie algebra and $\mathcal{L}$ is a lattice lying between the lattice of weights and the lattice of roots: $Q \subset \mathcal{L} \subset P$. The subgroup $\mathcal{L}$ is the lattice of weights of all rational representations of the group G.

Let us outline a construction of a Hopf algebra of regular functions on G. Set $\mathcal{L}_+ = \mathcal{L} \cap P_+$. The subspaces $(\mathrm{End}\, L(\lambda))^*$ are embedded into the Hopf algebra $(U\mathfrak{g})^*$ and are linearly independent. One can show that the Hopf subalgebra

$$\mathbb{C}[G] = \bigoplus_{\lambda \in \mathcal{L}_+} (\mathrm{End}\, L(\lambda))^* \tag{2.32}$$

is a finitely generated domain and thus is the algebra of regular functions on the connected affine algebraic group.

This construction implicitly uses the reductivity of the Lie algebra $\mathfrak{g}$ and therefore the existence of its compact real form. The equality (2.32) is the Peter–Weil decomposition for the corresponding compact real algebraic group [**327**, §5.2].

A description of the algebra $\mathbb{C}[G]$ in terms of generators and relations can be found in Popov's paper [**256**].

Let us introduce a quantum analog of the Hopf algebra $\mathbb{C}[G]$. To each representation $\pi : U_q\mathfrak{g} \to \mathrm{End}\, V$ and each linear functional $\varphi : \mathrm{End}\, V \to \mathbb{C}$, we assign the linear functional $\varphi \circ \pi : U_q\mathfrak{g} \to \mathbb{C}$. Such linear functionals are called matrix elements of the representation π. Given an irreducible finite-dimensional representation π, the dual linear map $(\mathrm{End}\, V)^* \to (U_q\mathfrak{g})^*$ is injective due to the Burnside theorem [**65**, p. 177], [**348**, p. 100].

Consider the full subcategory of all $\mathcal{L}$-weight finite-dimensional $U_q\mathfrak{g}$-modules. Once V, V', V'' belong to this category, then $V' \oplus V''$, $V'' \otimes V'$, V^* are in it. The trivial $U_q\mathfrak{g}$-module belongs also to this category. Thus the subspace of the vector space of linear functionals on $U_q\mathfrak{g}$ spanned by the matrix elements of $\mathcal{L}$-weight finite-dimensional representations of $U_q\mathfrak{g}$ is a Hopf algebra. It is denoted by $\mathbb{C}[G]_q$ and is called the algebra of regular functions on the quantum group G; see [**199**, p. 96].[3]

The following result is an analog of the Peter–Weil theorem.

[3]It is also called the coordinate ring of the quantum group G and denoted by $\mathcal{O}_q(G)$.

PROPOSITION 2.40. $\mathbb{C}[G]_q \cong \bigoplus_{\lambda\in\mathcal{L}_+}(\mathrm{End}\, L(\lambda))^*$.

PROOF. Every finite-dimensional weight $U_q\mathfrak{g}$-module is isomorphic to a sum of submodules $L(\lambda)$, $\lambda \in P_+$, with multiplicities; see Subsection 2.1.8. Therefore $\mathbb{C}[G]_q$ is a sum of subspaces $(\mathrm{End}\, L(\lambda))^*$ naturally embedded to $\mathbb{C}[G]_q$. It remains to verify that this sum is direct.

Let $U_q\mathfrak{g}^{\mathrm{op}}$ be a Hopf algebra obtained from $U_q\mathfrak{g}$ by replacing the multiplication with the opposite one. We endow $\mathbb{C}[G]_q$ with the structure of a $U_q\mathfrak{g}$-bimodule by setting the respective actions of $U_q\mathfrak{g}$ and $U_q\mathfrak{g}^{\mathrm{op}}$ as follows:

$$R_{\mathrm{reg}}(\xi): \; f(\eta) \mapsto f(\eta\xi), \qquad L_{\mathrm{reg}}(\xi): \; f(\eta) \mapsto f(\xi\eta),$$

where $f \in \mathbb{C}[G]_q$, $\xi, \eta \in U_q\mathfrak{g}$.

The subspaces $(\mathrm{End}\, L(\lambda))^* \hookrightarrow \mathbb{C}[G]_q$ are simple finite-dimensional pairwise nonisomorphic $U_q\mathfrak{g}^{\mathrm{op}} \otimes U_q\mathfrak{g}$-modules. Therefore, the natural map of their external direct sum to $\mathbb{C}[G]_q$ is injective (see Remark 2.25, which allows us to substitute the noncommutative algebra $U_q\mathfrak{g}$ by its center throughout our proof of the injectivity). □

REMARK 2.41. $\mathbb{C}[G]_q$ is a $U_q\mathfrak{g}^{\mathrm{op}} \otimes U_q\mathfrak{g}$-module algebra.

REMARK 2.42. In what follows the representation R_{reg} is more substantial than L_{reg}. We write ξf instead of $R_{\mathrm{reg}}(\xi) f$ in the cases where this does not lead to ambiguities.

Let us introduce the adjoint representation Ad_ξ of the algebra $U_q\mathfrak{g}$ in the space $\mathbb{C}[G]_q$ by

$$\mathrm{Ad}_\xi : f(\eta) \mapsto \sum_i f(S^{-1}(\xi_i')\eta\xi_i''), \qquad \xi, \eta \in U_q\mathfrak{g},$$

where $\Delta(\xi) = \sum_i \xi_i' \otimes \xi_i''$.

Let π_λ be the representation of $U_q\mathfrak{g}$ corresponding to the simple $\mathcal{L}$-weight finite-dimensional $U_q\mathfrak{g}$-module $L(\lambda)$. The element

$$\chi_\lambda(\xi) = \mathrm{tr}\; \pi_\lambda(\xi \cdot K_{-2\rho}), \quad \xi \in U_q\mathfrak{g}, \tag{2.33}$$

of the algebra $\mathbb{C}[G]_q$ is called the character of the representation π_λ.

Due to results of Subsection 2.1.8, $\dim \mathrm{Hom}_{U_q\mathfrak{g}}(\mathrm{End}(L(\lambda)^*, \mathbb{C}) = 1$, just as in the classical case $q = 1$. Thus Proposition 1.64 and the Peter–Weil decomposition yield

COROLLARY 2.43. *The characters χ_λ, $\lambda \in \mathcal{L}_+$, belong to the subspaces $(\mathrm{End}\, L(\lambda))^* \hookrightarrow \mathbb{C}[G]_q$; they are invariant under the adjoint action*

$$\mathrm{Ad}_\xi\, \chi_\lambda = \varepsilon(\chi_\lambda) f, \qquad \xi \in U_q\mathfrak{g},$$

and form a basis of the vector space of invariant elements.

REMARK 2.44. 1. Proposition 2.29 implies that the natural pairing of Hopf algebras

$$\mathbb{C}[G]_q \times U_q\mathfrak{g} \to \mathbb{C}, \qquad f \times \xi \mapsto f(\xi), \tag{2.34}$$

is nondegenerate. Thus $U_q\mathfrak{g} \hookrightarrow \mathbb{C}[G]_q^*$. It is an analog of the embedding of the universal enveloping algebra to the space of distributions concentrated at the unit of G. This embedding is well known in the theory of real Lie groups; see [**168**, p. 167].

2. The algebra $\mathbb{C}[G]_q$ is finitely generated. Indeed, one can easily prove that the semigroup $\mathcal{L}_+$ is generated by a finite number of elements, and the smallest subalgebra containing the corresponding subspaces $(\operatorname{End} L(\lambda))^*$ coincides with $\mathbb{C}[G]_q$; see [**49**, p. 61].

The fundamental group $\pi_1(G)$ is naturally isomorphic to $P/\mathcal{L}$ [**327**, p. 204]. In the sequel we assume that $\mathcal{L} = P$ (i.e., the group G is simply connected) unless the contrary is stated explicitly.

REMARK 2.45. In the classical case $q = 1$ to each subset $\mathbb{S} \subset \{1, 2, \ldots, l\}$ one can associate the Lie subalgebras $\mathfrak{q}_{\mathbb{S}}^{\pm} \subset \mathfrak{g}$ generated by the set $\{H_j, E_j\}_{j=1,2,\ldots,l} \cup \{F_i\}_{i\in\mathbb{S}}$ and the set $\{H_j, E_j\}_{j=1,2,\ldots,l} \cup \{F_i\}_{i\in\mathbb{S}}$, respectively [**130**, p. 291]. The constructions of the present subsection are fully applicable to the connected reductive subgroup $L_{\mathbb{S}} \subset G$ with the Lie algebra $\mathfrak{l}_{\mathbb{S}} = \mathfrak{q}_{\mathbb{S}}^{+} \cap \mathfrak{q}_{\mathbb{S}}^{-}$. Thus we get a Hopf algebra $\mathbb{C}[L_{\mathbb{S}}]_q$, which is a quantum analog of the Hopf algebra $\mathbb{C}[L_{\mathbb{S}}]$ of regular functions on $L_{\mathbb{S}}$.

In quantum group theory, it is hard or even impossible to point out a quantum analog of a closed subgroup. On the other hand one often can find a quantum analog of an algebraic G-space with the stabilizer of a point isomorphic to this subgroup. The quantum cone introduced in Subsection 1.6.3 serves as an example. Let us list some well-known results on subgroups of affine algebraic groups and on algebraic G-spaces. Certainly, the Zariski topology will be used and all G-modules are assumed to be finite-dimensional.

A homomorphic image of an affine algebraic group is a closed subgroup.

A homogeneous space X with a marked point x is called a quotient G/H if the stabilizer H satisfies the following universality requirement. For any homogeneous space Y with a marked point y and the stabilizer H there exists a unique morphism of homogeneous spaces $X \to Y$ that maps x to y. A quotient exists and is unique up to an isomorphism.

Each closed subgroup $H \subset G$ is a stabilizer of a one-dimensional subspace of a certain G-module; it endows the orbit with the structure of quasi-projective variety isomorphic to G/H [**130**, pp. 135–146].

In the sequel all subgroups are assumed closed. A homogeneous space G/H is affine if and only if H is reductive [**328**, p. 216]. A subgroup $H \subset G$ is called parabolic if G/H is a projective variety.

Let us consider the G-orbit of the highest weight vector of a simple finite-dimensional G-module. Kostant showed that its projectivization is closed and its homogeneous coordinate ring is a quadratic algebra, i.e., the ideal of polynomials that vanish on the G-orbit of the highest weight vector is generated by elements of degree 2. A detailed proof can be found, for example, in [**105**].

Consider a G-space X and a point $x \in X$. The orbit Gx is a nonsingular subvariety, and its boundary is a union of orbits of strictly lower dimensions [**130**, p. 108]. In particular, there exists a closed G-orbit in the closure of the orbit Gx.

Further results on this topic can be found in the surveys [**297**], [**328**] and in the lectures [**298**].

2.2.2. Commutation relations. Consider a finite-dimensional P-weight $U_q\mathfrak{g}$-module V. To each pair $v \in V$, $l \in V^*$, we assign the linear functional on $U_q\mathfrak{g}$:

$$c_{l,v}^{V}(\xi) \stackrel{\text{def}}{=} l(\xi v), \quad \xi \in U_q\mathfrak{g}.$$

The algebra $\mathbb{C}[G]_q$ admits a finite number of generators $c_{l,v}^{L(\Lambda)}$, the linear functional l and the vector v can be chosen weight.

In the sequel we use the brief notation $c_{\lambda,\mu}^{\Lambda}$ for the element $c_{l,v}^{L(\Lambda)}$, where $v \in L(\Lambda)_\mu$, $l \in (L(\Lambda)_\lambda)^*$, if it does not lead to ambiguities. It is implicit that $(L(\Lambda)_\lambda)^* \hookrightarrow L(\Lambda)^*$.

One can obtain commutation relations between the elements $c_{\lambda,\mu}^{\Lambda}$ using the morphisms of $U_q\mathfrak{g}$-modules $\check{R}_{V,V'} : V \otimes V' \to V' \otimes V$ described in Subsection 2.1.4. Namely, let V, V' be finite-dimensional P-weight $U_q\mathfrak{g}$-modules and π, π' the corresponding representations of the algebra $U_q\mathfrak{g}$. Then

$$(\pi' \otimes \pi)(\xi)\check{R}_{V,V'} = \check{R}_{V,V'}(\pi \otimes \pi')(\xi), \qquad \xi \in U_q\mathfrak{g}. \tag{2.35}$$

It remains to pass from (2.35) to the equality of matrix elements of the operators involved:

$$c^{V'\otimes V}_{l'\otimes l,\check{R}(v\otimes v')} = c^{V\otimes V'}_{\check{R}^*(l'\otimes l),v\otimes v'}, \tag{2.36}$$

where $\check{R} = \check{R}_{V,V'}$ and $v \in V$, $l \in V^*$, $v' \in V'$, $l' \in (V')^*$. The equalities (2.36) are commutation relations between the elements $c_{l,v}^{V}$, $c_{l',v'}^{V'}$ because matrix elements of the tensor product of representations are products of matrix elements of tensor factors:

$$c^{V\otimes V'}_{l\otimes l',v\otimes v'} = c^{V}_{l,v} \cdot c^{V'}_{l',v'}, \qquad c^{V'\otimes V}_{l'\otimes l,v'\otimes v} = c^{V'}_{l',v'} \cdot c^{V}_{l,v}.$$

In [**79**] Drinfeld established the approach for producing commutation relations using the universal R-matrix described above. His approach goes back to the quantum method of the inverse scattering problem developed by Faddeev and his students [**294**]. The following statement can be proved using commutation relations deduced from (2.36).

PROPOSITION 2.46. [**49**, p. 78], [**148**, p. 266] *The algebra $\mathbb{C}[G]_q$ is (both right and left) Noetherian domain.*

Recall the standard notation w_0 for the longest element of the Weyl group W.

PROPOSITION 2.47. [**199**, p. 98] *Let*

$$\Lambda, \Lambda' \in P_+, \qquad \lambda, \lambda', \mu, \mu' \in P$$

and the following two conditions hold:

$$1.\quad \mu = \Lambda \quad or \quad \mu' = w_0\Lambda'; \qquad 2.\quad \lambda = w_0\Lambda \quad or \quad \lambda' = \Lambda'.$$

Then we have

$$c^{\Lambda'}_{\lambda',\mu'} c^{\Lambda}_{\lambda,\mu} = q^{(\mu,\mu')-(\lambda,\lambda')} c^{\Lambda}_{\lambda,\mu} c^{\Lambda'}_{\lambda',\mu'}. \tag{2.37}$$

PROOF. It follows from (2.35), (2.36) and (2.13) that it suffices to verify the equalities

$$R_{L(\Lambda),L(\Lambda')}(v \otimes v') = q^{-t_0} v \otimes v', \qquad v \otimes v' \in L(\Lambda)_\mu \otimes L(\Lambda')_{\mu'}, \tag{2.38}$$

$$\left(R_{L(\Lambda),L(\Lambda')}\right)^* (l \otimes l') = (q^{-t_0})^* l \otimes l', \qquad l \otimes l' \in (L(\Lambda)_\lambda)^* \otimes (L(\Lambda')_{\lambda'})^*.$$

The last one means that

$$R_{L(\Lambda)^*,L(\Lambda')^*}(l \otimes l') = q^{-t_0} l \otimes l', \qquad l \otimes l' \in (L(\Lambda)_\lambda)^* \otimes (L(\Lambda')_{\lambda'})^* \tag{2.39}$$

since $(S \otimes S)R = R$; see (2.19). Note that the highest weights of the $U_q\mathfrak{g}$-modules $L(\Lambda)$, $L(\Lambda')$, $L(\Lambda)^*$, $L(\Lambda')^*$ are equal to Λ, Λ', $-w_0\Lambda$, $-w_0\Lambda'$, respectively, and their lowest weights are $w_0\Lambda$, $w_0\Lambda'$, $-\Lambda$, $-\Lambda'$.

Now we use a decomposition of $U_q\mathfrak{n}^+ \otimes U_q\mathfrak{n}^-$ into a sum of subspaces,

$$U_q\mathfrak{n}^+ \otimes U_q\mathfrak{n}^- = \bigoplus_{\mu\in Q_+} \bigoplus_{\nu\in Q_+} (U_q\mathfrak{n}^+)_\mu \otimes (U_q\mathfrak{n}^-)_{-\nu}.$$

The explicit form of the universal R-matrix (2.11) shows that in the space of formal series one has

$$Rq^{\mathbf{t}_0} - 1 \in \bigoplus_{\substack{\nu\in Q_+\\ \nu\neq 0}} (U_q\mathfrak{n}^+)_\nu \otimes (U_q\mathfrak{n}^-)_{-\nu}.$$

The equality (2.13) implies that (2.38) holds if $\mu = \Lambda$ or $\mu' = w_0\Lambda'$, and (2.39) holds if $\lambda = w_0\Lambda$ or $\lambda' = \Lambda'$. □

COROLLARY 2.48. *The subalgebra generated by the elements*

$$c^{\Lambda}_{w_0\Lambda,\Lambda},\ c^{\Lambda}_{\Lambda,w_0\Lambda}, \qquad \Lambda \in P_+, \tag{2.40}$$

is commutative and its generators (2.40) *quasi-commute with all* $c^{\Lambda'}_{\lambda',\mu'} \in \mathbb{C}[G]_q$*:*

$$c^{\Lambda'}_{\lambda',\mu'} c^{\Lambda}_{w_0\Lambda,\Lambda} = q^{(\Lambda,\mu')-(w_0\Lambda,\lambda')} c^{\Lambda}_{w_0\Lambda,\Lambda} c^{\Lambda'}_{\lambda',\mu'},$$
$$c^{\Lambda}_{\Lambda,w_0\Lambda} c^{\Lambda'}_{\lambda',\mu'} = q^{(\mu',w_0\Lambda)-(\lambda',\Lambda)} c^{\Lambda'}_{\lambda',\mu'} c^{\Lambda}_{\Lambda,w_0\Lambda}.$$

REMARK 2.49. Let us drop the requirement $\lambda \in P$ and replace the finite-dimensional weight $U_q\mathfrak{g}$-modules $L(\lambda)$ by the Verma modules $M(\lambda)$. The matrix elements $c^{\Lambda}_{\lambda,\mu}$, $c^{\Lambda'}_{\lambda',\mu'}$ of the corresponding representations belong to the algebra $(U_q\mathfrak{g})^*$. The proof of Proposition 2.47 shows that the commutation relations (2.37) hold even in this more general case if $\mu = \Lambda$ and $\lambda' = \Lambda'$.

2.2.3. Compact quantum groups. Let us obtain a quantum analog of the compact form K of a complex reductive group G [**327**]. A real Lie algebra is called compact if it has a positive definite nondegenerate symmetric bilinear form which is invariant with respect to the adjoint action. Such subalgebras are conjugate [**47**, p. 26]. The real subalgebra of a simple complex Lie algebra $\mathfrak{g}$ generated by $\{iH_j,\ E_j - F_j,\ i(E_j + F_j)\}_{j=1,2,\dots,l}$ serves as an example; see [**326**, p. 139].

Consider the Hopf $*$-algebra $(U_q\mathfrak{g}, \star)$ introduced in Subsection 2.1.5 by (2.21) and the dual Hopf $*$-algebra $(U_q\mathfrak{g})^\star$:

$$f^\star(\xi) \stackrel{\text{def}}{=} \overline{f((S(\xi))^\star)}, \qquad \xi \in U_q\mathfrak{g}, \quad f \in (U_q\mathfrak{g})^\star;$$

see (1.20).

Recall that it is formed by matrix elements of finite-dimensional representations of $U_q\mathfrak{g}$, not necessarily weight. The Hopf subalgebra $\mathbb{C}[G]_q \hookrightarrow (U_q\mathfrak{g})^\star$ inherits the above involution since once f is a matrix element of a finite-dimensional weight representation π, $f^\star$ is a matrix element of $\pi^\star$. (In the sequel we will specify the last claim; see (2.46).)

Following Drinfeld, we call the Hopf $*$-algebra

$$\mathbb{C}[K]_q \stackrel{\text{def}}{=} (\mathbb{C}[G]_q, \star) \tag{2.41}$$

the algebra of regular functions on the real affine algebraic group K.

REMARK 2.50. In the classical case $q = 1$, the involution $\star$ in $\mathbb{C}[G]$ is introduced in a similar way, and one has

$$K = \left\{ g \in G \,\middle|\, f^\star(g) = \overline{f(g)} \ \text{ for all } \ f \in \mathbb{C}[G] \right\}.$$

In Subsection 2.1.8 the finite-dimensional weight $U_q\mathfrak{g}$-modules $L(\Lambda)$ are endowed with the invariant positive-definite Hermitian forms $(\cdot,\cdot)$ such that $(v(\lambda), v(\lambda)) = 1$. Choose an orthonormal basis of weight vectors $\{v^\Lambda_{\mu,j}\}$ in each finite-dimensional weight subspace $L(\Lambda)_\mu$. Once the subspace $L(\Lambda)_\mu$ is one dimensional, we write v^Λ_μ instead of $v^\Lambda_{\mu,1}$ for brevity.

Use the notation

$$c^\Lambda_{\lambda,i;\mu,j}(\xi) \stackrel{\text{def}}{=} \left(\xi v^\Lambda_{\mu,j}, v^\Lambda_{\lambda,i}\right)$$

for the matrix elements of the representations π_Λ corresponding to the $U_q\mathfrak{g}$-modules $L(\Lambda)$. To match with the notation of Subsection 2.2.2, we write $c^\Lambda_{\lambda,\mu}$ instead of $c^\Lambda_{\lambda,i;\mu,j}$ if it does not lead to ambiguities.

The set $\left\{c^\Lambda_{\lambda,i;\mu,j}\right\}$ is a basis of weight vectors of the $U_q\mathfrak{g}^{\mathrm{op}} \otimes U_q\mathfrak{g}$-module $\mathbb{C}[K]_q$:

$$R_{\mathrm{reg}}\left(K_m^{\pm 1}\right) c^\Lambda_{\lambda,i;\mu,j} = q_m^{\pm\mu_m} c^\Lambda_{\lambda,i;\mu,j},$$
$$L_{\mathrm{reg}}\left(K_m^{\pm 1}\right) c^\Lambda_{\lambda,i;\mu,j} = q_m^{\pm\lambda_m} c^\Lambda_{\lambda,i;\mu,j},$$

where $\lambda = \sum_k \lambda_k \overline{\omega}_k$, $\mu = \sum_k \mu_k \overline{\omega}_k$.

It follows from the definitions that the coproduct Δ and the counit ε act on the matrix elements $c^\Lambda_{\lambda,i;\mu,j}$ as follows:

$$\Delta\left(c^\Lambda_{\lambda,i;\mu,j}\right) = \sum_{\nu,m} c^\Lambda_{\lambda,i;\nu,m} \otimes c^\Lambda_{\nu,m;\mu,j}, \tag{2.42}$$

$$\varepsilon\left(c^\Lambda_{\lambda,i;\mu,j}\right) = \delta_{\lambda,\mu}\delta_{i,j}. \tag{2.43}$$

To find an explicit form of the elements $S\left(c^\Lambda_{\lambda,i;\mu,j}\right)$, $\left(c^\Lambda_{\lambda,i;\mu,j}\right)^\star$, we endow $U_q\mathfrak{g}$ with the *antilinear* involution τ

$$\tau\left(K_j^{\pm 1}\right) = S\left(K_j^{\pm 1}\right), \qquad \tau(E_j) = q_j S(E_j), \qquad \tau(F_j) = q_j^{-1} S(F_j), \tag{2.44}$$

where $j = 1, 2, \ldots, l$. To each $\Lambda \in P_+$ one can assign the $U_q\mathfrak{g}$-module $L(\Lambda)$ and the $*$-representation π_Λ in the finite-dimensional Hilbert space $L(\Lambda)$. Let π^τ_Λ be the representation of $U_q\mathfrak{g}$ in $L(\Lambda)$ defined by

$$(\pi^\tau_\Lambda(\xi)v_1, v_2) = (v_1, \pi_\Lambda(\tau(\xi))v_2), \qquad \xi \in U_q\mathfrak{g},\ v_1, v_2 \in L(\Lambda).$$

Since $\star\tau = \tau\star$, one has

$$\begin{aligned}(\pi^\tau_\Lambda(\xi)v_1, v_2) = (v_1, \pi_\Lambda(\tau(\xi))v_2) &= (\pi_\Lambda((\tau(\xi))^\star)v_1, v_2) \\ &= (\pi_\Lambda(\tau(\xi^\star))v_1, v_2) = (v_1, \pi^\tau_\Lambda(\xi^\star)v_2).\end{aligned}$$

Thus π^τ_Λ is the $*$-representation with highest weight $-w_0\Lambda$.

The following statement differs insignificantly from the Soibelman result [**199**, p. 100].

PROPOSITION 2.51. *Let $U : L(\Lambda) \to L(-w_0\Lambda)$ be a unique (up to a scalar factor) unitary operator such that $\pi^\tau_\Lambda(\xi) = U^{-1}\pi_{-w_0\Lambda}(\xi)U$ for all $\xi \in U_q\mathfrak{g}$. In the $*$-algebra $\mathbb{C}[K]_q$ we have*

$$S\left(c^\Lambda_{\lambda,i;\mu,j}\right) = q^{(\mu-\lambda,\rho)} \cdot \widetilde{c}^{\,-w_0\Lambda}_{-\mu,j;-\lambda,i}, \tag{2.45}$$

$$\left(c^\Lambda_{\lambda,i;\mu,j}\right)^\star = q^{(\lambda-\mu,\rho)} \cdot \widetilde{c}^{\,-w_0\Lambda}_{-\lambda,i;-\mu,j}, \tag{2.46}$$

where $\left\{c^\Lambda_{\lambda,i;\mu,j}\right\}$, $\left\{\widetilde{c}^{\,-w_0\Lambda}_{\lambda,i;\mu,j}\right\}$ are the matrix elements of the representations π_Λ, $\pi_{-w_0\Lambda}$ in the bases $\left\{v^\Lambda_{\mu,j}\right\}$, $\left\{U\,v^\Lambda_{-\mu,j}\right\}$.

PROOF. Consider the automorphism $\xi \mapsto K_\rho^{-1}\xi K_\rho$ of the algebra $U_q\mathfrak{g}$. It is well defined; see Subsection 2.1.2. By the same arguments as in the proof of (2.6), formula (2.44) shows that $S(\xi) = K_\rho^{-1}\tau(\xi)K_\rho$ for each element ξ of the smallest real subalgebra which contains all the generators $K_i^{\pm 1}$, E_i, F_i, $i = 1, 2, \dots, l$. Thus one has for such ξ

$$c^\Lambda_{\lambda,i;\mu,j}(S(\xi)) = q^{(\mu-\lambda,\rho)} c^\Lambda_{\lambda,i;\mu,j}(\tau(\xi)).$$

The last equality implies (2.45). For all $\xi \in U_q\mathfrak{g}$ we have

$$c^\Lambda_{\lambda,i;\mu,j}(S(\xi)) = \overline{c^\Lambda_{\mu,j;\lambda,i}((S(\xi))^\star)} = \left(c^\Lambda_{\mu,j;\lambda,i}\right)^\star(\xi).$$

Thus,

$$S\left(c^\Lambda_{\lambda,i;\mu,j}\right) = \left(c^\Lambda_{\mu,j;\lambda,i}\right)^\star, \tag{2.47}$$

and (2.46) follows from (2.45), (2.47). □

The following equalities can be derived from the definition of the antipode of a Hopf algebra and from (2.42), (2.43), (2.47):

$$\sum_{\mu,j} c^\Lambda_{\lambda,i;\mu,j} \left(c^\Lambda_{\lambda,i;\mu,j}\right)^\star = 1, \tag{2.48}$$

$$\sum_{\lambda,i} \left(c^\Lambda_{\lambda,i;\mu,j}\right)^\star c^\Lambda_{\lambda,i;\mu,j} = 1. \tag{2.49}$$

These equalities yield that for each element $f \in \mathbb{C}[K]_q$, one has

$$\|f\|_\infty \stackrel{\text{def}}{=} \sup_T \|T(f)\| < \infty, \tag{2.50}$$

where the supremum is taken over all classes T of unitary equivalent $*$-representations of the algebra $\mathbb{C}[K]_q$ in Hilbert spaces. Let us construct a faithful $*$-representation of $\mathbb{C}[K]_q$. In that way we prove that $\|f\|_\infty$ is a norm and $\mathbb{C}[K]_q$ with this norm is a normed algebra.

REMARK 2.52. It is well known that the same norm can be obtained if one restricts oneself to irreducible $*$-representations in (2.50); see [**73**, p. 53].

Introduce a linear functional $\nu : \mathbb{C}[K]_q \to \mathbb{C}$ that vanishes on the hyperplane $\bigoplus_{\lambda\neq 0}(\operatorname{End} L(\lambda))^* \subset \mathbb{C}[K]_q$. Evidently,

$$(\nu \otimes \mathrm{id})\Delta(f) = (\mathrm{id} \otimes \nu)\Delta(f) = \nu(f)\cdot 1, \qquad f \in \mathbb{C}[K]_q, \tag{2.51}$$

and ν is a unique (up to a scalar factor) $(U_q\mathfrak{g})^{\mathrm{op}} \otimes U_q\mathfrak{g}$-invariant linear functional. Fix it by the requirement $\nu(1) = 1$. It is a quantum analog of the integral over the Haar measure on K. Thus we also use the notation $\nu(f) = \int_{K_q} f\,d\nu$.

The equality (2.6) shows that the representations $\pi_\Lambda(\xi)$ and $\pi_\Lambda(S^2(\xi))$ of $U_q\mathfrak{g}$ in $L(\Lambda)$ are equivalent. Namely, the linear operator

$$F_\Lambda : L(\Lambda) \to L(\Lambda), \quad F_\Lambda : v^\Lambda_{\mu,j} \mapsto q^{-2(\mu,\rho)}\, v^\Lambda_{\mu,j},$$

is intertwining:

$$\pi_\Lambda(S^2(\xi))F_\Lambda = F_\Lambda \pi_\Lambda(\xi), \qquad \xi \in U_q\mathfrak{g}.$$

It is explained in [**295**, p. 26] that once we know an explicit form of an intertwining operator, it is not hard to get the following orthogonality relations.

PROPOSITION 2.53.

$$\int\limits_{K_q} (c^{\Lambda}_{\lambda,i;\mu,j})^{\star} c^{\Lambda'}_{\lambda',i';\mu',j'} d\nu = \frac{\delta_{\Lambda,\Lambda'}\delta_{\lambda,\lambda'}\delta_{i,i'}\delta_{\mu,\mu'}\delta_{j,j'}q^{2(\lambda,\rho)}}{\dim_q L(\Lambda)}, \tag{2.52}$$

where

$$\dim_q L(\Lambda) \stackrel{\text{def}}{=} \sum_{\lambda\in P} \dim L(\Lambda)_{\lambda} q^{2(\lambda,\rho)}.$$

COROLLARY 2.54. *The sesquilinear form*

$$(\psi_1, \psi_2) = \int\limits_{K_q} \psi_2^{\star}\psi_1 d\nu, \qquad \psi_1, \psi_2 \in \mathbb{C}[K]_q,$$

is positive definite.

The following claim can be deduced from (2.52), (2.47) and the definition of character (2.33) under which $\chi_{\Lambda} = \sum_{\lambda,i} c^{\Lambda}_{\lambda,i;\lambda,i} q^{-2(\lambda,\rho)}$.

COROLLARY 2.55. *The projection P_{Λ} in $\mathbb{C}[K]_q$ to $(\operatorname{End} L(\Lambda))^*$ parallel to $\bigoplus_{\lambda\neq\Lambda}(\operatorname{End} L(\lambda))^*$ is the integral operator with kernel*

$$\mathcal{P}_{\Lambda} = \dim_q L(\Lambda) \cdot (\mathrm{id} \otimes S)\Delta(\chi_{\Lambda}).$$

We complete $\mathbb{C}[K]_q$ with respect to the norm $\|\psi\|_2 = (\int_{K_q} \psi^{\star}\psi d\nu)^{1/2}$ and obtain a Hilbert space $L^2(d\nu)_q$, which is a quantum analog of the space $L^2(d\nu)$.

REMARK 2.56. 1. Since $w_0\rho = -\rho$, one has

$$\dim_q L(-w_0\Lambda) = \sum_{\lambda\in P} \dim(L(-w_0\Lambda)_{-w_0\lambda}) q^{2(\lambda,\rho)} = \dim_q L(\Lambda),$$

where w_0 is the longest element of the Weyl group W.

2. $\|\chi_{\Lambda}\|_2 = 1$ in $L^2(d\nu)_q$.

Proposition 2.46 and (2.49) imply

PROPOSITION 2.57. *The representation of $\mathbb{C}[K]_q$ in the pre-Hilbert space $\mathbb{C}[K]_q$*

$$\widehat{f} : \mathbb{C}[K]_q \to \mathbb{C}[K]_q, \qquad \widehat{f} : \psi \mapsto f\psi,$$

is a faithful $*$-representation, and all operators of this representation are bounded.

COROLLARY 2.58. *This representation can be extended by continuity to a faithful $*$-representation of $\mathbb{C}[K]_q$ in the Hilbert space $L^2(d\nu)_q$.*

The completion $C(K)_q$ of the $*$-algebra $\mathbb{C}[K]_q$ with respect to the norm $\|f\|_{\infty}$ is a C^*-algebra; it is called the algebra of continuous functions on the quantum group K.

Consider the tensor product $C(K)_q \otimes C(K)_q$ of C^*-algebras as the vector spaces [**73**, p. 71]. Due to the definitions, the bialgebra structure extends by continuity from the dense subalgebra $\mathbb{C}[K]_q$ to $C(K)_q$:

$$\Delta : C(K)_q \to C(K)_q \otimes C(K)_q; \qquad \varepsilon : C(K)_q \to \mathbb{C}.$$

The invariant integral ν admits an extension by continuity to $C(K)_q$. Indeed, $\|f\|_2 \le \|f\|_\infty$ since the definition of $\|\cdot\|_\infty$ yields the inequality $\|f\varphi\|_2 \le \|f\|_\infty \|\varphi\|_2$, and $\|1\|_2 = 1$. Therefore $|\int_{K_q} f d\nu| \le \|f\|_2 \|1\|_2 = \|f\|_2 \le \|f\|_\infty$.

Obviously, (2.51) holds for all $f \in C(K)_q$.

REMARK 2.59. The subset $\left\{c^\rho_{\lambda,i;\mu,j}\right\} \subset \mathbb{C}[K]_q$ and the C^*-algebra $C(K)_q$ satisfy all the requirements that Woronowicz included in the definition of a compact matrix pseudogroup [**342**, p. 616].

2.2.4. Example: the algebra of functions on the quantum group SU_2. Consider the Hopf algebra $U_q\mathfrak{sl}_2$, the Hopf $*$-algebra $U_q\mathfrak{su}_2 = (U_q\mathfrak{sl}_2, \star)$ and the Hopf algebra $\mathbb{C}[SL_2]_q$ introduced in Subsections 1.3.1, 1.3.2, and 1.3.6.

We endow $\mathbb{C}[SL_2]_q$ with an involution by duality (see the previous subsection) and obtain the Hopf $*$-algebra $\mathbb{C}[SU_2]_q$, which is a quantum analog of the algebra of regular functions on SU_2. By the definitions, one has

$$\begin{pmatrix} t_{11}^\star & t_{12}^\star \\ t_{21}^\star & t_{22}^\star \end{pmatrix} = \begin{pmatrix} t_{22} & -qt_{21} \\ -q^{-1}t_{12} & t_{11} \end{pmatrix}. \tag{2.53}$$

Let us use the description of the algebra $\mathbb{C}[SL_2]_q$ in terms of generators and relations given in Subsection 1.3.6 to find irreducible $*$-representations of the algebra $\mathbb{C}[SU_2]_q$.

EXAMPLE 2.60. The algebra $\mathbb{C}[SU_2]_q$ has one-dimensional $*$-representations

$$\chi_\varphi(t_{11}) = e^{i\varphi}, \quad \chi_\varphi(t_{12}) = \chi_\varphi(t_{21}) = 0, \quad \chi_\varphi(t_{22}) = e^{-i\varphi}, \qquad \varphi \in \mathbb{R}/(2\pi\mathbb{Z}).$$

EXAMPLE 2.61. Let $\{e_j\}_{j=0}^\infty$ be the standard basis of the Hilbert space $l^2(\mathbb{Z}_+)$. The following equalities determine an irreducible infinite-dimensional $*$-representation Π of the algebra $\mathbb{C}[SU_2]_q$ in $l^2(\mathbb{Z}_+)$:

$$\Pi(t_{11})e_j = \begin{cases} \sqrt{1-q^{2j}}\, e_{j-1}, & j > 0, \\ 0, & j = 0, \end{cases} \qquad \Pi(t_{12})e_j = q^{j+1}e_j,$$

$$\Pi(t_{21})e_j = q^j e_j, \qquad \Pi(t_{22})e_j = \sqrt{1-q^{2(j+1)}}e_{j+1}.$$

REMARK 2.62. For all $f \in C(SU_2)_q$, one has

$$\|\Pi(f)\| \ge \max_{\varphi \in \mathbb{R}/(2\pi\mathbb{Z})} |\chi_\varphi(f)|.$$

Indeed, let $\mathcal{B}$ be the algebra of all bounded operators in $l^2(\mathbb{Z}_+)$, $\mathcal{K}$ the two-sided ideal of compact operators, and $j : \mathcal{B} \to \mathcal{B}/\mathcal{K}$ the canonical factorization map. Then in the Calkin algebra $\mathcal{B}/\mathcal{K}$ we have

$$\|j \circ \Pi(f)\| = \max_{\varphi \in \mathbb{R}/(2\pi\mathbb{Z})} |\chi_\varphi(f)|, \qquad f \in \mathbb{C}[SU_2]_q.$$

It remains to use the fact that $\|j\| \le 1$.

The first statement of the following proposition uses the coproduct in $\mathbb{C}[SU_2]_q$ and is nearly obvious. The second claim can be proved by the method described in Subsection 1.2.3. We have the commutation relations

$$t_{11}x = q^2xt_{11}, \qquad t_{12}x = xt_{12}, \qquad t_{21}x = xt_{21}, \qquad t_{22}x = q^{-2}xt_{22},$$

instead of the commutation relations (1.10) (here $x = t_{12}t_{12}^\star$).

PROPOSITION 2.63. [**321**, p. 5] *The $*$-representations of $\mathbb{C}[SU_2]_q$*

$$\chi_\varphi, \qquad \Pi \otimes \chi_\varphi, \qquad \varphi \in \mathbb{R}/(2\pi\mathbb{Z}), \tag{2.54}$$

are irreducible and pairwise nonequivalent. Each irreducible $*$-representation of $\mathbb{C}[SU_2]_q$ is unitary equivalent to one of the representations (2.54).

COROLLARY 2.64. *In the C^*-algebra $C(SU_2)_q$ one has*

$$\|f\| = \max_{\varphi \in \mathbb{R}/(2\pi\mathbb{Z})} \|(\Pi \otimes \chi_\varphi)(f)\|, \qquad f \in \mathbb{C}[SU_2]_q. \tag{2.55}$$

In the classical case $q = 1$, the spectrum $\operatorname{spec}(x)$ of the element $x = |t_{12}|^2$ of the Banach algebra $C(SU_2)$ is $[0,1]$. In the quantum case the spectrum of x appears to be a countable set.

COROLLARY 2.65. *In the C^*-algebra $C(SU_2)_q$ one has*

$$\operatorname{spec}(x) = q^{2\mathbb{N}} \cup \{0\}.$$

We reproduce below an explicit form of an invariant integral that equals 1 at the unit element.

PROPOSITION 2.66. [**321**, p. 11] *For all $f \in \mathbb{C}[SU_2]_q$, one has*

$$\int_{(SU_2)_q} f d\nu = \frac{q^{-2}-1}{2\pi} \int_0^{2\pi} \operatorname{tr}((\Pi \otimes \chi_\varphi)(f \cdot x)) d\varphi. \tag{2.56}$$

Now we turn to more general results.

2.2.5. Irreducible $*$-representations of the algebra $\mathbb{C}[K]_q$ and an invariant integral. Construction of irreducible $*$-representations of the algebra $\mathbb{C}[K]_q$ is realized by reducing to the special case $K = SU_2$. Let us choose $i \in \{1, 2, \ldots, l\}$ and consider the embedding of Hopf $*$-algebras

$$U_{q_i}\mathfrak{sl}_2 \hookrightarrow U_q\mathfrak{g}, \qquad K^{\pm 1} \mapsto K_i^{\pm 1}, \quad E \mapsto E_i, \quad F \mapsto F_i.$$

Its image will be denoted by $(U_{q_i}\mathfrak{sl}_2)_i$.

The dual linear map is a homomorphism of the Hopf $*$-algebras $\mathbb{C}[K]_q \to \mathbb{C}[SU_2]_{q_i}$ since the restriction of a finite-dimensional weight representation of the algebra $U_q\mathfrak{g}$ to its subalgebra $(U_{q_i}\mathfrak{sl}_2)_i$ is a finite-dimensional weight representation of this subalgebra. The morphisms of Hopf $*$-algebras

$$\psi_i : \mathbb{C}[K]_q \to \mathbb{C}[SU_2]_{q_i}, \qquad i = 1, 2, \ldots, l,$$

and irreducible $*$-representations of $\mathbb{C}[K]_q$

$$\Pi_i = \Pi \circ \psi_i, \qquad \chi_{\varphi,i} = \chi_\varphi \circ \psi_i, \quad \varphi \in \mathbb{R}/(2\pi\mathbb{Z}),$$

appear this way. Thus we have the l-dimensional torus $\mathbb{T}^l = (\mathbb{R}/(2\pi\mathbb{Z}))^l$ of one-dimensional $*$-representations

$$\chi_\varphi = \chi_{\varphi_1,1} \otimes \chi_{\varphi_2,2} \otimes \cdots \otimes \chi_{\varphi_l,l}, \quad \varphi = (\varphi_1, \varphi_2, \ldots, \varphi_l) \in \mathbb{T}^l.$$

It is natural to expect that some of the tensor products $\Pi_{i_1} \otimes \Pi_{i_2} \otimes \cdots \otimes \Pi_{i_N}$ are also irreducible. Let us emphasize that the operators of representations are bounded (see (2.48), (2.49)), and we deal with tensor products in the category of Hilbert spaces. Irreducibility in this context means the absence of common nontrivial invariant subspaces, not the absence of common nontrivial invariant linear subvarieties of the operators of representations.

PROPOSITION 2.67. [**199**, p. 121] *Let $w = s_{i_1} s_{i_2} \cdots s_{i_N}$ be a reduced expression of an element $w \in W$ and $\varphi \in \mathbb{T}^l$. Consider the $*$-representation of $\mathbb{C}[K]_q$*

$$\Pi_{i_1} \otimes \Pi_{i_2} \otimes \cdots \otimes \Pi_{i_N} \otimes \chi_\varphi. \tag{2.57}$$

(1) *The $*$-representation (2.57) is irreducible and each irreducible $*$-representation of the algebra $\mathbb{C}[K]_q$ is unitary equivalent to one of the representations (2.57).*
(2) *Two $*$-representations (2.57) are unitary equivalent if and only if the corresponding elements $w \in W$ and $\varphi \in (\mathbb{R}/(2\pi\mathbb{Z}))^l$ coincide.*

Using (2.55) and Proposition 2.67, we obtain

COROLLARY 2.68. *Let $w_0 = s_{i_1} s_{i_2} \cdots s_{i_M}$ be a reduced expression of the longest element and*

$$\Pi_{w_0} = \Pi_{i_1} \otimes \Pi_{i_2} \otimes \cdots \otimes \Pi_{i_M}.$$

Then for each $f \in \mathbb{C}[K]_q$ one has

$$\|f\|_\infty = \max_{\varphi \in \mathbb{T}^l} \|(\Pi_{w_0} \otimes \chi_\varphi)(f)\|. \tag{2.58}$$

Proposition 2.67 also yields that the $*$-representation Π_w, $w \in W$, of $\mathbb{C}[K]_q$ is determined by w up to a unitary equivalence, that is, it does not depend on the choice of reduced expression.

Stokman shows in [**299**] that one can obtain a similar description of $*$-representations for an important class of homogeneous spaces of quantum groups using the results of [**300**].

Let us turn to the classical case $q = 1$. Consider a subset $\mathbb{S} \subset \{1, 2, \ldots, l\}$ and an algebraic torus $\mathbb{T} \subset G$ with the Lie algebra

$$\{\xi \in \mathfrak{h} |\ \alpha_j(\xi) = 0 \quad \text{for} \quad j \in \mathbb{S}\};$$

see [**327**, p. 133]. The centralizer $L_\mathbb{S} \subset G$ of this torus is a connected reductive group with the Lie algebra $\mathfrak{l}_\mathbb{S} \subset \mathfrak{g}$ generated by the set $\{F_i, E_j |\ i \in \mathbb{S}\} \cup \{H_j \mid j = 1, 2, \ldots, l\}$ [**130**, pp. 202, 254]. Thus $L_\mathbb{S}$-invariant functions on G are distinguished by the requirement of their $U\mathfrak{l}_\mathbb{S}$-invariance, and the following constructions become well substantiated.

Let $\mathbb{S} \subset \{1, 2, \ldots, l\}$ and $U_q\mathfrak{l}_\mathbb{S} \subset U_q\mathfrak{g}$ be the Hopf subalgebra generated by the set

$$\{K_i^{\pm 1}\}_{i=1,2,\ldots,l} \cup \{E_j, F_j\}_{j \in \mathbb{S}}.$$

Consider the subalgebra

$$\mathbb{C}[L_\mathbb{S} \backslash G]_q \stackrel{\text{def}}{=} \{f \in \mathbb{C}[G]_q |\ L_{\text{reg}}(\xi) f = \varepsilon(\xi) f, \quad \xi \in U_q\mathfrak{l}_\mathbb{S}\}.$$

It inherits the structure of locally finite-dimensional $U_q\mathfrak{g}$-module algebra since

$$L_{\text{reg}}(\xi) \cdot R_{\text{reg}}(\eta) = R_{\text{reg}}(\eta) \cdot L_{\text{reg}}(\xi), \qquad \xi, \eta \in U_q\mathfrak{g}.$$

The subalgebra $\mathbb{C}[L_\mathbb{S} \backslash G]_q$ inherits the involution $\star$. It is known from [**299**, p. 474] and [**117**] that this algebra is generated by the set $\{c^{\overline{\omega}_i}_{\overline{\omega}_i;\lambda,j} (c^{\overline{\omega}_i}_{\overline{\omega}_i;\mu,k})^\star |\ i \in \mathbb{S}\}$.

Let us study representations of the algebra $\mathbb{C}[L_\mathbb{S} \backslash G]_q$ with involution. Let $W_\mathbb{S} \subset W$ be the subgroup generated by the set $\{s_i\}_{i \in \mathbb{S}}$, and

$${}^\mathbb{S}W = \{w \in W \mid l(ws_i) > l(w), \quad i \in \mathbb{S}\}.$$

The following result was obtained by Kostant in [**201**].

LEMMA 2.69. [**131**, p. 19]

(1) *Each element* $w \in W$ *is uniquely decomposable into a product*

$$w = {}^{\mathbb{S}}w \cdot w_{\mathbb{S}}, \qquad w_{\mathbb{S}} \in W_{\mathbb{S}}, \quad {}^{\mathbb{S}}w \in {}^{\mathbb{S}}W;$$

(2) ${}^{\mathbb{S}}w$ *is a unique element of the smallest length in* $\{w\, W_{\mathbb{S}}\}$*;*

(3) $l(w) = l(w_{\mathbb{S}}) + l({}^{\mathbb{S}}w)$.

PROPOSITION 2.70. [**299**, p. 480] *Let* $w \in {}^{\mathbb{S}}W$. *The restrictions of* Π_w *to the subalgebra* $\mathbb{C}[L_{\mathbb{S}}\backslash G]_q$ *are irreducible* $*$*-representations. Every irreducible* $*$*-representation of the algebra* $\mathbb{C}[L_{\mathbb{S}}\backslash G]_q$ *is unitary equivalent to exactly one of these restrictions.*

The equality (2.56) delivers an explicit form of invariant integral on $\mathbb{C}[SU_2]_q$. Let us give a generalization of this result to the case of the Hopf $*$-algebra $\mathbb{C}[K]_q$.

For all $\Lambda \in P_+$, $w \in W$, the weight subspace $L(\Lambda)_{w\Lambda}$ is one dimensional. Thus a normalized basis vector $v_{w\Lambda}^{\Lambda} \in L(\Lambda)_{w\Lambda}$ is defined up to a scalar factor whose modulus is 1,[4] and there is a uniquely determined element

$$x_{\Lambda,w} = c^{\Lambda}_{\Lambda,\ w\Lambda} \cdot (c^{\Lambda}_{\Lambda,\ w\Lambda})^{\star}$$

of the algebra $\mathbb{C}[K]_q$.

Given $w \in W$ and a reduced expression $w = s_{i_1} s_{i_2} \cdots s_{i_N}$, we introduce the elements

$$w_j = s_{i_{j+1}} s_{i_{j+2}} \cdots s_{i_N}, \quad j = 0, 1, \ldots, N-1; \qquad w_N = 1.$$

LEMMA 2.71. (cf. [**261**, p. 410]) The bounded linear operators

$$(\Pi_w \otimes \chi_\varphi)(x_{\Lambda,w}), \qquad \Lambda \in P_+,$$

in $l^2(\mathbb{Z}_+)^{\otimes N}$ pairwise commute and

$$(\Pi_w \otimes \chi_\varphi)(x_{\Lambda,w}) e_{k_1} \otimes e_{k_2} \otimes \cdots \otimes e_{k_N} = \prod_{j=1}^{N} q^{2(k_j+1)(w_j\Lambda, \alpha_{i_j})} e_{k_1} \otimes e_{k_2} \otimes \cdots \otimes e_{k_N},$$

where $\{e_{k_1} \otimes e_{k_2} \otimes \cdots \otimes e_{k_N}\}$ is the standard basis of the Hilbert space $l^2(\mathbb{Z}_+)^{\otimes N}$.

COROLLARY 2.72. *The nonnegative linear operator* $(\Pi_{w_0} \otimes \chi_\varphi)(x_{\rho,w_0})$ *is of trace class and has zero kernel.*

PROPOSITION 2.73. [**261**, p. 410] *For each* $f \in \mathbb{C}[K]_q$,

$$\int\limits_{K_q} f d\nu = \mathrm{const}(q) \int\limits_{\mathbb{T}^l} \mathrm{tr}((\Pi_{w_0} \otimes \chi_\varphi)(f \cdot x_{\rho,w_0})) d\varphi,$$

where $d\varphi$ *is the Haar measure on the group* $\mathbb{T}^l$.

REMARK 2.74. An explicit form of the positive numerical factor $\mathrm{const}(q)$ is given in [**261**].

The invariant integral admits an extension by continuity from $\mathbb{C}[K]_q$ to $C(K)_q$.

---

[4] Recall that the invariant Hermitian form in $L(\Lambda)$ given in Subsection 2.1.5 is positive definite for all $\lambda \in P_+$ (see Subsection 2.1.8).

COROLLARY 2.75. *The state ν is faithful, i.e., for all nonzero $f \in C(K)_q$ one has*

$$\int\limits_{K_q} f^\star f d\nu > 0.$$

PROOF. Let us assume all $*$-representations of the algebra $\mathbb{C}[K]_q$ are extended by continuity to $C(K)_q$. Let $f \in C(K)_q$ and $\int_{K_q} f^\star f d\nu = 0$. The continuity of the operator function $(\Pi_{w_0} \otimes \chi_\varphi)(f)$ on the torus $\mathbb{T}^l$ and the fact that the operator $(\Pi_{w_0} \otimes \chi_\varphi)(x_{\rho,w_0}) \geq 0$ has no kernel yield that $\Pi_{w_0} \otimes \chi_\varphi(f) = 0$ for all $\varphi \in \mathbb{T}^l$. We extend both sides of (2.58) by continuity to $C(K)_q$ and obtain

$$\|f\|_\infty = \max_{\varphi \in \mathbb{T}^l} \|(\Pi_{w_0} \otimes \chi_\varphi)(f)\| = 0, \qquad f \in C(K)_q.$$

Thus $f = 0$. □

2.3. Quantum vector spaces and Harish-Chandra modules

2.3.1. Generalized Verma modules. Just as in Subsection 2.1.1, consider a Cartan matrix (a_{ij}), $i, j = 1, 2, \ldots, l$, of simple complex Lie algebra $\mathfrak{g}$, together with the associated Hopf algebras $U\mathfrak{g}$ and $U_q\mathfrak{g}$ with the generators

$$\{H_i, E_i, F_i\}, \quad i = 1, 2, \ldots, l; \qquad \{K_i, K_i^{-1}, E_i, F_i\}, \quad i = 1, 2, \ldots, l,$$

respectively.

Choose a subset $\mathbb{S} \subset \{1, 2, \ldots, l\}$ and endow the Lie algebra $\mathfrak{g}$ with a grading associated to the element $H_\mathbb{S} \in \mathfrak{h}$ given by

$$\alpha_j(H_\mathbb{S}) = \begin{cases} 2, & j \notin \mathbb{S}, \\ 0, & j \in \mathbb{S}, \end{cases}$$

namely,

$$\mathfrak{g} = \bigoplus_j \mathfrak{g}_j, \qquad \mathfrak{g}_j = \{\xi \in \mathfrak{g} | \ \mathrm{ad}_{H_\mathbb{S}} \xi = 2j\xi\}.$$

Obviously,

$$\mathfrak{g} = \mathfrak{u}^- \oplus \mathfrak{l} \oplus \mathfrak{u}^+,$$

where $\mathfrak{l} = \mathfrak{g}_0$, $\mathfrak{u}^- = \bigoplus_{j<0} \mathfrak{g}_j$, $\mathfrak{u}^+ = \bigoplus_{j>0} \mathfrak{g}_j$, and $\mathfrak{q}^\pm = \mathfrak{l} \oplus \mathfrak{u}^\pm$. Of course, $\mathfrak{q}^\pm \supset \mathfrak{b}^\pm$; that is, the Lie subalgebras $\mathfrak{q}^\pm$ are parabolic.

In the next subsection we endow the algebra $\mathbb{C}[\mathfrak{u}^-]$ of polynomials on $\mathfrak{u}^-$ with a structure of $U\mathfrak{g}$-module algebra and then obtain its quantum analog $\mathbb{C}[\mathfrak{u}^-]_q$. Here are necessary definitions.

First, we endow every weight $U_q\mathfrak{g}$-module V with the grading as follows:

$$V = \bigoplus_r V_r, \qquad V_r = \{v \in V | \ H_\mathbb{S} v = 2rv\}.$$

Next, consider the Hopf subalgebras $U_q\mathfrak{q}^+$, $U_q\mathfrak{q}^-$, $U_q\mathfrak{l}$, $U_q\mathfrak{l}_{ss}$ generated by

$$\{K_i^{\pm 1}, E_i\}_{i=1,2,\ldots,l} \cup \{F_j\}_{j \in \mathbb{S}}, \qquad \{K_i^{\pm 1}, F_i\}_{i=1,2,\ldots,l} \cup \{E_j\}_{j \in \mathbb{S}},$$
$$\{K_i^{\pm 1}\}_{i=1,2,\ldots,l} \cup \{E_j, F_j\}_{j \in \mathbb{S}}, \qquad \{K_j^{\pm 1}, E_j, F_j\}_{j \in \mathbb{S}},$$

respectively. Note that the Hopf subalgebra $U_q\mathfrak{l}_{ss} \subset U_q\mathfrak{l}$ is a quantum analog of the universal enveloping algebra $U\mathfrak{l}_{ss}$ of the "semisimple part" $\mathfrak{l}_{ss} = [\mathfrak{l}, \mathfrak{l}]$ of the Lie algebra $\mathfrak{l}$.

Evidently, $U_q\mathfrak{l} = U_q\mathfrak{q}^+ \cap U_q\mathfrak{q}^-$, so that one has the well-defined surjective morphisms of Hopf algebras

$$\pi^+ : U_q\mathfrak{q}^+ \to U_q\mathfrak{l}, \qquad \pi^- : U_q\mathfrak{q}^- \to U_q\mathfrak{l}$$

such that

$$\pi^\pm|_{U_q\mathfrak{l}} = \mathrm{id}, \qquad \pi^+(E_j) = \pi^-(F_j) = 0, \qquad j \notin \mathbb{S}.$$

The morphisms $\pi^\pm$ frequently work as good substitutes for quantum analogs of the Hopf subalgebras $U\mathfrak{u}^\pm \subset U\mathfrak{g}$. For example, q-analogs of spaces of $\mathfrak{u}^\pm$-invariants in a $U_q\mathfrak{g}$-module V can be introduced as follows:

$$V^{\mathfrak{u}^+} \stackrel{\mathrm{def}}{=} \{v \in V |\ \mathrm{Ker}\,\pi^+ \cdot v = 0\}, \qquad V^{\mathfrak{u}^-} \stackrel{\mathrm{def}}{=} \{v \in V |\ \mathrm{Ker}\,\pi^- \cdot v = 0\}.$$

Let us turn to generalized Verma modules; see [**212**]. Let

$$P_+^{\mathbb{S}} = \left\{ \lambda = \sum_{j=1}^{l} \lambda_j \overline{\omega}_j \in \mathfrak{h}_{\mathbb{R}}^* \,\middle|\, \lambda_j \in \mathbb{Z}_+ \text{ for } j \in \mathbb{S} \right\}.$$

To every $\lambda \in P_+^{\mathbb{S}}$ we associate a simple weight $U_q\mathfrak{l}$-module $L(\mathfrak{l}, \lambda)$ with a generator $v(\mathfrak{l}, \lambda)$ and the defining relations

$$K_j^{\pm 1} v(\mathfrak{l}, \lambda) = q_j^{\pm \lambda_j} v(\mathfrak{l}, \lambda), \qquad j = 1, 2, \ldots, l,$$
$$E_j v(\mathfrak{l}, \lambda) = 0, \qquad F_j^{\lambda_j + 1} v(\mathfrak{l}, \lambda) = 0, \qquad j \in \mathbb{S};$$

cf. (2.30). The morphisms $\pi^\pm : U_q\mathfrak{q}^\pm \to U_q\mathfrak{l}$ allow one to introduce $U_q\mathfrak{q}^\pm$-modules $L(\mathfrak{q}^\pm, \lambda)$ in the same vector space.

The $U_q\mathfrak{g}$-module

$$N(\mathfrak{q}^+, \lambda) = U_q\mathfrak{g} \otimes_{U_q\mathfrak{q}^+} L(\mathfrak{q}^+, \lambda)$$

is called the generalized Verma module with highest weight $\lambda \in P_+^{\mathbb{S}}$. We are about to produce a basis of the vector space $N(\mathfrak{q}^+, \lambda)$ using the Poincaré–Birkhoff–Witt basis of the algebra $U_q\mathfrak{n}^-$; the latter basis will correspond to a special reduced expression of the longest element $w_0 \in W$.

Consider the subgroup $W_{\mathbb{S}} \subset W$ generated by the simple reflections $\{s_i\}_{i \in \mathbb{S}}$. Also let $w_0^{\mathbb{S}}$ be the shortest element in the coset $W_{\mathbb{S}} w_0$. It follows from Lemma 2.69 that such element $w_0^{\mathbb{S}}$ is unique (namely, $w_0^{\mathbb{S}} = ({}^{\mathbb{S}}w_0)^{-1}$) and

$$w_0 = w_{0,\mathbb{S}} \cdot w_0^{\mathbb{S}},$$

where $w_{0,\mathbb{S}}$ is the longest element of $W_{\mathbb{S}}$.

Choose the reduced expressions

$$w_{0,\mathbb{S}} = s_{i_1} s_{i_2} \cdots s_{i_{M'}}, \qquad w_0^{\mathbb{S}} = s_{i_{M'+1}} s_{i_{M'+2}} \cdots s_{i_M}.$$

Their concatenation $w_0 = s_{i_1} s_{i_2} \cdots s_{i_M}$ is a reduced expression of w_0. Proposition 2.3 implies that $U_q\mathfrak{g}$ is a free right module over $U_q\mathfrak{q}^+$ with the basis

$$F_{\beta_M}^{j_M} F_{\beta_{M-1}}^{j_{M-1}} \cdots F_{\beta_{M'+1}}^{j_{M'+1}}, \qquad (j_M, j_{M-1}, \ldots, j_{M'+1}) \in \mathbb{Z}_+^{M-M'}.$$

PROPOSITION 2.76. *Let $\lambda \in P_+^{\mathbb{S}}$ and $\{v_1, v_2, \ldots, v_d\}$ be a basis of the vector space $L(\mathfrak{q}^+, \lambda)$. Then the vectors*

$$F_{\beta_M}^{j_M} F_{\beta_{M-1}}^{j_{M-1}} \cdots F_{\beta_{M'+1}}^{j_{M'+1}} v_i, \qquad j_k \in \mathbb{Z}_+, \qquad i \in \{1, 2, \ldots, d\}, \tag{2.59}$$

form a basis of homogeneous elements of the graded vector space $N(\mathfrak{q}^+, \lambda)$.

REMARK 2.77. The elements of the basis (2.59) are homogeneous because F_β are weight vectors of the $U_q\mathfrak{g}$-module $U_q\mathfrak{g}$, whose weights are just their classical counterparts, and all v_i have the same degree of homogeneity: $\deg(v_j) = \frac{1}{2}\sum_{k\notin\mathbb{S}}\lambda_k$.

As a consequence of Propositions 2.76 and 2.30 we have

PROPOSITION 2.78. *The homogeneous components of the graded vector space $N(\mathfrak{q}^+,\lambda)$ are finite-dimensional weight $U_q\mathfrak{l}$-modules. The multiplicities of weights and dimensions of these $U_q\mathfrak{l}$-modules coincide with their classical values, i.e., those at $q=1$.*

Note that $N(\mathfrak{q}^+,\lambda)$ is a locally $U_q\mathfrak{b}^+$-finite-dimensional module, where $\mathfrak{b}^+\subset\mathfrak{g}$ is the standard Borel subalgebra. Just as Proposition 2.76, one can prove

LEMMA 2.79. *The homogeneous components of the $U_q\mathfrak{l}$-bimodules $U_q\mathfrak{q}^\pm$ are both left and right free $U_q\mathfrak{l}$-modules of finite rank.*

The following universality property of the generalized Verma module $N(\mathfrak{q}^+,\lambda)$ is a consequence of the definition and involves the natural embedding of $U_q\mathfrak{q}^+$-modules

$$L(\mathfrak{q}^+,\lambda)\hookrightarrow U_q\mathfrak{g}\otimes_{U_q\mathfrak{q}^+}L(\mathfrak{q}^+,\lambda),\qquad v\mapsto 1\otimes v.$$

LEMMA 2.80. *Consider a $U_q\mathfrak{g}$-module V. Every morphism of $U_q\mathfrak{q}^+$-modules $f: L(\mathfrak{q}^+,\lambda)\to V$ can be extended in a unique way to a morphism of $U_q\mathfrak{g}$-modules*

$$N(\mathfrak{q}^+,\lambda)\to V,\qquad \xi\otimes v\mapsto \xi f(v).$$

Let $v(\mathfrak{q}^+,\lambda)=1\otimes v(\mathfrak{l},\lambda)$ be the highest weight vector of $N(\mathfrak{q}^+,\lambda)$. Just as in the case $q=1$ [**212**], one can prove

PROPOSITION 2.81. *Consider the Verma module $M(\lambda)$ with highest weight $\lambda\in P_+^{\mathbb{S}}$ and the standard projection of $U_q\mathfrak{g}$-modules*

$$p_\lambda: M(\lambda)\to N(\mathfrak{q}^+,\lambda),\qquad p_\lambda: v(\lambda)\mapsto v(\mathfrak{q}^+,\lambda).$$

The kernel $\operatorname{Ker}p_\lambda$ *is the smallest submodule of the $U_q\mathfrak{g}$-module $M(\lambda)$ among those containing $\left\{F_i^{\lambda_i+1}v(\lambda)\right\}_{i\in\mathbb{S}}$.*

REMARK 2.82. Let $L\subset M(\lambda)$ be the submodule generated by $\left\{F_i^{\lambda_i+1}v(\lambda)\right\}_{i\in\mathbb{S}}$. In [**212**] the relation $L=\operatorname{Ker}p_\lambda$ is deduced from the facts that the exact sequence of $U\mathfrak{g}$-modules

$$0\to\operatorname{Ker}p_\lambda/L\to M(\lambda)/L\to M(\lambda)/\operatorname{Ker}p_\lambda\to 0 \tag{2.60}$$

splits and the $U\mathfrak{g}$-module $M(\lambda)/L$ is generated by its highest weight vector. On the other hand, the sequence (2.60) splits by Lemma 2.80 on the universality of $N(\mathfrak{q}^+,\lambda)\cong M(\lambda)/\operatorname{Ker}p_\lambda$ and commutativity of the diagram

$$\begin{array}{ccc} U\mathfrak{q}^+\cdot v(\lambda) & \xrightarrow{\ \approx\ } & U\mathfrak{q}^+\cdot v(\mathfrak{q}^+,\lambda)\\ \big\downarrow & & \big\downarrow\\ M(\lambda) & \xrightarrow{\ p_\lambda\ } & N(\mathfrak{q}^+,\lambda)\end{array}$$

in the category of $U\mathfrak{q}^+$-modules.

COROLLARY 2.83. *The generalized Verma module $N(\mathfrak{q}^+,\lambda)$ is canonically isomorphic to the $U_q\mathfrak{g}$-module with the generator $v(\mathfrak{q}^+,\lambda)$ and the defining relations*

$$(2.61)\qquad \begin{gathered} E_j v(\mathfrak{q}^+,\lambda)=0, \qquad K_j^{\pm 1} v(\mathfrak{q}^+,\lambda)=q_j^{\pm\lambda_j} v(\mathfrak{q}^+,\lambda), \qquad j=1,2,\ldots,l; \\ F_j^{\lambda_j+1} v(\mathfrak{q}^+,\lambda)=0, \qquad j\in\mathbb{S}. \end{gathered}$$

REMARK 2.84. It follows from the definitions that there exists the largest proper submodule $K(\mathfrak{q}^+,\lambda)\subsetneqq N(\mathfrak{q}^+,\lambda)$. If $\lambda\in P_+$, then $\dim(N(\mathfrak{q}^+,\lambda)/K(\mathfrak{q}^+,\lambda))<\infty$ and $\dim(N(\mathfrak{q}^+,\lambda)/K)=\infty$ for each proper submodule $K\subsetneqq K(\mathfrak{q}^+,\lambda)$. The latter claim can be proved in the same way as a similar claim in Subsection 2.1.5.

To conclude, we observe that a slight variation of proofs would allow one to produce similar results for the $U_q\mathfrak{g}$-modules

$$N(\mathfrak{q}^-,\lambda)=U_q\mathfrak{g}\otimes_{U_q\mathfrak{q}^-}L(\mathfrak{q}^-,w_{0,\mathbb{S}}\lambda), \qquad \lambda\in -P_+^{\mathbb{S}},$$

which are generalized Verma modules with lowest weight λ. Here, as above, $w_{0,\mathbb{S}}$ denotes the longest element of the subgroup $W_{\mathbb{S}}$ of the Weyl group W.

2.3.2. The quantum vector space $\mathfrak{u}^-$. Let us turn to a construction of the $U_q\mathfrak{g}$-module algebra $\mathbb{C}[\mathfrak{u}^-]_q$. The method used here goes back to V. Drinfeld and is based on producing the dual coalgebra first; see [**79**]. As a background of the method, one should mention the following fact, which results from the definitions.

PROPOSITION 2.85. *Let A be a Hopf algebra. Consider the Hopf algebra A^{cop} which differs from A by replacing its coproduct with the opposite one. If C is an A^{cop}-module coalgebra, then the dual vector space C^* is an A-module algebra.*

REMARK 2.86. It seems reasonable to explain why the coproduct should be replaced by its opposite. Normally, the vector space dual to a coalgebra is an algebra in such a way that the tensor factors under dualization are not flipped. On the contrary, in the category of $U_q\mathfrak{g}$-modules one has $V_1^*\otimes V_2^*\hookrightarrow(V_2\otimes V_1)^*$, that is, the tensor factors flip. The passage to the opposite coproduct allows one to bypass this discrepancy.

Here is an example of $U_q\mathfrak{g}^{\mathrm{cop}}$-module coalgebra.

PROPOSITION 2.87. *The maps*

$$\Delta^+: v(\mathfrak{q}^+,0)\mapsto v(\mathfrak{q}^+,0)\otimes v(\mathfrak{q}^+,0), \qquad \varepsilon^+: v(\mathfrak{q}^+,0)\mapsto 1$$

can be uniquely extended to morphisms of $U_q\mathfrak{g}^{\mathrm{cop}}$-modules

$$\Delta^+: N(\mathfrak{q}^+,0)\to N(\mathfrak{q}^+,0)\otimes N(\mathfrak{q}^+,0), \qquad \varepsilon^+: N(\mathfrak{q}^+,0)\to\mathbb{C}.$$

These morphisms endow $N(\mathfrak{q}^+,0)$ with a structure of $U_q\mathfrak{g}^{\mathrm{cop}}$-module coalgebra.

PROOF. The uniqueness of Δ^+ follows from the fact that $v(\mathfrak{q}^+,0)$ generates the $U_q\mathfrak{g}$-module $N(\mathfrak{q}^+,0)$. The existence follows from relations

$$\begin{gathered} E_j(v(\mathfrak{q}^+,0)\otimes v(\mathfrak{q}^+,0))=(K_j^{\pm 1}-1)(v(\mathfrak{q}^+,0)\otimes v(\mathfrak{q}^+,0))=0, \qquad j=1,2,\ldots,l, \\ F_j(v(\mathfrak{q}^+,0)\otimes v(\mathfrak{q}^+,0))=0, \qquad j\in\mathbb{S}, \end{gathered}$$

which themselves follow from (2.2) and (2.61). The existence and uniqueness of ε are obvious. All the required properties of Δ^+ except coassociativity follow from definitions. It suffices to verify the coassociativity on the generator $v(\mathfrak{q}^+,0)$:

$$(v(\mathfrak{q}^+,0)\otimes v(\mathfrak{q}^+,0))\otimes v(\mathfrak{q}^+,0)=v(\mathfrak{q}^+,0)\otimes(v(\mathfrak{q}^+,0)\otimes v(\mathfrak{q}^+,0)).$$ □

□

By Proposition 2.85, the algebra $\mathbb{C}[[\mathfrak{u}^-]]_q$ of all linear functionals on $N(\mathfrak{q}^+,0)$ is a $U_q\mathfrak{g}$-module algebra. It is a q-analog of the $U\mathfrak{g}$-module algebra $\mathbb{C}[[\mathfrak{u}^-]]$ of formal series on the complex vector space $\mathfrak{u}^- \cong \mathfrak{g}/\mathfrak{q}^+$. To make the passage from formal series to polynomials in the above construction, it suffices to replace the tensor category of vector spaces by the tensor category of graded vector spaces. Namely, direct sums and tensor products of graded vector spaces are endowed with the standard gradings

$$(V' \oplus V'')_r = V'_r \oplus V''_r, \qquad (V' \otimes V'')_r = \bigoplus_{r'+r''=r} (V'_{r'} \otimes V''_{r''}).$$

Note that every graded vector space

$$V = \bigoplus_r V_r, \qquad V_r = \{v \in V|\ \deg v = r\}, \tag{2.62}$$

determines the graded vector space $V^* \overset{\text{def}}{=} \bigoplus_r V_r^*$. It is just the space of all the linear functionals on V iff (2.62) contains only finitely many summands.

A bialgebra A is said to be graded if it is a graded vector space such that the multiplication $m : A \otimes A \to A$ and the coproduct $\Delta : A \to A \otimes A$ are compatible with the grading, i.e., they are linear maps of degree zero. The notions of graded algebra and graded module, graded coalgebra and graded comodule, as well as that of graded Hopf algebra, are introduced in a similar way.

In the previous subsection every weight $U_q\mathfrak{g}$-module was endowed with the grading in terms of the element $H_{\mathbb{S}}$.

Let us treat the generalized Verma modules $N(\mathfrak{q}^+,\lambda)$ as objects of the tensor category of $U_q\mathfrak{g}^{\mathrm{cop}}$-modules, and the dual graded $U_q\mathfrak{g}$-modules as objects of the tensor category of $U_q\mathfrak{g}$-modules.

Note that $U_q\mathfrak{g}$, $U_q\mathfrak{g}^{\mathrm{cop}}$ are graded Hopf algebras

$$\deg(K_j^{\pm 1}) = 0, \qquad \deg(E_j) = \begin{cases} 1, & j \in \mathbb{S}, \\ 0, & j \notin \mathbb{S}, \end{cases} \qquad \deg(F_j) = \begin{cases} -1, & j \in \mathbb{S}, \\ 0, & j \notin \mathbb{S}, \end{cases}$$

and $N(\mathfrak{q}^+,0)$ is a graded $U_q\mathfrak{g}^{\mathrm{cop}}$-module coalgebra. The dual graded $U_q\mathfrak{g}$-module algebra

$$\mathbb{C}[\mathfrak{u}^-]_q = \bigoplus_{j \geq 0} \mathbb{C}[\mathfrak{u}^-]_{q,j}, \qquad \mathbb{C}[\mathfrak{u}^-]_{q,j} \overset{\text{def}}{=} N(\mathfrak{q}^+,0)^*_{-j},$$

is called the polynomial algebra on the quantum complex vector space $\mathfrak{u}^-$. A structure of $U_q\mathfrak{g}$-module algebra is imposed by duality:

$$\xi f(v) = f(S^{-1}(\xi)v), \qquad \xi \in U_q\mathfrak{g}, \quad f \in \mathbb{C}[\mathfrak{u}^-]_q, \quad v \in N(\mathfrak{q}^+,0), \tag{2.63}$$

where S is the antipode of $U_q\mathfrak{g}$. Such $U_q\mathfrak{g}$-module algebras were considered by Kébé [**163, 164**].

PROPOSITION 2.88. *The algebra $\mathbb{C}[\mathfrak{u}^-]_q$ is a domain.*

PROOF. Consider the natural projection

$$(U_q\mathfrak{b}^-)^{\mathrm{cop}} \to N(\mathfrak{q}^+,0) \to 0, \qquad \xi \mapsto \xi\, v(\mathfrak{q}^+,0),$$

which is certainly a morphism of coalgebras. It follows from Propositions 2.7 and 2.8 that the dual (in the category of graded vector spaces) linear map is an embedding of algebras $\mathbb{C}[\mathfrak{u}^-]_q \hookrightarrow U_q\mathfrak{b}^+$. It remains to apply Corollary 2.5, which claims that the algebra $U_q\mathfrak{g}$ is a domain. □

It follows from Proposition 2.78 that the homogeneous components $\mathbb{C}[\mathfrak{u}^-]_{q,j}$ of the graded vector space $\mathbb{C}[\mathfrak{u}^-]_q$ are $U_q\mathfrak{l}$ -modules of the same dimension as in the classical case $q = 1$:

$$\dim \mathbb{C}[\mathfrak{u}^-]_{q,j} = \binom{j + \dim \mathfrak{u}^- - 1}{\dim \mathfrak{u}^- - 1} = \binom{j + \dim \mathfrak{u}^- - 1}{j}. \tag{2.64}$$

Consider the simplest example $\mathfrak{g} = \mathfrak{sl}_2$. In this case $N(\mathfrak{q}^+, 0)$ is the Verma module $M(0)$ with highest weight zero, and $\{F^j v(0)\}_{j\in\mathbb{Z}_+}$ is a basis of the vector space $M(0)$.

Obviously, $M(0) = \bigoplus_{j\geq 0} M(0)_{-j}$, $M(0)_{-j} = \mathbb{C}F^j v(0)$. The unit of the $U_q\mathfrak{sl}_2$-module algebra $\mathbb{C}[\mathfrak{u}^-]_q$ coincides with the linear functional which takes the value 1 at $v(0)$ and the value 0 on $\bigoplus_{j\neq 0} M(0)_j$. Consider the element $z \in M(0)^*$ determined by

$$z(S(F^j)v(0)) = \begin{cases} q^{1/2}, & j = 1, \\ 0, & j \neq 1, \end{cases}$$

where S is the antipode of $U_q\mathfrak{sl}_2$. One can verify that $z^j(F^j v(0)) \neq 0$ by induction in j. Hence z^j is a nonzero element of the one-dimensional subspace $M(0)^*_{-j} \subset M(0)^*$, and $\mathbb{C}[\mathfrak{u}^-]_q$ is isomorphic to the algebra $\mathbb{C}[z]$ of polynomials in one variable. The structure of the $U_q\mathfrak{sl}_2$-module algebra on $\mathbb{C}[\mathfrak{u}^-]_q$ is completely determined by the action of the generators $K^{\pm 1}$, E, F on z. An application of (2.63) and the explicit form of S makes it easy to obtain the relations

$$K^{\pm 1} z = q^{\pm 2} z, \qquad Fz = q^{1/2}, \qquad Ez = -q^{1/2} z^2$$

(the factor $-q^{1/2}$ in the latter one is found using $(EF - FE)z = \frac{K - K^{-1}}{q - q^{-1}} z)$.

The above simplest example can be generalized in a number of ways. The case of minimal parabolic subalgebra $\mathbb{S} = \varnothing$, $\mathfrak{q}^+ = \mathfrak{b}^+$ was studied by Joseph and his collaborators [**147**]. In the opposite case of maximal parabolic subalgebra, $\mathbb{S}$ contains all the elements $1, 2, \ldots, l$ except just one of those, to be denoted by l_0 in the sequel. Let $\alpha_1, \alpha_2, \ldots, \alpha_l$ be the simple roots of $\mathfrak{g}$ (see Subsection 2.1.1) and $\delta = \sum_{j=1} n_j \alpha_j$ its highest root. The most interesting cases are $n_{l_0} = 1$ and $n_{l_0} = 2$. The first case leads to q-analogs of prehomogeneous vector spaces of commutative parabolic type (see Subsection 2.3.8), and the second case is related to the Borel–De Siebenthal discrete series of representations of real reductive groups [**90, 96**].

REMARK 2.89. The generalized Verma module $N(\mathfrak{q}^-, 0) = \bigoplus_{j\geq 0} N(\mathfrak{q}^-, 0)_j$ is endowed with the structure of a graded $U_q\mathfrak{g}$-module coalgebra in the same way as $N(\mathfrak{q}^+, 0)$: $\Delta^- v(\mathfrak{q}^-, 0) = v(\mathfrak{q}^-, 0) \otimes v(\mathfrak{q}^-, 0)$. The dual graded vector space

$$\mathbb{C}[\mathfrak{u}^+]_q = \bigoplus_{j\geq 0} \mathbb{C}[\mathfrak{u}^+]_{q,-j}, \qquad \mathbb{C}[\mathfrak{u}^+]_{q,-j} = N(\mathfrak{q}^-, 0)^*_j,$$

is a $U_q\mathfrak{g}$-module algebra of *polynomials on the quantum complex vector space* $\mathfrak{u}^+$. It will be shown in Subsection 2.3.3 that the algebras $\mathbb{C}[\mathfrak{u}^+]_q$ and $\mathbb{C}[\mathfrak{u}^-]_q$ are anti-isomorphic. Hence $\mathbb{C}[\mathfrak{u}^+]_q$ is a domain.

2.3.3. The $(U_q\mathfrak{g}, *)$-module $*$-algebra $\mathrm{Pol}(\mathfrak{u}^-)_q$. In Subsection 1.3.5 the universal R-matrix was used to obtain the $U_q\mathfrak{su}_{1,1}$-module algebra $\mathrm{Pol}(\mathbb{C})_q$. We are about to demonstrate that a very similar argument allows one to get a significantly more general result.

Consider the complex vector space $\mathfrak{u}^- \oplus \mathfrak{u}^+$ and the algebra $\mathbb{C}[\mathfrak{u}^- \oplus \mathfrak{u}^+]$ of polynomials on $\mathfrak{u}^- \oplus \mathfrak{u}^+$. We are going to obtain its q-analog by endowing the $U_q\mathfrak{g}$-module

$$\mathbb{C}[\mathfrak{u}^- \oplus \mathfrak{u}^+]_q \stackrel{\mathrm{def}}{=} \mathbb{C}[\mathfrak{u}^-]_q \otimes \mathbb{C}[\mathfrak{u}^+]_q$$

with the structure of a $U_q\mathfrak{g}$-module algebra.

Recall that the $U_q\mathfrak{g}$-modules $\mathbb{C}[\mathfrak{u}^\pm]_q$ are weight modules and the $U_q\mathfrak{g}$-module $\mathbb{C}[\mathfrak{u}^-]_q$ is locally $U_q\mathfrak{b}^-$-finite dimensional, while the $U_q\mathfrak{g}$-module $\mathbb{C}[\mathfrak{u}^+]_q$ is locally $U_q\mathfrak{b}^+$-finite dimensional. It was explained in Subsection 2.1.4 that this allows one to get a well-defined morphism of $U_q\mathfrak{g}$-modules

$$\check{R} = \check{R}_{\mathbb{C}[\mathfrak{u}^+]_q, \mathbb{C}[\mathfrak{u}^-]_q} : \mathbb{C}[\mathfrak{u}^+]_q \otimes \mathbb{C}[\mathfrak{u}^-]_q \to \mathbb{C}[\mathfrak{u}^-]_q \otimes \mathbb{C}[\mathfrak{u}^+]_q.$$

We endow $\mathbb{C}[\mathfrak{u}^- \oplus \mathfrak{u}^+]_q$ with an algebra structure by defining a multiplication

$$m : \mathbb{C}[\mathfrak{u}^- \oplus \mathfrak{u}^+]_q \otimes \mathbb{C}[\mathfrak{u}^- \oplus \mathfrak{u}^+]_q \to \mathbb{C}[\mathfrak{u}^- \oplus \mathfrak{u}^+]_q$$

in terms of the multiplications

$$m^\pm : \mathbb{C}[\mathfrak{u}^\pm]_q \otimes \mathbb{C}[\mathfrak{u}^\pm]_q \to \mathbb{C}[\mathfrak{u}^\pm]_q, \qquad m^\pm : f_1 \otimes f_2 \mapsto f_1 f_2,$$

in $\mathbb{C}[\mathfrak{u}^\pm]_q$. Namely, set

$$m = (m^- \otimes m^+)(\mathrm{id}_{\mathbb{C}[\mathfrak{u}^-]_q} \otimes \check{R} \otimes \mathrm{id}_{\mathbb{C}[\mathfrak{u}^+]_q});$$

see a similar definition in Subsection 1.3.5.

Proposition 2.9 implies the following relations between the morphisms of $U_q\mathfrak{g}$-modules:

$$(m^- \otimes \mathrm{id})\check{R}_{\mathbb{C}[\mathfrak{u}^+]_q, \mathbb{C}[\mathfrak{u}^-]_q \otimes \mathbb{C}[\mathfrak{u}^-]_q} = \check{R}_{\mathbb{C}[\mathfrak{u}^+]_q, \mathbb{C}[\mathfrak{u}^-]_q}(\mathrm{id} \otimes m^-),$$
$$(\mathrm{id} \otimes m^+)\check{R}_{\mathbb{C}[\mathfrak{u}^+]_q \otimes \mathbb{C}[\mathfrak{u}^+]_q, \mathbb{C}[\mathfrak{u}^-]_q} = \check{R}_{\mathbb{C}[\mathfrak{u}^+]_q, \mathbb{C}[\mathfrak{u}^-]_q}(m^+ \otimes \mathrm{id}),$$

and Proposition 2.10 implies

$$\check{R}_{\mathbb{C}[\mathfrak{u}^+]_q, \mathbb{C}[\mathfrak{u}^-]_q \otimes \mathbb{C}[\mathfrak{u}^-]_q} = (\mathrm{id}_{\mathbb{C}[\mathfrak{u}^-]_q} \otimes \check{R}_{\mathbb{C}[\mathfrak{u}^+]_q, \mathbb{C}[\mathfrak{u}^-]_q}) \cdot (\check{R}_{\mathbb{C}[\mathfrak{u}^+]_q, \mathbb{C}[\mathfrak{u}^-]_q} \otimes \mathrm{id}_{\mathbb{C}[\mathfrak{u}^-]_q}),$$
$$\check{R}_{\mathbb{C}[\mathfrak{u}^+]_q \otimes \mathbb{C}[\mathfrak{u}^+]_q, \mathbb{C}[\mathfrak{u}^-]_q} = (\check{R}_{\mathbb{C}[\mathfrak{u}^+]_q, \mathbb{C}[\mathfrak{u}^-]_q} \otimes \mathrm{id}_{\mathbb{C}[\mathfrak{u}^+]_q}) \cdot (\mathrm{id}_{\mathbb{C}[\mathfrak{u}^+]_q} \otimes \check{R}_{\mathbb{C}[\mathfrak{u}^+]_q, \mathbb{C}[\mathfrak{u}^-]_q}).$$

These relations, together with the associativity of $m^\pm$ in $\mathbb{C}[\mathfrak{u}^\pm]_q$, will be used to prove that the multiplication m in $\mathbb{C}[\mathfrak{u}^- \oplus \mathfrak{u}^+]_q$ is associative. The rest of the claims of the proposition below follow from Proposition 2.9.

PROPOSITION 2.90. *The multiplication m endows the $U_q\mathfrak{g}$-module $\mathbb{C}[\mathfrak{u}^- \oplus \mathfrak{u}^+]_q$ with a structure of the $U_q\mathfrak{g}$-module algebra whose unit is $1 \otimes 1$.*

REMARK 2.91. Proposition 2.10 implies that the embeddings of vector spaces

$$\mathbb{C}[\mathfrak{u}^-]_q \hookrightarrow \mathbb{C}[\mathfrak{u}^- \oplus \mathfrak{u}^+]_q, \qquad f \mapsto f \otimes 1,$$
$$\mathbb{C}[\mathfrak{u}^+]_q \hookrightarrow \mathbb{C}[\mathfrak{u}^- \oplus \mathfrak{u}^+]_q, \qquad f \mapsto 1 \otimes f,$$

are morphisms of $U_q\mathfrak{g}$-module algebras.

Consider the graded Lie algebra $\mathfrak{g} = \bigoplus_i \mathfrak{g}_i$ introduced in Subsection 2.3.1, together with its involutive automorphism θ,

$$\theta|_{\mathfrak{g}_i} = \begin{cases} 1, & i \in 2\mathbb{Z}, \\ -1, & i \in 2\mathbb{Z}+1. \end{cases}$$

Consider also an extension of this automorphism to an automorphism of the universal enveloping algebra $U\mathfrak{g}$. Its q-analog is the automorphism θ_q of the Hopf algebra $U_q\mathfrak{g}$ given by

$$\theta_q(K_i^{\pm 1}) = K_i^{\pm 1}, \qquad \theta_q(E_i) = \begin{cases} E_i, & i \in \mathbb{S}, \\ -E_i, & i \notin \mathbb{S}, \end{cases} \qquad \theta_q(F_i) = \begin{cases} F_i, & i \in \mathbb{S}, \\ -F_i, & i \notin \mathbb{S}. \end{cases}$$

In Section 2.2 the Hopf $*$-algebra $(U_q\mathfrak{g}, \star)$ is essentially involved in the study of compact quantum groups; here $\star$ is the involutive antilinear antiautomorphism (2.21). The right substitute for it in the context of noncompact quantum groups is the Hopf $*$-algebra $(U_q\mathfrak{g}, *)$, where

$$* = \theta_q \cdot \star = \star \cdot \theta_q; \tag{2.65}$$

that is,

$$(K_j^{\pm 1})^* = K_j^{\pm 1}, \qquad j = 1, 2, \ldots, l,$$

$$E_j^* = \begin{cases} K_j F_j, & j \in \mathbb{S}, \\ -K_j F_j, & j \notin \mathbb{S}, \end{cases} \qquad F_j^* = \begin{cases} E_j K_j^{-1}, & j \in \mathbb{S}, \\ -E_j K_j^{-1}, & j \notin \mathbb{S}. \end{cases}$$

Note that this kind of involution was initially considered in [**94**, p. 188].

We endow $\mathbb{C}[\mathfrak{u}^- \oplus \mathfrak{u}^+]_q$ with the structure of a $(U_q\mathfrak{g}, *)$-module algebra using the natural embeddings

$$\mathbb{C}[\mathfrak{u}^\pm]_q \hookrightarrow \mathbb{C}[\mathfrak{u}^-]_q \otimes \mathbb{C}[\mathfrak{u}^+]_q, \qquad f_- \mapsto f_- \otimes 1, \quad f_+ \mapsto 1 \otimes f_+.$$

PROPOSITION 2.92. *There exists a unique involutive antilinear antiautomorphism $*$ of the algebra $\mathbb{C}[\mathfrak{u}^- \oplus \mathfrak{u}^+]_q$ such that*

(1) $* : \mathbb{C}[\mathfrak{u}^\pm]_q \to \mathbb{C}[\mathfrak{u}^\mp]_q$,
(2) $(\xi f)^* = (S(\xi))^* f^*$ *for all* $\xi \in U_q\mathfrak{g}$, $f \in \mathbb{C}[\mathfrak{u}^- \oplus \mathfrak{u}^+]_q$.

PROOF. Let us start with proving the uniqueness. Consider an involution with the properties (1) and (2), together with its restriction to the subalgebras $\mathbb{C}[\mathfrak{u}^\pm]_q$. It follows from (2) and the definition of gradings in $\mathbb{C}[\mathfrak{u}^\pm]_{q,j}$ that

$$* : \mathbb{C}[\mathfrak{u}^\pm]_{q,j} \to \mathbb{C}[\mathfrak{u}^\mp]_{q,-j}.$$

This allows one to introduce the dual antilinear maps $* : N(\mathfrak{q}^\pm, 0) \to N(\mathfrak{q}^\mp, 0)$ given by

$$f^*(v) = \overline{f(v^*)}, \qquad f \in \mathbb{C}[\mathfrak{u}^\pm]_q, \quad v \in N(\mathfrak{q}^\pm, 0). \tag{2.66}$$

Note that $* : N(\mathfrak{q}^\pm, 0)_j \to N(\mathfrak{q}^\mp, 0)_{-j}$. Since $1^* = 1$ in $\mathbb{C}[\mathfrak{u}^- \oplus \mathfrak{u}^+]_q$ one has

$$* : v(\mathfrak{q}^\pm, 0) \mapsto v(\mathfrak{q}^\mp, 0). \tag{2.67}$$

Using (2), (2.66)–(2.67), together with $S * S* = \mathrm{id}$, we proceed as follows:

$$\begin{aligned} (\xi f)^*(v(\mathfrak{q}^\pm,0)) &= ((S(\xi))^* f^*)(v(\mathfrak{q}^\pm,0)), \\ \overline{(\xi f)(v(\mathfrak{q}^\mp,0))} &= f^*(S((S(\xi))^*)v(\mathfrak{q}^\pm,0)), \\ \overline{(\xi f)(v(\mathfrak{q}^\mp,0))} &= f^*(\xi^* v(\mathfrak{q}^\pm,0)), \\ \overline{f(S(\xi)v(\mathfrak{q}^\mp,0))} &= \overline{f((\xi^* v(\mathfrak{q}^\pm,0))^*)}, \\ f((S^{-1}(\xi))^* v(\mathfrak{q}^\mp,0))7 &= f((\xi v(\mathfrak{q}^\pm))^*). \end{aligned}$$

Hence,

$$* : \xi v(\mathfrak{q}^\pm,0) \mapsto (S^{-1}(\xi))^* v(\mathfrak{q}^\mp,0), \qquad \xi \in U_q\mathfrak{g}. \tag{2.68}$$

The involutions $* : \mathbb{C}[\mathfrak{u}^\pm]_q \to \mathbb{C}[\mathfrak{u}^\mp]_q$, $* : N(\mathfrak{q}^\pm,0) \to N(\mathfrak{q}^\mp,0)$ are determined uniquely by (2.66) and (2.68). Therefore, an involution with desired properties in $\mathbb{C}[\mathfrak{u}^- \oplus \mathfrak{u}^+]_q$ is unique.

To construct the involution in question, we are going to revert the above arguments. The rule (2.68) for $*$ allows one to define the antilinear maps $* : N(\mathfrak{q}^\pm,0) \to N(\mathfrak{q}^\mp,0)$ (these are well defined, as one can conclude from the definition of the $U_q\mathfrak{g}$-modules $N(\mathfrak{q}^\pm,0)$). Now (2.66) determines the antilinear maps $* : \mathbb{C}[\mathfrak{u}^\pm]_q \to \mathbb{C}[\mathfrak{u}^\mp]_q$. It remains to introduce the antilinear map in $\mathbb{C}[\mathfrak{u}^- \oplus \mathfrak{u}^+]_q = \mathbb{C}[\mathfrak{u}^-]_q \otimes \mathbb{C}[\mathfrak{u}^+]_q$ by setting

$$(f_- \otimes f_+)^* = f_+^* \otimes f_-^*, \qquad f_\pm \in \mathbb{C}[\mathfrak{u}^\pm]_q,$$

and to prove that it endows $\mathbb{C}[\mathfrak{u}^- \oplus \mathfrak{u}^+]_q$ with a structure of $(U_q\mathfrak{g}, *)$-module algebra.

First, the antilinear maps $* : N(\mathfrak{q}^\pm,0) \to N(\mathfrak{q}^\mp,0)$ are antihomomorphisms of coalgebras, because the map $\xi \mapsto (S^{-1}(\xi))^*$ is an antihomomorphism of the coalgebra $U_q\mathfrak{g}$. Second, these antilinear maps are mutually inverse since we have $*S^{-1} * S^{-1} = \mathrm{id}$. Hence $* : \mathbb{C}[\mathfrak{u}^\pm]_q \to \mathbb{C}[\mathfrak{u}^\mp]_q$ are antilinear antihomomorphisms of algebras inverse to each other. Third, one deduces from (2.19) that

$$(S \otimes S)(R) = R, \qquad R^{*\otimes *} = R^{21}, \tag{2.69}$$

where R^{21} differs from R only by the flip of tensor factors. By (2.69) we easily establish that

$$(f_1 f_2)^* = f_2^* f_1^*, \qquad f_1, f_2 \in \mathbb{C}[\mathfrak{u}^- \oplus \mathfrak{u}^+]_q;$$

see [**289**, Sec. 16]. Thus $*$ is an involution of the $U_q\mathfrak{g}$-module algebra $\mathbb{C}[\mathfrak{u}^- \oplus \mathfrak{u}^+]_q$.

It remains to show that the involutions in $U_q\mathfrak{g}$ and $\mathbb{C}[\mathfrak{u}^- \oplus \mathfrak{u}^+]_q$ are compatible:

$$(\xi f)^* = (S(\xi))^* f^*, \qquad \xi \in U_q\mathfrak{g}, \quad f \in \mathbb{C}[\mathfrak{u}^- \oplus \mathfrak{u}^+]_q.$$

We start with the special case $f \in \mathbb{C}[\mathfrak{u}^\pm]_q$. Using (2.66), (2.68) and $S * S = *$, one can pass through the above sequence of relations in the reverse direction:

$$f((\xi v)^*) = f((S^{-1}(\xi))^* v^*), \qquad \overline{f(S(\xi)v^*)} = \overline{f((\xi^* v)^*)},$$
$$\overline{(\xi f)(v^*)} = f^*(\xi^* v), \qquad \overline{(\xi f)(v^*)} = f^*(S((S(\xi))^*)v),$$
$$(\xi f)^*(v) = ((S(\xi))^* f^*)(v),$$

where $v \in N(\mathfrak{q}^\pm,0)$, $f \in N(\mathfrak{q}^\pm,0)^*$. The general case reduces to this special case. In fact, let $f = f_- f_+$, where $f_\pm \in \mathbb{C}[\mathfrak{u}^\pm]_q$. Then $(\xi(f_- f_+))^* = \sum_i (\xi_i'' f_+)^* (\xi_i' f_-)^*$, where $\Delta(\xi) = \sum_i \xi_i' \otimes \xi_i''$ and

$$(S(\xi))^*(f_- f_+)^* = (S(\xi))^*(f_+^* f_-^*) = \sum_i ((S(\xi_i''))^* f_+^*)(S(\xi_i')^* f_-^*). \qquad \square$$

The presence of the antilinear anti-isomorphism $* : \mathbb{C}[\mathfrak{u}^-]_q \to \mathbb{C}[\mathfrak{u}^+]_q$ allows one to treat the $U_q\mathfrak{g}$-module algebra $\mathbb{C}[\mathfrak{u}^+]_q$ as a quantum analog (q-analog) of the algebra of antiholomorphic polynomials on $\mathfrak{u}^-$.

Let us impose the notation $\mathrm{Pol}(\mathfrak{u}^-)_q = (\mathbb{C}[\mathfrak{u}^- \oplus \mathfrak{u}^+]_q, *)$ for the $(U_q\mathfrak{g}, *)$-module algebra we have just produced. The example of quantum disc, which corresponds to $\mathfrak{g} = \mathfrak{sl}_2$, demonstrates that $\mathrm{Pol}(\mathfrak{u}^-)_q$ is a q-analog of the polynomial algebra on the real vector space $(\mathfrak{u}^-)_\mathbb{R}$.

The vector space $\mathrm{Pol}(\mathfrak{u}^-)_q$ admits the decomposition

$$\mathrm{Pol}(\mathfrak{u}^-)_q = \bigoplus_{i,j=0}^{\infty} \mathbb{C}[\mathfrak{u}^-]_{q,i} \cdot \mathbb{C}[\mathfrak{u}^+]_{q,-j}, \tag{2.70}$$

so that $\mathbb{C}[\mathfrak{u}^-]_{q,0} \cdot \mathbb{C}[\mathfrak{u}^+]_{q,0} = \mathbb{C} \cdot 1$. In other words, given $f \in \mathrm{Pol}(\mathfrak{u}^-)_q$, there exists a unique decomposition

$$f = \langle f \rangle \cdot 1 + \sum_{\{(i,j) \in \mathbb{Z}_+^2 | (i,j) \neq (0,0)\}} f_{i,j}, \qquad f_{i,j} \in \mathbb{C}[\mathfrak{u}^-]_{q,i} \cdot \mathbb{C}[\mathfrak{u}^+]_{q,-j}, \tag{2.71}$$

where $\langle \cdot \rangle$ is the linear functional on $\mathrm{Pol}(\mathfrak{u}^-)_q$ defined by the latter relation.

PROPOSITION 2.93. *The sesquilinear form* $(f_1, f_2) \stackrel{\mathrm{def}}{=} \langle f_2^* f_1 \rangle$ *in* $\mathbb{C}[\mathfrak{u}^-]_q$ *is non-degenerate and*

$$(f_1, f_2) = \overline{(f_2, f_1)}, \qquad f_1, f_2 \in \mathbb{C}[\mathfrak{u}^-]_q. \tag{2.72}$$

PROOF. We start with proving that

$$\langle f_2^* f_1 \rangle = (f_2^* \otimes f_1)(R(v(\mathfrak{q}^-, 0) \otimes v(\mathfrak{q}^+, 0))), \qquad f_1, f_2 \in \mathbb{C}[\mathfrak{u}^-]_q, \tag{2.73}$$

where R is the universal R-matrix (2.11). Recall that

$$N(\mathfrak{q}^-, 0) \otimes N(\mathfrak{q}^+, 0) = \bigoplus_{i,j=0}^{\infty} N(\mathfrak{q}^-, 0)_i \otimes N(\mathfrak{q}^+, 0)_{-j}.$$

We endow the finite-dimensional vector spaces $N(\mathfrak{q}^-, 0)_i \otimes N(\mathfrak{q}^+, 0)_{-j}$ with their standard topology and $N(\mathfrak{q}^-, 0) \otimes N(\mathfrak{q}^+, 0)$ with the weakest topology that makes continuous all the "projections" onto $N(\mathfrak{q}^-, 0)_i \otimes N(\mathfrak{q}^+, 0)_{-j}$ parallel to $\bigoplus_{i' \neq i, j' \neq j} N(\mathfrak{q}^-, 0)_{i'} \otimes N(\mathfrak{q}^+, 0)_{-j'}$. The space of formal series of the form

$$\sum_{i,j} f_{i,j}, \qquad f_{i,j} \in N(\mathfrak{q}^-, 0)_i \otimes N(\mathfrak{q}^+, 0)_{-j},$$

is a completion of the topological vector space $N(\mathfrak{q}^-, 0) \otimes N(\mathfrak{q}^+, 0)$ as above; it will be denoted by $N(\mathfrak{q}^-, 0) \widehat{\otimes} N(\mathfrak{q}^+, 0)$. The pairing

$$(\mathbb{C}[\mathfrak{u}^+]_q \otimes \mathbb{C}[\mathfrak{u}^-]_q) \times (N(\mathfrak{q}^-, 0) \otimes N(\mathfrak{q}^+, 0)) \to \mathbb{C}$$

extends by continuity to a pairing

$$(\mathbb{C}[\mathfrak{u}^+]_q \otimes \mathbb{C}[\mathfrak{u}^-]_q) \times (N(\mathfrak{q}^-, 0) \widehat{\otimes} N(\mathfrak{q}^+, 0)) \to \mathbb{C}.$$

If

$$\varphi = \sum_i \varphi_i, \quad \varphi_i \in \mathbb{C}[\mathfrak{u}^-]_{q,i}, \qquad \psi = \sum_i \psi_j, \quad \psi_j \in \mathbb{C}[\mathfrak{u}^+]_{q,-j},$$

then $\varphi_0 = \varphi(v(\mathfrak{q}^+,0))$, $\psi_0 = \psi(v(\mathfrak{q}^-,0))$. Hence the definition of multiplication in $\mathbb{C}[\mathfrak{u}^- \oplus \mathfrak{u}^+]_q$ and $(S \otimes S)(R) = R$ imply (2.73).

Now we are in a position to finish the proof of our proposition. The action of $U_q\mathfrak{g} \otimes U_q\mathfrak{g}$ on $N(\mathfrak{q}^-,0) \otimes N(\mathfrak{q}^+,0)$ and the antilinear map $v_1 \otimes v_2 \mapsto v_2^* \otimes v_1^*$ in this space admit extensions by continuity to $N(\mathfrak{q}^-,0)\widehat{\otimes} N(\mathfrak{q}^+,0)$.

Due to (2.73), one can deduce (2.72) from

$$R^{21}(v(\mathfrak{q}^+,0) \otimes v(\mathfrak{q}^-,0)) = \left(R\left(v(\mathfrak{q}^-,0) \otimes v(\mathfrak{q}^+,0)\right)\right)^{*\otimes *}.$$

Furthermore, the latter relation follows from (2.66), (2.68) and (2.69).

The sesquilinear form $(\cdot,\cdot)$ in $\mathbb{C}[\mathfrak{u}^-]_q$ is nondegenerate, as one can deduce from (2.73), the multiplicative formula for the universal R-matrix (2.11), and the explicit form of the bases in $N(\mathfrak{q}^\pm,0)$ obtained in Subsection 2.3.1. □

2.3.4. The category $C(\mathfrak{g},\mathfrak{l})_q$ and Hermitian symmetric pairs. Consider the full subcategory $C(\mathfrak{g},\mathfrak{l})_q$ of the category of $U_q\mathfrak{g}$-modules formed by P-weight locally $U_q\mathfrak{l}$-finite-dimensional modules. It is closed under the operations $\oplus$, $\otimes$, the passages to submodules and quotient modules. Obviously, $\mathrm{Pol}(\mathfrak{u}^-)_q \in C(\mathfrak{g},\mathfrak{l})_q$.

Each weight $U_q\mathfrak{g}$-module has been endowed with the grading determined by the element $H_{\mathbb{S}} \in \mathfrak{h}$; see Subsection 2.3.1. By the definition of this grading, the Hopf algebras $U_q\mathfrak{q}^\pm$ are graded

$$U_q\mathfrak{q}^+ = \bigoplus_{j=0}^{\infty}(U_q\mathfrak{q}^+)_j, \qquad U_q\mathfrak{q}^- = \bigoplus_{j=0}^{\infty}(U_q\mathfrak{q}^-)_{-j},$$

so that $(U_q\mathfrak{q}^+)_0 = (U_q\mathfrak{q}^-)_0 = U_q\mathfrak{l}$.

PROPOSITION 2.94. *Each $U_q\mathfrak{g}$-module V possesses the largest submodule $V_\mathfrak{l}$ which is in $C(\mathfrak{g},\mathfrak{l})_q$.*

PROOF. Obviously, each $U_q\mathfrak{g}$-module V contains the largest P-weight submodule $V_\mathfrak{h}$. It remains to prove that the subspace $V' = \{v \in V_\mathfrak{h} | \dim(U_q\mathfrak{l}v) < \infty\}$ is a submodule of the $U_q\mathfrak{g}$-module $V_\mathfrak{h}$. By Lemma 2.79, for every j there exist finite subsets $\{\eta_1^\pm,\eta_2^\pm,\ldots\}$ such that $(U_q\mathfrak{q}^\pm)_j = \sum_{k=1}^{N^\pm(j)} \eta_k^\pm U_q\mathfrak{l}$. Hence,

$$\dim(U_q\mathfrak{l}\,\xi v) \le \dim\left((U_q\mathfrak{q}^\pm)_j\, v\right) = \dim\left(\sum_{k=1}^{N^\pm(j)} \eta_k^\pm U_q\mathfrak{l}\, v\right) < \infty$$

for all $\xi \in (U_q\mathfrak{q}^\pm)_j$, $v \in V'$; that is, $\xi v \in V'$. □

Thus we obtain a functor Γ from the category of all $U_q\mathfrak{g}$-modules to $C(\mathfrak{g},\mathfrak{l})_q$ which assigns to each $U_q\mathfrak{g}$-module V its submodule $V_\mathfrak{l}$, and to each morphism $V \to U$ its restriction to $V_\mathfrak{l}$. In the representation theory of real reductive groups these kinds of functors are called the Zuckerman functors [**175**].

Let $V \in C(\mathfrak{g},\mathfrak{l})_q$. The vector space of all linear functionals on V is a $U_q\mathfrak{g}$-module. An application of Γ to this module yields a $U_q\mathfrak{g}$-module $V^c \in C(\mathfrak{g},\mathfrak{l})_q$, to be called the dual object in $C(\mathfrak{g},\mathfrak{l})_q$. Every morphism $f : V_1 \to V_2$ in $C(\mathfrak{g},\mathfrak{l})_q$ determines the dual morphism $f^* : V_2^c \to V_1^c$ in an obvious way. Note that in the special case $\mathbb{S} = \varnothing$, the category $C(\mathfrak{g},\mathfrak{l})_q$ coincides with the category of P-weight $U_q\mathfrak{g}$-modules, and the above notion is compatible with that introduced before in Subsection 2.1.2.

Let us obtain a decomposition of a $U_q\mathfrak{g}$-module from $C(\mathfrak{g},\mathfrak{l})_q$ as a sum of its $U_q\mathfrak{l}$-isotypic components. The following claim can be proved in the same way as Proposition 2.34.

PROPOSITION 2.95. *Let $V \in C(\mathfrak{g},\mathfrak{l})_q$. Then $V = \bigoplus_{\lambda\in P_+^{\mathbb{S}}} V(\lambda)$ in the category of $U_q\mathfrak{l}$-modules, where the $U_q\mathfrak{l}$-modules $V(\lambda)$ are multiples of $L(\mathfrak{l},\lambda)$.*

As in [**116, 158**], we call a finitely generated module $V \in C(\mathfrak{g},\mathfrak{l})_q$ a Harish-Chandra module if the multiplicities of the simple $U_q\mathfrak{l}$-modules $L(\mathfrak{l},\lambda)$ in V are finite:

$$\dim_{U_q\mathfrak{l}}(L(\mathfrak{l},\lambda),V) < \infty, \qquad \lambda \in P_+^{\mathbb{S}}$$

(equivalently, if $\dim V(\lambda) < \infty$ for all $\lambda \in P_+^{\mathbb{S}}$).

REMARK 2.96. The study of Harish-Chandra modules is an important trend in representation theory. One has a number of works treating the special case $q = 1$, $\mathbb{S} = \varnothing$. For example, a classification of simple Harish-Chandra modules in this case was obtained by O. Mathieu [**230**]. A vast literature is dedicated to the Harish-Chandra modules related to representations of real reductive groups [**270**]. A classification of simple Harish-Chandra modules in this case is among the best known advances of the theory [**175**, pp. 741, 747]. The classification problem for simple Harish-Chandra modules in the general case is discussed in the works of Penkov, Zuckerman, and Serganova [**250, 249, 248**].

It should be noted that one could never expect that all the objects of the theory of real reductive groups admit quantization. Among the reasons for this, one should mention that the Lie subalgebra $\mathfrak{g}_0$ coincides with the subalgebra of fixed elements for the involution θ (as in Subsection 2.3.3) if and only if

$$\mathfrak{g} = \mathfrak{g}_{-1} \oplus \mathfrak{g}_0 \oplus \mathfrak{g}_1.$$

In what follows we need to make sure that the grading terminates sufficiently fast. To make it possible, we assume that a simple root α_{l_0} is chosen so that its coefficient in the decomposition of the highest root δ of the Lie algebra $\mathfrak{g}$ is 1 and

$$\mathbb{S} = \{1, 2, \ldots, l\} \setminus \{l_0\}.$$

The tables in [**44**] can be readily used to distinguish all such simple roots; see Subsection 2.3.8.

It is conventional in this important particular case to use the notation

$$\mathfrak{k} = \mathfrak{g}_0, \qquad \mathfrak{p}^{\pm} = \mathfrak{g}_{\pm 1}, \qquad \mathfrak{q}^{\pm} = \mathfrak{k} \oplus \mathfrak{p}^{\pm},$$

and to call the pair $(\mathfrak{g},\mathfrak{k})$ a Hermitian symmetric pair [**123**].

REMARK 2.97. 1. In our context we deal with the complexifications of real Lie subalgebras considered in the theory of Hermitian symmetric spaces [**123**].

2. Although the algebra $\mathfrak{k}$ is not simple, the results on quantum universal enveloping algebras listed in Section 2.1 are valid for $U_q\mathfrak{k}$.

3. The scalar product in $\mathfrak{h}_{\mathbb{R}}$ was chosen in Subsection 2.1.3 so that the squares of lengths for short roots are 2, while it is easy to show that α_{l_0} is a long root. Hence the scalar product in $\mathfrak{h}_{\mathbb{R}}$ chosen for $\mathfrak{g}$ works also for $\mathfrak{k}$.

Recall that the Hermitian form $(f_1, f_2) = \langle f_2^* f_1\rangle$ in $\mathbb{C}[\mathfrak{p}^-]_q$ was introduced in Subsection 2.3.3.

PROPOSITION 2.98. *If* $(\mathfrak{g},\mathfrak{k})$ *is a Hermitian symmetric pair, then* $(f,f)>0$ *for all nonzero* $f\in\mathbb{C}[\mathfrak{p}^-]_q$.

PROOF. The $U_q\mathfrak{h}$-invariance of the form $(\cdot,\cdot)$ implies that the homogeneous components $\mathbb{C}[\mathfrak{p}^-]_{q,j}$ are pairwise orthogonal. It remains to prove that the Hermitian form in question has positive restrictions to the subspaces $\mathbb{C}[\mathfrak{p}^-]_{q,j}$.

Consider the basis of the vector space $N(\mathfrak{q}^+,0)_{-j}$:

$$\left\{F_{\beta_M}^{j_M}F_{\beta_{M-1}}^{j_{M-1}}\cdots F_{\beta_{M'+1}}^{j_{M'+1}}v(\mathfrak{q}^+,0)\,\middle|\,\sum_{k=M'+1}^{M}j_k=j\right\}$$

introduced in Subsection 2.3.1, together with the biorthogonal basis in $\mathbb{C}[\mathfrak{p}^-]_{q,j}$. Proposition 2.93 and the results of Subsection 2.1.3 imply that the matrix of Hermitian form $(q^{-1}-q)^{-j}(\cdot,\cdot)$ with respect to this basis is nondegenerate and continuous in $q\in(0,1)$. Relations (2.73), (2.11) and Proposition 2.93 can be readily used to prove that as $q\to1$ one has for $f\in\mathbb{C}[\mathfrak{p}^-]_{q,j}$

$$\begin{aligned}&(q^{-1}-q)^{-j}(f,f)\\&\quad\sim(f^*\otimes f)\left(\overset{\curvearrowleft}{\prod_{k>M'}}\exp_{q^2_{\beta_k}}\left(\frac{q_{\beta_k}^{-1}-q_{\beta_k}}{q^{-1}-q}E_{\beta_k}\otimes F_{\beta_k}\right)(v(\mathfrak{q}^-,0)\otimes v(\mathfrak{q}^+,0))\right).\end{aligned}$$

Hence the matrix of Hermitian form has a limit as $q\to1$. It remains to prove positivity for the limit matrix. Note that for the Hermitian symmetric pair $(\mathfrak{g},\mathfrak{k})$ in the limit case $q=1$, one has the relations (2.66), (2.68), and for $k>M'$ one has additionally $E_{\beta_k}^*=-F_{\beta_k}$. Therefore, the limit Hermitian form is just the sum of squares of modules for the coefficients of the Taylor series for f at the origin, modulo some inessential positive factors. □

The study of Harish-Chandra modules corresponding to the Hermitian symmetric pairs $(\mathfrak{g},\mathfrak{k})$ and geometric realizations for those Harish-Chandra modules deserves special attention. This is the way to obtain remarkable results in noncommutative complex analysis in quantum bounded symmetric domains.

2.3.5. Quantum analogs of the subalgebras $U\mathfrak{u}^\pm\subset U\mathfrak{g}$. The results of this and the next subsections were obtained by D. Shklyarov, cf. [**276**]. In Subsection 2.3.1 the graded Lie algebra $\mathfrak{g}=\mathfrak{u}^-\bigoplus\mathfrak{l}\bigoplus\mathfrak{u}^+$ was considered. In the present subsection we are going to introduce q-analogs of universal enveloping algebras $U\mathfrak{u}^\pm$. More precisely, we are about to consider $U_q\mathfrak{g}$ together with the adjoint action of the subalgebra $U_q\mathfrak{l}$, and then we construct embeddings of $U_q\mathfrak{l}$-module algebras $i^\pm:\mathbb{C}[\mathfrak{u}^\mp]_q\hookrightarrow U_q\mathfrak{g}$. After that we set $U_q\mathfrak{u}^\pm\overset{\mathrm{def}}{=}i^\pm\mathbb{C}[\mathfrak{u}^\mp]_q$.

Here are our arguments are intended to persuade the reader that the subalgebras $U_q\mathfrak{u}^\pm$ produced this way can be treated as q-analogs of the universal enveloping algebras $U\mathfrak{u}^\pm$. First, these are $U_q\mathfrak{l}$-module subalgebras. Second, $U_q\mathfrak{u}^\pm\subset U_q\mathfrak{b}^\pm$. Third, the generators of $U_q\mathfrak{u}^\pm$ turn into the generators of $U\mathfrak{u}^\pm$ under the formal passage to a limit as $q\to1$; see Proposition 2.103. Finally, it will be demonstrated in the next subsection that the subalgebras $U_q\mathfrak{u}^\pm$ can be used to obtain a q-analog of the decomposition $U\mathfrak{g}=U\mathfrak{u}^-\cdot U\mathfrak{l}\cdot U\mathfrak{u}^+$.

We use the brief Sweedler notation for the coproduct: $\Delta(\xi)=\xi_1\otimes\xi_2$, cf. [**161**, p. 55]. The double indices are used to label the iterated application of the coproduct, for example, $(1\otimes\Delta)\Delta(\xi)=\xi_1\otimes\xi_{21}\otimes\xi_{22}$. Under this notation, the coassociativity for the coproduct is to be written as follows: $\xi_1 \otimes \xi_{21} \otimes \xi_{22} = \xi_{11} \otimes \xi_{12} \otimes \xi_2$.

Let us start with constructing the embedding i^+. Consider the generalized Verma module $N(\mathfrak{q}^+,0) = \bigoplus_{j=0}^{\infty} N(\mathfrak{q}^+,0)_{-j}$ endowed with its standard grading and the vector space $U_q\mathfrak{g}\widehat{\otimes}N(\mathfrak{q}^+,0)$ of the formal series

$$\sum_{j=0}^{\infty} c_j, \qquad c_j \in U_q\mathfrak{g} \otimes N(\mathfrak{q}^+,0)_{-j}.$$

Obviously, $U_q\mathfrak{g} \otimes N(\mathfrak{q}^+,0) \hookrightarrow U_q\mathfrak{g}\widehat{\otimes}N(\mathfrak{q}^+,0)$. Let

$$\mathcal{K} = \sum_{j=0}^{\infty} a_j \otimes b_j v(\mathfrak{q}^+,0) \in U_q\mathfrak{g}\widehat{\otimes}N(\mathfrak{q}^+,0),$$

where $\sum_{j=0}^{\infty} a_j \otimes b_j$ is the universal R-matrix. It is implicit that the latter decomposition is derived from the multiplicative formula (2.11) by multiplying series for q-exponents, and the reduced expression of w_0 is chosen in the same way as in Subsection 2.3.1. The factor q^{-t_0} may be omitted since $v(\mathfrak{q}^+,0)$ is the weight vector of weight zero.

PROPOSITION 2.99. *The linear map*

$$i^+ : \mathbb{C}[\mathfrak{u}^-]_q \to U_q\mathfrak{g}, \qquad i^+ : f \mapsto (\mathrm{id} \otimes f)(\mathcal{K}),$$

is an embedding of $U_q\mathfrak{l}$-module algebras.

PROOF. Let us prove the injectivity of i^+. We are going to use the fact that the generator $v(\mathfrak{q}^+,0)$ is $U_q\mathfrak{l}$-invariant:

$$\xi v(\mathfrak{q}^+,0) = \varepsilon(\xi) \cdot v(\mathfrak{q}^+,0), \qquad \xi \in U_q\mathfrak{l}. \tag{2.74}$$

Since the reduced expression for w_0 was chosen in Subsection 2.3.1 in such a way that the elements $F_{\beta_1}, F_{\beta_2}, \ldots, F_{\beta_{M'}}$ belong to $U_q\mathfrak{l}$, one has

$$\mathcal{K} = \sum c_{j_M, j_{M-1}, \ldots, j_{M'+1}} \cdot E_{\beta_M}^{j_M} E_{\beta_{M-1}}^{j_{M-1}} \cdots E_{\beta_{M'+1}}^{j_{M'+1}} \otimes F_{\beta_M}^{j_M} F_{\beta_{M-1}}^{j_{M-1}} \cdots F_{\beta_{M'+1}}^{j_{M'+1}} v(\mathfrak{q}^+,0). \tag{2.75}$$

Furthermore, all the coefficients

$$c_{j_M, j_{M-1}, \ldots, j_{M'+1}}, \qquad (j_M, j_{M-1}, \ldots, j_{M'+1}) \in \mathbb{Z}_+^{M-M'},$$

are nonzero, see (2.11). It remains to use the fact that the elements $E_{\beta_M}^{j_M} E_{\beta_{M-1}}^{j_{M-1}} \cdots E_{\beta_{M'+1}}^{j_{M'+1}}$ are linearly independent, and the elements $F_{\beta_M}^{j_M} F_{\beta_{M-1}}^{j_{M-1}} \cdots F_{\beta_{M'+1}}^{j_{M'+1}} v(\mathfrak{q}^+,0)$ form a basis in $N(\mathfrak{q}^+,0)$; see Corollary 2.6 and Proposition 2.76.

Let us show that i^+ is a morphism of algebras. Let $f_1, f_2 \in \mathbb{C}[\mathfrak{u}^-]_q$. Then

$$\begin{aligned} i^+(f_1 f_2) &= (\mathrm{id} \otimes f_1 f_2) \left(\sum_{j=0}^{\infty} a_j \otimes b_j v(\mathfrak{q}^+,0) \right) \\ &= (\mathrm{id} \otimes f_1 \otimes f_2) \left(\sum_{j=0}^{\infty} a_j \otimes \Delta^+(b_j v(\mathfrak{q}^+,0)) \right), \end{aligned}$$

where Δ^+ is the coproduct introduced in Subsection 2.3.2. Its definition implies

$$(\mathrm{id} \otimes f_1 \otimes f_2)\left(\sum_{j=0}^{\infty} a_j \otimes \Delta^+(b_j v(\mathfrak{q}^+,0))\right)$$
$$= (\mathrm{id} \otimes f_1 \otimes f_2)\left(\sum_{j=0}^{\infty} a_j \otimes \Delta^{\mathrm{cop}}(b_j)(v(\mathfrak{q}^+,0) \otimes v(\mathfrak{q}^+,0))\right).$$

It follows from the property (2.18) of the universal R-matrix that

$$\sum_{j=0}^{\infty} a_j \otimes \Delta^{\mathrm{cop}}(b_j) = \sum_{i,j=0}^{\infty} a_i a_j \otimes b_i \otimes b_j.$$

It remains to observe that

$$(\mathrm{id} \otimes f_1 \otimes f_2)\left(\sum_{i,j=0}^{\infty} a_i a_j \otimes b_i v(\mathfrak{q}^+,0) \otimes b_j v(\mathfrak{q}^+,0)\right) = i^+(f_1)\, i^+(f_2).$$

To complete the proof of our proposition, we show that i^+ is a morphism of $U_q\mathfrak{l}$-modules. Let us identify the space $U_q\mathfrak{g}\widehat{\otimes}N(\mathfrak{q}^+,0)$ with $\mathrm{Hom}(\mathbb{C}[\mathfrak{u}^-]_q, U_q\mathfrak{g})$ by assigning to $\sum_j \eta_j \otimes v_j \in U_q\mathfrak{g}\widehat{\otimes}N(\mathfrak{q}^+,0)$ the linear map $f \mapsto (\mathrm{id} \otimes f)(\sum_j \eta_j \otimes v_j)$. We endow the tensor product $U_q\mathfrak{g}\widehat{\otimes}N(\mathfrak{q}^+,0)$ with a structure of $U_q\mathfrak{l}$-module using the opposite coproduct $\Delta^{\mathrm{cop}} : U_q\mathfrak{l} \to U_q\mathfrak{l} \otimes U_q\mathfrak{l}$:

$$\xi\left(\sum_j \eta_j \otimes v_j\right) = \sum_j \mathrm{ad}(\xi_2)\eta_j \otimes \xi_1 v_j.$$

LEMMA 2.100. *If an element of the $U_q\mathfrak{l}$-module $U_q\mathfrak{g}\widehat{\otimes}N(\mathfrak{q}^+,0)$ is $U_q\mathfrak{l}$-invariant, then the associated linear map from $\mathbb{C}[\mathfrak{u}^-]_q$ to $U_q\mathfrak{g}$ is a morphism of $U_q\mathfrak{l}$-modules.*

PROOF OF THE LEMMA. Let A be a Hopf algebra and let V_1, V_2 be some A-modules. The canonical map

$$V_1^* \otimes V_1 \otimes V_2 \to V_2, \qquad f_1 \otimes v_1 \otimes v_2 \mapsto f_1(v_1)v_2,$$

is a morphism of A-modules. We thus get a map from the space of A-invariant elements $(V_1 \otimes V_2)^A$ to $\mathrm{Hom}_A(V_1^*, V_2)$. It remains to apply this argument to $A = U_q\mathfrak{l}$, $V_1 = N(\mathfrak{q}^+,0)$, $V_2 = U_q\mathfrak{g}$. □

To complete the proof of Proposition 2.99, it suffices to demonstrate that $\mathcal{K}$ is $U_q\mathfrak{l}$-invariant. Let $\xi \in U_q\mathfrak{l}$. Then

$$\xi(\mathcal{K}) = \xi\left(\sum_j a_j \otimes b_j v(\mathfrak{q}^+,0)\right) = \sum_j \mathrm{ad}(\xi_2)a_j \otimes \xi_1 b_j v(\mathfrak{q}^+,0)$$
$$= \sum_j \xi_{21} a_j S(\xi_{22}) \otimes \xi_1 b_j v(\mathfrak{q}^+,0).$$

By coassociativity,

$$\sum_j \xi_{21} a_j S(\xi_{22}) \otimes \xi_1 b_j v(\mathfrak{q}^+,0) = \sum_j \xi_{12} a_j S(\xi_2) \otimes \xi_{11} b_j v(\mathfrak{q}^+,0).$$

In view of the property (2.17) of the universal R-matrix,

$$\sum_j \xi_{12}a_j \otimes \xi_{11}b_j = \sum_j a_j\,\xi_{11} \otimes b_j\xi_{12}.$$

Hence $\sum_j \xi_{12}a_jS(\xi_2) \otimes \xi_{11}b_jv(\mathfrak{q}^+,0) = \sum_j a_j\,\xi_{11}S(\xi_2) \otimes b_j\xi_{12}v(\mathfrak{q}^+,0)$. The $U_q\mathfrak{l}$-invariance of the generator $v(\mathfrak{q}^+,0)$ (see (2.74)) implies

$$\sum_j a_j\xi_{11}S(\xi_2) \otimes b_j\xi_{12}v(\mathfrak{q}^+,0) = \sum_j a_j\xi_{11}S(\xi_2) \otimes b_j\varepsilon(\xi_{12})v(\mathfrak{q}^+,0).$$

The definition of counit and antipode in a Hopf algebra implies

$$\sum_j a_j\,\xi_{11}S(\xi_2) \otimes b_j\varepsilon(\xi_{12})v(\mathfrak{q}^+,0) = \sum_j a_j\xi_{11}\varepsilon(\xi_{12})S(\xi_2) \otimes b_jv(\mathfrak{q}^+,0)$$

$$= \sum_j a_j\xi_1S(\xi_2) \otimes b_jv(\mathfrak{q}^+,0) = \varepsilon(\xi)\sum_j a_j \otimes b_jv(\mathfrak{q}^+,0).$$

Hence $\mathcal{K}$ is $U_q\mathfrak{l}$-invariant, and the proof of Proposition 2.99 is completed. □

Let $U_q\mathfrak{u}^+$ be the image of the linear map i^+.

COROLLARY 2.101. *$U_q\mathfrak{u}^+$ is a $U_q\mathfrak{l}$-module subalgebra in $U_q\mathfrak{g}$.*

Formula (2.75) and the arguments following it imply

PROPOSITION 2.102. *The elements*

$$E_{\beta_M}^{j_M}E_{\beta_{M-1}}^{j_{M-1}}\cdots E_{\beta_{M'+1}}^{j_{M'+1}}, \qquad j_{M'+1},\ldots,j_{M-1},j_M \in \mathbb{Z}_+, \tag{2.76}$$

form a basis of the vector space $U_q\mathfrak{u}^+$.

This fact, together with Proposition 2.4 allow one to deduce

PROPOSITION 2.103. *$U_q\mathfrak{u}^+$ is a unital subalgebra in the algebra $U_q\mathfrak{g}$ generated by $E_{\beta_{M'+1}},\ldots,E_{\beta_{M-1}},E_{\beta_M}$.*

The latter claim and the results of Subsection 2.1.3 allow one to permute the factors in (2.76).

COROLLARY 2.104. *The elements*

$$E_{\beta_{M'+1}}^{j_{M'+1}}\cdots E_{\beta_{M-1}}^{j_{M-1}}E_{\beta_M}^{j_M}, \qquad j_{M'+1},\ldots,j_{M-1},j_M \in \mathbb{Z}_+,$$

form a basis of the vector space $U_q\mathfrak{u}^+$.

Let us turn to producing a q-analog of the universal enveloping algebra $U\mathfrak{u}^-$.

Recall that $U_q\mathfrak{g}$ is a $U_q\mathfrak{g}$-module algebra with respect to the adjoint action, and the involution (2.65) endows $U_q\mathfrak{g}$ with a structure of Hopf $*$-algebra. The following claim is well known.

PROPOSITION 2.105. *The $*$-algebra $(U_q\mathfrak{g},*)$ is a $(U_q\mathfrak{g},*)$-module algebra:*

$$(\mathrm{ad}(\xi)\eta)^* = \mathrm{ad}((S(\xi))^*)\eta^*, \qquad \xi,\eta \in U_q\mathfrak{g}.$$

PROOF. Obviously,

$$(\mathrm{ad}(\xi)\eta)^* = (\xi_1\eta S(\xi_2))^* = (S(\xi_2))^*\eta^*\xi_1^*. \tag{2.77}$$

On the other hand, $*$ is an automorphism and S is an antiautomorphism of the coalgebra $U_q\mathfrak{g}$. Therefore, $\triangle((S(\xi))^*) = (S(\xi_2))^* \otimes (S(\xi_1))^*$. Hence,

$$\mathrm{ad}((S(\xi))^*)\eta^* = (S(\xi_2))^*\eta^*S((S(\xi_1))^*). \tag{2.78}$$

Since $S((S(\xi))^*) = \xi^*$ for each $\xi \in U_q\mathfrak{g}$, the right-hand side of (2.77) coincides with the right hand side of (2.78). □

The latter claim readily implies that the subalgebra

$$U_q\mathfrak{u}^- \stackrel{\text{def}}{=} \{\xi^* \,|\, \xi \in U_q\mathfrak{u}^+\} \subset U_q\mathfrak{g}$$

is a $U_q\mathfrak{l}$-module algebra.

REMARK 2.106. Consider the linear map[5]

$$i^- : \mathbb{C}[\mathfrak{u}^+]_q \to U_q\mathfrak{g}, \qquad i^- : \; f \mapsto (i^+(f^*))^*.$$

It is an embedding of $U_q\mathfrak{l}$-module algebras due to Propositions 2.105, 2.99, and the relation $i^-\mathbb{C}[\mathfrak{u}^+]_q = U_q\mathfrak{u}^-$.

The maps $i^\pm$ allow one to produce an embedding of $U_q\mathfrak{l}$-modules

$$i : \mathbb{C}[\mathfrak{u}^-]_q \otimes \mathbb{C}[\mathfrak{u}^+]_q \to U_q\mathfrak{g}, \qquad i : f_1 \otimes f_2 \mapsto i^+(f_1) \cdot i^-(f_2).$$

This embedding respects the involutions in $\mathrm{Pol}(\mathfrak{u}^-)_q = \mathbb{C}[\mathfrak{u}^-]_q \otimes \mathbb{C}[\mathfrak{u}^+]_q$ and $U_q\mathfrak{g}$:

$$(i(f))^* = i(f^*), \qquad f \in \mathrm{Pol}(\mathfrak{u}^-)_q,$$

but it fails to be a morphism of algebras.

From now until the end of this subsection, we use the assumption of Subsection 2.3.4 that $(\mathfrak{g}, \mathfrak{k})$ is a Hermitian symmetric pair and we use the notation $\mathfrak{k}$, $\mathfrak{p}^\pm$ instead of $\mathfrak{l}$, $\mathfrak{u}^\pm$. Recall that in this case $\mathbb{S} = \{1, 2, \ldots, l\} \setminus \{l_0\}$,

$$\mathfrak{k} = \mathfrak{g}_0, \qquad \mathfrak{p}^\pm = \mathfrak{g}_{\pm 1}, \qquad \mathfrak{q}^\pm = \mathfrak{g}_0 \oplus \mathfrak{g}_{\pm 1}.$$

PROPOSITION 2.107. *Let $(\mathfrak{g}, \mathfrak{k})$ be a Hermitian symmetric pair. Then the subspace $\mathbb{C}[\mathfrak{p}^-]_{q,1}$ generates the subalgebra $\mathbb{C}[\mathfrak{p}^-]_q$, and the subspace $\mathbb{C}[\mathfrak{p}^+]_{q,-1}$ generates the subalgebra $\mathbb{C}[\mathfrak{p}^+]_q$.*

PROOF. It suffices to demonstrate that $\mathbb{C}[\mathfrak{p}^-]_{q,1}$ generates $\mathbb{C}[\mathfrak{p}^-]_q$. This subspace and this algebra can be replaced by their images under the embedding i^+ in $U_q\mathfrak{g}$. It remains to use the fact that the elements (2.76) form a basis in $i^+\mathbb{C}[\mathfrak{p}^-]_q$, and the subspace $i^+\mathbb{C}[\mathfrak{p}^-]_{q,1}$ contains all the *root vectors*

$$E_{\beta_{M'+1}}, E_{\beta_{M'+2}}, \ldots, E_{\beta_{M-1}}, E_{\beta_M}, \tag{2.79}$$

because the pair $(\mathfrak{g}, \mathfrak{k})$ is Hermitian symmetric, and the vectors (2.79) are homogeneous of degree 1 in this case. □

Let us emphasize that in the Hermitian symmetric case we use the notation $U_q\mathfrak{p}^\pm$ instead of $U_q\mathfrak{u}^\pm$.

PROPOSITION 2.108. *$U_q\mathfrak{p}^+ \subset U_q\mathfrak{g}$ is the smallest unital $U_q\mathfrak{k}$-module subalgebra which contains E_{l_0}.*

PROOF. Let $\mathcal{U}^+ \subset U_q\mathfrak{g}$ be the smallest unital $U_q\mathfrak{k}$-module subalgebra which contains E_{l_0}. We have to prove that $\mathcal{U}^+ = U_q\mathfrak{p}^+$.

Let us prove the inclusion $\mathcal{U}^+ \subset U_q\mathfrak{p}^+$. For that, it suffices to show $E_{l_0} \in U_q\mathfrak{p}^+$. But among the roots $\beta_{M'+1}, \ldots, \beta_{M-1}, \beta_M$ there exists a distinguished simple root α_{l_0}. To be precise, set $\beta_N = \alpha_{l_0}$, $M'+1 \le N \le M$. The elements E_{β_N} and E_{l_0} differ only by a (nonzero) factor since both belong to $(U_q\mathfrak{g})_{\alpha_{l_0}} \cap U_q\mathfrak{n}^+$, which is

[5] Inside the parentheses we use the involution on the algebra $\mathbb{C}[\mathfrak{u}^- \oplus \mathfrak{u}^+]_q$ defined in Proposition 2.92, while the involution outside the parentheses is just that on $U_q\mathfrak{g}$ given by (2.65).

one dimensional (see Proposition 2.3). It remains to use the fact that $E_{\beta_N} \in U_q\mathfrak{p}^+$, which follows from Proposition 2.102.

Now we prove the inclusion $\mathcal{U}^+ \supset U_q\mathfrak{p}^+$. Recall that all the weight $U_q\mathfrak{g}$-modules, in particular $U_q\mathfrak{g}$ with the adjoint action, have been endowed with a grading; see Subsection 2.3.1. The subspaces $\mathcal{U}^+$, $U_q\mathfrak{p}^+$ inherit the grading, and their homogeneous components are $U_q\mathfrak{k}$-submodules. Since $\mathcal{U}^+ \subset U_q\mathfrak{p}^+$, one has $(\mathcal{U}^+)_1 \subset (U_q\mathfrak{p}^+)_1$. It remains to prove the inequality

$$\dim(\mathcal{U}^+)_1 \geq \dim(U_q\mathfrak{p}^+)_1.$$

In fact, this will imply the inclusion $(\mathcal{U}^+)_1 \supset (U_q\mathfrak{p}^+)_1$, hence also the target claim $\mathcal{U}^+ \supset U_q\mathfrak{p}^+$, because the $U_q\mathfrak{k}$-submodule $(U_q\mathfrak{p}^+)_1$ generates the algebra $U_q\mathfrak{p}^+$.

First of all, note that the dimension of $(U_q\mathfrak{p}^+)_1$ is the same as in the limit case $q = 1$, which is just $M - M'$, because the elements $E_{\beta_{M'+1}}, \ldots, E_{\beta_{M-1}}, E_{\beta_M}$ have the degree of homogeneity 1 and form a basis in $(U_q\mathfrak{p}^+)_1$; see Proposition 2.103.

Also E_{l_0} is a homogeneous element of degree 1, hence $(\mathcal{U}^+)_1 \supset \mathrm{ad}(U_q\mathfrak{k})E_{l_0}$. In addition, E_{l_0} is a singular vector of the $U_q\mathfrak{k}$-module $\mathrm{ad}(U_q\mathfrak{k})E_{l_0}$, that is

$$\mathrm{ad}(F_j)E_{l_0} = F_j E_{l_0} K_j - E_{l_0} F_j K_j = 0, \qquad j \neq l_0,$$

and is a weight vector with weight α_{l_0}. By Proposition 2.30, the dimensions of simple weight modules do not change under the passage from the classical universal enveloping algebras to their quantum counterparts.

Thus, just as in the case $q = 1$,

$$\dim(\mathrm{ad}(U_q\mathfrak{k})E_{l_0}) \geq \dim((U_q\mathfrak{p}^+)_1).$$

Hence,

$$\dim(\mathcal{U}^+)_1 \geq \dim(\mathrm{ad}(U_q\mathfrak{k})E_{l_0}) \geq \dim(U_q\mathfrak{p}^+)_1. \qquad \square$$

The description of the algebra $U_q\mathfrak{p}^+$ given in Proposition 2.108 is due to Jakobsen's paper [**136**], where it was used as a definition of quantum analog of the universal enveloping algebra $U\mathfrak{p}^+$ in the Hermitian symmetric case. One of the principal goals of [**136**] was to produce an analog of the algebra $\mathbb{C}[\mathfrak{p}^-]$ of polynomials on the vector space $\mathfrak{p}^-$, and this analog is *defined* there to be $\mathbb{C}[\mathfrak{p}^-]_q = U_q\mathfrak{p}^+$. In the Hermitian symmetric case this definition can be motivated as follows: for $q = 1$, the Killing form on $\mathfrak{g}$ determines an isomorphism of $U\mathfrak{k}$-modules $\mathfrak{p}^- \cong (\mathfrak{p}^+)^*$, hence an isomorphism of $U\mathfrak{k}$-module algebras $\mathbb{C}[\mathfrak{p}^-] \cong S(\mathfrak{p}^+)$. A close approach to producing a quantum analog of the algebra $\mathbb{C}[\mathfrak{p}^-]$ is used in [**156, 21**].

Proposition 2.108 and the isomorphism $i^+ : \mathbb{C}[\mathfrak{p}^-]_q \to U_q\mathfrak{p}^+$ introduced in this subsection actually set up an equivalence of our approach and that of [**136**] to producing an analog of $\mathbb{C}[\mathfrak{p}^-]$.

2.3.6. A quantum analog of the decomposition $U\mathfrak{g} = U\mathfrak{u}^- \cdot U\mathfrak{l} \cdot U\mathfrak{u}^+$. It follows from $\mathfrak{g} = \mathfrak{u}^- \oplus \mathfrak{l} \oplus \mathfrak{u}^+$, together with the Poincaré–Birkhoff–Witt theorem, that one has a decomposition

$$U\mathfrak{g} = U\mathfrak{u}^- \cdot U\mathfrak{l} \cdot U\mathfrak{u}^+. \tag{2.80}$$

More precisely, the linear map

$$m : U\mathfrak{u}^- \otimes U\mathfrak{l} \otimes U\mathfrak{u}^+ \to U\mathfrak{g}, \qquad m : \xi \otimes \eta \otimes \zeta \mapsto \xi\eta\zeta,$$

determined by the multiplication in $U\mathfrak{g}$, is bijective. This is how the term *decomposition* will be understood in this subsection.

We want to obtain a quantum analog of (2.80). To be more precise, we are going to demonstrate that there exist the following six decompositions:

$$U_q\mathfrak{g} = U_q\mathfrak{u}^- \cdot U_q\mathfrak{l} \cdot U_q\mathfrak{u}^+ = U_q\mathfrak{u}^- \cdot U_q\mathfrak{u}^+ \cdot U_q\mathfrak{l} = U_q\mathfrak{l} \cdot U_q\mathfrak{u}^- \cdot U_q\mathfrak{u}^+ \\ = U_q\mathfrak{l} \cdot U_q\mathfrak{u}^+ \cdot U_q\mathfrak{u}^- = U_q\mathfrak{u}^+ \cdot U_q\mathfrak{u}^- \cdot U_q\mathfrak{l} = U_q\mathfrak{u}^+ \cdot U_q\mathfrak{l} \cdot U_q\mathfrak{u}^-. \tag{2.81}$$

PROPOSITION 2.109. *The following decompositions hold:*

$$U_q\mathfrak{q}^+ = U_q\mathfrak{l} \cdot U_q\mathfrak{u}^+, \qquad U_q\mathfrak{q}^+ = U_q\mathfrak{u}^+ \cdot U_q\mathfrak{l}. \tag{2.82}$$

PROOF. Let us obtain the first of those. Using the decomposition $U_q\mathfrak{g} = U_q\mathfrak{n}^- \cdot U_q\mathfrak{h} \cdot U_q\mathfrak{n}^+$ (see Subsection 2.1.3), we get

$$U_q\mathfrak{q}^+ = (U_q\mathfrak{q}^+ \cap U_q\mathfrak{n}^-) \cdot U_q\mathfrak{h} \cdot U_q\mathfrak{n}^+, \qquad U_q\mathfrak{l} = (U_q\mathfrak{l} \cap U_q\mathfrak{n}^-) \cdot U_q\mathfrak{h} \cdot (U_q\mathfrak{l} \cap U_q\mathfrak{n}^+).$$

Since $U_q\mathfrak{u}^+ \subset U_q\mathfrak{n}^+$ and $U_q\mathfrak{q}^+ \cap U_q\mathfrak{n}^- = U_q\mathfrak{l} \cap U_q\mathfrak{n}^-$, it suffices to obtain the decomposition $U_q\mathfrak{n}^+ = (U_q\mathfrak{l} \cap U_q\mathfrak{n}^+) \cdot U_q\mathfrak{u}^+$. It remains to use the fact that the elements

$$E_{\beta_1}^{j_1} E_{\beta_2}^{j_2} \cdots E_{\beta_{M'}}^{j_{M'}}, \qquad j_1, j_2, \ldots, j_{M'} \in \mathbb{Z}_+, \tag{2.83}$$

form a basis of the vector space $U_q\mathfrak{l} \cap U_q\mathfrak{n}^+$, together with Corollary 2.104 and Proposition 2.3.

The second decomposition in (2.82) can be proved similarly. One needs only to start with the decomposition $U_q\mathfrak{g} = U_q\mathfrak{n}^+ \cdot U_q\mathfrak{h} \cdot U_q\mathfrak{n}^-$, and then finally to apply Proposition 2.102 and Corollary 2.6, together with the fact that the elements

$$E_{\beta_{M'}}^{j_{M'}} \cdots E_{\beta_2}^{j_2} E_{\beta_1}^{j_1}, \qquad j_1, j_2, \ldots, j_{M'} \in \mathbb{Z}_+, \tag{2.84}$$

form a basis of the vector space $U_q\mathfrak{l} \cap U_q\mathfrak{n}^+$. □

An application of the involution $*$ to (2.82) yields

COROLLARY 2.110. *The following decompositions hold:*

$$U_q\mathfrak{q}^- = U_q\mathfrak{l} \cdot U_q\mathfrak{u}^-, \qquad U_q\mathfrak{q}^- = U_q\mathfrak{u}^- \cdot U_q\mathfrak{l}. \tag{2.85}$$

PROPOSITION 2.111. *The following decompositions hold:*

$$U_q\mathfrak{g} = U_q\mathfrak{q}^- \cdot U_q\mathfrak{u}^+, \qquad U_q\mathfrak{g} = U_q\mathfrak{u}^+ \cdot U_q\mathfrak{q}^-. \tag{2.86}$$

PROOF. This claim can be easily derived from the proof of Proposition 2.109. In fact,

$$U_q\mathfrak{g} = U_q\mathfrak{n}^- \cdot U_q\mathfrak{h} \cdot U_q\mathfrak{n}^+ = U_q\mathfrak{n}^- \cdot U_q\mathfrak{h} \cdot (U_q\mathfrak{l} \cap U_q\mathfrak{n}^+) \cdot U_q\mathfrak{u}^+ = U_q\mathfrak{q}^- \cdot U_q\mathfrak{u}^+.$$

In a similar way,

$$U_q\mathfrak{g} = U_q\mathfrak{n}^+ \cdot U_q\mathfrak{h} \cdot U_q\mathfrak{n}^- = U_q\mathfrak{u}^+ \cdot (U_q\mathfrak{l} \cap U_q\mathfrak{n}^+) \cdot U_q\mathfrak{h} \cdot U_q\mathfrak{n}^- = U_q\mathfrak{u}^+ \cdot U_q\mathfrak{q}^-. \quad \square$$

We apply the involution $*$ to (2.86) to obtain

COROLLARY 2.112. *The following decompositions hold:*

$$U_q\mathfrak{g} = U_q\mathfrak{q}^+ \cdot U_q\mathfrak{u}^-, \qquad U_q\mathfrak{g} = U_q\mathfrak{u}^- \cdot U_q\mathfrak{q}^+. \tag{2.87}$$

The decompositions (2.86) and (2.87), together with (2.82) and (2.85) imply the result announced at the beginning of this subsection.

PROPOSITION 2.113. *The decompositions* (2.81) *hold.*

To conclude, we describe a simple and natural generalization of the results of this and the previous subsections. Recall that our construction of Lie subalgebras $\mathfrak{l}$, $\mathfrak{q}^{\pm}$, $\mathfrak{u}^{\pm}$ involves a subset $\mathbb{S} \subset \{1, \ldots, l\}$; see Subsection 2.3.1. Our previous assumption was that this subset is fixed, and so it was missing in the notation. We alter the notation and write $\mathfrak{l}_{\mathbb{S}}$, $\mathfrak{q}^{\pm}_{\mathbb{S}}$, $\mathfrak{u}^{\pm}_{\mathbb{S}}$, $U_q\mathfrak{l}_{\mathbb{S}}$, $U_q\mathfrak{q}^{\pm}_{\mathbb{S}}$, $U_q\mathfrak{u}^{\pm}_{\mathbb{S}}$ instead of $\mathfrak{l}$, $\mathfrak{q}^{\pm}$, $\mathfrak{u}^{\pm}$, $U_q\mathfrak{l}$, $U_q\mathfrak{q}^{\pm}$, $U_q\mathfrak{u}^{\pm}$ until the end of this subsection so that the dependence on the choice of $\mathbb{S}$ is now explicit.

Let $\mathbb{K} \subset \{1, \ldots, l\}$ and $\mathbb{L} \subset \mathbb{K}$. One has the natural inclusions of Lie algebras $\mathfrak{l}_{\mathbb{L}} \subset \mathfrak{l}_{\mathbb{K}}$, $\mathfrak{q}^{\pm}_{\mathbb{L}} \subset \mathfrak{q}^{\pm}_{\mathbb{K}}$, together with the corresponding inclusions of the associated Hopf algebras $U_q\mathfrak{l}_{\mathbb{L}} \subset U_q\mathfrak{l}_{\mathbb{K}}$, $U_q\mathfrak{q}^{\pm}_{\mathbb{L}} \subset U_q\mathfrak{q}^{\pm}_{\mathbb{K}}$. Now one can apply the arguments of Subsection 2.3.1 used before to obtain the decomposition $\mathfrak{g} = \mathfrak{u}^- \oplus \mathfrak{l} \oplus \mathfrak{u}^+$, to the pair $(\mathfrak{l}_{\mathbb{K}}, \mathfrak{l}_{\mathbb{L}})$ of reductive Lie algebras. We thus get an ad $\mathfrak{l}_{\mathbb{L}}$-invariant decomposition

$$\mathfrak{l}_{\mathbb{K}} = \mathfrak{u}^-_{\mathbb{KL}} \oplus \mathfrak{l}_{\mathbb{L}} \oplus \mathfrak{u}^+_{\mathbb{KL}}.$$

Let us impose the notation

$$\mathfrak{q}^{\pm}_{\mathbb{KL}} = \mathfrak{l}_{\mathbb{L}} \oplus \mathfrak{u}^{\pm}_{\mathbb{KL}} = \mathfrak{q}^{\pm}_{\mathbb{L}} \cap \mathfrak{l}_{\mathbb{K}}, \qquad U_q\mathfrak{q}^{\pm}_{\mathbb{KL}} = U_q\mathfrak{q}^{\pm}_{\mathbb{L}} \cap U_q\mathfrak{l}_{\mathbb{K}}.$$

One can reproduce literally the constructions of Subsection 2.3.2 to introduce the algebras $\mathbb{C}[\mathfrak{u}^{\pm}_{\mathbb{KL}}]_q$, which are q-analogs of polynomial algebras on the vector spaces $\mathfrak{u}^{\pm}_{\mathbb{KL}}$. Note that $\mathbb{C}[\mathfrak{u}^{\pm}_{\mathbb{KL}}]_q$ are $U_q\mathfrak{l}_{\mathbb{K}}$-module algebras, and, in particular, $U_q\mathfrak{l}_{\mathbb{L}}$-module algebras. We apply the construction of Subsection 2.3.5 to introduce $U_q\mathfrak{l}_{\mathbb{L}}$-module subalgebras $U_q\mathfrak{u}^{\pm}_{\mathbb{KL}} \subset U_q\mathfrak{l}_{\mathbb{K}}$, which are quantum analogs of the universal enveloping algebras $U\mathfrak{u}^{\pm}_{\mathbb{KL}}$.[6]

One can use the same argument as in the proofs of Propositions 2.109, 2.111, 2.113, and Corollaries 2.110 and 2.112 to obtain the following result.

PROPOSITION 2.114. *The claims of Propositions* 2.109, 2.111, 2.113, *and Corollaries* 2.110 and 2.112 *are still valid if* $\mathfrak{g}$, $\mathfrak{l}$, $\mathfrak{q}^+$, $\mathfrak{q}^-$, $\mathfrak{u}^+$, $\mathfrak{u}^-$ *are replaced, respectively, by*

- $\mathfrak{l}_{\mathbb{K}}$, $\mathfrak{l}_{\mathbb{L}}$, $\mathfrak{q}^+_{\mathbb{KL}}$, $\mathfrak{q}^-_{\mathbb{KL}}$, $\mathfrak{u}^+_{\mathbb{KL}}$, $\mathfrak{u}^-_{\mathbb{KL}}$;
- $\mathfrak{q}^+_{\mathbb{K}}$, $\mathfrak{l}_{\mathbb{L}}$, $\mathfrak{q}^+_{\mathbb{L}}$, $\mathfrak{q}^-_{\mathbb{KL}}$, $\mathfrak{u}^+_{\mathbb{L}}$, $\mathfrak{u}^-_{\mathbb{KL}}$;
- $\mathfrak{q}^-_{\mathbb{K}}$, $\mathfrak{l}_{\mathbb{L}}$, $\mathfrak{q}^+_{\mathbb{KL}}$, $\mathfrak{q}^-_{\mathbb{L}}$, $\mathfrak{u}^+_{\mathbb{KL}}$, $\mathfrak{u}^-_{\mathbb{L}}$.

All the decompositions derived here are just products of $U_q\mathfrak{l}_{\mathbb{L}}$*-module subalgebras.*

2.3.7. The quadratic algebra $\mathbb{C}[\mathfrak{p}^-]_q$. It was demonstrated in Subsection 2.3.5 that in the Hermitian symmetric case the subspace $\mathbb{C}[\mathfrak{p}^-]_{q,1}$ generates the algebra $\mathbb{C}[\mathfrak{p}^-]_q$. Now we find defining relations.

Consider the linear map

$$m : \mathbb{C}[\mathfrak{p}^-]_q \otimes \mathbb{C}[\mathfrak{p}^-]_q \to \mathbb{C}[\mathfrak{p}^-]_q, \qquad m : f_1 \otimes f_2 \mapsto f_1 f_2,$$

together with the universal R-matrix of the Hopf algebra $U_q\mathfrak{g}$. The definition of the multiplication m implies that

$$\varphi\psi = m\check{R}_{\mathbb{C}[\mathfrak{p}^-]_q, \mathbb{C}[\mathfrak{p}^-]_q}(\varphi \otimes \psi), \qquad \varphi, \psi \in \mathbb{C}[\mathfrak{p}^-]_q, \tag{2.88}$$

which means that $\mathbb{C}[\mathfrak{p}^-]_q$ is a commutative algebra in the braided tensor category $\mathcal{C}^-$; see Subsection 1.3.8.

[6]The embeddings $\mathbb{C}[\mathfrak{u}^{\pm}_{\mathbb{KL}}]_q \hookrightarrow U_q\mathfrak{l}_{\mathbb{K}}$ are to be produced using the universal R-matrix for the quantum universal enveloping algebra $U_q\mathfrak{l}^{ss}_{\mathbb{K}}$ of the semisimple part $\mathfrak{l}^{ss}_{\mathbb{K}} = [\mathfrak{l}_{\mathbb{K}}, \mathfrak{l}_{\mathbb{K}}]$. Just as before, the multiplicative formula for the universal R-matrix should be written here using a special reduced expression for the longest element of the group $W_{\mathbb{K}}$ (see Subsection 2.3.1).

Let $\mathcal{L}$ be the kernel of the linear map

$$\mathbb{C}[\mathfrak{p}^-]_{q,1} \otimes \mathbb{C}[\mathfrak{p}^-]_{q,1} \to \mathbb{C}[\mathfrak{p}^-]_{q,2}, \qquad \varphi \otimes \psi \mapsto \varphi\psi - m\check{R}_{\mathbb{C}[\mathfrak{p}^-]_q,\mathbb{C}[\mathfrak{p}^-]_q}(\varphi \otimes \psi).$$

In the classical limit case $q = 1$ this is the subspace of all antisymmetric tensors.

Consider the subalgebra $U_q\mathfrak{k}_{ss}$ generated by $\{K_j^{\pm 1}, E_j, F_j\}_{j \neq l_0}$.

By (2.88), $m\mathcal{L} = 0$. Equivalently, the elements of $\mathcal{L}$ are quadratic relations. We will describe this subspace in terms of the universal R-matrix of the Hopf algebra $U_q\mathfrak{k}_{ss}$ and the associated morphism of $U_q\mathfrak{k}$-modules

$$\widetilde{R}_{\mathbb{C}[\mathfrak{p}^-]_{q,1},\mathbb{C}[\mathfrak{p}^-]_{q,1}} : \mathbb{C}[\mathfrak{p}^-]_{q,1} \otimes \mathbb{C}[\mathfrak{p}^-]_{q,1} \to \mathbb{C}[\mathfrak{p}^-]_{q,1} \otimes \mathbb{C}[\mathfrak{p}^-]_{q,1},$$

determined by the action of the universal R-matrix, followed by the "naive" flip of tensor factors.

LEMMA 2.115. *$\mathcal{L}$ contains all the eigenvectors of the operator $\widetilde{R}_{\mathbb{C}[\mathfrak{p}^-]_{q,1},\mathbb{C}[\mathfrak{p}^-]_{q,1}}$ with negative eigenvalues.*

PROOF. Let $\mathcal{L}'$ be the spectral subspace of $\widetilde{R}_{\mathbb{C}[\mathfrak{p}^-]_{q,1},\mathbb{C}[\mathfrak{p}^-]_{q,1}}$ associated to the negative semi-axis $(-\infty, 0)$. It suffices to prove that $\mathcal{L} \supset \mathcal{L}'$.

Obviously, $\mathcal{L}$ and $\mathcal{L}'$ are $U_q\mathfrak{k}$-submodules. In the classical case $q = 1$, the multiplicities of simple weight $U\mathfrak{k}$-modules in $(\mathfrak{p}^-)^* \otimes (\mathfrak{p}^-)^*$ do not exceed 1 because the weight subspaces of $U\mathfrak{k}$-module $\mathfrak{p}^-$ are one dimensional [**348**, pp. 591–592]. Hence by Proposition 2.30, these multiplicities are just 1 for all $q \in (0,1)$. Therefore, the subspaces $\mathcal{L}$ and $\mathcal{L}'$ are determined by their $U_q\mathfrak{k}$-spectra, that is, the collections of highest weights of their simple $U_q\mathfrak{k}$-submodules.

For $q = 1$ these collections of highest weights are the same. It remains to monitor their dependence on $q \in (0,1]$. By (2.31), the spectrum of the operator $\widetilde{R}_{\mathbb{C}[\mathfrak{p}^-]_{q,1},\mathbb{C}[\mathfrak{p}^-]_{q,1}}$ lies on the real axis and does not contain 0. Hence the analytic dependence of the spectral projection corresponding to the negative semi-axis follows from the analytic dependence of the operator $\widetilde{R}_{\mathbb{C}[\mathfrak{p}^-]_{q,1},\mathbb{C}[\mathfrak{p}^-]_{q,1}}$.

Let us monitor the dependence on q for the operators

$$E_j, F_j, \quad j \neq l_0, \qquad H_i, \quad i = 1, 2, \ldots, l,$$

the subspace $\mathcal{L}$, and the operator $\widetilde{R}_{\mathbb{C}[\mathfrak{p}^-]_{q,1},\mathbb{C}[\mathfrak{p}^-]_{q,1}}$. For that, choose a basis of weight vectors in $\mathbb{C}[\mathfrak{p}^-]_q$ which is biorthogonal to that described in Proposition 2.76. We thus obtain bases in $\mathbb{C}[\mathfrak{p}^-]_{q,1}$, $\mathbb{C}[\mathfrak{p}^-]_{q,1} \otimes \mathbb{C}[\mathfrak{p}^-]_{q,1}$, $\mathbb{C}[\mathfrak{p}^-]_{q,2}$. The target claim $\mathcal{L} \supset \mathcal{L}'$ now follows from the fact that the matrix elements of the operators m and $\widetilde{R}_{\mathbb{C}[\mathfrak{p}^-]_{q,1},\mathbb{C}[\mathfrak{p}^-]_{q,1}}$ in the above bases depend analytically on $q \in (0,1]$. □

PROPOSITION 2.116. *$\mathcal{L}$ is an eigenspace of $\widetilde{R}_{\mathbb{C}[\mathfrak{p}^-]_{q,1},\mathbb{C}[\mathfrak{p}^-]_{q,1}}$ corresponding to its unique negative eigenvalue $-q^{\frac{4}{(H_S,H_S)}}$. The multiplicity of this eigenvalue is $\frac{\dim\mathfrak{p}^-(\dim\mathfrak{p}^- - 1)}{2}$.*

PROOF. Clearly $\dim\mathcal{L} \le \frac{\dim\mathfrak{p}^-(\dim\mathfrak{p}^- - 1)}{2}$ since $\dim\mathbb{C}[\mathfrak{p}^-]_{q,2} = \frac{\dim\mathfrak{p}^-(\dim\mathfrak{p}^- + 1)}{2}$. It remains to apply Lemma 2.115 and the fact that the dimension of the eigenspace of $\widetilde{R}_{\mathbb{C}[\mathfrak{p}^-]_{q,1},\mathbb{C}[\mathfrak{p}^-]_{q,1}}$ corresponding to the eigenvalue $-q^{\frac{4}{(H_S,H_S)}}$, is at least $\frac{\dim\mathfrak{p}^-(\dim\mathfrak{p}^- - 1)}{2}$. Let us prove the latter inequality.

Consider the operator $\check{R}_{N(\mathfrak{q}^+,-\alpha_{l_0})^*,N(\mathfrak{q}^+,-\alpha_{l_0})^*}$ and its restriction to the subspace $N(\mathfrak{q}^+,-\alpha_{l_0})^*_{-1} \otimes N(\mathfrak{q}^+,-\alpha_{l_0})^*_{-1}$. It is easy to show that this restriction differs from the operator $\widetilde{R}_{\mathbb{C}[\mathfrak{p}^-]_{q,1},\mathbb{C}[\mathfrak{p}^-]_{q,1}}$ we are interested in only the scalar factor $q^{-\frac{4}{(H_{\mathbb{S}},H_{\mathbb{S}})}}$. (In fact, to compare the universal R-matrices of the Hopf algebras $U_q\mathfrak{g}$ and $U_q\mathfrak{k}_{ss}$, one can use the reduced expression for the element w_0 described in Subsection 2.3.1 by (2.11), (2.14), together with the compatibility of scalar products in the Cartan subalgebras; see Remark 2.97.)

To complete the proof of Proposition 2.116, we need the following auxiliary results. First, the images of morphisms of $U_q\mathfrak{g}$-modules

$$N(\mathfrak{q}^+,w\rho-\rho)\to N(\mathfrak{q}^+,-\alpha_{l_0})\otimes N(\mathfrak{q}^+,-\alpha_{l_0}),\qquad w\in W^{\mathbb{S}}\quad\text{and}\quad l(w)=2,$$

intersect with $N(\mathfrak{q}^+,-\alpha_{l_0})_{-1}\otimes N(\mathfrak{q}^+,-\alpha_{l_0})_{-1}$ by subspaces whose sum has dimension $\frac{\dim\mathfrak{p}^-(\dim\mathfrak{p}^- -1)}{2}$. Second, the restriction of $\check{R}_{N(\mathfrak{q}^+,-\alpha_{l_0}),N(\mathfrak{q}^+,-\alpha_{l_0})}$ to the image of every such morphism is just -1. These results will be obtained in Subsection 2.6.2. The last step of the proof is to refer to the relation

$$\left(\check{R}_{N(\mathfrak{q}^+,\lambda)^*N(\mathfrak{q}^+,\mu)^*}\right)^*=\left(\check{R}_{N(\mathfrak{q}^+,\lambda)N(\mathfrak{q}^+,\mu)}\right)^{-1};$$

see Subsection 2.1.4. □

REMARK 2.117. Consider the quadratic algebra defined by the spaces of generators $\mathbb{C}[\mathfrak{p}^-]_{q,1}$ and relations $\mathcal{L}\subset\mathbb{C}[\mathfrak{p}^-]_{q,1}\otimes\mathbb{C}[\mathfrak{p}^-]_{q,1}$. It is the quotient algebra $F=T(\mathbb{C}[\mathfrak{p}^-]_{q,1})/J$ of the tensor algebra $T(\mathbb{C}[\mathfrak{p}^-]_{q,1})$ by the two-sided ideal J generated by $\mathcal{L}\subset T(\mathbb{C}[\mathfrak{p}^-]_{q,1})$.

Let us produce a monomial basis of the vector space F. Just as in the classical case $q=1$, the weights of the $U_q\mathfrak{k}$-module $\mathbb{C}[\mathfrak{p}^-]_{q,1}$ are of the form $-\alpha_{l_0}-\sum_{i\neq l_0}n_i\alpha_i$, and all the weight subspaces are one dimensional. We endow the set of weights of this $U_q\mathfrak{k}$-module with the linear order determined by the lexicographic ordering on the strings $(-n_1,-n_2,\ldots,-n_{l_0-1},-1,-n_{l_0+1},\ldots,-n_l)$. Choose a basis $\{z_1,z_2,\ldots,z_{\dim\mathfrak{p}^-}\}$ of weight vectors in the space $\mathbb{C}[\mathfrak{p}^-]_{q,1}$ in the same way as in the proof of Lemma 2.115, and sort them in the ascending order of weights. Obviously, the free algebra $\mathbb{C}\langle z_1,z_2,\ldots,z_{\dim\mathfrak{p}^-}\rangle$ is isomorphic to the tensor algebra $T(\mathbb{C}[\mathfrak{p}^-]_{q,1})$, and (2.11) implies that the tensors

$$z_i\otimes z_j+q^{-\frac{4}{(H_{\mathbb{S}},H_{\mathbb{S}})}}\sum_{k<m}\widetilde{R}_{ij}^{km}z_k\otimes z_m,\qquad i>j,$$

form a Gröbner basis of the ideal J; see Subsection 2.3.9. Hence the set

$$\{z_1^{j_1}z_2^{j_2}z_3^{j_3}\cdots z_{\dim\mathfrak{p}^-}^{j_{\dim\mathfrak{p}^-}}\mid j_1<j_2<\cdots<j_{\dim\mathfrak{p}^-}\}\tag{2.89}$$

is a basis of the vector space F. The action of the generators E_i, F_i, $K_i^{\pm1}$ in this basis is given by the matrices whose elements belong to the field of rational functions $\mathbb{Q}(q)$ and have no poles at $q\in(0,1]$. This follows from the definitions and Proposition 2.4.

Due to the next result, the same formula (2.89) determines a basis of the vector space $\mathbb{C}[\mathfrak{p}^-]_q$.

PROPOSITION 2.118. *$\mathbb{C}[\mathfrak{p}^-]_q$ is a quadratic algebra with the space of generators $\mathbb{C}[\mathfrak{p}^-]_{q,1}$ and the space of relations $\mathcal{L}$.*

PROOF. Just as in Remark 2.117, consider the quadratic algebra F with the space of generators $\mathbb{C}[\mathfrak{p}^-]_{q,1}$ and the space of relations $\mathcal{L}$. One needs to prove that the natural morphism of graded algebras $F \to \mathbb{C}[\mathfrak{p}^-]_q$ is bijective. The surjectivity is due to the fact that the subspace $\mathbb{C}[\mathfrak{p}^-]_{q,1}$ generates the algebra $\mathbb{C}[\mathfrak{p}^-]_q$; see Proposition 2.107. It remains to apply the equalities

$$\dim \mathbb{C}[\mathfrak{p}^-]_{q,j} = \binom{j + \dim \mathfrak{p}^- - 1}{\dim \mathfrak{p}^- - 1} = \dim F_j,$$

where the latter can be easily obtained using the basis (2.89). □

REMARK 2.119. Proposition 2.107, together with the refining Proposition 2.118, are due to I. Heckenberger and S. Kolb [**117**]. Their proofs are significantly different from ours.

2.3.8. Appendix on prehomogeneous vector spaces. The following definitions and results are expounded in detail in the monograph of H. Rubenthaler [**265**].

Let V be the space of a finite-dimensional rational representation π of a connected complex affine algebraic group. It is called a prehomogeneous vector space if there exists an open orbit Ω. Such an orbit is evidently unique. If the representation π is irreducible, the prehomogeneous vector space V is also called irreducible. A complete list of irreducible prehomogeneous vector spaces, up to a natural isomorphism, was obtained by M. Sato and T. Kimura [**167**, pp. 228–229]. In what follows we assume the group in question to be reductive.

A prehomogeneous vector space V is said to be regular if the Zariski closed set $V \setminus \Omega$ is a hypersurface.

Let us introduce some extensively used classes of prehomogeneous vector spaces. Suppose $\mathfrak{g}$ is a simple complex Lie algebra, $\{\alpha_1, \alpha_2, \ldots, \alpha_l\}$ the system of its simple roots, and $\mathbb{S} \subset \{1, 2, \ldots, l\}$.

Consider the simply connected complex affine algebraic group G with Lie algebra $\mathfrak{g}$. The coordinate ring of G is a Hopf algebra generated by matrix elements of the fundamental representations; see [**326**, p. 94], [**327**, p. 204].

In Subsection 2.3.1 a Lie algebra was endowed with the grading

$$\mathfrak{g} = \bigoplus_i \mathfrak{g}_i.$$

Let $L = \{a \in G | \ \mathrm{Ad}_a H_{\mathbb{S}} = H_{\mathbb{S}}\}$ be the subgroup of all the elements which preserve the above grading; see [**130**, pp. 114–116]. This subgroup is connected [**132**, p. 33], [**327**, pp. 227–228].

Due to the Vinberg theorem [**265**, p. 119], [**324**], the vector spaces $\mathfrak{g}_{\pm 1}$ are prehomogeneous vector spaces with the symmetry group L. These $\mathfrak{g}_0$-modules are simple if and only if the set $\{1, 2, \ldots, l\} \setminus \mathbb{S}$ reduces to a single element [**265**, p. 126], [**326**, p. 113].

We follow the idea of H. Rubenthaler in distinguishing an important class of such prehomogeneous vector spaces. Assume that $\mathbb{S} = \{1, 2, \ldots, l\} \setminus \{l_0\}$ and that $n_{l_0} = 1$ in the decomposition of the highest root

$$\delta = \sum_j n_j \alpha_j. \tag{2.90}$$

In this case the grading of the Lie algebra $\mathfrak{g}$ terminates shortly:

$$\mathfrak{g} = \mathfrak{g}_{-1} \oplus \mathfrak{g}_0 \oplus \mathfrak{g}_{+1},$$

and its homogeneous components $\mathfrak{g}_{\pm 1}$ are Abelian Lie algebras. These are called *prehomogeneous vector spaces of commutative parabolic type* [**265**, p. 147]. A complete list of such prehomogeneous vector spaces can be easily produced using an explicit form of the decompositions (2.90) for irreducible root systems [**44**]. In the notation of Bourbaki [**44**] this list has the following form:

A_l, $l_0 \leq 1, 2, \ldots, l$, complex $l_0 \times (l + 1 - l_0)$-matrices;
C_l, $l_0 = l$, complex symmetric $l \times l$-matrices;
D_l, $l_0 = l$, complex antisymmetric $l \times l$-matrices;
B_l, $l_0 = 1$; D_l, $l_0 = 1$;
E_6, $l_0 = 6$;
E_7, $l_0 = 7$

(one can readily find the pairs of isomorphic prehomogeneous vector spaces of this list).

The following prehomogeneous vector spaces are nonregular [**234**]:

$$A_l, \text{ for } l_0 \neq \frac{l+1}{2};$$
$$D_l, \text{ for odd } l; \qquad E_6.$$

The elements of this list are frequently depicted as the Dynkin diagrams with marked vertex corresponding to the simple root α_{l_0}.

The list of prehomogeneous vector spaces
of commutative parabolic type

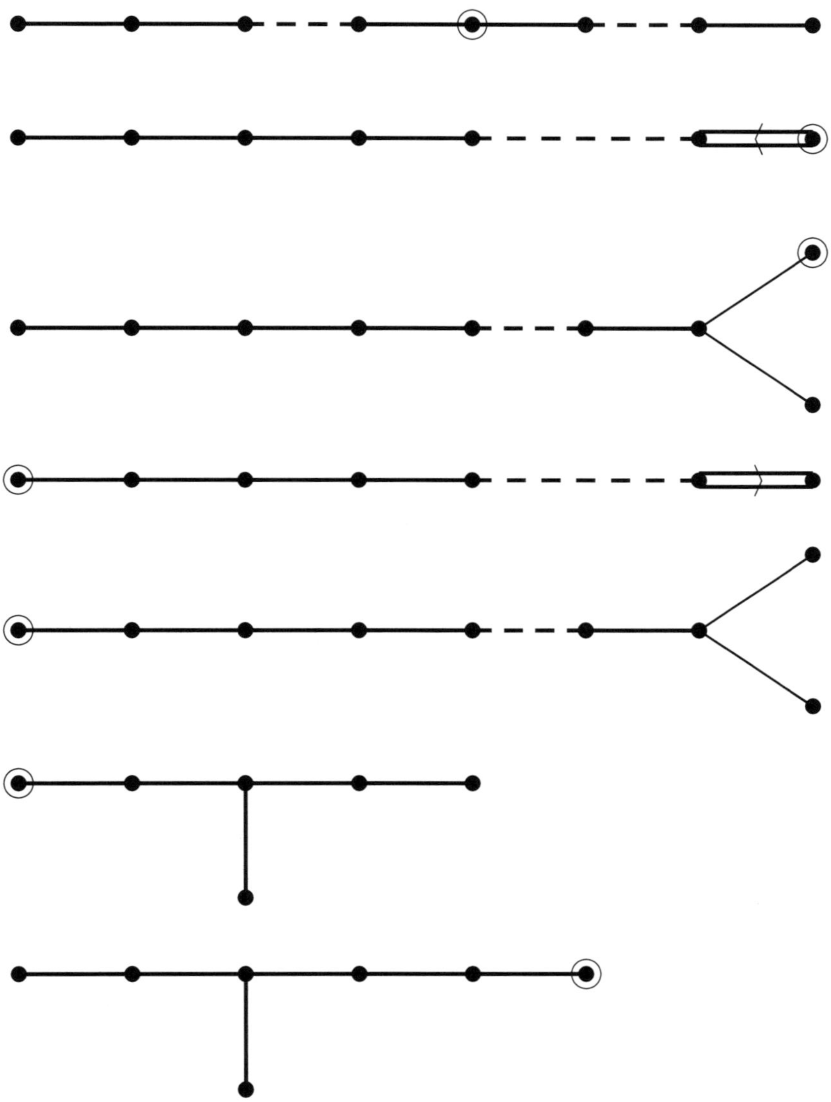

Every irreducible bounded symmetric domain admits a standard realization (the Harish-Chandra realization) as a centrally symmetric convex domain of a prehomogeneous vector space of commutative parabolic type [**123**, p. 342]. This prehomogeneous vector space is regular if and only if the bounded symmetric domain in question has the tube type, that is, $w_0\alpha_{l_0} = -\alpha_{l_0}$; see [**9**, p. 38].

2.3.9. Appendix on noncommutative Gröbner bases. Let us present the principal notions of the theory of Gröbner bases for noncommutative algebras. In doing this, we follow the survey paper by V. Ufnarovskii [**309**, pp. 31–34]. Note that to pass from the noncommutative case to the essentially more simple commutative case [**63, 18, 257**], it suffices to impose some minor changes to the definition of the Gröbner basis listed in [**309**, pp. 35, 36].

Choose a nonzero $q \in \mathbb{C}$. In this subsection the ground field $K \subset \mathbb{C}$ will be assumed to be the smallest subfield which contains q. Thus the field K should be either a finite extension of the field of rational numbers or its transcendental extension of degree 1.

Consider a finite set X, to be called an alphabet. Let $\langle X \rangle$ be the semigroup of words which contains the empty word $\varnothing$, with the operation of concatenation words. The formal linear combinations of the elements of $\langle X \rangle$ form a free algebra $K\langle X \rangle$ with the unit $1 \cdot \varnothing$. To every finite subset $R \subset K\langle X \rangle$ we associate the smallest two-sided ideal $I \subset K\langle X \rangle$ which contains R, together with the quotient algebra

$$A = K\langle X \rangle / I. \tag{2.91}$$

The algebras isomorphic to (2.91) are called finitely defined, X and R are said to be the sets of generators and relations, respectively.

EXAMPLE 2.120. [**64**, p. 97]. The group algebra of the symmetric group S_N admits a description in terms of the generators $s_i = (i, i+1)$, $i = 1, 2, \ldots, N-1$, and the defining relations

$$s_i^2 = 1, \qquad i = 1, 2, \ldots, N-1,$$
$$(s_i s_j)^2 = 1, \qquad |i-j| \neq 1, \quad i, j \in \{1, 2, \ldots, N-1\},$$
$$(s_j s_{j+1})^3 = 1, \qquad j \in \{1, 2, \ldots, N-2\}.$$

In what follows the alphabet X is assumed to be linearly ordered, and the semigroup of words $\langle X \rangle$ to be endowed with the so-called deglex order. Namely, the words are compared by their length, and in the case of equal lengths they are compared lexicographically. It follows from the definition that the set of words $\langle X \rangle$ is linearly ordered, and the order agrees with the semigroup operation: if $a_1 \geq a_2$ and $b_1 \geq b_2$, then $a_1 b_1 \geq a_2 b_2$. Obviously, $\varnothing$ is the smallest element of the semigroup of words.

The leading term of the element

$$v = \sum a_w w, \qquad v \in K\langle X \rangle, \quad a_w \in K, \quad w \in \langle X \rangle,$$

is the largest word w present in this decomposition with a nonzero coefficient a_w. The leading term of the element v is denoted by $L(v)$, and the set of leading terms of the elements of a two-sided ideal I by $L(I)$.

Let us use the above order in $\langle X \rangle$ to produce a subspace which is a complement to I in the vector space $K\langle X \rangle$. Denote by N the set of words which are not in $L(I)$. N is called the set of normal words.

PROPOSITION 2.121. *$K\langle X \rangle = KN \oplus I$, where KN is the subspace of the vector space $K\langle X \rangle$ spanned by N.*

PROOF. Obviously, $KN \cap I = 0$. Suppose that $KN \oplus I \neq K\langle X \rangle$. Consider the smallest word w which is not in $KN \oplus I$. Since $w \notin N$, there exists $v \in I$ such that its leading term is w. Equivalently, $v = aw + v'$, $a \in K$, and every word present in v' belongs to $KN \oplus I$. Hence $v' \in KN \oplus I$. Therefore, $w \in KN \oplus I$, which contradicts our choice of w. □

Consider the projection in the vector space $K\langle X\rangle$ onto the subspace KN parallel to I:

$$K\langle X\rangle \to KN, \qquad v \to \overline{v}.$$

It is called the operator of reduction to the normal form, and $\overline{v}$ the normal form of v. By Proposition 2.121, this operator determines an isomorphism of the vector spaces $A = K\langle X\rangle/I$ and KN. This isomorphism is helpful in the case when the set of normal words N is described explicitly.

We suggest combining the search for an explicit description of this set with the search for an algorithm of reducing to the normal form. It should be noted that for some pairs X, R such an algorithm does not exist. As an example, one may consider the group algebra $K[G]$ of the group G with finitely many generators and defining relations for which the famous word problem is not algorithmically solvable [**227**, p. 103].

A word w' is called a subword of a word w if $w = aw'b$ for some $a, b \in \langle X\rangle$. A subword w' is proper if $w' \neq w$. A subset M is called a two-sided ideal of a semigroup S, if $SM \subset M$ and $MS \subset M$.

Consider the subsets that generate the two-sided ideal $L(I)$ of the semigroup $\langle X\rangle$. There exists the smallest such subset O, it is formed by those words from $L(I)$ that have no proper subwords from $L(I)$. We follow the terminology of D. Anick who called such words obstructions. Since O generates the two-sided ideal $L(I)$ of the semigroup $\langle X\rangle$, the set $G = \{w - \overline{w} |\, w \in O\}$ generates the two-sided ideal I of the algebra $K\langle X\rangle$.

The above set G is called a Gröbner basis of the ideal I (the survey [**309**] uses the term "reduced Gröbner basis"). Of course, a Gröbner basis depends on the choice of a linear order in X.

LEMMA 2.122. *An obstruction $w \in O$ is the leading term of the associated element $w - \overline{w}$ of the Gröbner basis.*

PROOF. Obviously, $N \cap L(I) = \varnothing$, $N \cup L(I) = \langle X\rangle$. Hence, $v = v' + v''$, $v' \in KN$, $v'' \in I$ imply

$$L(v) = L(v' + v'') = L(L(v') + L(v'')) = \begin{cases} L(v'), & \text{if } L(v) \in N, \\ L(v''), & \text{if } L(v) \in L(I). \end{cases}$$

It remains to set $v = w$, $v' = \overline{w}$, $v'' = w - \overline{w}$ and use the fact that $w = L(w)$ for $w \in L(I)$. □

Suppose that a Gröbner basis is finite. This requirement is essential in the noncommutative case. Let us demonstrate that there exists an algorithm of reduction of an element $v \in K\langle X\rangle$ to the normal form, and describe this algorithm.

To every element $u \in I$, we associate the linear operator

$$\hat{u} : K\langle X\rangle \to K\langle X\rangle. \tag{2.92}$$

Let $u \in I$ and $a_{L(u)} \in K$ be the coefficient at the leading term $L(u)$ in the decomposition $u = \sum_{w\in\langle X\rangle} a_w w$. We introduce the linear operator (2.92) by defining its action on the elements of the basis $\langle X\rangle$ of the vector space $K\langle X\rangle$. By definition, $\hat{u}(w) = w$ if $L(u)$ never occurs as a subword of $w \in \langle X\rangle$. Otherwise, one can find the first occurrence of the subword $L(u)$ in w, and $\hat{u}(w)$ is obtained by replacing this subword with $L(u)-\frac{1}{a_{L(u)}}u$. In other words, the operator $\hat{u}$ realizes the substitution

$$L(u) \mapsto L(u) - \frac{1}{a_{L(u)}}u. \tag{2.93}$$

Let $v \in K\langle X\rangle$. An element $v' \in K\langle X\rangle$ is called a G-reduction of v, if, first, v' can be derived from v via a consecutive application of operators of the form $\hat{u}$, $u \in G$ and, second, v' is a common fixed vector of these linear operators.

PROPOSITION 2.123. *A G-reduction of an element $v \in K\langle X\rangle$ exists, is unique, and is precisely the normal form $\overline{v}$ of this element.*

COROLLARY 2.124. *A word $w \in \langle X\rangle$ is normal if and only if no subword of w is the leading term of an element of a Gröbner basis.*

REMARK 2.125. We do not discuss here the passage from a finite list of substitution rules (2.93) to an algorithm of reducing to the normal form because contemporary computing algebra software allows one to arrange this passage.

EXAMPLE 2.126. (cf. [**309**, p. 33]) Consider the algebra A with generators x, y and the defining relation $x^2+y^2=0$: $A = K\langle x,y\rangle/(x^2+y^2)$. Let us order the generators by setting $x>y$. One can show that in this case $G=\{x^2+y^2, xy^2-y^2x\}$, and the list of substitution rules has the form

$$x^2 \mapsto -y^2, \qquad xy^2 \mapsto y^2x.$$

A word w is normal if and only if it has no subwords x^2, xy^2:

$$N = \{y^n(xy)^k, \quad y^n(xy)^kx \mid n,k \in \mathbb{Z}_+\}.$$

Thus with a finite Gröbner basis at hand, it is easy to obtain an algorithm of reducing to the normal form. It remains to explain how to find a finite Gröbner basis of the two-sided ideal $I=(R)$, assuming that it exists. For that, we suggest transforming the set R

$$R \to R_1 \to R_2 \to \cdots,$$

using operations of the three types described below to pass from R_j to R_{j+1}.

(1) *Normalization.* Assume that $v \in R_j$ and the leading term $L(v)$ occurs in the decomposition $v = \sum a_w w$ with the coefficient $a_{L(v)} \neq 1$. The normalization here is then just the replacement of v with $a_{L(v)}^{-1}v$.
(2) *Reduction.* Let $v, v' \in R_j$, $v \neq v'$, and let $L(v')$ be a subword of $L(v)$: $L() = aL(v')b$, $a,b \in \langle X\rangle$. The reduction is just discarding v from R_j if $v = av'b$, and the replacement of v with $v-av'b$ if $v \neq av'b$.
(3) Consider two elements $u, v \in R_j$ whose coefficients at the leading terms $L(u)$, $L(v)$ are both 1. A word $w \in \langle X\rangle$ is called the composition of u and v if w starts with $L(u)$ and ends with $L(v)$, and these subwords of w have a nonempty intersection

$$w = xyz, \qquad L(u) = xy, \qquad L(v) = yz,$$

where y is a nonempty word. The presence of composition allows one to apply to w both substitutions $xy \mapsto L(u)-u$ and $yz \mapsto L(v)-v$. The results, in general, are different. In fact, their difference is $(L(u)-u)z - x(L(v)-v) = xv-uz$. An operation of type 3 attaches to R_j the element $xv-uz \in K\langle X\rangle$ if this element is not in R_j. The element $xv-uz$ is called the result of composition.

The order of applying operations is essential. The operations of type 1 are assigned the highest priority level. These are just those to be applied in the case when at least one of the elements of R_j has the coefficient at the leading term different from 1. The operations of type 3 are assigned the lowest priority level. Such operations are used only when no operations of type 1 or 2 are applicable.

REMARK 2.127. It may happen happen that the element attached to R_j via an operation of type 3, *reduces to zero* by operations of type 1 or 2 which are applied immediately afterwards. This is a source of *infinite loop*. To avoid this, one should impose an additional prohibition to apply "useless" operations of type 3.

If the above "algorithm" terminates after finitely many steps because none of the above operations is applicable to a current set (with Remark 2.127 being taken into account), then this set is a Gröbner basis for the ideal I. To prove this, it suffices to apply the Bergman Diamond Lemma [**32**, p. 181] in the following formulation (cf. [**309**, p. 33]):

LEMMA 2.128. *Let $G \subset K\langle X\rangle$ be a set to which no reduction or normalization operation is applicable. If the result of each composition reduces to zero in finitely many consecutive steps of reduction or normalization, then G is a Gröbner basis.*

2.4. Spaces of functions in quantum bounded symmetric domains

2.4.1. The Fock representation and finite functions. Consider a complex vector space $V \neq \{0\}$. A bounded domain $\mathbb{D}$ of this vector space is said to be symmetric if every $z \in \mathbb{D}$ is an isolated fixed point of some involutive automorphism s_z (with $s_z^2 = \mathrm{id}$) of the domain $\mathbb{D}$, the latter being treated as a complex manifold [**123**, p. 342], [**197**, p. 193]. Such a domain is called reducible if in the category of complex manifolds it is isomorphic to a Cartesian product of bounded symmetric domains. The unit disc $\mathbb{D} = \{z \in \mathbb{C} | \, |z| < 1\}$ is the simplest example of irreducible bounded symmetric domain.

Consider a Hermitian symmetric pair $(\mathfrak{g}, \mathfrak{k})$ (see Remark 2.97 in Subsection 2.3.4). Such pairs $(\mathfrak{g}, \mathfrak{k})$ are produced by complexifying the orthogonal involutive Lie algebras [**123**, p. 217], [**197**, p. 196]. Each such pair determines a decomposition of the simple complex Lie algebra $\mathfrak{g}$:

$$\mathfrak{g} = \mathfrak{p}^- \oplus \mathfrak{k} \oplus \mathfrak{p}^+.$$

It was demonstrated by Harish-Chandra [**123**, pp. 342–352], [**267**, p. 6] that every irreducible bounded symmetric domain is isomorphic to a domain $\mathbb{D}$ that is embedded into $\mathfrak{p}^-$ in a standard way. We intend to produce its q-analog.

Just as in the special case of quantum disc (see Section 1.4) we need a $U_q\mathfrak{g}$-module algebra $\mathcal{D}(\mathbb{D})_q$ of finite functions in the quantum bounded symmetric domain $\mathbb{D}$. To begin with, we introduce the Fock representation of the $*$-algebra $\mathrm{Pol}(\mathfrak{p}^-)_q = (\mathbb{C}[\mathfrak{p}^- \oplus \mathfrak{p}^+]_q, *)$.

Consider the $\mathbb{C}[\mathfrak{p}^- \oplus \mathfrak{p}^+]_q$-module $\mathcal{H}$ with a single generator v_0 and the defining relations

$$f v_0 = \langle f \rangle v_0, \qquad f \in \mathbb{C}[\mathfrak{p}^+]_q,$$

where $\langle f \rangle$ is the linear functional introduced in Subsection 2.3.3. Equivalently,

$$f v_0 = 0, \qquad f \in \bigoplus_{j=1}^{\infty} \mathbb{C}[\mathfrak{p}^+]_{q,-j}. \tag{2.94}$$

Proposition 2.107 implies that the set of defining relations (2.94) can be replaced with the subset

$$fv_0 = 0, \qquad f \in \mathbb{C}[\mathfrak{p}^+]_{q,-1}.$$

LEMMA 2.129. *The linear map*

$$\mathbb{C}[\mathfrak{p}^-]_q \to \mathcal{H}, \qquad f \mapsto fv_0,$$

is an isomorphism of vector spaces.

PROOF. The kernel of the linear map $\mathbb{C}[\mathfrak{p}^- \oplus \mathfrak{p}^+]_q \to \mathcal{H}$, $f \mapsto fv_0$, is the left ideal of the algebra $\mathbb{C}[\mathfrak{p}^- \oplus \mathfrak{p}^+]_q$ generated by $\bigoplus_{j=1}^{\infty} \mathbb{C}[\mathfrak{p}^+]_{q,-j}$. This ideal is just $\bigoplus_{i=0}^{\infty} \bigoplus_{j=1}^{\infty} \mathbb{C}[\mathfrak{p}^-]_{q,i} \otimes \mathbb{C}[\mathfrak{p}^+]_{q,-j}$. Hence $\mathcal{H} \cong \bigoplus_{i=0}^{\infty} \mathbb{C}[\mathfrak{p}^-]_{q,i} = \mathbb{C}[\mathfrak{p}^-]_q$. □

This lemma, together with Proposition 2.98, imply

PROPOSITION 2.130. *The Hermitian form in $\mathcal{H}$*

$$(f_1 v_0, f_2 v_0) = \langle f_2^* f_1 \rangle, \qquad f_1, f_2 \in \mathrm{Pol}(\mathfrak{p}^-)_q,$$

is well defined, positive, and one has

$$(fv', v'') = (v', f^* v''), \qquad v', v'' \in \mathcal{H},\ f \in \mathrm{Pol}(\mathfrak{p}^-)_q.$$

The latter proposition allows one to introduce a representation T_F of the $*$-algebra $\mathrm{Pol}(\mathfrak{p}^-)_q$ in the pre-Hilbert space $\mathcal{H}$. It is called the Fock representation, and v_0 is called the vacuum vector.

We turn to the special case $\mathfrak{g} = \mathfrak{sl}_2$. It follows from (1.10) that

$$(z^n v_0, z^n v_0) = ((1 - q^2 y) z^{n-1} v_0, z^{n-1} v_0) = (1 - q^{2n})(z^{n-1} v_0, z^{n-1} v_0) = (q^2; q^2)_n.$$

Hence the action of $T_F(z)$, $T_F(z^*)$ on the elements of the orthonormal basis $\left\{ \frac{z^n v_0}{\sqrt{(q^2;q^2)_n}} \right\}_{n \in \mathbb{Z}_+}$ is described by (1.9).

REMARK 2.131. Evidently, the Fock representation is irreducible. Indeed, let $\mathcal{H}_1 \subset \mathcal{H}$ be a nonzero common invariant subspace for the operators of this representation. Then $\mathcal{H}_1$ contains a nonzero vector v such that $T_F(f)v = 0$ for all $f \in \mathbb{C}[\mathfrak{p}^+]_{q,-j}$, $j \neq 0$. This vector is orthogonal to $\bigoplus_{j=0}^{\infty} \mathbb{C}[\mathfrak{p}^-]_{q,j} v_0$, hence it belongs to $\mathbb{C}v_0$. Therefore, $v_0 \in \mathcal{H}_1$ and $\mathcal{H}_1 = \mathcal{H}$.

Let us turn to a construction of the algebra $\mathcal{D}(\mathbb{D})_q$. In the special case of quantum disc, this algebra was introduced in Subsection 1.4.3.

We endow the vector space $\mathcal{H}$ with the grading

$$\mathcal{H} = \bigoplus_{j=0}^{\infty} \mathcal{H}_j, \qquad \mathcal{H}_j = \mathbb{C}[\mathfrak{p}^-]_{q,j} v_0.$$

The subspaces $\mathcal{H}_j \cong \mathbb{C}[\mathfrak{p}^-]_{q,j}$ are pairwise orthogonal and finite-dimensional. It follows from (2.64) that

$$\dim \mathcal{H}_j = \binom{j + \dim \mathfrak{p}^- - 1}{\dim \mathfrak{p}^- - 1}.$$

A linear operator A in $\mathcal{H}$ is called finite if $A\mathcal{H}_j = 0$ for all but finitely many $j \in \mathbb{Z}_+$. As an example, one can consider the one-dimensional orthogonal projection P_0 onto the vacuum subspace $\mathcal{H}_0 = \mathbb{C}v_0$. Finite linear operators form an algebra which appears to be canonically isomorphic to the algebra $\mathcal{D}(\mathbb{D})_q$ defined below.

We introduce an element f_0 which will correspond to the projection P_0 under this canonical isomorphism.

More precisely, consider the unital algebra $\mathrm{Fun}(\mathbb{D})_q$ generated by the subalgebra $\mathrm{Pol}(\mathfrak{p}^-)_q$ and the element f_0 which satisfies the following defining relations:

$$f_0^2 = f_0, \tag{2.95}$$

$$\psi f_0 = \langle\psi\rangle f_0, \quad \psi \in \mathbb{C}[\mathfrak{p}^+]_q; \qquad f_0\psi = \langle\psi\rangle f_0, \quad \psi \in \mathbb{C}[\mathfrak{p}^-]_q. \tag{2.96}$$

According to Proposition 2.107, the set of relations (2.96) in this definition can be replaced with the subset

$$\psi f_0 = 0, \quad \psi \in \mathbb{C}[\mathfrak{p}^+]_{q,-1}; \qquad f_0\psi = 0, \quad \psi \in \mathbb{C}[\mathfrak{p}^-]_{q,1}.$$

For example, in the case of quantum disc the elements z, z^*, and f_0 generate the algebra $\mathrm{Fun}(\mathbb{D})_q$, while the set of defining relations includes (1.8) and

$$z^* f_0 = f_0 z = 0, \qquad f_0^2 = f_0.$$

There exists a unique extension of the involution $*$ from $\mathrm{Pol}(\mathfrak{p}^-)_q$ to $\mathrm{Fun}(\mathbb{D})_q$ such that $f_0^* = f_0$. Obviously, one has

LEMMA 2.132. *There exists a unique extension of T_F to a $*$-representation of the $*$-algebra $\mathrm{Fun}(\mathbb{D})_q$ which takes f_0 to the orthogonal projection P_0 onto the vacuum subspace.*

We keep the current notation T_F for this extension of the Fock representation.

Let $\mathcal{D}(\mathbb{D})_q$ be the two-sided ideal of $\mathrm{Fun}(\mathbb{D})_q$ generated by f_0. It is called the algebra of finite functions on the quantum bounded symmetric domain.

Note that the algebra of finite linear operators in $\mathcal{H}$ is a $*$-algebra (with the adjoint operator defined in the standard way).

PROPOSITION 2.133. *The representation T_F yields an isomorphism between the $*$-algebra $\mathcal{D}(\mathbb{D})_q$ and the $*$-algebra of all finite linear operators in $\mathcal{H}$.*

PROOF. T_F is a $*$-representation. Hence it suffices to prove that the restriction of T_F to $\mathcal{D}(\mathbb{D})_q$ is a bijective map from $\mathcal{D}(\mathbb{D})_q$ to the algebra of all finite linear operators in $\mathcal{H}$.

Let $\mathcal{D}(\mathbb{D})_{q,i,j} = \mathbb{C}[\mathfrak{p}^-]_{q,i} \cdot f_0 \cdot \mathbb{C}[\mathfrak{p}^+]_{q,-j}$. If $f \in \mathcal{D}(\mathbb{D})_{q,i,j}$, then $T_F(f)$ maps $\mathcal{H}_j$ to $\mathcal{H}_i$ and vanishes on $\bigoplus_{k\neq j}\mathcal{H}_k$. We thus get a linear map from $\mathcal{D}(\mathbb{D})_{q,i,j}$ to $\mathrm{Hom}(\mathcal{H}_j,\mathcal{H}_i)$. This map is onto by Proposition 2.93, and one has

$$\dim \mathcal{D}(\mathbb{D})_{q,i,j} = \dim \mathrm{Hom}(\mathcal{H}_j,\mathcal{H}_i).$$

Therefore, T_F yields an isomorphism

$$\mathcal{D}(\mathbb{D})_{q,i,j} = \mathbb{C}[\mathfrak{p}^-]_{q,i} f_0 \mathbb{C}[\mathfrak{p}^+]_{q,-j} \cong \mathrm{Hom}(\mathcal{H}_j,\mathcal{H}_i).$$

On the other hand, $\mathcal{D}(\mathbb{D})_q = \bigoplus_{i,j=0}^{\infty}\mathcal{D}(\mathbb{D})_{q,i,j}$, while the vector space of finite linear operators coincides with $\bigoplus_{i,j=0}^{\infty}\mathrm{Hom}(\mathcal{H}_j,\mathcal{H}_i) \hookrightarrow \mathrm{End}\,\mathcal{H}$. □

Now we prove the faithfulness of the Fock representation of $\mathrm{Pol}(\mathfrak{p}^-)_q$. Just as in Proposition 2.133, one can prove

LEMMA 2.134. *For all $j,k \in \mathbb{Z}_+$ the map*

$$\mathbb{C}[\mathfrak{p}^-]_{q,k} \cdot \mathbb{C}[\mathfrak{p}^+]_{q,-j} \to \mathrm{Hom}(\mathcal{H}_j,\mathcal{H}_k), \qquad f \mapsto T_F(f)|_{\mathcal{H}_j},$$

is bijective.

PROPOSITION 2.135. *The Fock representation of* $\mathrm{Pol}(\mathfrak{p}^-)_q$ *is faithful.*

PROOF. We endow the vector space $\mathrm{Pol}(\mathfrak{p}^-)_q$ with the bigrading

$$\mathrm{Pol}(\mathfrak{p}^-)_q = \bigoplus_{k,j=0}^{\infty} \mathbb{C}[\mathfrak{p}^-]_{q,k}\, \mathbb{C}[\mathfrak{p}^+]_{q,-j}$$

and $\mathbb{Z}_+^2$ with the standard partial order by setting $(k_1,j_1) \leq (k_2,j_2)$ if and only if $k_1 \leq k_2$ and $j_1 \leq j_2$. Assume that the claim of our proposition fails, i.e., $T_F(f)=0$ for some $f \neq 0$. Apply the decomposition

$$f = \sum_{k,j=0}^{\infty} f_{k,j}, \qquad f_{k,j} \in \mathbb{C}[\mathfrak{p}^-]_{q,k}\mathbb{C}[\mathfrak{p}^+]_{q,-j},$$

and choose a nonzero component $f_{k_0j_0}$ with the minimal *index* (k_0,j_0) in this decomposition. Since $T_F(f)=0$, one has $T_F(f_{k_0j_0})|_{\mathcal{H}_{j_0}}=0$ because $T_F(f_{k_0j_0})|_{\mathcal{H}_{j_0}}$ is a composition of the linear operator $T_F(f)|_{\mathcal{H}_{j_0}}$ and the projection onto $\mathcal{H}_{k_0}$ parallel to $\bigoplus_{k\neq k_0}\mathcal{H}_k$. It remains to observe that

$$f_{k_0j_0} \neq 0, \qquad f_{k_0j_0} \in \mathbb{C}[\mathfrak{p}^-]_{q,k_0}\mathbb{C}[\mathfrak{p}^+]_{q,-j_0}, \qquad T_F(f_{k_0j_0})|_{\mathcal{H}_{j_0}} = 0.$$

Thus we get a contradiction with Lemma 2.134. □

REMARK 2.136. It will follow from Proposition 2.142 and Corollary 2.203 below that $T_F(\mathrm{Pol}(\mathfrak{p}^-)_q)$ does not contain nonzero elements of $T_F(\mathcal{D}(\mathbb{D})_q)$. Hence the Fock representation of $\mathrm{Fun}(\mathbb{D})_q = \mathrm{Pol}(\mathfrak{p}^-)_q \oplus \mathcal{D}(\mathbb{D})_q$ is faithful.

2.4.2. The Fock representation and distributions. Consider the finite-dimensional subspaces $\mathcal{H}_j \subset \mathcal{H}$, $j \in \mathbb{Z}_+$, introduced in Subsection 2.4.1. Their direct product $\overline{\mathcal{H}}$ is the vector space of formal series with terms from $\mathcal{H}_j$. Note that the vector spaces $\mathrm{Hom}(\mathcal{H}_j,\mathcal{H}_k)$ are finite-dimensional. Just as in Subsection 1.4.3, the linear maps from $\mathcal{H}$ to $\overline{\mathcal{H}}$ will be called generalized linear operators in $\mathcal{H}$. Each of them is a formal series

$$A = \sum_{k,j=0}^{\infty} A_{kj}, \qquad A_{kj} \in \mathrm{Hom}(\mathcal{H}_j,\mathcal{H}_k), \tag{2.97}$$

and the vector space of generalized linear operators in $\mathcal{H}$ is endowed with the topology of termwise convergence of formal series.

The topological vector space $\mathcal{D}(\mathbb{D})'_q$ of *distributions on the quantum bounded symmetric domain* defined below turns out to be canonically isomorphic to the topological vector space of generalized linear operators in $\mathcal{H}$, just as in the special case of quantum disc; see Subsection 1.4.3.

Recall that

$$\mathrm{Pol}(\mathfrak{p}^-)_q = \bigoplus_{k,j=0}^{\infty} \mathbb{C}[\mathfrak{p}^-]_{q,k} \cdot \mathbb{C}[\mathfrak{p}^+]_{q,-j} \cong \bigoplus_{k,j=0}^{\infty} \mathbb{C}[\mathfrak{p}^-]_{q,k} \otimes \mathbb{C}[\mathfrak{p}^+]_{q,-j},$$

so that $\dim(\mathbb{C}[\mathfrak{p}^-]_{q,k} \cdot \mathbb{C}[\mathfrak{g}^+]_{q,-j}) < \infty$. Equivalently, every $f \in \mathrm{Pol}(\mathfrak{p}^-)_q$ admits a unique decomposition as a finite sum

$$f = \sum_{k,j=0}^{\infty} f_{k,j}, \qquad f_{k,j} \in \mathbb{C}[\mathfrak{p}^-]_{q,k} \cdot \mathbb{C}[\mathfrak{p}^+]_{q,-j}. \tag{2.98}$$

Consider the vector space $\mathcal{D}(\mathbb{D})'_q$ of formal series of the form (2.98) with the topology of termwise convergence. The involution $*$ extends by continuity from the dense linear subvariety $\mathrm{Pol}(\mathfrak{p}^-)_q$ to $\mathcal{D}(\mathbb{D})'_q$:

$$* : \sum_{k,j=0}^{\infty} f_{k,j} \mapsto \sum_{k,j=0}^{\infty} f^*_{k,j}.$$

The elements of $\mathcal{D}(\mathbb{D})'_q$ are called *distributions on the quantum bounded symmetric domain.*

PROPOSITION 2.137. (1) *The Fock representation of* $\mathrm{Pol}(\mathfrak{p}^-)_q$ *admits an extension by continuity to a linear map from* $\mathcal{D}(\mathbb{D})'_q$ *to the space of generalized linear operators in* $\mathcal{H}$.

(2) *This linear map is an isomorphism of topological vector spaces.*

PROOF. For $k, j \in \mathbb{Z}_+$, we introduce linear maps

$$a_{k,j} : \mathrm{Pol}(\mathfrak{p}^-)_q \to \mathbb{C}[\mathfrak{p}^-]_{q,k} \cdot \mathbb{C}[\mathfrak{p}^+]_{q,-j}; \quad b_{k,j} : \mathrm{Pol}(\mathfrak{p}^-)_q \to \mathrm{Hom}(\mathcal{H}_j, \mathcal{H}_k).$$

Namely, the maps $a_{k,j}$ are introduced using the decomposition (2.98) by setting $a_{k,j}(f) = f_{k,j}$, and the maps $b_{k,j}$ using (2.97):

$$T_F(f) = \sum_{k,j=0}^{\infty} b_{k,j}(f), \qquad f \in \mathrm{Pol}(\mathfrak{p}^-)_q.$$

We endow the vector space $\mathrm{Pol}(\mathfrak{p}^-)_q$ with two topologies. Let $\mathcal{T}_a$ be the weakest topology which makes continuous all the maps $a_{k,j}$, and $\mathcal{T}_b$ the weakest topology which makes continuous all the maps $b_{k,j}$. The topologies $\mathcal{T}_a$, $\mathcal{T}_b$ are induced by the embeddings of $\mathrm{Pol}(\mathfrak{p}^-)_q$ to $\mathcal{D}(\mathbb{D})'_q$ and to the space of generalized linear operators in $\mathcal{H}$, respectively. Hence these topological vector spaces are completions of $\mathrm{Pol}(\mathfrak{p}^-)_q$ in the topologies $\mathcal{T}_a$, $\mathcal{T}_b$. Therefore, it suffices to prove that $\mathcal{T}_a = \mathcal{T}_b$. Obviously, $\mathrm{Ker}\, b_{k,j} \supset \bigcap_{i \le \min(k,j)} \mathrm{Ker}\, a_{k-i,j-i}$. In fact, $b_{k,j}(f)$ is a linear combination of $a_{k-i,j-i}(f)$ with coefficients

$$c_{k,j,i} : \mathbb{C}[\mathfrak{p}^-]_{q,k-i}\mathbb{C}[\mathfrak{g}^+]_{q,-j+i} \to \mathrm{Hom}(\mathcal{H}_j, \mathcal{H}_k),$$

which do not depend on $f \in \mathrm{Pol}(\mathfrak{p}^-)_q$:

$$b_{k,j}(f) = \sum_{i \le \min(k,j)} c_{k,j,i} a_{k-i,j-i}(f). \tag{2.99}$$

Hence the topology $\mathcal{T}_a$ is not weaker than $\mathcal{T}_b$. It remains to observe that Lemma 2.134 implies the invertibility of the linear maps $c_{k,j,0}$, $k, j \in \mathbb{Z}_+$. Therefore, the *triangular* system of equations (2.99) is solvable with respect to $a_{k,j}(f)$, and the topology $\mathcal{T}_b$ is not weaker than $\mathcal{T}_a$. □

Observe that the vector space of generalized linear operators is endowed with the involution $*$, and the above linear map from $\mathcal{D}(\mathbb{D})'_q$ to this vector space intertwines the involutions.

This proposition and Proposition 2.133 imply

COROLLARY 2.138. *The embedding of the vector space* $\mathrm{Pol}(\mathfrak{p}^-)_q$ *to* $\mathcal{D}(\mathbb{D})'_q$ *canonically extends to a linear map from the vector space* $\mathrm{Fun}(\mathbb{D})_q$ *to* $\mathcal{D}(\mathbb{D})'_q$ *and determines an embedding of* $\mathcal{D}(\mathbb{D})_q$ *into* $\mathcal{D}(\mathbb{D})'_q$ *as a dense linear subvariety.*

We will identify $\mathcal{D}(\mathbb{D})_q$ and $\mathrm{Pol}(\mathfrak{p}^-)_q$ with their images under the embedding into $\mathcal{D}(\mathbb{D})'_q$. Now Proposition 2.137 implies

COROLLARY 2.139. *For every* $f_1, f_2 \in \mathrm{Fun}(\mathbb{D})_q$ *the linear map*

$$\mathcal{D}(\mathbb{D})_q \to \mathcal{D}(\mathbb{D})_q, \qquad f \mapsto f_1 \cdot f \cdot f_2,$$

admits an extension by continuity to a linear operator in $\mathcal{D}(\mathbb{D})'_q$. *This way* $\mathcal{D}(\mathbb{D})'_q$ *is endowed with a structure of* $\mathrm{Fun}(\mathbb{D})_q$*-bimodule.*

REMARK 2.140. The structure of $\mathrm{Pol}(\mathfrak{p}^-)_q$-bimodule in $\mathcal{D}(\mathbb{D})'_q$ induced by the embedding $\mathrm{Pol}(\mathfrak{p}^-)_q \hookrightarrow \mathrm{Fun}(\mathbb{D})_q$ can be derived in a different way, namely, via extending by continuity from $\mathrm{Pol}(\mathfrak{p}^-)_q$:

$$f_1 \times f_2 : f \mapsto f_1 \cdot f \cdot f_2, \qquad f_1, f_2, f \in \mathrm{Pol}(\mathfrak{p}^-)_q.$$

COROLLARY 2.141. (1) *Every* $f \in \mathcal{D}(\mathbb{D})'_q$ *admits a unique expansion as a convergent series*

$$f = \sum_{k,j=0}^{\infty} f_{k,j}, \qquad f_{k,j} \in \mathbb{C}[\mathfrak{p}^-]_{q,k} f_0 \mathbb{C}[\mathfrak{p}^+]_{q,-j}. \tag{2.100}$$

(2) *An element* $f \in \mathcal{D}(\mathbb{D})'_q$ *belongs to* $\mathcal{D}(\mathbb{D})_q$ *if and only if all but finitely many terms of the series* (2.100) *are zero.*

In the special case of quantum disc, the following result can be derived from Proposition 1.76.

PROPOSITION 2.142. *An element* $f \in \mathcal{D}(\mathbb{D})'_q$ *belongs to the subvariety* $\mathcal{D}(\mathbb{D})_q$ *if and only if for some* N *one has*

$$\mathbb{C}[\mathfrak{p}^+]_{q,-n} f = f \mathbb{C}[\mathfrak{p}^-]_{q,n} = 0 \tag{2.101}$$

for all $n \geq N$.

PROOF. If $f \in \mathcal{D}(\mathbb{D})_q$, then for n big enough the relations (2.101) are due to the definitions of $\mathrm{Fun}(\mathbb{D})_q$ and $\mathcal{D}(\mathbb{D})_q$. On the other hand, (2.101) implies that

$$f_0 \mathbb{C}[\mathfrak{p}^+]_{q,-n} f = f \mathbb{C}[\mathfrak{p}^-]_{q,n} f_0 = 0$$

for n big enough. Hence by Proposition 2.93 all but finitely many terms of the series (2.100) are zero. It remains to apply Corollary 2.141. □

2.4.3. Action of $U_q\mathfrak{g}$ in spaces of test functions and distributions.

LEMMA 2.143. *For each* $\xi \in U_q\mathfrak{g}$, *there exists* $N \in \mathbb{N}$ *such that*

$$\xi\left(\mathbb{C}[\mathfrak{p}^-]_{q,k}\mathbb{C}[\mathfrak{p}^+]_{q,-j}\right) \subset \bigoplus_{|k'-k|\leq N \ \&\ |j'-j|\leq N} \mathbb{C}[\mathfrak{p}^-]_{q,k'}\mathbb{C}[\mathfrak{p}^+]_{q,-j'}$$

for all $k, j \in \mathbb{Z}_+$.

PROOF. The set of all ξ which satisfy the claim of the lemma is a subalgebra containing all the elements E_i, F_i, $K_i^{\pm}$, $i = 1, 2, \ldots, l$. □

Recall that $\mathrm{Pol}(\mathfrak{p}^-)_q$ is a $U_q\mathfrak{g}$-covariant $\mathrm{Pol}(\mathfrak{p}^-)_q$-bimodule. It was demonstrated in the previous subsection that the structure of $\mathrm{Pol}(\mathfrak{p}^-)_q$-bimodule is extended by continuity to $\mathcal{D}(\mathbb{D})'_q$. Lemma 2.143 implies that the structure of $U_q\mathfrak{g}$-module also extends by continuity to $\mathcal{D}(\mathbb{D})'_q$, and that these structures are compatible. More precisely, one has

PROPOSITION 2.144. (1) *For every $\xi \in U_q\mathfrak{g}$ the operator*

$$\mathrm{Pol}(\mathfrak{p}^-)_q \to \mathrm{Pol}(\mathfrak{p}^-)_q, \qquad f \mapsto \xi f,$$

admits an extension by continuity to the topological vector space $\mathcal{D}(\mathbb{D})'_q$. This way the vector space $\mathcal{D}(\mathbb{D})'_q$ is endowed with a structure of $U_q\mathfrak{g}$-covariant $\mathrm{Pol}(\mathfrak{p}^-)_q$-bimodule.

(2) *For all $\xi \in U_q\mathfrak{g}$, $f \in \mathcal{D}(\mathbb{D})'_q$ one has*

$$(\xi f)^* = (S(\xi))^* f^*. \tag{2.102}$$

An application of Proposition 2.142 yields

COROLLARY 2.145. *The linear subvariety $\mathcal{D}(\mathbb{D})_q \subset \mathcal{D}(\mathbb{D})'_q$ is a $U_q\mathfrak{g}$-covariant $\mathrm{Pol}(\mathfrak{p}^-)_q$-bimodule.*

Recall the notation

$$\mathfrak{q}^\pm = \mathfrak{k} \oplus \mathfrak{p}^\pm$$

introduced in Subsection 2.3.4.

COROLLARY 2.146. *Consider the $U_q\mathfrak{g}$-module $\mathcal{D}(\mathbb{D})_q$. The linear subvariety $\mathbb{C}[\mathfrak{p}^-]_q f_0$ is a $U_q\mathfrak{q}^+$-submodule of this module, and the linear subvariety $f_0\mathbb{C}[\mathfrak{p}^+]_q$ is a $U_q\mathfrak{q}^-$-submodule:*

$$U_q\mathfrak{q}^+(\mathbb{C}[\mathfrak{p}^-]_q f_0) \subset \mathbb{C}[\mathfrak{p}^-]_q f_0, \qquad U_q\mathfrak{q}^-(f_0\mathbb{C}[\mathfrak{p}^+]_q) \subset f_0\mathbb{C}[\mathfrak{p}^+]_q.$$

PROOF. A finite linear operator in $\mathcal{H} = \bigoplus_{j=0}^\infty \mathcal{H}_j$ belongs to $T_F(\mathbb{C}[\mathfrak{p}^-]_q f_0)$ if and only if its kernel contains $\bigoplus_{j=1}^\infty \mathcal{H}_j$. Thus by Proposition 2.93

$$\mathbb{C}[\mathfrak{p}^-]_q f_0 = \left\{ f \in \mathcal{D}(\mathbb{D})_q \,\middle|\, f\left(\bigoplus_{j=1}^\infty \mathbb{C}[\mathfrak{p}^-]_{q,j}\right) = 0 \right\}.$$

In a similar way one proves

$$f_0\mathbb{C}[\mathfrak{p}^+]_q = \left\{ f \in \mathcal{D}(\mathbb{D})_q \,\middle|\, \left(\bigoplus_{j=1}^\infty \mathbb{C}[\mathfrak{p}^+]_{q,-j}\right) f = 0 \right\}.$$

It remains to use the fact that $\mathcal{D}(\mathbb{D})_q$ is both $U_q\mathfrak{q}^+$- and $U_q\mathfrak{q}^-$-module algebra and the inclusions

$$U_q\mathfrak{q}^+\left(\bigoplus_{j=1}^\infty \mathbb{C}[\mathfrak{p}^-]_{q,j}\right) \subset \bigoplus_{j=1}^\infty \mathbb{C}[\mathfrak{p}^-]_{q,j}; \quad U_q\mathfrak{q}^-\left(\bigoplus_{j=1}^\infty \mathbb{C}[\mathfrak{p}^+]_{q,-j}\right) \subset \bigoplus_{j=1}^\infty \mathbb{C}[\mathfrak{p}^+]_{q,-j}. \quad \square$$

REMARK 2.147. The linear map

$$\mathcal{H} \xrightarrow{\sim} \mathbb{C}[\mathfrak{p}^-]_q f_0, \qquad \psi v_0 \mapsto \psi f_0, \quad \psi \in \mathbb{C}[\mathfrak{p}^-]_q,$$

is an isomorphism of $\mathrm{Pol}(\mathfrak{p}^-)_q$-modules and supplies a *geometric realization* of the Fock representation. By Corollary 2.146, this linear map endows the vector space $\mathcal{H}$ with a structure of $U_q\mathfrak{q}^+$-covariant $\mathrm{Pol}(\mathfrak{p}^-)_q$-module. We denote the corresponding representation of $U_q\mathfrak{q}^+$ in $\mathcal{H}$ by Γ.

Let us show that $\mathrm{Pol}(\mathbb{C})_{q_{l_0}}$ admits a natural embedding into $\mathrm{Pol}(\mathfrak{p}^-)_q$.

LEMMA 2.148. (1) *Consider the graded $U_q\mathfrak{g}$-module algebra $\mathbb{C}[\mathfrak{p}^-]_q$. There exists a unique element $z_{\mathrm{low}} \in \mathbb{C}[\mathfrak{p}^-]_{q,1}$ such that*

$$F_{l_0} z_{\mathrm{low}} = q_{l_0}^{1/2}, \qquad F_j z_{\mathrm{low}} = 0, \quad j \neq l_0. \tag{2.103}$$

(2) *The following relations are valid:*

$$K_i^{\pm 1} z_{\mathrm{low}} = q_i^{\pm a_{il_0}} z_{\mathrm{low}}, \qquad i = 1, 2, \ldots, l, \tag{2.104}$$

$$E_{l_0} z_{\mathrm{low}} = -q_{l_0}^{1/2} z_{\mathrm{low}}^2,$$

$$E_{l_0} z_{\mathrm{low}}^* = q_{l_0}^{-3/2}, \quad F_{l_0} z_{\mathrm{low}}^* = -q_{l_0}^{5/2} z_{\mathrm{low}}^{*2}, \quad K_i^{\pm 1} z_{\mathrm{low}}^* = q_i^{\mp a_{il_0}} z_{\mathrm{low}}^*, \ i = 1, 2, \ldots, l.$$

PROOF. 1. It was observed in Subsection 2.3.8 that the $U_q\mathfrak{k}$-module $\mathbb{C}[\mathfrak{p}^-]_{q,1}$ is simple. Thus its lowest weight vector $z_{\mathrm{low}} \neq 0$ is determined up to a nonzero scalar factor. Let us prove that

$$F_{l_0} z_{\mathrm{low}} \neq 0. \tag{2.105}$$

In fact, assuming the contrary, we have $U_q\mathfrak{n}^- z_{\mathrm{low}} = \mathbb{C} z_{\mathrm{low}}$. Hence for all $\xi \in U_q\mathfrak{n}^-$, the linear functional $\xi z_{\mathrm{low}} \in \mathbb{C}[\mathfrak{p}^-]_{q,1}$ annihilates $v(\mathfrak{q}^+, 0) \in N(\mathfrak{q}^+, 0)$. This already implies that z_{low} annihilates all the vectors from $U_q\mathfrak{n}^- v(\mathfrak{q}^+, 0)$. Thus $z_{\mathrm{low}} = 0$, which contradicts to our choice of z_{low}.

It follows from (2.105) that z_{low} can be chosen so that $F_{l_0} z_{\mathrm{low}} = q_{l_0}^{1/2}$, and it is a unique element with the properties (2.103).

2. The equality (2.104) is obvious, because the lowest weight of the $U_q\mathfrak{k}$-module $\mathbb{C}[\mathfrak{p}^-]_{q,1}$ differs by sign from the highest weight of the $U_q\mathfrak{k}$-module $N(\mathfrak{q}^+, 0)_{-1}$, and the highest weight vector of the latter module is $F_{l_0} v(\mathfrak{q}^+, 0)$, whose weight is $-\alpha_{l_0}$.

The relation $E_{l_0} z_{\mathrm{low}} = \mathrm{const} \cdot z_{\mathrm{low}}^2$ is due to the following observations. First, the lowest weight subspace of the $U_q\mathfrak{k}$-module $\mathbb{C}[\mathfrak{p}^-]_{q,2}$ is spanned by z_{low}^2. Second, the vector $E_{l_0} z_{\mathrm{low}} \in \mathbb{C}[\mathfrak{p}^-]_{q,2}$ belongs to this weight subspace because its weight is $2\alpha_{l_0}$. The equality $\mathrm{const} = -q_{\mathrm{low}}^{1/2}$ is due to the relation

$$(E_{l_0} F_{l_0} - F_{l_0} E_{l_0}) z_{\mathrm{low}} = \frac{K_{l_0} - K_{l_0}^{-1}}{q_{l_0} - q_{l_0}^{-1}} z_{\mathrm{low}}.$$

The rest of the relations follow from what has already been proved, together with the compatibility of involutions (2.102). □

LEMMA 2.149.

$$z_{\mathrm{low}}^* z_{\mathrm{low}} = q_{l_0}^2 z_{\mathrm{low}} z_{\mathrm{low}}^* + 1 - q_{l_0}^2. \tag{2.106}$$

PROOF. We are about to apply the multiplicative formula (2.11) for the universal R-matrix under the assumption that the reduced expression for $w_0 \in W$ is chosen in the same way as in Subsection 2.3.1: $w_0 = w_{0,\mathbb{S}} \cdot {}^{\mathbb{S}}w_0$, where

$$w_{0,\mathbb{S}} = s_{i_1} s_{i_2} \cdots s_{i_{M'}}, \qquad {}^{\mathbb{S}}w_0 = s_{i_{M'+1}} s_{i_{M'+2}} \cdots s_{i_M}.$$

The definition of multiplication $m : \mathrm{Pol}(\mathfrak{p}^-)_q^{\otimes 2} \to \mathrm{Pol}(\mathfrak{p}^-)_q$ implies that, up to the flip of the tensor factors, $m(z_{\mathrm{low}}^* \otimes z_{\mathrm{low}})$ equals

$$\exp_{q_{M'}^2}\left(\left(q_{M'} - q_{M'}^{-1}\right) E_{\beta_{M'}} \otimes F_{\beta_{M'}}\right) q^{-t_0} z_{\mathrm{low}}^* \otimes z_{\mathrm{low}}.$$

It is easy to show that

$$\beta_{M'} = \alpha_{l_0}, \qquad E_{\beta_{M'}} \otimes F_{\beta_{M'}} = c \cdot E_{l_0} \otimes F_{l_0}, \qquad c \in \mathbb{C} \backslash \{0\}.$$

Together with (2.13), the first relation implies that

$$q^{-t_0}(z_{\mathrm{low}}^* \otimes z_{\mathrm{low}}) = q_{l_0}^{-\frac{H_{l_0} \otimes H_{l_0}}{2}} (z_{\mathrm{low}}^* \otimes z_{\mathrm{low}}).$$

Hence,

$$z_{\mathrm{low}}^* z_{\mathrm{low}} = q_{l_0}^2 z_{\mathrm{low}} z_{\mathrm{low}}^* + c(1 - q_{l_0}^2), \qquad c \neq 0.$$

It remains to prove that $c = 1$. For that, it suffices to apply $F_{l_0} \in U_q\mathfrak{g}$ to both sides of the relation

$$z_{\mathrm{low}}^* z_{\mathrm{low}}^2 = q_{l_0}^4 z_{\mathrm{low}}^2 z_{\mathrm{low}}^* + c(1 - q_{l_0}^4) z_{\mathrm{low}},$$

which leads to a quadric equation with respect to c, whose solutions ($c = 0$, $c = 1$) are easy to guess. □

COROLLARY 2.150. *Suppose that* $\mathrm{Pol}(\mathfrak{p}^-)_q$ *is endowed with the structure of* $U_{q_{l_0}}\mathfrak{su}_{1,1}$*-module algebra induced by the embedding*

$$U_{q_{l_0}}\mathfrak{sl}_2 \hookrightarrow U_q\mathfrak{g}, \qquad K^{\pm 1} \mapsto K_{l_0}^{\pm 1}, \qquad E \mapsto E_{l_0}, \qquad F \mapsto F_{l_0}.$$

The map $z \mapsto z_{\mathrm{low}}$ *can be uniquely extended to an embedding of* $U_q\mathfrak{su}_{1,1}$*-module* $*$*-algebras*

$$\mathrm{Pol}(\mathbb{C})_{q_{l_0}} \hookrightarrow \mathrm{Pol}(\mathfrak{p}^-)_q.$$

Now we describe the action of the generators of $U_q\mathfrak{g}$ on f_0.

PROPOSITION 2.151. $K_i^{\pm 1} f_0 = f_0$ *for* $i = 1, 2, \ldots, l$,

$$F_j f_0 = \begin{cases} -\frac{q_{l_0}^{1/2}}{q_{l_0}^{-2} - 1} f_0 z_{\mathrm{low}}^*, & j = l_0, \\ 0, & j \neq l_0, \end{cases} \qquad E_j f_0 = \begin{cases} -\frac{q_{l_0}^{1/2}}{1 - q_{l_0}^2} z_{\mathrm{low}} f_0, & j = l_0, \\ 0, & j \neq l_0. \end{cases}$$

PROOF. Due to the definition of T_F and Lemma 2.134, all the terms $a_{j,k}$ of the decomposition

$$f_0 = \sum_{j,k=0}^{\infty} a_{j,k}, \qquad a_{j,k} \in \mathbb{C}[\mathfrak{p}^-]_{q,k} \mathbb{C}[\mathfrak{p}^+]_{q,-j},$$

are zero for $j \neq k$. Hence $H_{\mathbb{S}} f_0 = 0$. Thus Corollary 2.146 implies the $U_q\mathfrak{k}$-invariance of f_0:

$$(2.107) \qquad K_i^{\pm 1} f_0 = f_0, \quad i = 1, 2, \ldots, l, \qquad E_j f_0 = F_j f_0 = 0, \quad j \neq l_0.$$

It remains to prove the relations $F_{l_0} f_0 = -\frac{q_{l_0}^{1/2}}{q_{l_0}^{-2} - 1} f_0 z_{\mathrm{low}}^*$ and $E_{l_0} f_0 = -\frac{q_{l_0}^{1/2}}{1 - q_{l_0}^2} z_{\mathrm{low}} f_0$. Corollary 2.146 implies that $F_{l_0} f_0 \in f_0 \mathbb{C}[\mathfrak{p}^+]_q$ and $E_{l_0} f_0 \in \mathbb{C}[\mathfrak{p}^-]_q f_0$. This claim allows a refinement. By (2.107), the weights of $F_{l_0} f_0$ and $E_{l_0} f_0$ are $-\alpha_{l_0}$ and α_{l_0}, respectively. On the other hand, the weight subspaces of such weights in $f_0 \mathbb{C}[\mathfrak{p}^+]_q$ and $\mathbb{C}[\mathfrak{p}^-]_q f_0$ are one dimensional and are spanned by $f_0 z_{\mathrm{low}}^*$ and $z_{\mathrm{low}} f_0$, respectively. Hence $F_{l_0} f_0 = c^- f_0 z_{\mathrm{low}}^*$ and $E_{l_0} f_0 = c^+ z_{\mathrm{low}} f_0$.

It remains to find the scalar factors c^-, c^+. It follows from the relations

$$f_0 z_{\mathrm{low}} = 0, \qquad \Delta(F_{l_0}) = F_{l_0} \otimes K_{l_0}^{-1} + 1 \otimes F_{l_0}, \qquad z_{\mathrm{low}}^* z_{\mathrm{low}} = q_{l_0}^2 z_{\mathrm{low}} z_{\mathrm{low}}^* + 1 - q_{l_0}^2$$

that

$$0 = F_{l_0}(f_0 z_{\mathrm{low}}) = (c^- f_0 z_{\mathrm{low}}^*)\left(q_{l_0}^{-2} z_{\mathrm{low}}\right) + f_0 q_{l_0}^{1/2} = \left(c^- q_{l_0}^{-2}\left(1 - q_{l_0}^2\right) + q_{l_0}^{1/2}\right) f_0.$$

Hence $c^- = -q_{l_0}^{1/2} \Big/ \left(q_{l_0}^{-2} - 1\right)$. An application of the relation $(S(E_{l_0}))^* = q_{l_0}^{-2} F_{l_0}$ and (2.102) allows one to find c^+. □

Note that in the case of quantum disc, the latter claim was obtained earlier in Subsection 1.4.3.

Let us endow $\mathrm{Fun}(\mathbb{D})_q$ with a structure of $U_q\mathfrak{g}$-module using the decomposition $\mathrm{Fun}(\mathbb{D})_q = \mathrm{Pol}(\mathfrak{p}^-)_q \oplus \mathcal{D}(\mathbb{D})_q$.

PROPOSITION 2.152. *The algebra* $\mathrm{Fun}(\mathbb{D})_q$ *is a* $U_q\mathfrak{g}$*-module algebra.*

PROOF. We are about to use the description of $\mathrm{Fun}(\mathbb{D})_q$ in terms of generators and relations as follows. The element f_0 and nonzero elements of $\mathrm{Pol}(\mathfrak{p}^-)_q$ are to be used as generators, and the set of defining relations is formed by (2.95) and (2.96), and all the relations that exist between the nonzero elements of the algebra $\mathrm{Pol}(\mathfrak{p}^-)_q$. The action of elements $K_i^{\pm 1}$, E_i, F_i, $i = 1, 2, \ldots, l$, of $U_q\mathfrak{g}$ on the above generators of $\mathrm{Fun}(\mathbb{D})_q$ was described in this subsection and Subsections 2.3.2 and 2.3.3. We use this description to endow the free algebra with the same generators as $\mathrm{Fun}(\mathbb{D})_q$ with a structure of $U_q\mathfrak{g}$-module algebra.

It remains to be persuaded that the two-sided ideal J of the free algebra, which corresponds to the defining relations of $\mathrm{Fun}(\mathbb{D})_q$, is a $U_q\mathfrak{g}$-submodule. Equivalently, it remains to verify that the action of generators of $U_q\mathfrak{g}$ on the generators of the two-sided ideal J does not take them out of J.

We start with the element $f_0^2 - f_0 \in J$. Proposition 2.151 yields

$$K_i^{\pm 1}(f_0 f_0 - f_0) = \left(K_i^{\pm 1} f_0\right)\left(K_i^{\pm 1} f_0\right) - K_i^{\pm 1} f_0 = f_0 f_0 - f_0, \quad i = 1, 2, \ldots, l,$$
$$F_j(f_0 f_0 - f_0) = E_j(f_0 f_0 - f_0) = 0, \quad j \neq l_0,$$
$$F_{l_0}(f_0 f_0 - f_0) = -\frac{q_{l_0}^{1/2}}{q_{l_0}^{-2} - 1}\left(f_0(z_{\mathrm{low}}^* f_0) + (f_0 f_0 - f_0) z_{\mathrm{low}}^*\right),$$
$$E_{l_0}(f_0 f_0 - f_0) = -\frac{q_{l_0}^{1/2}}{1 - q_{l_0}^2}\left(z_{\mathrm{low}}(f_0 f_0 - f_0) + (f_0 z_{\mathrm{low}}) f_0\right).$$

Obviously, all the right-hand sides belong to J.

The action of $K_i^{\pm 1}$, E_i, F_i, $i = 1, 2, \ldots, l$, on the elements corresponding to the rest of the defining relations does not take out of J, because $\mathrm{Pol}(\mathfrak{p}^-)_q$ is a $U_q\mathfrak{g}$-module algebra, and $\mathcal{D}(\mathbb{D})_q$ is a $U_q\mathfrak{g}$-covariant $\mathrm{Pol}(\mathfrak{p}^-)_q$-bimodule. □

COROLLARY 2.153. *The* $*$*-algebra* $\mathcal{D}(\mathbb{D})_q$ *is a* $(U_q\mathfrak{g}, *)$*-module algebra.*

2.4.4. Existence, uniqueness, and an explicit form of an invariant integral on $\mathcal{D}(\mathbb{D})_q$. It was demonstrated in the previous subsection that $\mathcal{D}(\mathbb{D})_q$ is a $U_q\mathfrak{g}$-module algebra. Let us prove the existence and uniqueness (up to a scalar factor) of an invariant integral $\nu : \mathcal{D}(\mathbb{D})_q \to \mathbb{C}$. We follow the ideas of Subsection 1.4.4 where these results were obtained in the special case of quantum disc.

Consider the representation Γ of $U_q\mathfrak{q}^+$ in the Fock space $\mathcal{H}$ as in Subsection 2.4.3. To every $\lambda = \sum_{i=1}^{l} n_i \alpha_i \in Q$, one can assign the element $K_\lambda = K_1^{n_1} K_2^{n_2} \cdots K_l^{n_l}$ of $U_q\mathfrak{h}$ and the linear operator $\Gamma(K_\lambda)$.

Since the linear operators $T_F(f)$, $f \in \mathcal{D}(\mathbb{D})_q$, are finite-dimensional, one can introduce the q-trace

$$\mathrm{tr}_q\, T_F(f) \stackrel{\mathrm{def}}{=} \mathrm{tr}(T_F(f)\Gamma(K_{-2\rho})), \qquad f \in \mathcal{D}(\mathbb{D})_q,$$

where ρ is half the sum of positive roots of the Lie algebra $\mathfrak{g}$.

PROPOSITION 2.154. *The linear functional*

$$\int_{\mathbb{D}_q} f d\nu \overset{\text{def}}{=} \left(1 - q_{l_0}^2\right)^{\dim \mathfrak{p}^-} \operatorname{tr}_q T_F(f) \tag{2.108}$$

is a positive $U_q\mathfrak{g}$-invariant integral on $\mathcal{D}(\mathbb{D})_q$.

PROOF. The restriction of $\Gamma(K_{-2\rho})$ to every weight subspace of $\mathcal{H}$ is a positive scalar operator. Hence $\operatorname{tr}_q T_F(f^*f) = \operatorname{tr}(T_F(f)\Gamma(K_{-2\rho})T_F(f)^*) > 0$ for all $f \in \mathcal{D}(\mathbb{D})_q$, and so the positivity of the linear functional (2.108) now follows from Proposition 2.133. Of course,

$$\int_{\mathbb{D}_q} f^* d\nu = \overline{\int_{\mathbb{D}_q} f d\nu}. \tag{2.109}$$

By Proposition 1.64 and (2.6), the linear functional (2.108) is a $U_q\mathfrak{q}^+$-invariant integral. Its $U_q\mathfrak{g}$-invariance follows from the $U_q\mathfrak{q}^+$-invariance, (2.109), and (2.102). □

Let us turn to proving uniqueness for the invariant integral.

LEMMA 2.155. *The element $f_0 \in \mathcal{D}(\mathbb{D})_q$ generates the $U_q\mathfrak{b}^+$-module $\mathbb{C}[\mathfrak{p}^-]_q f_0$ and the $U_q\mathfrak{b}^-$-module $f_0\mathbb{C}[\mathfrak{p}^+]_q$.*

PROOF. By (2.102), it suffices to prove that $U_q\mathfrak{b}^+ f_0 = \mathbb{C}[\mathfrak{p}^-]_q f_0$. Proposition 2.93 allows one to prove easily that the $U_q\mathfrak{b}^+$-invariant pairing

$$\mathbb{C}[\mathfrak{p}^+]_q \times \mathbb{C}[\mathfrak{p}^-] f_0 \to \mathbb{C}, \qquad f_+ \times f_- f_0 \mapsto \int_{\mathbb{D}_q} f_+ f_- f_0 d\nu,$$

is nondegenerate.

Hence in the category of graded $U_q\mathfrak{b}^+$-modules, one has $\mathbb{C}[\mathfrak{p}^-]_q f_0 \cong \mathbb{C}[\mathfrak{p}^+]_q^*$. Also, in this category $N(\mathfrak{q}^-, 0) \cong \mathbb{C}[\mathfrak{p}^+]_q^*$, which is due to the definition of the $U_q\mathfrak{g}$-module $\mathbb{C}[\mathfrak{p}^+]_q$. Therefore, $\mathbb{C}[\mathfrak{p}^-]_q f_0 \cong N(\mathfrak{q}^-, 0)$, and the latter isomorphism maps the homogeneous component $\mathbb{C}f_0$ onto the homogeneous component $\mathbb{C}v(\mathfrak{q}^-, 0)$. It remains to use the fact that $v(\mathfrak{q}^-, 0)$ generates the $U_q\mathfrak{b}^+$-module $N(\mathfrak{q}^-, 0)$. □

The following claim generalizes Proposition 1.82.

PROPOSITION 2.156. $U_q\mathfrak{g} f_0 = \mathcal{D}(\mathbb{D})_q$.

PROOF. By Lemma 2.155, it suffices to show that

$$(E_{j_1} E_{j_2} \cdots E_{j_r} f_0)\mathbb{C}[\mathfrak{p}^+]_q \subset U_q\mathfrak{g} f_0 \tag{2.110}$$

for all $r \in \mathbb{Z}_+$, $j_1, j_2, \ldots, j_r \in \{1, 2, \ldots, l\}$. We proceed by induction. For $r = 0$ the inclusion (2.110) follows from Lemma 2.155. Let us describe the passage from $r - 1$ to r.

Let $\varphi = E_{j_2} E_{j_3} \cdots E_{j_r} f_0$, $\psi \in \mathbb{C}[\mathfrak{p}^+]_q$. By the induction hypotheses one has $\varphi\psi \in U_q\mathfrak{g} f_0$. Thus Proposition 2.152 implies that

$$(E_{j_1} E_{j_2} \cdots E_{j_r} f_0)\psi = (E_{j_1}\varphi)\psi = E_{j_1}(\varphi\psi) - (K_{j_1}\varphi)(E_{j_1}\psi),$$

since $\Delta(E_{j_1}) = E_{j_1} \otimes 1 + K_{j_1} \otimes E_{j_1}$.

It remains to observe that $E_{j_1}(\varphi\psi)$ and $(K_{j_1}\varphi)(E_{j_1}\psi)$ belong to $U_q\mathfrak{g} f_0$ by the induction hypothesis. □

COROLLARY 2.157. *The $U_q\mathfrak{g}$-invariant integral on $\mathcal{D}(\mathbb{D})_q$ is unique up to a scalar factor.*

Since the $U_q\mathfrak{g}$-invariant integral (2.108) is positive, one can endow the vector space $\mathcal{D}(\mathbb{D})_q$ with the norm $\|f\|_2 = (\int_{\mathbb{D}_q} f^* f d\nu)^{\frac{1}{2}}$. The completion of $\mathcal{D}(\mathbb{D})_q$ with respect to the above norm is a Hilbert space. It is to be denoted by $L^2(d\nu)_q$ and is a q-analog of the space $L^2(d\nu)$ of square summable functions in the invariant measure $d\nu$ on the bounded symmetric domain $\mathbb{D}$.

It follows from (2.108) that for each homogeneous normalized vector $v_i \in \mathcal{H}_i$, $v_j \in \mathcal{H}_j$ of the graded pre-Hilbert space $\mathcal{H} = \bigoplus_{k=0}^{\infty} \mathcal{H}_k$, one has the inequality

$$|(T_F(f)v_i, v_j)| \leq \mathrm{const}(i,j)\|f\|_2, \qquad f \in \mathcal{D}(\mathbb{D})_q.$$

Hence the embedding $\mathcal{D}(\mathbb{D})_q \hookrightarrow \mathcal{D}(\mathbb{D})'_q$ described in Subsection 2.4.2 admits an extension by continuity to an embedding $L^2(d\nu)_q \hookrightarrow \mathcal{D}(\mathbb{D})'_q$ (the injectivity is easily derivable from the definition of $L^2(d\nu)_q$ and (2.108)). In what follows, the Hilbert space $L^2(d\nu)_q$ will be identified with its image under the embedding into $\mathcal{D}(\mathbb{D})'_q$.

For every distribution $f \in \mathcal{D}(\mathbb{D})'_q$ on the quantum bounded symmetric domain $\mathbb{D}$ one has the well-defined sesquilinear form $(T_F(f)v', v'')$ in $\mathcal{H}$. A distribution $f \in \mathcal{D}(\mathbb{D})'_q$ will be called bounded if

$$\|f\|_\infty = \sup_{v',v'' \in \mathcal{H}\setminus\{0\}} \frac{(T_F(f)v', v'')}{\|v'\|\|v''\|} < \infty.$$

In this case there exists a unique bounded linear operator $\overline{T}$ in the completion $\overline{\mathcal{H}}$ of the pre-Hilbert space $\mathcal{H}$ such that $(\overline{T}v', v'') = (T_F(f)v', v'')$ for all $v', v'' \in \mathcal{H}$. Bounded distributions form a C^*-algebra $L^\infty(\mathbb{D})_q$ with the norm $\|\cdot\|_\infty$. This C^*-algebra is naturally isomorphic to the C^*-algebra of all bounded linear operators in the Hilbert space $\overline{\mathcal{H}}$. If $\varphi \in L^\infty(\mathbb{D})_q$, $\psi \in L^2(d\nu)_q$, then $\varphi\psi \in L^2(d\nu)_q$ and

$$\|\varphi\psi\|_2 \leq \|\varphi\|_\infty \|\psi\|_2.$$

Indeed,

$$\begin{aligned}\|\varphi\psi\|_2^2 &= \left(1 - q_{l_0}^2\right)^{\dim \mathfrak{p}^-} \mathrm{tr}(T_F(\varphi)T_F(\psi)\Gamma(K_{-2\rho})T_F(\psi)^* T_F(\varphi)^*) \\ &\leq \|\varphi\|_\infty^2 \left(1 - q_{l_0}^2\right)^{\dim \mathfrak{p}^-} \mathrm{tr}(T_F(\psi)\Gamma(K_{-2\rho})T_F(\psi)^*) \leq \|\varphi\|_\infty^2 \|\psi\|_2^2.\end{aligned}$$

One can obtain a natural representation of the C^*-algebra $L^\infty(\mathbb{D})_q$ in the Hilbert space $L^2(d\nu)_q$ by assigning to every $\varphi \in L^\infty(\mathbb{D})_q$ the linear operator $\hat{\varphi} : \psi \mapsto \varphi\psi$.

2.4.5. Isomorphism of the $U_q\mathfrak{g}$-modules $\mathcal{D}(\mathbb{D})_q$ and $U_q\mathfrak{g} \otimes_{U_q\mathfrak{k}} \mathbb{C}$. To every weight locally finite-dimensional $U_q\mathfrak{k}$-module V we assign a $U_q\mathfrak{g}$-module $P(V) = U_q\mathfrak{g} \otimes_{U_q\mathfrak{k}} V$. It is easy to show that the latter module is a projective object in the category $C(\mathfrak{g}, \mathfrak{k})_q$. Such projective objects are called standard, cf. [**175**, p. 104]. Let us split $P(\mathbb{C}) = U_q\mathfrak{g} \otimes_{U_q\mathfrak{k}} \mathbb{C}$ as a product of generalized Verma modules in the tensor category of $U_q\mathfrak{g}^{\mathrm{cop}}$-modules.

The vector $v(\mathfrak{q}^+, 0) \otimes v(\mathfrak{q}^-, 0) \in N(\mathfrak{q}^+, 0) \otimes N(\mathfrak{q}^-, 0)$ is $U_q\mathfrak{k}$-invariant, that allows one to introduce the morphism of $U_q\mathfrak{g}$-modules

$$\mathcal{J} : P(\mathbb{C}) \to N(\mathfrak{q}^+, 0) \otimes N(\mathfrak{q}^-, 0), \qquad \mathcal{J} : 1 \otimes 1 \mapsto v(\mathfrak{q}^+, 0) \otimes v(\mathfrak{q}^-, 0). \tag{2.111}$$

PROPOSITION 2.158. *The morphism of $U_q\mathfrak{g}$-modules (2.111) is bijective.*

We precede the proof with the following

LEMMA 2.159. *Let $P_{\mathfrak{h}}(\mathbb{C}) = U_q\mathfrak{g} \otimes_{U_q\mathfrak{h}} \mathbb{C}$. In the category of $U_q\mathfrak{g}^{\mathrm{cop}}$-modules the morphism of $U_q\mathfrak{g}$-modules*

$$\mathcal{J}_{\mathfrak{h}} : P_{\mathfrak{h}}(\mathbb{C}) \to N(\mathfrak{b}^+, 0) \otimes N(\mathfrak{b}^-, 0), \qquad \mathcal{J}_{\mathfrak{h}} : 1 \otimes 1 \mapsto v(\mathfrak{b}^+, 0) \otimes v(\mathfrak{b}^-, 0),$$

is an isomorphism.

PROOF OF THE LEMMA. We endow the $U_q\mathfrak{h}$-modules $N(\mathfrak{b}^\pm, 0)$ with the filtrations related to the weight subspaces

$$F_{\mu^+}N(\mathfrak{b}^+, 0) = \bigoplus_{\substack{\nu^+ \in Q_+ \\ 0 \le \nu^+ \le \mu^+}} N(\mathfrak{b}^+, 0)_{-\nu^+}, \qquad F_{\mu^-}N(\mathfrak{b}^-, 0) = \bigoplus_{\substack{\nu^- \in Q_+ \\ 0 \le \nu^- \le \mu^-}} N(\mathfrak{b}^-, 0)_{\nu^-}.$$

The tensor product $N(\mathfrak{b}^+, 0) \otimes N(\mathfrak{b}^-, 0)$ is endowed with a similar filtration

$$F_{\mu^+,\mu^-}\left(N(\mathfrak{b}^+, 0) \otimes N(\mathfrak{b}^-, 0)\right) = \bigoplus_{\substack{\nu^+ \in Q_+ \\ 0 \le \nu^+ \le \mu^+}} \bigoplus_{\substack{\nu^- \in Q_+ \\ 0 \le \nu^- \le \mu^-}} N(\mathfrak{b}^+, 0)_{-\nu^+} \otimes N(\mathfrak{b}^-, 0)_{\nu^-}.$$

Finally, we use the adjoint action of $U_q\mathfrak{h}$ in $U_q\mathfrak{n}^\pm$ to endow the $U_q\mathfrak{h}$-module $P_{\mathfrak{h}}(\mathbb{C})$ with the filtration as follows:

$$F_{\mu^+,\mu^-}(P_{\mathfrak{h}}(\mathbb{C})) = \sum_{\substack{\nu^+ \in Q_+ \\ \nu^+ \le \mu^+}} \sum_{\substack{\nu^- \in Q_+ \\ \nu^- \le \mu^-}} \text{linear span of} \left((U_q\mathfrak{n}^-)_{-\nu^+}(U_q\mathfrak{n}^+)_{\nu^-} \otimes 1\right),$$

where $\mu^\pm$ are elements of the weight lattice P.

Let us use the above filtrations to prove the lemma. First, it is easy to show that

$$\mathcal{J}_{\mathfrak{h}} F_{\mu^+,\mu^-}(P_{\mathfrak{h}}(\mathbb{C})) \supset F_{\mu^+,\mu^-}\left(N(\mathfrak{b}^+, 0) \otimes N(\mathfrak{b}^-, 0)\right). \tag{2.112}$$

In fact, if one assumes the contrary, there are minimal pairs among the pairs μ^+, μ^- which do not satisfy (2.112) (a partial order on the set of such pairs is introduced in an obvious way). We get a contradiction because if

$$\xi = F_{i_1}^{j_1} F_{i_2}^{j_2} F_{i_3}^{j_3} \cdots, \qquad \eta = E_{k_1}^{l_1} E_{k_2}^{l_2} E_{k_3}^{l_3} \cdots, \qquad i_1, i_2, \ldots; k_1, k_2 \cdots \in \{1, 2, \ldots, l\},$$

then up to lower degree terms one has

$$\mathcal{J}_{\mathfrak{h}}(\xi\eta(1 \otimes 1)) = \mathrm{const} \cdot \xi v(\mathfrak{b}^+, 0) \otimes \eta v(\mathfrak{b}^-, 0),$$

where const is a nonzero constant.

Second, it follows from the results of Subsection 2.1.3 that the subspaces $F_{\mu^+,\mu^-}(P_{\mathfrak{h}}(\mathbb{C}))$, $F_{\mu^+,\mu^-}(N(\mathfrak{b}^+, 0) \otimes N(\mathfrak{b}^-, 0))$ are finite-dimensional, and their dimensions are the same. Hence,

$$\mathcal{J}_{\mathfrak{h}} : F_{\mu^+,\mu^-}(P_{\mathfrak{h}}(\mathbb{C})) \xrightarrow{\approx} F_{\mu^+,\mu^-}\left(N(\mathfrak{b}^+, 0) \otimes N(\mathfrak{b}^-, 0)\right),$$

and so it remains to apply the equalities $\bigcup_{\mu^+,\mu^-} F_{\mu^+,\mu^-}(P_{\mathfrak{h}}(\mathbb{C})) = P_{\mathfrak{h}}(\mathbb{C})$,

$$\bigcup_{\mu^+,\mu^-} F_{\mu^+,\mu^-}\left(N(\mathfrak{b}^+, 0) \otimes N(\mathfrak{b}^-, 0)\right) = N(\mathfrak{b}^+, 0) \otimes N(\mathfrak{b}^-, 0). \qquad \square$$

PROOF OF THE PROPOSITION. Let us argue as in the proof of the lemma. Recall that the weight $U_q\mathfrak{k}$-modules are endowed with the gradings determined by the element $H_{\mathbb{S}} \in \mathfrak{h}$; see Subsection 2.1.2. In particular,

$$N(\mathfrak{q}^+, 0) = \bigoplus_{j=0}^{\infty} N(\mathfrak{q}^+, 0)_{-j}, \qquad N(\mathfrak{q}^-, 0) = \bigoplus_{j=0}^{\infty} N(\mathfrak{q}^-, 0)_j.$$

We endow $N(\mathfrak{q}^+,0)\otimes N(\mathfrak{q}^-,0)$ with the filtration

$$F_{i^+i^-}\left(N(\mathfrak{q}^+,0)\otimes N(\mathfrak{q}^-,0)\right)=\bigoplus_{0\le j^+\le i^+}\bigoplus_{0\le j^-\le i^-}N(\mathfrak{q}^+,0)_{-j^+}\otimes N(\mathfrak{q}^-,0)_{j^-}.$$

Then using the isomorphism

$$U_q\mathfrak{p}^-\otimes U_q\mathfrak{p}^+\otimes U_q\mathfrak{k}\xrightarrow{\approx}U_q\mathfrak{g},\qquad \xi^-\otimes\xi^+\otimes\eta\mapsto\xi^-\xi^+\eta,$$

and the adjoint action of $U_q\mathfrak{k}$ in $U_q\mathfrak{p}^\pm$, we endow the $U_q\mathfrak{k}$-module $P(\mathbb{C})$ with the filtration

$$F_{i^+i^-}\left(P(\mathbb{C})\right)=\sum_{0\le j^+\le i^+}\sum_{0\le j^-\le i^-}\text{linear span of}\left((U_q\mathfrak{p}^-)_{-j^+}(U_q\mathfrak{p}^+)_{j^-}\otimes 1\right).$$

Due to the definitions, $P(\mathbb{C})=\bigcup_{i^+,i^-}F_{i^+i^-}\left(P(\mathbb{C})\right)$,

$$N(\mathfrak{q}^+,0)\otimes N(\mathfrak{q}^-,0)=\bigcup_{i^+,i^-}F_{i^+i^-}\left(N(\mathfrak{q}^+,0)\otimes N(\mathfrak{q}^-,0)\right).$$

In addition, the subspaces $F_{i^+i^-}\left(N(\mathfrak{q}^+,0)\otimes N(\mathfrak{q}^-,0)\right)$ and $F_{i^+i^-}\left(P(\mathbb{C})\right)$ are finite-dimensional, and their dimensions are the same; see Subsections 2.3.1 and 2.3.5.

It remains to prove the inclusion

$$\mathcal{J}F_{i^+i^-}\left(P(\mathbb{C})\right)\supset F_{i^+i^-}\left(N(\mathfrak{q}^+,0)\otimes N(\mathfrak{q}^-,0)\right),\qquad i^+,i^-\in\mathbb{Z}_+. \tag{2.113}$$

The commutative diagram

$$\begin{array}{ccc} P_{\mathfrak{h}}(\mathbb{C}) & \xrightarrow{\mathcal{J}_{\mathfrak{h}}} & N(\mathfrak{b}^+,0)\otimes N(\mathfrak{b}^-,0)\\ \varphi\Big\downarrow & & \Big\downarrow\psi\\ P(\mathbb{C}) & \xrightarrow{\mathcal{J}} & N(\mathfrak{q}^+,0)\otimes N(\mathfrak{q}^-,0)\end{array}$$

in the category of $U_q\mathfrak{g}^{\mathrm{cop}}$-modules can be readily used to derive (2.113) from (2.112) and the following claims.

First of all, the subspace $N(\mathfrak{q}^+,0)_{-j^+}$ is the image of the sum of weight subspaces $N(\mathfrak{b}^+,0)_{-\nu^+}$ with $\nu^+(H_{\mathbb{S}})=j^+$, under the canonical projection morphism $N(\mathfrak{b}^+,0)\to N(\mathfrak{q}^+,0)$. Similarly, $N(\mathfrak{q}^-,0)_{j^-}$ is the image of the sum of weight subspaces $N(\mathfrak{b}^-,0)_{\nu^-}$ with $\nu^-(H_{\mathbb{S}})=j^-$. This allows one to arrange an upper estimate for the preimage of the subspace $F_{i^+i^-}\left(N(\mathfrak{q}^+,0)\otimes N(\mathfrak{q}^-,0)\right)$ under ψ.

Finally, $(U_q\mathfrak{n}^-)_{-\nu^+}(U_q\mathfrak{n}^+)_{\nu^-}\otimes 1\subset(U_q\mathfrak{q}^-)_{-\nu^+}(U_q\mathfrak{q}^+)_{\nu^-}\otimes 1$ and

$$\mathcal{J}\left((U_q\mathfrak{q}^-)_{-\nu^+}(U_q\mathfrak{q}^+)_{\nu^-}\otimes 1\right)\subset(U_q\mathfrak{p}^-)_{-j^+}(U_q\mathfrak{p}^+)_{j^-}(v(\mathfrak{q}^+,0)\otimes v(\mathfrak{q}^-,0)),$$

if $j^\pm=\nu^\pm(H_{\mathbb{S}})$.

This claim gives a lower estimate for the preimage of $F_{i^+i^-}(P(\mathbb{C}))$ under φ. □

Let us prove that $P(\mathbb{C})\approx\mathcal{D}(\mathbb{D})_q$. Consider the vector space

$$\mathrm{Pol}(\mathfrak{p}^-)_q=\mathbb{C}[\mathfrak{p}^-]_q\otimes\mathbb{C}[\mathfrak{p}^+]_q=N(\mathfrak{q}^+,0)^*\otimes N(\mathfrak{q}^-,0)^*$$

together with its descending filtration by the subspaces $\bigoplus_{j+k\ge n}\mathbb{C}[\mathfrak{p}^-]_{q,j}\mathbb{C}[\mathfrak{p}^+]_{q,-k}$. Their annihilators in the dual space form an ascending sequence. Due to Proposition 2.158, the union of annihilators is a $U_q\mathfrak{g}$-submodule which is naturally isomorphic to $P(\mathbb{C})$.

On the other hand, it is easy to show, using the principal properties of the algebra of finite functions and the invariant integral, that this union of annihilators is naturally isomorphic to $\mathcal{D}(\mathbb{D})_q$.

We thus obtain the isomorphism of $U_q\mathfrak{g}$-modules

$$P(\mathbb{C}) \overset{\approx}{\to} \mathcal{D}(\mathbb{D})_q, \qquad 1 \otimes 1 \mapsto \frac{1}{\left(1-q_{l_0}^2\right)^{\dim \mathfrak{p}^-}} f_0.$$

This proves the following refinement of Proposition 2.156, which also generalizes Proposition 1.84.

PROPOSITION 2.160. *f_0 is a generator of the $U_q\mathfrak{g}$-module $\mathcal{D}(\mathbb{D})_q$, and*

$$\xi f_0 = \varepsilon(\xi) f_0, \qquad \xi \in U_q\mathfrak{k},$$

is the defining relation.

REMARK 2.161. The duality between the $U_q\mathfrak{g}$-module algebra $\mathrm{Pol}(\mathfrak{p}^-)_q$ and the $U_q\mathfrak{g}$-module coalgebra $P(\mathbb{C})$ used in the proof allows one to view $P(\mathbb{C})$ as a quantum analog of the space of distributions concentrated at $\mathbf{z} = 0$. A close statement is present in the work by Heckenberger and Kolb [**117**]. There the elements of $P(\mathbb{C})$ are viewed as q-analogs of distributions on a Hermitian symmetric space which are concentrated at a point corresponding to the unit of a group. The principal difference of our results from those of [**117**] is that the evaluation at $\mathbf{z} = 0$ is not a morphism of algebras, while the evaluation at the unit of quantum group (i.e., the counit) is a morphism of algebras.

REMARK 2.162. Let $\lambda^+ \in P_+^{\mathbb{S}}$, $\lambda^- \in -P_-^{\mathbb{S}}$ and $w_{0,\mathbb{S}}$ be the longest element of the Weyl group $W_{\mathbb{S}} \subset W$. Consider the generalized Verma module $N(\mathfrak{q}^+, \lambda^+)$ with highest weight λ^+ and the generalized Verma module $N(\mathfrak{q}^-, \lambda^-)$ with highest weight λ^-. The proof of Proposition 2.158 admits quite literal transfer to the case of nonzero weights $\lambda^\pm$. So $P(L(\mathfrak{k}, w_{0,\mathbb{S}}\lambda^-) \otimes L(\mathfrak{k}, \lambda^+))$ and $N(\mathfrak{q}^+, \lambda^+) \otimes N(\mathfrak{q}^-, \lambda^-)$ are isomorphic in the tensor category of $U_q\mathfrak{g}^{\mathrm{cop}}$-modules.

2.4.6. Boundedness of the Fock representation operators of the $*$-algebra $\mathrm{Pol}(\mathfrak{p}^-)_q$. In the classical case $q = 1$, the polynomials on the vector space $\mathfrak{p}^-$ belong to $L^\infty(\mathbb{D})$ since the domain $\mathbb{D}$ is bounded. We intend to obtain a similar result in the case $0 < q < 1$, which will substantiate the idea of *boundedness of quantum domains* we deal with.

Let us start with proving that the operator $T_F(z_{\mathrm{low}})$ is bounded. In the case of a quantum disc this is evident: $\|T_F(z)\| = 1$. Consider the embedding of $*$-algebras

$$i : \mathrm{Pol}(\mathbb{C})_{q_{l_0}} \hookrightarrow \mathrm{Pol}(\mathfrak{p}^-)_q, \qquad i : z \mapsto z_{\mathrm{low}};$$

see Corollary 2.150.

LEMMA 2.163. (1) *The representation $T_F \cdot i$ in the pre-Hilbert space $\mathcal{H}$ is multiple to the Fock representation of the $*$-algebra $\mathrm{Pol}(\mathbb{C})_{q_{l_0}}$.*

(2) $\|T_F(z_{\mathrm{low}})\| = 1$.

PROOF. It suffices to prove the first claim. Let $T = T_F \cdot i$. Consider the subspace $\mathcal{L} = \sum_{j=0}^{\infty}(\operatorname{Ker} T(z^*) \cap \mathcal{H}_j)$. It will be enough to prove that the linear span $\mathcal{M}$ of $\{T(z^j)v|\, j \in \mathbb{Z}_+,\ v \in \mathcal{L}\}$ coincides with $\mathcal{H}$.

Evidently, $\mathcal{M} = \sum_{j=0}^{\infty}(\mathcal{M} \cap \mathcal{H}_j)$. Suppose that $\mathcal{M} \neq \mathcal{H}$. Then $\mathcal{M} \cap \mathcal{H}_j \neq \mathcal{H}_j$ for some $j \in \mathbb{Z}_+$, hence there exists a nonzero vector $v \in \mathcal{H}_j$ orthogonal to $\mathcal{M}$. For some $N \in \mathbb{Z}_+$ we have $T(z^*)^N v \neq 0$ and $T(z^*)^{(N+1)} v = 0$, because the operator $T(z^*)$ reduces the degree of homogeneity. Hence $T(z)^N T(z^*)^N v \in \mathcal{M}$ and $(T(z^*)^N v, T(z^*)^N v) = 0$, which contradicts the choice of N. □

COROLLARY 2.164. $\|T_F(z_{\mathrm{low}} z_{\mathrm{low}}^*)\| = 1$.

Consider the linear functional $\langle \cdot \rangle$ on $\mathrm{Pol}(\mathfrak{p}^-)_q$ introduced by (2.71). Let $(\cdot, \cdot)$ be the standard Hermitian form on $\mathbb{C}[\mathfrak{p}^-]_q$:

$$(\psi_1, \psi_2) = \langle \psi_2^* \psi_1 \rangle, \qquad \psi_1, \psi_2 \in \mathbb{C}[\mathfrak{p}^-]_q; \tag{2.114}$$

see Subsection 2.3.3.

LEMMA 2.165. (1) *The standard Hermitian form on* $\mathbb{C}[\mathfrak{p}^-]_{q,1}$ *is positive,* $U_q\mathfrak{k}$*-invariant, and*

$$(z_{\mathrm{low}}, z_{\mathrm{low}}) = 1 - q_{l_0}^2. \tag{2.115}$$

(2) *If* $\{z_j\}$ *is an orthogonal basis in* $\mathbb{C}[\mathfrak{p}^-]_{q,1}$, *then the element*

$$\Psi = \sum_{j=1}^{\dim \mathfrak{p}^-} \frac{z_j z_j^*}{(z_j, z_j)}$$

does not depend on the choice of this basis and is $U_q\mathfrak{k}$*-invariant.*

(3) *Consider the* $U_q\mathfrak{k}$*-module*

$$\mathbb{C}[\mathfrak{p}^-]_{q,1} \cdot \mathbb{C}[\mathfrak{p}^+]_{q,-1} = l.s. \left\{ \psi^- \cdot \psi^+ \in \mathrm{Pol}(\mathfrak{p}^-)_q \, \middle| \, \psi^- \in \mathbb{C}[\mathfrak{p}^-]_{q,1}, \ \psi^+ \in \mathbb{C}[\mathfrak{p}^+]_{q,-1} \right\}.$$

Its subspace of $U_q\mathfrak{k}$*-invariant elements is one dimensional and is spanned by* Ψ.

PROOF. The positivity of standard Hermitian form follows from Proposition 2.98, and (2.115) is due to the relations $f_0\mathbb{C}[\mathfrak{p}^-]_{q,1} = \mathbb{C}[\mathfrak{p}^+]_{q,-1} f_0 = 0$ and (2.106), (2.114). The definition of linear functional $\langle \cdot \rangle$ implies its $U_q\mathfrak{k}$-invariance, whence the $U_q\mathfrak{k}$-invariance of the Hermitian form (2.114). The element Ψ obviously does not depend on the choice of orthogonal basis $\{z_j\}$.

To prove the remaining claims, to every $\mathcal{K} \in \mathbb{C}[\mathfrak{p}^-]_{q,1}\mathbb{C}[\mathfrak{p}^+]_{q,-1}$, we assign the linear operator of multiplication by $\mathcal{K}$ in the space $\mathbb{C}[\mathfrak{p}^-]_{q,1} f_0$:

$$\mathbb{C}[\mathfrak{p}^-]_{q,1} f_0 \mapsto \mathbb{C}[\mathfrak{p}^-]_{q,1} f_0, \qquad \psi \mapsto \mathcal{K}\psi.$$

Due to Proposition 2.93, the linear map

$$\mathbb{C}[\mathfrak{p}^-]_{q,1}\mathbb{C}[\mathfrak{p}^+]_{q,-1} \to \mathrm{End}(\mathbb{C}[\mathfrak{p}^-]_{q,1} f_0) \tag{2.116}$$

we get this way is bijective. Also, the $U_q\mathfrak{k}$-invariance of f_0 implies that it is an isomorphism of $U_q\mathfrak{k}$-modules. Hence it provides a one-to-one map of the spaces of $U_q\mathfrak{k}$-invariant elements. The map (2.116) sends Ψ to the identity operator in $\mathbb{C}[\mathfrak{p}^-]_{q,1} f_0$. Hence this element is $U_q\mathfrak{k}$-invariant.

It was established in Subsection 2.3.2 that the $U_q\mathfrak{k}$-module $\mathbb{C}[\mathfrak{p}^-]_{q,1} \cong \mathbb{C}[\mathfrak{p}^-]_{q,1} f_0$ is simple. This implies that the subspace of $U_q\mathfrak{k}$-invariant elements of $\mathrm{End}(\mathbb{C}[\mathfrak{p}^-]_{q,1} f_0)$ is one dimensional. Hence the subspace of $U_q\mathfrak{k}$-invariant elements of the $U_q\mathfrak{k}$-module $\mathbb{C}[\mathfrak{p}^-]_{q,1}\mathbb{C}[\mathfrak{p}^+]_{q,-1}$ is one dimensional. It contains the nonzero $U_q\mathfrak{k}$-invariant element Ψ, hence is spanned by this element. □

PROPOSITION 2.166. *The linear operator* $T_F(\Psi)$ *in the pre-Hilbert space* $\mathcal{H}$ *is bounded.*

PROOF. As in Subsections 2.2.1 and 2.2.3, we consider the Hopf $*$-algebra $\mathbb{C}[K]_q$ of matrix elements of P-weight finite-dimensional $U_q\mathfrak{k}$-modules. Let $\nu : \mathbb{C}[K]_q \to \mathbb{C}$ be the invariant integral on the quantum group K normalized by $\int_{K_q} 1 d\nu = 1$.

Since the $U_q\mathfrak{k}$-action in $\mathrm{Pol}(\mathfrak{p}^-)_q$ is locally finite-dimensional, we have the well-defined coaction

$$\Delta : \mathrm{Pol}(\mathfrak{p}^-)_q \to \mathrm{Pol}(\mathfrak{p}^-)_q \otimes \mathbb{C}[K]_q;$$

see Subsection 1.3.6. Consider the linear operator of *averaging with respect to the K-action*

$$\mathrm{av} : \mathrm{Pol}(\mathfrak{p}^-)_q \to \mathrm{Pol}(\mathfrak{p}^-)_q, \qquad \mathrm{av} : f \mapsto (\mathrm{id} \otimes \nu)\Delta(f).$$

Obviously, $\Delta(\mathbb{C}[\mathfrak{p}^-]_{q,1}\mathbb{C}[\mathfrak{p}^+]_{q,-1}) \to (\mathbb{C}[\mathfrak{p}^-]_{q,1}\mathbb{C}[\mathfrak{p}^+]_{q,-1}) \otimes \mathbb{C}[K]_q$,

$$\mathrm{av}(\mathbb{C}[\mathfrak{p}^-]_{q,1}\mathbb{C}[\mathfrak{p}^+]_{q,-1}) \subset (\mathbb{C}[\mathfrak{p}^-]_{q,1}\mathbb{C}[\mathfrak{p}^+]_{q,-1}).$$

It follows from the $U_q\mathfrak{k}$-invariance of $\mathrm{av}(z_{\mathrm{low}} z_{\mathrm{low}}^*)$ and Lemma 2.165 that

$$\mathrm{av}(z_{\mathrm{low}} z_{\mathrm{low}}^*) = \mathrm{const} \cdot \Psi. \tag{2.117}$$

Let us show that in the latter relation const $\neq 0$. Using the morphism of $U_q\mathfrak{k}$-modules (2.116) and the q-trace $\mathrm{tr}_q : \mathrm{End}(\mathbb{C}[\mathfrak{p}^-]_{q,1} f_0) \to \mathbb{C}$ introduced in Subsection 2.1.2, it is easy to find a nonzero $U_q\mathfrak{k}$-invariant integral μ on $\mathbb{C}[\mathfrak{p}^-]_{q,1}\mathbb{C}[\mathfrak{p}^+]_{q,-1}$ and then to prove that $\mu(z_{\mathrm{low}} z_{\mathrm{low}}^*) \neq 0$. We have

$$(\mu \otimes \mathrm{id})\Delta(\psi) = \mu(\psi) \otimes 1, \qquad \psi \in \mathbb{C}[\mathfrak{p}^-]_{q,1}\mathbb{C}[\mathfrak{p}^+]_{q,-1}.$$

Hence $\mu(\mathrm{av}(z_{\mathrm{low}} z_{\mathrm{low}}^*)) = \mu(z_{\mathrm{low}} z_{\mathrm{low}}^*) \neq 0$. Therefore $\mathrm{const} \cdot \mu(\Psi) \neq 0$.

One can observe from (2.117) that in order to prove Proposition 2.166, it suffices to obtain the inequality $\|T_F(\mathrm{av}(z_{\mathrm{low}} z_{\mathrm{low}}^*))\| \leq 1$. Recall that the homogeneous components $\mathcal{H}_k$ of the graded vector space $\mathcal{H} = \bigoplus_{k=0}^{\infty} \mathcal{H}_k$ are finite-dimensional and pairwise orthogonal. The finite-dimensional C^*-algebras $\mathrm{End}(\mathcal{H}_k)$ are $U_q\mathfrak{k}$-module algebras, which allow one to introduce the linear operators

$$\Delta_k : \mathrm{End}(\mathcal{H}_k) \to \mathrm{End}(\mathcal{H}_k), \qquad \mathrm{av}_k : \mathrm{End}(\mathcal{H}_k) \to \mathrm{End}(\mathcal{H}_k)$$

in the same way as Δ, av. It follows from the definitions that

$$\|T_F(\mathrm{av}(z_{\mathrm{low}} z_{\mathrm{low}}^*))\| = \sup_k \|T_F(\mathrm{av}(z_{\mathrm{low}} z_{\mathrm{low}}^*))|_{\mathcal{H}_k}\| = \sup_k \|\,\mathrm{av}_k(T_F(z_{\mathrm{low}} z_{\mathrm{low}}^*)|_{\mathcal{H}_k})\|,$$

because the linear map

$$\mathbb{C}[\mathfrak{p}^-]_{q,1}\mathbb{C}[\mathfrak{p}^+]_{q,-1} \to \mathrm{End}(\mathcal{H}_k), \qquad \psi \mapsto T_F(\psi)|_{\mathcal{H}_k},$$

is a morphism of $U_q\mathfrak{k}$-module algebras. By Lemma 2.163 one has $\|T_F(z_{\mathrm{low}} z_{\mathrm{low}}^*)|_{\mathcal{H}_k}\| \leq 1$. It remains to prove that the linear operator av_k of *averaging with respect to the action of the quantum group K* is norm decreasing on the C^*-algebra $\mathrm{End}\,\mathcal{H}_k$. On the other hand, $\Delta_k : \mathrm{End}\,\mathcal{H}_k \to \mathrm{End}\,\mathcal{H}_k \otimes C(K)_q$ is a contraction, just like any morphism of C^*-algebras. Hence Proposition 2.166 follows from Lemma 2.167 below. □

Consider the invariant integral $\nu : C(K)_q \to \mathbb{C}$; see Subsection 2.2.5. Let Mat_n be the algebra of complex $n \times n$ matrices. It is well known that Mat_n admits a unique structure of C^*-algebra.

LEMMA 2.167. *For all $n \in \mathbb{N}$ the linear map of C^*-algebras*

$$\mathrm{id} \otimes \nu : \mathrm{Mat}_n \otimes C(K)_q \to \mathrm{Mat}_n$$

is a contraction.

PROOF. Let us start with the case $n = 1$. The results of Subsection 2.2.3 imply that the C^*-algebra $C(K)_q$ possesses the representation in the Hilbert space $L^2(d\nu)_q$

$$\widehat{\psi} : f \mapsto \psi f, \qquad \psi \in C(K)_q, \quad f \in L^2(d\nu)_q,$$

such that $\nu(\psi) = (\widehat{\psi}1, 1)$. Hence $|\nu(\psi)| = |(\widehat{\psi}1, 1)| \leq \|\widehat{\psi}\| \leq \|\psi\|$. This argument is still valid in the general case.

Namely, the C^*-algebra $\mathrm{Mat}_n \otimes C(K)_q$ acts naturally in the Hilbert space $\mathbb{C}^n \otimes L^2(d\nu)_q$. We introduce the corresponding representation $\mathcal{R}$.

Let $\mathcal{L} = \mathbb{C}^n \otimes 1 \hookrightarrow \mathbb{C}^n \otimes L^2(d\nu)_q$ and $P_{\mathcal{L}}$ be the orthogonal projection onto $\mathcal{L}$. The statement of the lemma follows from the obvious inequality

$$\| \mathrm{id} \otimes \nu(\psi)\| = \|P_{\mathcal{L}}\mathcal{R}(\psi)|_{\mathcal{L}}\| \leq \|\psi\|, \qquad \psi \in \mathrm{Mat}_n \otimes C(K)_q. \qquad \square$$

REMARK 2.168. The final part of the proof of the latter lemma can be replaced by a reference to well-known results of the theory of C^*-algebras [**17**, pp. 148, 153]. It allows one to pass from the special case $n = 1$ to the general case $n \in \mathbb{Z}_+$ while estimating the norm of the linear map $\mathrm{id} \otimes \nu$.

Proposition 2.166 is proved completely. It implies the boundedness of the linear operators $T_F(f)$ for $f \in \mathbb{C}[\mathfrak{p}^-]_{q,1}$. Besides, such elements generate the $*$-algebra $\mathrm{Pol}(\mathfrak{p}^-)_q$ by Proposition 2.107. Thus we obtain

COROLLARY 2.169. *All the operators of T_F in the pre-Hilbert space $\mathcal{H}$ are bounded.*

Equivalently, $\mathrm{Fun}(\mathbb{D})_q \subset L^\infty(\mathbb{D})_q$. Denote by $C(\overline{\mathbb{D}})_q$ the closure of $\mathrm{Fun}(\mathbb{D})_q$ in $L^\infty(\mathbb{D})_q$. This C^*-algebra is a quantum analog of the commutative C^*-algebra $C(\overline{\mathbb{D}})$ of continuous functions on $\overline{\mathbb{D}}$. In the special case of a quantum disc, it was considered in Section 1.2.

2.5. The canonical embedding

2.5.1. Introduction. In what follows, all pairs $(\mathfrak{g}, \mathfrak{k})$ are assumed to be Hermitian symmetric. Recall the Cartan involution θ_q in $U_q\mathfrak{g}$

$$\theta_q(K_i^{\pm 1}) = K_i^{\pm 1}, \qquad \theta_q(E_i) = \begin{cases} E_i, & i \neq l_0, \\ -E_i, & i = l_0, \end{cases} \qquad \theta_q(F_i) = \begin{cases} F_i, & i \neq l_0, \\ -F_i, & i = l_0, \end{cases}$$

introduced in Subsection 2.3.3, together with the associated Hopf-$*$-algebra $(U_q\mathfrak{g}, *)$, where $* = \star\theta_q = \theta_q\star$. A duality argument (see (1.20)) allows one to define a Hopf $*$-algebra $((U_q\mathfrak{g})^\star, \natural)$:

$$f^\natural(\xi) = \overline{f((S(\xi))^*)}, \qquad f \in (U_q\mathfrak{g})^\star, \; \xi \in U_q\mathfrak{g}. \tag{2.118}$$

Burnside's theorem implies that the homomorphism of algebras $U_q\mathfrak{g} \to \mathrm{End}\, L(\Lambda)$ is onto for all $\Lambda \in P_+$, hence $(\mathrm{End}\, L(\Lambda))^* \hookrightarrow (U_q\mathfrak{g})^\star$. Since

$$\natural : (\mathrm{End}\, L(\Lambda))^* \to (\mathrm{End}\, L(-w_0\Lambda))^*,$$

the involution $\natural$ admits a restriction to the subalgebra

$$\mathbb{C}[G]_q = \sum_{\Lambda \in P_+} (\mathrm{End}\, L(\Lambda))^* \subset (U_q\mathfrak{g})^\star.$$

Consider the Hopf $*$-algebra $\mathbb{C}[G_0]_q = (\mathbb{C}[G]_q, \sharp)$. It serves as a q-analog of the algebra of regular functions on the real affine algebraic group

$$G_0 = \{g \in G |\ f^\sharp(g) = \overline{f(g)},\ f \in \mathbb{C}[G]\},$$

cf. (2.41). It is well known that G_0 acts by automorphisms of a bounded symmetric domain $\mathbb{D}$ which admits a standard embedding into $\mathfrak{p}^-$; see [**197**].

The presence of a distinguished point $0 \in \mathbb{D}$ allows one to associate to every function $f(z)$ on $\mathbb{D}$ a function $f(g^{-1} \cdot 0)$ on G_0. We are interested in quantum analogs of function algebras on $\mathbb{D}$ and their images under the above embedding into algebras of functions on G_0.

In the special case $\mathfrak{g} = \mathfrak{sl}_2$, G_0 is the real affine algebraic group $SU_{1,1}$; it is easy to verify that

$$t_{11}^\sharp = t_{22}, \qquad t_{12}^\sharp = q t_{21}, \tag{2.119}$$

where $\{t_{ij}\}_{i,j=1,2}$ are the standard generators of $\mathbb{C}[SL_2]_q$. In fact, consider the automorphism θ_q^* of the Hopf algebra $\mathbb{C}[SL_2]_q$ which is dual to the automorphism

$$\theta_q : U_q\mathfrak{sl}_2 \to U_q\mathfrak{sl}_2, \qquad \theta_q : K^{\pm 1} \mapsto K^{\pm 1}, \quad \theta_q : E \mapsto -E, \quad \theta_q : F \mapsto -F.$$

Its action on the generators t_{ij} is given by

$$\begin{pmatrix} \theta_q^*(t_{11}) & \theta_q^*(t_{12}) \\ \theta_q^*(t_{21}) & \theta_q^*(t_{22}) \end{pmatrix} = \begin{pmatrix} t_{11} & -t_{12} \\ -t_{21} & t_{22} \end{pmatrix}.$$

It remains to use the relation (2.65), together with the explicit formulas (2.53) which describe the action of $\star$ on the generators t_{ij}.

Recall that the canonical embedding in the $\mathfrak{sl}_2$ case, i.e., $\mathbb{D} = \{z \in \mathbb{C} | |z| < 1\}$, was described in detail in Subsection 1.7.1.

2.5.2. A quantum analog of the longest element w_0 of the Weyl group. The construction of the canonical embedding becomes considerably more intricate under the passage from the special case $\mathfrak{g} = \mathfrak{sl}_2$ to the general case, and the passage requires the notion of a quantum Weyl group described below.

PROPOSITION 2.170. *There exists a unique linear functional $\overline{s}$ on $\mathbb{C}[SL_2]_q$ such that*

$$\overline{s}(t_{22} f) = \overline{s}(f t_{11}) = 0,$$
$$\overline{s}(t_{12} f) = \overline{s}(f t_{12}) = q\overline{s}(f),$$
$$\overline{s}(t_{21} f) = \overline{s}(f t_{21}) = -\overline{s}(f),$$

for all $f \in \mathbb{C}[SL_2]_q$ with the property $f(1) = 1$.

PROOF. Let $\{e_j\}$, $j \in \mathbb{Z}_+$, be the standard basis of the Hilbert space $l^2(\mathbb{Z}_+)$ and Π the $*$-representation of the algebra $\mathbb{C}[SU_2]_q = (\mathbb{C}[SL_2]_q, \star)$ in this space described in Subsection 2.2.4.

The linear functional $\overline{s}(f) = (\Pi(f)e_0, e_0)$ possesses all the required properties. The uniqueness of $\overline{s}$ follows from the fact that $\mathbb{C}[SL_2]_q$ is a linear span of the monomials

$$\{t_{22}^d t_{21}^c t_{12}^b t_{11}^a \mid a, b, c, d \in \mathbb{Z}_+\}.$$

□

REMARK 2.171. It is easy to prove that every element of $\mathbb{C}[SL_2]_q$ admits a unique decomposition

$$f = \sum_{j=1}^{\infty} t_{22}^j f_j(t_{12}, t_{21}) + f_0(t_{12}, t_{21}) + \sum_{j=1}^{\infty} f_{-j}(t_{12}, t_{21}) t_{11}^j,$$

where each f_j is a polynomial in two variables; see Subsection 2.2.4. This observation provides another proof of existence for $\overline{s}$:

$$\overline{s}(f) = f_0(q, -1), \qquad f \in \mathbb{C}[SL_2]_q.$$

The element $\overline{s} \in \mathbb{C}[SL_2]_q^*$ is a q-analog of the element $\left(\begin{smallmatrix} 0 & 1 \\ -1 & 0 \end{smallmatrix}\right)$ of SL_2 since

$$\begin{pmatrix} \overline{s}(t_{11}) & \overline{s}(t_{12}) \\ \overline{s}(t_{21}) & \overline{s}(t_{22}) \end{pmatrix} = \begin{pmatrix} 0 & q \\ -1 & 0 \end{pmatrix}. \tag{2.120}$$

Let us endow the vector space $U_q\mathfrak{sl}_2$ with the weakest topology which makes continuous all the irreducible finite-dimensional weight representations

$$\pi_\lambda : U_q\mathfrak{sl}_2 \to \operatorname{End}(L(\lambda)), \qquad \lambda \in \mathbb{Z}_+\overline{\omega}.$$

These representations separate elements of $U_q\mathfrak{sl}_2$. Then the above topology is Hausdorff and the completion of $U_q\mathfrak{sl}_2$ is canonically isomorphic to the direct product of algebras $\times_j \operatorname{End} L(j\overline{\omega})$, that is, to the algebra $\mathbb{C}[SL_2]_q^*$ with the weak-$*$ topology. The following claim is a direct consequence of the definitions.

LEMMA 2.172. *Every finite-dimensional weight representation of $U_q\mathfrak{sl}_2$ in a vector space L extends by continuity to a representation of $\mathbb{C}[SL_2]_q^*$, thus L is endowed with the structure of a $\mathbb{C}[SL_2]_q^*$-module.*

Recall that given a weight $U_q\mathfrak{sl}_2$-module V, one has the well-defined linear operators H and $X^\pm$ in V such that

$$K^{\pm 1} = q^{\pm H}, \quad X^+ = Eq^{-\frac{1}{2}H}, \quad X^- = q^{\frac{1}{2}H}F; \tag{2.121}$$

see Subsection 2.1.2 and the relations (2.15) and (2.16).

Let $j \in \mathbb{Z}_+$. We follow the conventions of the quantum theory of angular momentum in introducing the standard basis $\{e_m^j\}_{m=j,j-1,\ldots,-j}$ of the vector space $L(2j\overline{\omega})$ by setting

$$He_m^j = 2me_m^j, \qquad X^+e_m^j = \begin{cases} ([j-m]_q[j+m+1]_q)^{1/2} e_{m+1}^j, & m \neq j, \\ 0, & m = j, \end{cases}$$

$$X^-e_m^j = \begin{cases} ([j-m+1]_q[j+m]_q)^{1/2} e_{m-1}^j, & m \neq -j, \\ 0, & m = -j, \end{cases}$$

with $[n]_q = \frac{q^n - q^{-n}}{q - q^{-1}}$.

LEMMA 2.173. [**321**, p. 8] *For all j, m one has*

$$\overline{s}e_m^j = (-1)^{j-m} q^{j^2+j} q^{-(m^2+m)} e_{-m}^j,$$

hence,

$$\overline{s}K\overline{s}^{-1} = K^{-1}.$$

Let us turn from the special case $G = SL_2$ to the general case; see Subsection 2.2.1. Recall that the group G is assumed to be simply connected. Let us endow the vector space $\mathbb{C}[G]_q^* \cong \times_{\lambda \in P_+} \operatorname{End}(L(\lambda))$ with the direct product topology, that is, the weak-$*$ topology. Consider the standard embeddings of Hopf $*$-algebras

$$\varphi_i : U_{q_i}\mathfrak{sl}_2 \hookrightarrow U_q\mathfrak{g},$$
$$\varphi_i : K^{\pm 1} \mapsto K_i^{\pm 1}, \quad \varphi_i : E \mapsto E_i, \quad \varphi_i : F \mapsto F_i, \qquad i = 1, 2, \ldots, l.$$

PROPOSITION 2.174. *The embeddings of algebras* $\varphi_i : U_{q_i}\mathfrak{sl}_2 \hookrightarrow U_q\mathfrak{g}$ *extend by continuity to embeddings of algebras* $\overline{\varphi}_i : \mathbb{C}[SL_2]_{q_i}^* \hookrightarrow \mathbb{C}[G]_q^*$.

PROOF. Lemma 2.172 allows us to extend, while the injectivity is due to

$$\dim \operatorname{Hom}_{U_{q_i}\mathfrak{sl}_2}(L(k\overline{\omega}), L(k\overline{\omega}_i)) > 0, \qquad k \in \mathbb{Z}_+,$$

where the $U_q\mathfrak{g}$-module $L(k\overline{\omega}_i)$ is considered here as a $U_{q_i}\mathfrak{sl}_2$-module. □

Let $\overline{s}_i = \overline{\varphi}_i(\overline{s})$, $i = 1, 2, \ldots, l$. It is well known [**216**, p. 157], [**171**, p. 427] that the elements $\overline{s}_1, \overline{s}_2, \ldots, \overline{s}_l$ of $\mathbb{C}[G]_q^*$ satisfy the braid relations

$$\overline{s}_i\overline{s}_j\overline{s}_i \cdots = \overline{s}_j\overline{s}_i\overline{s}_j \cdots , \tag{2.122}$$

where the number of factors in each side of (2.122) is two if $a_{ij}a_{ji} = 0$, three if $a_{ij}a_{ji} = 1$, four if $a_{ij}a_{ji} = 2$, and six if $a_{ij}a_{ji} = 3$.[7] Let $w = s_{i_1}s_{i_2}\cdots s_{i_M}$ be a reduced expression of an element $w \in W$, and set

$$\overline{w} = \overline{s}_{i_1}\overline{s}_{i_2}\cdots\overline{s}_{i_M}.$$

This element of $\mathbb{C}[G]_q^*$ does not depend on the choice of reduced expression, as is clear from (2.122), together with a well-known property of Coxeter groups [**44**, p. 19].

We get a quantum analog $\overline{w}$ of $w \in W$ and, in particular, a quantum analog of the longest element w_0 of the Weyl group W.

Lemma 2.173 allows one to prove

PROPOSITION 2.175. [**199**, p. 138] *The elements* $\overline{s}_i \in \mathbb{C}[G]_q^*$ *are invertible, and*

$$\overline{s}_i K_j \overline{s}_i^{-1} = K_j K_i^{-a_{ij}}, \qquad i, j = 1, 2, \ldots, l.$$

COROLLARY 2.176. *Let* $V = \bigoplus_\lambda V_\lambda$ *be a finite-dimensional* $U_q\mathfrak{g}$*-module which admits a decomposition as a direct sum of its weight subspaces. Then*

$$\overline{w}V_\lambda = V_{w\lambda}, \qquad w \in W. \tag{2.123}$$

Now let us claim that the relation

$$\widetilde{w}_0 = \overline{w}_0^{-1} \cdot q^{-\frac{1}{2}\sum\limits_{i,j=1}^{l} c_{ij}d_id_jH_iH_j} \tag{2.124}$$

determines an element $\widetilde{w}_0$ of $\mathbb{C}[G]_q^*$. First, by Proposition 2.175, the $\overline{s}_i$ are invertible, hence so is their product $\overline{w}_0$. Second, $\mathbb{C}[G]_q^*$ is a direct product of the finite-dimensional algebras $\operatorname{End} L(\lambda)$ and all $L(\lambda)$ are weight $U_q\mathfrak{g}$-modules. Hence the second factor in (2.124), which is normally called the *Cartan correcting multiplier*, determines an element of $\mathbb{C}[G]_q^*$.

[7] The latter possibility does not occur in the Hermitian symmetric case we consider here.

In the case $\mathfrak{g}=\mathfrak{sl}_2$ one has $\widetilde{w}_0=\bar{s}^{-1}\cdot q^{-\frac{H^2}{4}}$. It can be verified that

$$\widetilde{w}_0^{-1}X^+\widetilde{w}_0=-q^{-1}X^-,\qquad \widetilde{w}_0^{-1}X^-\widetilde{w}_0=-q\,X^+,$$
$$\Delta\widetilde{w}_0=(\widetilde{w}_0\otimes\widetilde{w}_0)R,$$

where $\Delta:\mathbb{C}[SL_2]_q^*\to(\mathbb{C}[SL_2]_q^{\otimes 2})^*$ and $R\in(\mathbb{C}[SL_2]_q^{\otimes 2})^*$ corresponds to the universal R-matrix of the Hopf algebra $U_q\mathfrak{sl}_2$; see [**171**, p. 425].

In a similar way, in the general case an element R of $(\mathbb{C}[G]_q^{\otimes 2})^*$ will correspond to the universal R-matrix of the Hopf algebra $U_q\mathfrak{g}$, cf. [**148**, p. 289].

PROPOSITION 2.177. [**216**, p. 252] *In* $(\mathbb{C}[G]_q^{\otimes 2})^*$ *the following relation is valid:*

$$\Delta\widetilde{w}_0=(\widetilde{w}_0\otimes\widetilde{w}_0)R, \tag{2.125}$$

where $\Delta:\mathbb{C}[G]_q^*\to(\mathbb{C}[G]_q^{\otimes 2})^*$.

REMARK 2.178. In [**216, 171**] $\widetilde{w}_0^{-1}$ is used instead of $\widetilde{w}_0$.

Consider the elements $X_i^{\pm}$ of $\mathbb{C}[G]_q^*$ given by

$$X_i^+=E_i q_i^{-\frac{H_i}{2}},\qquad X_i^-=q_i^{\frac{H_i}{2}}F_i,$$

cf. (2.121) and Remark 2.13.

The following result is due to Joseph; see [**149**].

PROPOSITION 2.179. *In* $\mathbb{C}[G]_q^*$ *one has*

$$\widetilde{w}_0^{-1}X_i^{\pm}\widetilde{w}_0=-q_i^{\mp 1}X_{i'}^{\mp},\qquad i=1,2,\dots,l,$$

where i' *is fixed by* $\alpha_{i'}=-w_0\alpha_i$.

REMARK 2.180. Our choice of the Cartan correcting multiplier in (2.124) is conventional (see, e.g., [**216**, p. 252]) and differs a little from that used by Joseph; see [**149**, p. 422].

2.5.3. Fundamental representations and special bases. The simple finite-dimensional $U_q\mathfrak{g}$-weight modules $L(\Lambda)$, $\Lambda\in P_+$, and matrix elements of their representations π_Λ are substantial in the study of algebras of functions on compact quantum groups; see Subsection 2.2.2. In this approach, some inconvenience is related to an ambiguity in choosing an orthonormal basis of weight vectors in $L(\Lambda)$.

Even in the special case of a fundamental weight Λ and $\lambda,\mu\in W\Lambda$, $\lambda\neq\mu$, with one-dimensional weight subspaces, the matrix element $c_{\lambda,\mu}^{\Lambda}$ is determined only up to a scalar factor of modulus one. We are going to remove this ambiguity by removing the arbitrariness in the choice of an orthonormal basis of weight vectors.

Let $\mathbb{S}=\{1,2,\dots,l\}\setminus\{l_0\}$. Denote by $W_{\mathbb{S}}$ the stabilizer of the fundamental weight $\overline{\omega}=\overline{\omega}_{l_0}$ [**131**, p. 22], that is, $W_{\mathbb{S}}=\{\sigma\in W\,|\,\sigma\overline{\omega}=\overline{\omega}\}$. Hence there is a natural one-to-one correspondence between $W\overline{\omega}$ and the set ${}^{\mathbb{S}}W$ of shortest elements in the cosets $\sigma W_{\mathbb{S}}$, $\sigma\in W$; see Lemma 2.69.

Recall that the representation $\pi_{\overline{\omega}}$ of the $*$-algebra $(U_q\mathfrak{g},\star)$ in $L(\overline{\omega})$ is unitarizable and the Hermitian form $(\cdot,\cdot)$ in question is uniquely determined by the normalization condition $(v(\overline{\omega}),v(\overline{\omega}))=1$; see Subsection 2.1.8.

An orthonormal basis formed by weight vectors of the $U_q\mathfrak{g}$-module $L(\overline{\omega})$ will be called special if the weight vectors with weights in $W\overline{\omega}$ are given by

$$v_{\sigma\overline{\omega}}^{\overline{\omega}}=\frac{\overline{\sigma}^{-1}v(\overline{\omega})}{\|\overline{\sigma}^{-1}v(\overline{\omega})\|},\qquad \sigma\in{}^{\mathbb{S}}W.$$

Here $\overline{\sigma} \in \mathbb{C}[G]_q^*$ is the quantum analog of σ introduced in the previous section (its invertibility and (2.123) are used).

We now give an equivalent definition of a special basis, cf. [**209**, p. 153]. Let $\sigma = s_{i_L} s_{i_{L-1}} \cdots s_{i_1}$ be a reduced expression for $\sigma \in {}^{\mathbb{S}}W$. Set $u_j = s_{i_j} s_{i_{j-1}} \cdots s_{i_1}$,

$$\lambda_j = \begin{cases} u_j \overline{\omega}, & j = 1, 2, \ldots, L, \\ \overline{\omega}, & j = 0, \end{cases} \qquad m_j = \frac{2(\lambda_j, \lambda_{j-1})}{(\lambda_j, \lambda_j)}.$$

One has $m_j \in \mathbb{N}$ [**44**, p. 93]. The validity of the following claim is a consequence of Lemma 2.173.

PROPOSITION 2.181. *An orthonormal basis of weight vectors is special if and only if*

$$v_{\sigma\overline{\omega}}^{\overline{\omega}} = \frac{F_{i_L}^{m_L} F_{i_{L-1}}^{m_{L-1}} \cdots F_{i_1}^{m_1} v(\overline{\omega})}{\left\| F_{i_L}^{m_L} F_{i_{L-1}}^{m_{L-1}} \cdots F_{i_1}^{m_1} v(\overline{\omega}) \right\|}$$

for all $\sigma \in {}^{\mathbb{S}}W$.

PROPOSITION 2.182. *The following relations for the matrix elements of the fundamental representations in the special bases are valid:*

$$S(c_{\sigma'\overline{\omega},\sigma''\overline{\omega}}^{\overline{\omega}}) = (-1)^{l(\sigma')-l(\sigma'')} q^{(\sigma''\overline{\omega}-\sigma'\overline{\omega},\rho)} c_{-\sigma''\overline{\omega},-\sigma'\overline{\omega}}^{-w_0\overline{\omega}}, \tag{2.126}$$

$$(c_{\sigma'\overline{\omega},\sigma''\overline{\omega}}^{\overline{\omega}})^\star = (-1)^{l(\sigma')-l(\sigma'')} q^{(\sigma''\overline{\omega}-\sigma'\overline{\omega},\rho)} c_{-\sigma'\overline{\omega},-\sigma''\overline{\omega}}^{-w_0\overline{\omega}}, \tag{2.127}$$

where $\sigma', \sigma'' \in {}^{\mathbb{S}}W$, *and* ρ *is half the sum of positive roots.*

PROOF. Let $\pi_{\overline{\omega}}$, $\pi_{-w_0\overline{\omega}}$ be the fundamental representations of $U_q\mathfrak{g}$ corresponding to the $U_q\mathfrak{g}$-modules $L(\overline{\omega})$ and $L(-w_0\overline{\omega})$, respectively. Consider the involutive antilinear antiautomorphism τ of $U_q\mathfrak{g}$, together with the representation $\pi_{\overline{\omega}}^{\tau}$ of that algebra in $L(\overline{\omega})$ described in Subsection 2.2.3. It is demonstrated there that there is a unique isometric linear map

$$U : L(\overline{\omega}) \to L(-w_0\overline{\omega}), \qquad U : v_{\overline{\omega}}^{\overline{\omega}} \mapsto v_{-\overline{\omega}}^{-w_0\overline{\omega}},$$

and this establishes the equivalence of $\pi_{\overline{\omega}}^{\tau}$ and $\pi_{-w_0\overline{\omega}}$. Proposition 2.181 and the definition of τ imply that

$$U v_{\sigma\overline{\omega}}^{\overline{\omega}} = (-1)^{l(\sigma)} v_{-\sigma\overline{\omega}}^{-w_0\overline{\omega}}, \qquad \sigma \in^{\mathbb{S}} W.$$

It remains to apply Proposition 2.51. □

To conclude, consider an important special case of the fundamental weight $\overline{\omega}$ corresponding to a simple root whose coefficient in the decomposition of the highest root is 1. Even more, we restrict ourselves to the simply laced Lie algebras (the series A, D, E). Under these assumptions $\overline{\omega}$ is a microweight; i.e., all the weights of the $U_q\mathfrak{g}$-module $L(\overline{\omega})$ belong to its W-orbit [**44**, p. 164]. Now (2.126) and (2.127) take the form

$$S(c_{\lambda,\mu}^{\overline{\omega}}) = (-q)^{(\mu-\lambda,\rho)} c_{-\mu,-\lambda}^{-w_0\overline{\omega}}, \qquad \lambda, \mu \in W\overline{\omega},$$

$$(c_{\lambda,\mu}^{\overline{\omega}})^\star = (-q)^{(\lambda-\mu,\rho)} c_{-\lambda,-\mu}^{-w_0\overline{\omega}}, \qquad \lambda, \mu \in W\overline{\omega},$$

because $l(s_i\sigma) = l(\sigma) + 1$, $s_i\sigma \in {}^{\mathbb{S}}W$, implies that one has

$$(s_i \sigma\overline{\omega} - \sigma\overline{\omega}, \rho) = -\left(\alpha_i, \sum_{j=1}^{l} \overline{\omega}_j\right) = -1.$$

2.5.4. The canonical embedding $\mathbb{C}[\mathfrak{p}^-]_q \hookrightarrow \mathbb{C}[X_{\mathbb{S}}^-]_{q,t}$. We maintain our assumptions that $\mathbb{S} = \{1, 2, \ldots, l\} \setminus \{l_0\}$ and that the simple root α_{l_0} has coefficient 1 in the decomposition of the highest root. Consider the fundamental weight

$$\overline{\omega}_{l_1} = -w_0 \overline{\omega}_{l_0}$$

together with the associated fundamental representation π_{l_1} of $U_q\mathfrak{g}$ in the vector space $L(\overline{\omega}_{l_1})$.

Pick a special basis in $L(\overline{\omega}_{l_1})$ and maintain the notation for matrix elements of representations of $U_q\mathfrak{g}$ as in Subsection 2.2.2. Consider the following distinguished matrix element of π_{l_1}:

$$t = c^{\overline{\omega}_{l_1}}_{\overline{\omega}_{l_1}, -\overline{\omega}_{l_0}}.$$

Corollary 2.48 implies

PROPOSITION 2.183. (1) $tt^\star = t^\star t$.

(2) *Let $\Lambda \in P_+$ and λ, μ be weights of $L(\Lambda)$. The matrix elements $c^{L(\Lambda)}_{\lambda,j;\mu,k}$ quasi-commute with t, $t^\star$. Namely,*

$$t c^{L(\Lambda)}_{\lambda,j;\mu,k} = \mathrm{const}_1 \cdot c^{L(\Lambda)}_{\lambda,j;\mu,k} t, \qquad t^\star c^{L(\Lambda)}_{\lambda,j;\mu,k} = \mathrm{const}_2 \cdot c^{L(\Lambda)}_{\lambda,j;\mu,k} t^\star,$$

where $\mathrm{const}_1, \mathrm{const}_2 \in q^{\frac{1}{s}\mathbb{Z}}$ depend on Λ, λ, μ, q, and s is the cardinality of the quotient group P/Q.

Set $x = tt^\star = t^\star t$.

COROLLARY 2.184. *If $\Lambda \in P_+$ and λ, μ are weights of $L(\Lambda)$, then*

$$x\, c^{L(\Lambda)}_{\lambda,j;\mu,k} = \mathrm{const} \cdot c^{L(\Lambda)}_{\lambda,j;\mu,k}\, x, \tag{2.128}$$

where $\mathrm{const} \in q^{\frac{1}{s}\mathbb{Z}}$ depends on Λ, λ, μ, q.

The multiplicative system $x^{\mathbb{Z}_+}$ is an Ore set; see (2.128). Let $\mathbb{C}[G]_{q,x}$ be the localization of $\mathbb{C}[G]_q$ with respect to this multiplicative system. It follows from Proposition 2.46 that $\mathbb{C}[G]_q \hookrightarrow \mathbb{C}[G]_{q,x}$.

PROPOSITION 2.185. (1) *The structure of $U_q\mathfrak{g}$-module algebra admits a unique extension from $\mathbb{C}[G]_q$ to $\mathbb{C}[G]_{q,x}$.*

(2) *For all $\xi \in U_q\mathfrak{g}$ and $f \in \mathbb{C}[G]_{q,x}$ there exists a unique Laurent polynomial $p_{f,\xi}$ in one variable with coefficients from $\mathbb{C}[G]_{q,x}$ such that*

$$\xi(f \cdot x^n) = p_{f,\xi}(q^{\frac{n}{s}}) x^n$$

for all $n \in \mathbb{Z}$.

PROOF. First consider the second claim. Since $\mathbb{C}[G]_q$ is a $U_q\mathfrak{g}$-module algebra, one can reduce the general case $\xi \in U_q\mathfrak{g}$ to the special case $\xi \in \{K_j^{\pm 1}, E_j, F_j\}_{j=1,2,\ldots,l}$. Within this special case, the existence of a Laurent polynomial $p_{f,\xi}$ follows from Corollary 2.184 and the explicit formulas (2.2) for $\triangle(K_j^{\pm 1})$, $\triangle(E_j)$, $\triangle(F_j)$. The rest of the statements of Proposition 2.185 can be proved in the same way as the corresponding claims in Proposition 1.164. □

We thus get a well-known statement.

COROLLARY 2.186. *The multiplicative system $t^{\mathbb{Z}_+}$ is an Ore set in $\mathbb{C}[G]_q$, so that there exists an embedding of $\mathbb{C}[G]_q$ into the localization $\mathbb{C}[G]_{q,t}$, and the structure of $U_q\mathfrak{g}$-module algebra can be uniquely extended from $\mathbb{C}[G]_q$ to $\mathbb{C}[G]_{q,t}$.*

REMARK 2.187. A different approach to proving that the structure of $U_q\mathfrak{g}$-module algebra admits an extension to a localization is presented in works by Lunts and Rosenberg [**219**], [**218**].

Now introduce the matrix element

$$t' = c^{\overline{\omega}_{l_1}}_{\overline{\omega}_{l_1}, -\overline{\omega}_{l_0}+\alpha_{l_0}}$$

of the fundamental representation π_{l_1} of $U_q\mathfrak{g}$ in the special basis of the vector space $L(\overline{\omega}_{l_1})$. Note that the weights $-\overline{\omega}_{l_0}$, $-\overline{\omega}_{l_0}+\alpha_{l_0}$ belong to the W-orbit of the highest weight $\overline{\omega}_{l_1}$, hence t, t' do not depend on a choice of a special basis.

The definitions yield that for all $j \in \{1, 2, \dots, l\}$,

$$F_j t = 0, \qquad K_j^{\pm 1} t = \begin{cases} q_j^{\mp 1} t, & j = l_0, \\ 0, & j \neq l_0, \end{cases} \qquad E_j t = \begin{cases} t', & j = l_0, \\ 0, & j \neq l_0. \end{cases} \tag{2.129}$$

PROPOSITION 2.188. *The map $i : z_{\text{low}} \mapsto t^{-1}t'$ can be uniquely extended to an embedding of $U_q\mathfrak{g}$-module algebras*

$$i : \mathbb{C}[\mathfrak{p}^-]_q \hookrightarrow \mathbb{C}[G]_{q,x}. \tag{2.130}$$

PROOF. The pair $(\mathfrak{g}, \mathfrak{k})$ in question is Hermitian symmetric. Hence the uniqueness of the homomorphism (2.130) is due to the fact that the $U_q\mathfrak{k}$-module $\mathbb{C}[\mathfrak{p}^-]_{q,1}$ is simple (see Subsection 2.3.2), together with Proposition 2.107 which claims that the subspace $\mathbb{C}[\mathfrak{p}^-]_{q,1}$ generates the unital algebra $\mathbb{C}[\mathfrak{p}^-]_q$.

Let us now turn to proving existence of the morphism (2.130). Consider the $U_q\mathfrak{g}$-module $\widetilde{w}_0 U_q\mathfrak{g}$, the dual $U_q\mathfrak{g}$-module $(\widetilde{w}_0 U_q\mathfrak{g})^*$, and its largest weight submodule $\mathbb{F} \subset (\widetilde{w}_0 U_q\mathfrak{g})^*$ formed by $U_q\mathfrak{b}^-$-finite elements. Here $U_q\mathfrak{b}^-$ is the standard notation for the Hopf subalgebra generated by $\{K_j^{\pm}, F_j\}_{j=1,2,\dots,l}$. That is, $\mathbb{F} = \bigoplus_{\lambda \in P} \mathbb{F}_\lambda$, with

$$\mathbb{F}_\lambda = \left\{ f \in (\widetilde{w}_0 U_q\mathfrak{g})^* \middle|\ \dim(U_q\mathfrak{b}^- f) < \infty,\ H_i f = \lambda(H_i) f,\ i = 1, \dots, l \right\}.$$

The vector subvariety $\mathbb{C}[G]_q$ is dense in $\mathbb{F}$ with respect to the weak topology. An application of (2.125) makes it easy to prove that the structure of $U_q\mathfrak{g}$-module algebra on $\mathbb{C}[G]_q$ admits an extension by continuity to $\mathbb{F}$. In fact,

$$\begin{aligned} \langle f_1 f_2, \widetilde{w}_0 \xi \rangle &= \langle R_{\mathbb{F},\mathbb{F}} \Delta(\xi)(f_1 \otimes f_2), \widetilde{w}_0 \otimes \widetilde{w}_0 \rangle, \\ \langle \xi f, \widetilde{w}_0 \eta \rangle &= \langle f, \widetilde{w}_0 \eta \xi \rangle, \qquad \xi, \eta \in U_q\mathfrak{g}, \qquad f, f_1, f_2 \in \mathbb{F}. \end{aligned}$$

Here $R_{\mathbb{F},\mathbb{F}}$ is the linear map in $\mathbb{F} \otimes \mathbb{F}$ determined by the action of the universal R-matrix.

It was noted in the proof of uniqueness of the morphism (2.130) that z_{low} generates the $U_q\mathfrak{k}$-module algebra $\mathbb{C}[\mathfrak{p}^-]_q$. Thus, the existence of the embedding (2.130) will be proved if one produces the embeddings of $U_q\mathfrak{g}$-module algebras

$$i_1 : \mathbb{C}[G]_{q,t} \hookrightarrow \mathbb{F}, \qquad i_2 : \mathbb{C}[\mathfrak{p}^-]_q \hookrightarrow \mathbb{F},$$

such that $i_1(t^{-1}t') = i_2(z_{\text{low}})$.

To produce an embedding i_1, we extend the natural pairing $\mathbb{C}[G]_q \times \widetilde{w}_0 U_q\mathfrak{g} \to \mathbb{C}$ to a pairing $\mathbb{C}[G]_{q,t} \times \widetilde{w}_0 U_q\mathfrak{g} \to \mathbb{C}$. In the classical case $q = 1$, such extension is possible because w_0 belongs to the domain $t \neq 0$ where all the functions from $\mathbb{C}[G]_{q,t}$ are regular. In the quantum case we are going to use

$$\langle t, \widetilde{w}_0 \rangle \neq 0, \tag{2.131}$$

which follows from (2.123).

Let $f \in \mathbb{C}[G]_{q,\lambda}$ be a weight vector with weight $\lambda \in P$ and set $c_k(f) = \langle ft^k, \widetilde{w}_0 \rangle$. An application of the equations

$$\left\langle ft^k, \widetilde{w}_0 \right\rangle = \left\langle R_{\mathbb{C}[G]_q,\mathbb{C}[G]_q}(ft^{k-1} \otimes t), \widetilde{w}_0 \otimes \widetilde{w}_0 \right\rangle,$$

(2.129), (2.11), (2.13) makes it easy to prove that the sequence of numbers $\{c_k(f)\}$, $k \in \mathbb{Z}_+$, is a solution of a first-order difference equation whose coefficients do not depend on f. Namely,

$$c_k(f) = q^{\mathrm{const}_1 \cdot (k-1) + \mathrm{const}_2 \cdot \lambda} \cdot \langle t, \widetilde{w}_0 \rangle c_{k-1}(t), \qquad k \in \mathbb{N}.$$

Using the last formula and (2.131), we can introduce $\{c_k(f)\}$ for all $k \in \mathbb{Z}$. Thus we get a "natural" extension of the linear functional $f \mapsto \langle f, \widetilde{w}_0 \rangle$ from $\mathbb{C}[G]_q$ to $\mathbb{C}[G]_{q,t}$.

Proposition 2.185 implies that the subalgebra $\mathbb{C}[G]_{q,t} \subset \mathbb{C}[G]_{q,x}$ is a $U_q\mathfrak{g}$-module algebra. Consider the pairing

$$\mathbb{C}[G]_{q,t} \times \widetilde{w}_0 U_q\mathfrak{g} \to \mathbb{C}, \qquad f \times \widetilde{w}_0\xi \mapsto \langle \xi f, \widetilde{w}_0 \rangle,$$

together with the associated morphism of $U_q\mathfrak{g}$-modules

$$i_1 : \mathbb{C}[G]_{q,t} \to (\widetilde{w}_0 U_q\mathfrak{g})^*, \qquad i_1 : f \mapsto \langle f, \cdot \rangle.$$

Let us prove that i_1 is injective. Since $\widetilde{w}_0 \in \mathbb{C}[G]_q^*$ is invertible, the vector subvariety $\widetilde{w}_0 U_q\mathfrak{g}$ is dense in $\mathbb{C}[G]_q^*$ with respect to the weak topology. Thus $\operatorname{Ker} i_1 \cap \mathbb{C}[G]_q = 0$. It remains to observe that $i_1(f) = 0$ implies $i_1(f \cdot t^j) = 0$ for all $j \in \mathbb{N}$ since

$$\langle f \cdot t^j, \widetilde{w}_0\xi \rangle = \left\langle R_{\mathbb{C}[G]_{q,t},\mathbb{C}[G]_{q,t}} \triangle(\xi)(f \otimes t^j), \widetilde{w}_0 \otimes \widetilde{w}_0 \right\rangle.$$

Note that $i_1\mathbb{C}[G]_{q,t} \subset \mathbb{F}$ since i_1 is a morphism of $U_q\mathfrak{g}$-modules and all $f \in \mathbb{C}[G]_{q,t}$ are $U_q\mathfrak{b}^-$-finite.[8]

We now prove that the embedding $i_1 : \mathbb{C}[G]_{q,t} \hookrightarrow \mathbb{F}$ we have just obtained is a morphism of algebras. It suffices to show that given a pair of weight vectors $f_1, f_2 \in \mathbb{C}[G]_q$, the equation

$$\langle (f_1 \cdot t^k)(f_2 \cdot t^k), \widetilde{w}_0 \rangle = \left\langle R_{\mathbb{C}[G]_{q,t},\mathbb{C}[G]_{q,t}}(f_1 \cdot t^k \otimes f_2 \cdot t^k), \widetilde{w}_0 \otimes \widetilde{w}_0 \right\rangle \tag{2.132}$$

holds for all $k \in \mathbb{Z}$. As the sequences of numbers in both sides of (2.132) are linear combinations of sequences of the form q^{ak^2+bk}, $a, b \in \mathbb{Q}$, it suffices to prove (2.132) for large k. For example, one may restrict considerations to the special case $f_1, f_2 \in \mathbb{C}[G]_q$. In this case (2.132) follows from (2.125).

The embedding i_1 is thus produced. Now let us turn to a construction of i_2. Let $i_2 : \mathbb{C}[\mathfrak{p}^-]_q \to (\widetilde{w}_0 U_q\mathfrak{g})^*$ be the linear map dual to

$$\mathcal{I} : \widetilde{w}_0 U_q\mathfrak{g} \to N(\mathfrak{q}^+, 0), \qquad \mathcal{I} : \widetilde{w}_0\xi \mapsto S(\xi)v(\mathfrak{q}^+, 0).$$

The definitions imply that i_2 is an injective morphism of $U_q\mathfrak{g}$-modules. Hence $i_2 : \mathbb{C}[\mathfrak{p}^-]_q \hookrightarrow \mathbb{F}$ since the $U_q\mathfrak{g}$-module $\mathbb{C}[\mathfrak{p}^-]_q$ is locally $U_q\mathfrak{b}^-$-finite-dimensional.

Now (2.11) and the fact that $v(\mathfrak{q}^+, 0)$ is the highest vector with zero weight imply that

$$\begin{aligned}\langle \xi(f_1 f_2), v(\mathfrak{q}^+, 0) \rangle &= \langle \triangle(\xi)(f_1 \otimes f_2), v(\mathfrak{q}^+, 0) \otimes v(\mathfrak{q}^+, 0) \rangle \\ &= \langle R_{\mathbb{C}[\mathfrak{p}^-]_q \mathbb{C}[\mathfrak{p}^-]_q} \triangle(\xi)(f_1 \otimes f_2), v(\mathfrak{q}^+, 0) \otimes v(\mathfrak{q}^+, 0) \rangle \\ &= \langle R_{\mathbb{F}\mathbb{F}} \triangle(\xi)(i_2(f_1) \otimes i_2(f_2)), \widetilde{w} \otimes \widetilde{w} \rangle = \langle \xi(i_2(f_1) i_2(f_2), \widetilde{w} \rangle,\end{aligned}$$

[8] It suffices to prove the last statement for $f \in \mathbb{C}[G]_q$ and $f = t^{-1}$. On the other hand, $\mathbb{C}[G]_q = \bigoplus_{\lambda \in P_+} \operatorname{End} L(\lambda)$, and $\dim(U_q\mathfrak{b}^- \cdot t^{-1}) = 1$, as one can see from (2.129).

with $f_1, f_2 \in \mathbb{C}[\mathfrak{p}^-]_q$, $\xi \in U_q\mathfrak{g}$. Thus i_2 is a morphism of algebras.

We now prove that $i_1(t^{-1}t') \in i_2\mathbb{C}[\mathfrak{p}^-]_q$. It suffices to demonstrate that $i_1(t^{-1}t')$ is orthogonal to the subspace $\operatorname{Ker} \mathcal{J} = \widetilde{w}_0(U_q\mathfrak{q}^+)_0$, where $(U_q\mathfrak{q}^+)_0 = U_q\mathfrak{q}^+ \cap \operatorname{Ker}\varepsilon$, and ε is the counit of the Hopf algebra $U_q\mathfrak{q}^+$. Consider the "regular" representation L_{reg} of the algebra $U_q\mathfrak{g}^{\mathrm{op}}$ in the space $\mathbb{C}[G]_q$, see Subsection 2.2.1. Applying the ideas of the proof of Proposition 2.185, we may extend L_{reg} to $\mathbb{C}[G]_{q,t}$. In this way, $\mathbb{C}[G]_{q,t}$ is endowed with a structure of $U_q\mathfrak{g}^{\mathrm{op}}$-module. Given $\xi \in (U_q\mathfrak{q}^+)_0$, the element $\eta = \widetilde{w}_0 \cdot \xi \cdot (\widetilde{w}_0)^{-1}$ of $(\mathbb{C}[G]_q)^*$ belongs to the subalgebra $U_q\mathfrak{g}$, and

$$L_{\mathrm{reg}}(\eta)(t^{-1}t') = 0. \tag{2.133}$$

In fact, while proving this relation, ξ may be assumed homogeneous of positive degree. Let us endow the $U_q\mathfrak{g}^{\mathrm{op}}$-module $V = \{L_{\mathrm{reg}}(\zeta)(t^{-1}t') | \ \zeta \in U_q\mathfrak{g}^{\mathrm{op}}\}$ with the grading corresponding to the subset $\mathbb{S}_1 = \{1, 2, \dots, l\} \setminus \{l_1\}$. It is easy to demonstrate that the element $L_{\mathrm{reg}}(\eta)(t^{-1}t') \in V$ has incompatible homogeneity properties; hence it is zero.

Now (2.133) implies the required orthogonality.

To complete the proof of Proposition 2.188, one has to establish that the element z with the property $i_1(t^{-1}t') = i_2(z)$ is just z_{low}. Recall that $\mathbb{S} = \{1, 2, \dots, l\} \setminus \{l_0\}$ and $\mathbb{C}[G]_{q,t}$ is a weight $U_q\mathfrak{g}$-module. Hence we have the well-defined linear map $H_{\mathbb{S}}$ in $\mathbb{C}[G]_{q,t}$. The relations

$$\begin{aligned} H_{\mathbb{S}}t &= -\overline{\omega}_{l_0}(H_{\mathbb{S}})t, & F_jt &= 0, \quad j = 1, 2, \dots, l, \\ H_{\mathbb{S}}t' &= (-\overline{\omega}_{l_0}(H_{\mathbb{S}}) + \alpha_{l_0}(H_{\mathbb{S}}))t', & F_jt &= 0, \quad j \neq l_0, \end{aligned}$$

imply that

$$H_{\mathbb{S}}(t^{-1}t') = 2t^{-1}t', \qquad F_j(t^{-1}t') = 0, \quad j \neq l_0.$$

Thus the element $z \in \mathbb{C}[\mathfrak{p}^-]_q$ with the property $i_1(t^{-1}t') = i_2(z)$ is a lowest weight vector of the $U_q\mathfrak{k}$-module $\mathbb{C}[\mathfrak{p}^-]_{q,1}$. A straightforward computation shows that

$$t' = q_{l_0}^{\frac{1}{2}} E_{l_0} t,$$

$F_{l_0}(t^{-1}t') = q_{l_0}^{\frac{1}{2}}$. Hence $F_{l_0}(z) = q_{l_0}^{1/2}$, and the relation $z = z_{\mathrm{low}}$ now follows from Lemma 2.148. □

The embedding $i : \mathbb{C}[\mathfrak{p}^-]_q \hookrightarrow \mathbb{C}[G]_{q,x}$ just produced will be called the canonical embedding.

PROPOSITION 2.189. *The image of $\mathbb{C}[\mathfrak{p}^-]_q$ under the canonical embedding into $\mathbb{C}[G]_{q,x}$ is the subalgebra generated by $\{t^{-1} c_{\overline{\omega}_{l_1},\lambda}^{\overline{\omega}_{l_1}}\}$, where $c_{\overline{\omega}_{l_1},\lambda}^{\overline{\omega}_{l_1}}$ are the elements of the "top row" of the matrix of the fundamental representation π_{l_1}.*

PROOF. Let $\mathcal{F} \subset \mathbb{C}[G]_{q,x}$ be the subalgebra generated by $\{t^{-1}c_{\overline{\omega}_{l_1},\lambda}^{\overline{\omega}_{l_1}}\}$. The definitions yield that it is a $U_q\mathfrak{g}$-module subalgebra.

We now establish that $i\mathbb{C}[\mathfrak{p}^-]_q = \mathcal{F}$. To begin with, observe that $i\mathbb{C}[\mathfrak{p}^-]_q \subset \mathcal{F}$ because $i(z_{\mathrm{low}}) \in \mathcal{F}$, hence $i\mathbb{C}[\mathfrak{p}^-]_{q,1} \subset \mathcal{F}$. The opposite inclusion $i\mathbb{C}[\mathfrak{p}^-]_q \supset \mathcal{F}$ is due to the fact that the linear span $\mathcal{L}$ of the set $\{c_{\overline{\omega}_{l_1},\lambda}^{\overline{\omega}_{l_1}}\}$ is just $U_q\mathfrak{b}^+ \cdot t$, since t is the lowest weight vector of the $U_q\mathfrak{g}$-module $\mathcal{L} \approx L(\overline{\omega}_{l_0})$. □

REMARK 2.190. An application of the canonical embedding, together with the well-known Hua–Schmid theorem (see, e.g., Subsection 2.5.10), allows one easily to establish that

$$\dim \mathrm{Hom}_{U_q\mathfrak{k}}(L(\mathfrak{k},\lambda), L(j\overline{\omega}_{l_0})) \leq 1$$

for all $\lambda \in P_+^{\mathbb{S}}$, $j \in \mathbb{N}$.

Let $\mathbb{C}[X_{\mathbb{S}}^-]_q \subset \mathbb{C}[G]_q$ be the smallest $U_q\mathfrak{g}$-module subalgebra which contains t, and $\mathbb{C}[X_{\mathbb{S}}^+]_q \subset \mathbb{C}[G]_q$ the smallest $U_q\mathfrak{g}$-module subalgebra which contains $t^\star$. This can also be formulated as follows: the subalgebra $\mathbb{C}[X_{\mathbb{S}}^-]_q$ is generated by the matrix elements $\left\{c^{\overline{\omega}_{l_1}}_{\overline{\omega}_{l_1};\lambda,j}\right\}$, and the subalgebra $\mathbb{C}[X_{\mathbb{S}}^+]_q$ is generated by $\left\{c^{\overline{\omega}_{l_0}}_{w_0\overline{\omega}_{l_0};\lambda,j}\right\}$; see (2.127).

We endow the algebra $\mathbb{C}[X_{\mathbb{S}}^-]_q$ with a grading by setting $\deg\left(c^{\overline{\omega}_{l_1}}_{\overline{\omega}_{l_1};\lambda,j}\right) = 1$. This grading is well defined since it can be imposed in a different way by using the linear map $L_{\mathrm{reg}}(H_{\mathbb{S}_1})$. Namely, $\deg f = j$ if and only if $L_{\mathrm{reg}}(H_{\mathbb{S}_1})f = jf$.

The above results imply the following statements, which are well known in quantum group theory.

First, the multiplicative system $t^{\mathbb{Z}_+}$ is an Ore set in $\mathbb{C}[X_{\mathbb{S}}^-]_q$, and this algebra admits the natural embedding into its localization: $\mathbb{C}[X_{\mathbb{S}}^-]_q \hookrightarrow \mathbb{C}[X_{\mathbb{S}}^-]_{q,t}$.

Second, $\mathbb{C}[X_{\mathbb{S}}^-]_q$ is a $U_q\mathfrak{g}$-module subalgebra.

Third, the structure of $U_q\mathfrak{g}$-module algebra on $\mathbb{C}[X_{\mathbb{S}}^-]_q$ admits an extension to $\mathbb{C}[X_{\mathbb{S}}^-]_{q,t}$, and this extension is unique.

Now Proposition 2.188 can be used to obtain

COROLLARY 2.191. *The map $i : z_{\mathrm{low}} \mapsto t^{-1}t'$ can be uniquely extended to an embedding of $U_q\mathfrak{g}$-module algebras $i : \mathbb{C}[\mathfrak{p}^-]_q \hookrightarrow \mathbb{C}[X_{\mathbb{S}}^-]_{q,t}$.*

REMARK 2.192. In a similar way one can endow the localization $\mathbb{C}[X_{\mathbb{S}}^+]_{q,t^\star}$ of $\mathbb{C}[X_{\mathbb{S}}^+]_q$ with respect to the multiplicative system $(t^\star)^{\mathbb{Z}_+}$ with the structure of a $U_q\mathfrak{g}$-module algebra and obtain an embedding $\mathbb{C}[\mathfrak{p}^+]_q \hookrightarrow \mathbb{C}[X_{\mathbb{S}}^+]_{q,t^\star}$.

Consider the classical case $q = 1$. We are about to use an isomorphism of $U\mathfrak{g}$-modules $L(\overline{\omega}_{l_1})^* \approx L(\overline{\omega}_{l_0})$. Consider the cone $X_{\mathbb{S}}^- \subset L(\overline{\omega}_{l_0})$ generated by the G-orbit of the highest weight vector $v(\overline{\omega}_{l_0}) \in L(\overline{\omega}_{l_0})$. Its projectivisation has been discussed in Subsection 2.2.1. It is a homogeneous space of the group G, and the stabilizer of the point $\mathbb{C}v(\overline{\omega}_{l_0})$ is a standard parabolic subgroup [**130**, p. 291].

We produced a quantum analog of the embedding of $\mathfrak{p}^-$ into the above homogeneous space, which is called a generalized flag variety. In the classical case $q = 1$, such an embedding is well known and is extensively investigated [**296**, pp. 34, 35]. In the special case $\mathfrak{g} = \mathfrak{sl}_N$ a result close to ours was obtained by R. Fioresi [**95**].

2.5.5. The $*$-algebra $(\mathbb{C}[\mathbb{X}_{\mathbb{S}}]_q, *)$. Consider the smallest unital $U_q\mathfrak{g}$-module subalgebra $\mathbb{C}[\mathbb{X}_{\mathbb{S}}]_q \subset \mathbb{C}[G]_q$ which contains t, $t^\star$.

It is known [**299**, p. 480] that $\mathbb{C}[\mathbb{X}_{\mathbb{S}}]_q$ consists of functions $f \in \mathbb{C}[G]_q$ such that

$$L_{\mathrm{reg}}(E_j)f = L_{\mathrm{reg}}(F_j)f = L_{\mathrm{reg}}(K_j^{\pm 1} - 1)f = 0, \qquad j \neq l_1.$$

EXAMPLE 2.193. If $\mathfrak{g} = \mathfrak{sl}_2$, then

$$\mathbb{C}[\mathbb{X}_{\mathbb{S}}]_q = \mathbb{C}[SL_2]_q.$$

PROPOSITION 2.194. *There exists a unique antilinear involution $*$ in $\mathbb{C}[\mathbb{X}_{\mathbb{S}}]_q$ such that $(\mathbb{C}[\mathbb{X}_{\mathbb{S}}]_q, *)$ is a $(U_q\mathfrak{g}, *)$-module algebra and $t^* = t^\star$.*

PROOF. The uniqueness of the required involution is obvious. We prove its existence by endowing $\mathbb{C}[G]_q$ with a $(U_q\mathfrak{g}, *)$-module algebra structure such that $t^* = t^\star$.

First, we consider the following natural embeddings:

$$U_q\mathfrak{g} \hookrightarrow \mathbb{C}[G]_q^*, \qquad \widetilde{w}_0 U_q\mathfrak{g} \hookrightarrow \mathbb{C}[G]_q^*.$$

The image of $\widetilde{w}_0 U_q\mathfrak{g}$ under the embedding is dense in $\mathbb{C}[G]_q^* \cong \times_{\lambda\in P_+} \operatorname{End} L(\lambda)$ in the product topology since $\widetilde{w}_0$ is invertible. Therefore, the canonical pairing $\langle\cdot,\cdot\rangle$

$$\mathbb{C}[G]_q \times \widetilde{w}_0 U_q\mathfrak{g} \to \mathbb{C}$$

is nondegenerate. Now we need some auxiliary observations.

LEMMA 2.195. *There exists a unique antilinear map* $*$ *in* $\mathbb{C}[G]_q$ *such that*

$$\langle f^*, \widetilde{w}_0\xi\rangle = \overline{\langle f, \widetilde{w}_0(S(\xi))^*\rangle} \tag{2.134}$$

for each $f \in \mathbb{C}[G]_q$, $\xi \in U_q\mathfrak{g}$.

PROOF OF THE LEMMA. The uniqueness is obvious. We establish existence by using the involution $\sharp$ in $\mathbb{C}[G_0]_q$; see Subsection 2.5.1.

The representation L_{reg} admits an extension by continuity to a representation $\overline{L}_{\text{reg}}$ of $\mathbb{C}[G]_q^{*\text{op}}$. Comparing (2.134) with (2.118), one has that for all $f \in \mathbb{C}[G]_q$, $\xi \in U_q\mathfrak{g}$

$$\langle \overline{L}_{\text{reg}}(\widetilde{w}_0) f^*, \xi\rangle = \langle \overline{L}_{\text{reg}}(\widetilde{w}_0) f, (S(\xi))^*\rangle = \left\langle \left(\overline{L}_{\text{reg}}(\widetilde{w}_0) f\right)^\sharp, \xi\right\rangle.$$

Hence,

$$\overline{L}_{\text{reg}}(\widetilde{w}_0)\cdot * = \sharp \cdot \overline{L}_{\text{reg}}(\widetilde{w}_0). \tag{2.135}$$

Finally,

$$f^* = L_{\text{reg}}(\widetilde{w}_0^{-1})\left(L_{\text{reg}}(\widetilde{w}_0) f\right)^\sharp.$$

The last equation gives an antilinear operator with the required properties. □

Now we prove that $*$ is an involution of $\mathbb{C}[G]_q$.

LEMMA 2.196. (1) $f^{**} = f$ *for each* $f \in \mathbb{C}[G]_q$.
(2) $(f_1 f_2)^* = f_2^* f_1^*$ *for each* $f_1, f_2 \in \mathbb{C}[G]_q$.

PROOF OF THE LEMMA. The square of the antilinear operator $*S : U_q\mathfrak{g} \to U_q\mathfrak{g}$ is the identity map. This proves the first statement:

$$\langle f^{**}, \widetilde{w}_0\xi\rangle = \langle f, \widetilde{w}_0(S((S(\xi))^*))^*\rangle = \langle f, \xi\rangle$$

for each $f \in \mathbb{C}[G]_q$, $\xi \in U_q\mathfrak{g}$.

Consider the second statement. Strictly speaking, the universal R-matrix should be substituted by the corresponding map $R_{\mathbb{C}[G]_q,\mathbb{C}[G]_q}$ in the coming proof. For clarity, let us keep just R. We use some properties of $*S$, which come from the following equalities:

$$\triangle(S(\xi)) = (S\otimes S)\triangle^{\text{cop}}(\xi), \qquad \triangle(\xi^*) = (\triangle(\xi))^{*\otimes *},$$
$$S\otimes S(R) = R, \qquad R^{*\otimes *} = R^{21},$$

where R^{21} is obtained from R via a flip of tensor factors. The last equality follows from the well-known properties of the universal R-matrix; see, for example Proposition 2.3.1 in [**199**].

For each $f_1, f_2 \in \mathbb{C}[G]_q$, $\xi \in U_q\mathfrak{g}$

$$\begin{aligned}\langle (f_1 f_2)^*, \widetilde{w}_0 \xi\rangle &= \overline{\langle f_1 f_2, \widetilde{w}_0 (S(\xi))^*\rangle} = \overline{\langle f_1 \otimes f_2, (\widetilde{w}_0 \otimes \widetilde{w}_0) R\triangle(S(\xi))^*\rangle},\\ \langle f_2^* f_1^*, \widetilde{w}_0 \xi\rangle &= \langle f_2^* \otimes f_1^*, (\widetilde{w}_0 \otimes \widetilde{w}_0) R\triangle(\xi)\rangle \\ &= \overline{\langle f_2 \otimes f_1, (\widetilde{w}_0 \otimes \widetilde{w}_0)(*S \otimes *S)(R\triangle(\xi))\rangle} \\ &= \overline{\langle f_2 \otimes f_1, (\widetilde{w}_0 \otimes \widetilde{w}_0) R^{21} \triangle^{\mathrm{cop}}((S(\xi))^*)\rangle} \\ &= \overline{\langle f_1 \otimes f_2, (\widetilde{w}_0 \otimes \widetilde{w}_0) R\triangle((S(\xi))^*)\rangle}.\end{aligned}$$

Therefore $\langle (f_1 f_2)^*, \widetilde{w}_0 \xi\rangle = \langle f_2^* f_1^*, \widetilde{w}_0 \xi\rangle$ for each $f_1, f_2 \in \mathbb{C}[G]_q$, $\xi \in U_q\mathfrak{g}$. □

Consider the $*$-algebra $\mathbb{C}[w_0 G_0]_q \overset{\mathrm{def}}{=} (\mathbb{C}[G]_q, *)$. It is a q-analog of the algebra $\mathbb{C}[w_0 G_0]$ of regular functions on the real affine algebraic variety $w_0 G_0$, where G_0 is a noncompact real form of G.

EXAMPLE 2.197. If $\mathfrak{g} = \mathfrak{sl}_2$, then $(\alpha, \alpha) = 2$, $(\overline{\omega}, \overline{\omega}) = \frac{1}{2}$, $(H, H) = 2$. Using (2.120), one can explicitly find the action of $\widetilde{w}_0$ in the fundamental representation $\overline{\pi}$ of the algebra $\mathbb{C}[SL_2]_q^*$:

$$\overline{\pi}(\widetilde{w}_0) = q^{-1/4}\begin{pmatrix} 0 & -1 \\ q^{-1} & 0 \end{pmatrix}.$$

Hence,

$$\begin{pmatrix} \overline{L}_{\mathrm{reg}}(\widetilde{w}_0) t_{11} & \overline{L}_{\mathrm{reg}}(\widetilde{w}_0) t_{12} \\ \overline{L}_{\mathrm{reg}}(\widetilde{w}_0) t_{21} & \overline{L}_{\mathrm{reg}}(\widetilde{w}_0) t_{22} \end{pmatrix} = q^{-1/4}\begin{pmatrix} 0 & -1 \\ q^{-1} & 0 \end{pmatrix}\begin{pmatrix} t_{11} & t_{12} \\ t_{21} & t_{22} \end{pmatrix}.$$

One can easily check that the involutions (1.186) and (2.119) satisfy (2.135). For example,

$$\overline{L}_{\mathrm{reg}}(\widetilde{w}_0) t_{11}^* = \overline{L}_{\mathrm{reg}}(\widetilde{w}_0)(-t_{22}) = -q^{-5/4} t_{12} = (-t_{21})^\sharp = \left(\overline{L}_{\mathrm{reg}}(\widetilde{w}_0) t_{11}\right)^\sharp.$$

Hence $\overline{L}_{\mathrm{reg}}(\widetilde{w}_0) f^* = \left(\overline{L}_{\mathrm{reg}}(\widetilde{w}_0) f\right)^\sharp$ for $f = t_{11}$.

Therefore, the new definition of the involution $*$ generalizes the previous one (1.186), which is applicable only in the special case $\mathfrak{g} = \mathfrak{sl}_2$.

LEMMA 2.198. *The $*$-algebra $\mathbb{C}[w_0 G_0]_q$ is a $(U_q\mathfrak{g}, *)$-module algebra.*

PROOF. For each $f \in \mathbb{C}[G]_q$, $\xi, \eta \in U_q\mathfrak{g}$, we have that

$$\begin{aligned}\langle (\xi f)^*, \widetilde{w}_0 \eta\rangle &= \overline{\langle \xi f, \widetilde{w}_0 (S(\eta))^*\rangle} = \overline{\langle f, \widetilde{w}_0 (S(\eta))^* \xi\rangle},\\ \langle (S(\xi))^* f^*, \widetilde{w}_0 \eta\rangle &= \langle f^*, \widetilde{w}_0 \eta (S(\xi))^*\rangle = \overline{\langle f, \widetilde{w}_0 (S(\eta(S(\xi))^*))^*\rangle} \\ &= \overline{\langle f, \widetilde{w}_0 (S(\eta))^* (S((S(\xi))^*))^*\rangle} = \overline{\langle f, \widetilde{w}_0 (S(\eta))^* \xi\rangle}.\end{aligned}$$

Hence for each $f \in \mathbb{C}[G]_q$, $\xi \in U_q\mathfrak{g}$ one has

$$(\xi f)^* = (S(\xi))^* f^*.$$ □

Now we prove that $t^* = t^\star$.

LEMMA 2.199.

$$t^* = \mathrm{const} \cdot c^{\overline{\omega}_{l_0}}_{-\overline{\omega}_{l_1}, \overline{\omega}_{l_0}}.$$

PROOF. Consider the fundamental $U_q\mathfrak{g}$-module $L(\overline{\omega}_{l_0})$ and the corresponding representation $\pi_{\omega_{l_0}}$ together with its dual $\pi^*_{\omega_{l_0}}$. The elements t^*, $c^{\overline{\omega}_{l_0}}_{-\overline{\omega}_{l_1},\overline{\omega}_{l_0}}$ belong to the same weight subspace of the simple $U_q\mathfrak{g}^{\mathrm{op}} \otimes U_q\mathfrak{g}$-module

$$\left\{ f \in (U_q\mathfrak{g})^* \,\middle|\, \operatorname{Ker} f \supset \operatorname{Ker} \pi^*_{\omega_{l_0}} \right\} \subset \mathbb{C}[G]_q,$$

which is one dimensional. □

LEMMA 2.200. $t^* = t^\star$.

PROOF. Due to (2.127), it is sufficient to prove that in the previous lemma const $= (-q)^{(\overline{\omega}_{l_1}+\overline{\omega}_{l_0},\rho)}$. Therefore, one should verify the following

$$\left\langle (c^{\overline{\omega}_{l_1}}_{\overline{\omega}_{l_1},-\overline{\omega}_{l_0}})^*, \widetilde{w}_0 \right\rangle = \overline{\left\langle c^{\overline{\omega}_{l_1}}_{\overline{\omega}_{l_1},-\overline{\omega}_{l_0}}, \widetilde{w}_0 \right\rangle} = (-q)^{(\overline{\omega}_{l_1}-w_0\overline{\omega}_{l_1},\rho)} \left\langle c^{\overline{\omega}_{l_0}}_{-\overline{\omega}_{l_1},\overline{\omega}_{l_0}}, \widetilde{w}_0 \right\rangle,$$

or, equivalently (see (2.124)),

$$\overline{\left\langle c^{\overline{\omega}_{l_1}}_{\overline{\omega}_{l_1},-\overline{\omega}_{l_0}}, \overline{w}_0^{-1} \right\rangle} = (-q)^{(\overline{\omega}_{l_1}-u_1\overline{\omega}_{l_1},\rho)} \left\langle c^{\overline{\omega}_{l_0}}_{-\overline{\omega}_{l_1},\overline{\omega}_{l_0}}, \overline{w}_0^{-1} \right\rangle. \tag{2.136}$$

Here u_1 is the shortest element in the coset $w_0 W_{\mathbb{S}_1}$ of the subgroup $W_{\mathbb{S}_1} \subset W$ generated by $\{s_j \,|\, j \neq l_1\}$.

If $\overline{\omega}_{l_1} - u_1\overline{\omega}_{l_1} = \sum_k n_k\alpha_k$, then $(\overline{\omega}_{l_1} - u_1\overline{\omega}_{l_1}, \rho) = \sum_k n_k d_k$ since $(\alpha_k, \overline{\omega}_j) = d_j\delta_{jk}$. Hence for a reduced expression $u_1 = s_{i_1}s_{i_2}\cdots s_{i_M}$ one has

$$(-q)^{(\overline{\omega}_{l_1}-u_1\overline{\omega}_{l_1},\rho)} = \prod_{j=1}^{M}(-q_{i_j}).$$

Due to the last equality and the definition of $\overline{w}_0^{-1}$, it suffices to prove (2.136) in the simplest special case $\mathfrak{g} = \mathfrak{sl}_2$. Here we can just use Proposition 2.170. □

Proposition 2.194 follows from the lemmas above. □

In the sequel we use the notation $\mathbb{S}_1 = \{1, 2, \ldots, l\} \setminus \{l_1\}$ as in the proof of Lemma 2.200.

REMARK 2.201. One has the following equation for the matrix elements in a special basis of $L(\overline{\omega}_{l_1})$:

$$\left(c^{\overline{\omega}_{l_1}}_{\overline{\omega}_{l_1},\lambda}\right)^* = (-1)^{\lambda(H_{\mathbb{S}_1})+\overline{\omega}_{l_0}(H_{\mathbb{S}_1})} \left(c^{\overline{\omega}_{l_1}}_{\overline{\omega}_{l_1},\lambda}\right)^\star, \qquad \lambda \in W\overline{\omega}_{l_0}.$$

In particular, for $l = m+n-1$, $l_0 = n$, $\mathfrak{g} = \mathfrak{sl}_{m+n}$, the involution $*$ can be written explicitly [**283**, p. 859]:

$$t^*_{ij} = \operatorname{sign}\left(\left(i - m - \frac{1}{2}\right)\left(j - n - \frac{1}{2}\right)\right) t^\star_{ij}.$$

2.5.6. The canonical embedding of $\mathrm{Pol}(\mathfrak{p}^-)_q$. Consider the subalgebra $\mathbb{C}[\mathbb{X}_\mathbb{S}]_{q,x} \subset \mathbb{C}[G]_{q,x}$ generated by $\mathbb{C}[\mathbb{X}_\mathbb{S}]_q$ and x^{-1}. It is a localization of the algebra $\mathbb{C}[\mathbb{X}_\mathbb{S}]_q$ with respect to the multiplicative system $x^{\mathbb{Z}_+}$.

The involution $*$ is uniquely extendable from $\mathbb{C}[\mathbb{X}_\mathbb{S}]_q$ to $\mathbb{C}[\mathbb{X}_\mathbb{S}]_{q,x}$, and we have $(x^{-1})^* = x^{-1}$. It is easy to prove that the structure of $U_q\mathfrak{g}$-module algebra and the involution are compatible, i.e., we have the $(U_q\mathfrak{g}, *)$-module algebra $(\mathbb{C}[\mathbb{X}_\mathbb{S}]_{q,x}, *)$; see Proposition 2.185.

Similarly, one can introduce the $(U_q\mathfrak{g}, *)$-module algebra

$$\mathbb{C}[w_0G_0]_{q,x} \stackrel{\mathrm{def}}{=} (\mathbb{C}[G]_{q,x}, *).$$

THEOREM 2.202. (1) *There exists a unique homomorphism of* $(U_q\mathfrak{g}, *)$-*module algebras* $\mathcal{I} : \mathrm{Pol}(\mathfrak{p}^-)_q \to \mathbb{C}[\mathbb{X}_\mathbb{S}]_{q,x}$ *such that* $\mathcal{I} : z_{\mathrm{low}} \mapsto t^{-1}t'$.
(2) $\mathcal{I}$ *is injective.*

PROOF. Let $\mathbb{C}[\mathbb{X}_\mathbb{S}]_{q,t}$ be the subalgebra in $\mathbb{C}[\mathbb{X}_\mathbb{S}]_{q,x}$ generated by $\mathbb{C}[\mathbb{X}_\mathbb{S}]_q$ and t^{-1}.

Proposition 2.188 implies that the map $i : z_{\mathrm{low}} \mapsto t^{-1}t'$ can be uniquely extended to an embedding of $U_q\mathfrak{g}$-module algebras $i : \mathbb{C}[\mathfrak{p}^-]_q \hookrightarrow \mathbb{C}[\mathbb{X}_\mathbb{S}]_{q,t}$.

Furthermore, $\mathrm{Pol}(\mathfrak{p}^-)_q = (\mathbb{C}[\mathfrak{p}^- \oplus \mathfrak{p}^+], *)$, and the multiplication

$$m : \mathbb{C}[\mathfrak{p}^-]_q \otimes \mathbb{C}[\mathfrak{p}^+]_q \to \mathbb{C}[\mathfrak{p}^- \oplus \mathfrak{p}^+]_q, \qquad m : f_- \otimes f_+ \mapsto f_- f_+,$$

is a bijection.

Hence the homomorphism $\mathcal{I}$ is unique, and, if it exists, one has

$$\mathcal{I}(f_- f_+) = i(f_-)(i(f_+))^*, \qquad f_\pm \in \mathbb{C}[\mathfrak{p}^\pm]_q. \tag{2.137}$$

Now we prove existence. Consider the linear map $\mathcal{I} : \mathrm{Pol}(\mathfrak{p}^-)_q \to \mathbb{C}[\mathbb{X}_\mathbb{S}]_{q,x}$ defined by (2.137). Here, $\mathcal{I}$ is a morphism of $U_q\mathfrak{g}$-modules since $\mathrm{Pol}(\mathfrak{p}^-)_q$ and $\mathbb{C}[\mathbb{X}_\mathbb{S}]_{q,x}$ are $(U_q\mathfrak{g}, *)$-module algebras. It remains to prove that $\mathcal{I}$ is an algebra homomorphism.

Consider the embedding $\mathbb{C}[\mathbb{X}_\mathbb{S}]_{q,x} \subset \mathbb{C}[G]_{q,x}$. Due to the definition of multiplication in $\mathbb{C}[\mathfrak{p}^- \oplus \mathfrak{p}^+]_q$, it is sufficient to obtain the equality

$$(\mathcal{I}f_+)(\mathcal{I}f_-) = m\check{R}_{\mathcal{I}\mathbb{C}[\mathfrak{p}^+]_q, \mathcal{I}\mathbb{C}[\mathfrak{p}^-]_q}(\mathcal{I}f_+ \otimes \mathcal{I}f_-), \tag{2.138}$$

where $f_\pm \in \mathcal{I}\mathbb{C}[\mathfrak{p}^\pm]_{q,\pm 1}$, and m is the multiplication in $\mathbb{C}[G]_{q,x}$. Instead of (2.138), we prove a more general statement:

$$q^{ck^2}\mathcal{I}(f_+)t^{*k} \cdot t^k \cdot \mathcal{I}(f_-) = m\check{R}_{\mathcal{I}\mathbb{C}[G]_{q,t^*}, \mathcal{I}\mathbb{C}[G]_{q,t}}(\mathcal{I}(f_+)t^{*k} \otimes t^k \mathcal{I}(f_-)), \tag{2.139}$$

where $k \in \mathbb{Z}$, $\mathbb{C}[G]_{q,t^*} = \{f^* | \, f \in \mathbb{C}[G]_{q,t}\}$ and c is a rational number that depends on f_-, f_+. Just as in Subsection 2.5.4, one proves that it is sufficient to obtain the last equality for k big enough. Namely, the last equality can be proved under the additional assumption that

$$\mathcal{I}(f_+)t^{*k} \in \mathbb{C}[X_\mathbb{S}^+]_q, \qquad \mathcal{I}(f_-)t^k \in \mathbb{C}[X_\mathbb{S}^-]_q, \tag{2.140}$$

where $\mathbb{C}[X_\mathbb{S}^-]_q$, $\mathbb{C}[X_\mathbb{S}^+]_q \subset \mathbb{C}[\mathbb{X}_\mathbb{S}]_q$ are the $U_q\mathfrak{g}$-module subalgebras generated by t, t^*, respectively; see also Subsection 2.5.4.

Note that $\mathbb{C}[X_\mathbb{S}^+]_q = (\mathbb{C}[X_\mathbb{S}^-]_q)^*$.

We endow $\mathbb{C}[\mathbb{X}_\mathbb{S}]_q$ with the grading via

$$\deg\left(c^{\varpi_{l_1}}_{\varpi_{l_1};\lambda,j}\right) = 1, \qquad \deg\left(c^{\varpi_{l_0}}_{-\varpi_{l_1};\mu,k}\right) = -1.$$

As in Subsection 2.5.4, one can prove that the grading is well defined.

The commutation relations (2.35) (see also (2.14) and the definition of the grading in $\mathbb{C}[\mathbb{X}_\mathbb{S}]_q$) and the equalities

$$L_{\mathrm{reg}}(F_j)\psi_+ = L_{\mathrm{reg}}(E_j)\psi_- = 0, \qquad j = 1, 2, \ldots, l,$$
$$L_{\mathrm{reg}}(H_j)\psi_\pm = 0, \quad j \neq l_1, \qquad (H_\mathbb{S}, H_\mathbb{S}) = (H_{\mathbb{S}_1}, H_{\mathbb{S}_1})$$

imply that

$$q^{-\deg(\psi_+)\deg(\psi_-)/(H_\mathbb{S}, H_\mathbb{S})}\psi_+\psi_- = m\check{R}_{\mathbb{C}[G]_q, \mathbb{C}[G]_q}(\psi_+ \otimes \psi_-), \tag{2.141}$$

where $\psi_+ \in \mathbb{C}[X_\mathbb{S}^+]_q]_q$, $\psi_- \in \mathbb{C}[X_\mathbb{S}^-]_q$.

Using (2.140), one can prove (2.139) from (2.141).

The existence and uniqueness of the homomorphism $\mathcal{I}$ are now proved. Let us prove its injectivity. We intend to prove this without assuming q to be transcendental. Our proof requires some auxiliary results.

The proof consists of a construction of a representation $\mathcal{T}$ of $\mathbb{C}[G]_{q,x}$ such that the corresponding representation $\mathcal{T} \circ \mathcal{I}$ of $\mathbb{C}[\mathfrak{p}^- \oplus \mathfrak{p}^+]_q$ is faithful. This will prove the injectivity of $\mathcal{I}$.

To construct of $\mathcal{T}$ we follow the standard approach to producing $*$-representations of $\mathbb{C}[K]_q = (\mathbb{C}[G]_q, \star)$ which is described in Subsection 2.2.5. Indeed, let $w_0^{\mathbb{S}}$ be the shortest element in the coset $W_{\mathbb{S}}\, w_0 \subset W$ of the subgroup $W_{\mathbb{S}}$ generated by the simple reflections s_i, $i \neq l_0$; see Lemma 2.69. Choose a reduced expression

$$w_0^{\mathbb{S}} = s_{i_1} s_{i_2} \cdots s_{i_N},$$

and introduce the irreducible representation of the Hopf $*$-algebra $\mathbb{C}[K]_q$ in $l^2(\mathbb{Z}_+)^{\otimes N}$

$$\Pi = \Pi_{i_1} \otimes \Pi_{i_2} \otimes \cdots \otimes \Pi_{i_N},$$

where Π_{i_j} are the $*$-representations of $(\mathbb{C}[G]_q, \star)$ in $l^2(\mathbb{Z}_+)$; see Subsection 2.2.5.

The definitions of Π_{i_j} and Π imply that the operators $\Pi(f)$, $f \in \mathbb{C}[G]_q$, admit the well-defined restriction to the linear span $\mathcal{L}$ of elements of the standard basis $\{e_{k_1} \otimes e_{k_2} \otimes \cdots \otimes e_{k_N}\}$ of the Hilbert space $l^2(\mathbb{Z}_+)^{\otimes N}$. It is known [**300**, pp. 314, 315] that these vectors are eigenvectors of the bounded self-adjoint compact linear operator $\Pi(x)$ and the eigenvalues are positive. Besides, the eigenspace of the highest eigenvalue is one dimensional and is spanned by $\mathbf{e} = e_0 \otimes e_0 \otimes \cdots \otimes e_0$. Hence the linear operator $\Pi(x)|_{\mathcal{L}}$ is invertible in the pre-Hilbert space $\mathcal{L}$. Therefore, the representation $\Pi|_{\mathcal{L}}$ of $\mathbb{C}[G]_q$ can be extended in a canonic way to a representation $\mathcal{T}$ of its localization $\mathbb{C}[G]_{q,x}$ such that

$$\mathcal{T}(f)v = \Pi(f)v, \qquad f \in \mathbb{C}[G]_q, \quad v \in \mathcal{L}.$$

It remains to prove the faithfulness of the representation $\mathcal{T}_{\mathcal{I}} = \mathcal{T} \cdot \mathcal{I}$ of $\mathbb{C}[\mathfrak{p}^- \oplus \mathfrak{p}^+]_q$. First of all, we establish the existence of an involution $\star$ in $\mathbb{C}[\mathfrak{p}^- \oplus \mathfrak{p}^+]_q$ such that

$$\mathcal{I}(f^\star) = \mathcal{I}(f)^\star, \qquad f \in \mathbb{C}[\mathfrak{p}^- \oplus \mathfrak{p}^+]_q. \tag{2.142}$$

Let us use the standard grading

$$\mathbb{C}[\mathfrak{p}^- \oplus \mathfrak{p}^+]_{q,j} = \left\{ f \in \mathbb{C}[\mathfrak{p}^- \oplus \mathfrak{p}^+]_q \middle|\, H_{\mathbb{S}} f = 2jf \right\}, \qquad j \in \mathbb{Z},$$

and the automorphism $f \mapsto (-1)^{\deg f} f$ of $\mathbb{C}[\mathfrak{p}^- \oplus \mathfrak{p}^+]_q$, related to this grading. Indeed, put

$$f^\star \stackrel{\text{def}}{=} (-1)^{\deg f} f^*.$$

We introduce the notation $\mathrm{Pol}(U)_q = (\mathbb{C}[\mathfrak{p}^- \oplus \mathfrak{p}^+]_q, \star)$. Note that $\mathrm{Pol}(U)_q$ is a $(U_q\mathfrak{g}, \star)$-module algebra, since $\mathrm{Pol}(\mathfrak{p}^-)_q$ is a $(U_q\mathfrak{g}, *)$-module algebra.

Now (2.142) follows from the fact that $\mathcal{I}(f^*) = \mathcal{I}(f)^*$ for each $f \in \mathrm{Pol}(\mathfrak{p}^-)_q$ and the definitions of the involutions. Hence $\mathcal{T}_{\mathcal{I}}$ is a $*$-representation of $\mathrm{Pol}(U)_q$ in the pre-Hilbert space $\mathcal{L}$. The eigenvalue of the eigenvector $\mathbf{e}$ is maximal, so the commutation relations (2.37) lead to the following:

$$\mathcal{T}\left(c^{\overline{\omega}_{l_1}}_{\overline{\omega}_{l_1},\lambda}\right)\mathbf{e} = 0, \quad \lambda \neq w_0\overline{\omega}_{l_1}; \qquad \mathcal{T}_{\mathcal{I}}(z)\mathbf{e} = 0, \quad z \in \mathbb{C}[\mathfrak{p}^-]_{q,1}. \tag{2.143}$$

The second equality points out a plan of action. We intend to introduce a Fock representation $\mathcal{T}_F$ of the $*$-algebra $\mathrm{Pol}(U)_q$ and a nonzero intertwining operator between $\mathcal{T}_F$ and $\mathcal{T}_{\mathcal{I}}$. Then the faithfulness of $\mathcal{T}_{\mathcal{I}}$ will follow from the faithfulness and irreducibility of $\mathcal{T}_F$.

The well-known equality $(S \otimes \mathrm{id})R = (\mathrm{id} \otimes S^{-1})R = R^{-1}$ [**80**, p. 326] leads to the invertibility of the linear map $\check{R}_{\mathbb{C}[\mathfrak{p}^+]_q, \mathbb{C}[\mathfrak{p}^-]_q}$; see Subsection 2.1.4. This allows us to turn from the expansion (2.70) to the expansion

$$\mathrm{Pol}(U)_q = \bigoplus_{i,j=0}^{\infty} \mathbb{C}[\mathfrak{p}^+]_{q,-j} \mathbb{C}[\mathfrak{p}^-]_{q,i}.$$

Obviously, $\mathbb{C}[\mathfrak{p}^+]_{q,0} = \mathbb{C}[\mathfrak{p}^-]_{q,0} = \mathbb{C} \cdot 1$. Hence for each $f \in \mathrm{Pol}(U)_q$ there exists a unique decomposition similar to (2.71):

$$f = \langle f \rangle \cdot 1 + \sum_{(i,j) \neq (0,0)} f_{ij}, \qquad f_{ij} \in \mathbb{C}[\mathfrak{p}^+]_{q,-j} \mathbb{C}[\mathfrak{p}^-]_{q,i},$$

where $\langle f \rangle$ is a linear functional on $\mathrm{Pol}(U)_q$.

Following Subsection 2.4.1, we introduce the $\mathrm{Pol}(U)_q$-module $\mathcal{H}$ with a generator v_0 and defining relations $f v_0 = 0$ for each $f \in \mathbb{C}[\mathfrak{p}^-]_{q,1}$, i.e., $f v_0 = \langle f \rangle v_0$ for $f \in \mathbb{C}[\mathfrak{p}^-]_q$. Evidently,

$$\mathbb{C}[\mathfrak{p}^+]_q \overset{\approx}{\to} \mathcal{H}, \qquad f \mapsto f v_0,$$

which allows us to endow $\mathcal{H}$ with the sesquilinear form

$$(f_1 v_0, f_2 v_0) = \langle f_2^\star f_1 \rangle, \qquad f_1, f_2 \in \mathbb{C}[\mathfrak{p}^+]_q. \tag{2.144}$$

Due to (2.143),

$$\langle f \rangle = (\mathcal{T}_{\mathcal{I}}(f)\mathbf{e}, \mathbf{e}), \qquad f \in \mathrm{Pol}(U)_q.$$

Hence (2.144) is a nonnegative Hermitian form. Its positivity can be proved using methods from function theory on compact quantum groups. In fact, if $\langle f^\star f \rangle = 0$ and $f \in \mathbb{C}[\mathfrak{p}^+]_q$, then for $k \in \mathbb{N}$ large enough the element $\mathcal{I}(f) t^{\star k}$ belongs to $(\mathbb{C}[G]_q, \star)$ and is annihilated by all its irreducible representations; see Proposition 2.67. Hence the L^2-norm of $\mathcal{I}(f) t^{\star k}$ is equal to 0. The orthogonality relations imply that $\mathcal{I}(f) t^{\star k} = 0$, and Proposition 2.188 leads to $f = 0$.

Thus $\mathcal{H}$ is a pre-Hilbert space and a $\mathrm{Pol}(U)_q$-module. Denote by $\mathcal{T}_F$ the corresponding $*$-representation of $\mathrm{Pol}(U)_q$ in $\mathcal{H}$; it is the mentioned Fock representation.

By obvious reasons, there exists a unique linear operator $\mathcal{J} : \mathcal{H} \to \mathcal{L}$, $\mathcal{J}(v_0) = \mathbf{e}$, which intertwines the representations $\mathcal{T}_F$ and $\mathcal{T}_{\mathcal{I}}$.

The irreducibility of $\mathcal{T}_F$ can be proved similarly to the irreducibility of T_F in Subsection 2.4.1. Indeed, if $\mathcal{H}' \neq 0$ is a common invariant subspace of $\mathcal{T}_F(f)$, $f \in \mathrm{Pol}(U)_q$, then $\mathcal{H}'$ has a nonzero vector v such that

$$\mathcal{T}_F(f) v = 0, \qquad f \in \bigoplus_{j=1}^{\infty} \mathbb{C}[\mathfrak{p}^-]_{q,j}.$$

Hence v is orthogonal to the subspace $(\bigoplus_{j=1}^{\infty} \mathbb{C}[\mathfrak{p}^+]_{q,-j}) v_0$ and, therefore, $\mathbb{C}v = \mathbb{C}v_0$, $\mathcal{H}' = \mathcal{H}$. The irreducibility of $\mathcal{T}_F$ is thus proved.

Let us endow $\mathcal{H}$ with the grading

$$\mathcal{H} = \bigoplus_{j=0}^{\infty} \mathcal{H}_j, \qquad \mathcal{H}_j = \mathbb{C}[\mathfrak{p}^+]_{q,-j} v_0.$$

The subspaces $\mathcal{H}_j$ are pairwise orthogonal and finite dimensional. As in Subsection 2.4.1, one can prove that for each $j, k \in \mathbb{Z}_+$ the map

$$\mathbb{C}[\mathfrak{p}^+]_{q,-j} \otimes \mathbb{C}[\mathfrak{p}^-]_{q,k} \overset{m}{\to} \mathbb{C}[\mathfrak{p}^+]_{q,-j} \mathbb{C}[\mathfrak{p}^-]_{q,k} \to \mathrm{Hom}(\mathcal{H}_k, \mathcal{H}_j), \qquad f \mapsto \mathcal{T}_F(f)|_{\mathcal{H}_k},$$

is bijective (see Lemma 2.134). Now use the positivity of the Hermitian form $(\cdot,\cdot)$ in $\mathcal{H}$ to obtain the faithfulness of $\mathcal{T}_F$; see Lemma 2.135.

The irreducibility and faithfulness of $\mathcal{T}_F$, hence the faithfulness of $\mathcal{T}_{\mathcal{I}}$ and the injectivity of $\mathcal{I}$, are now proved. □

The following statement is due to the fact that $\mathbb{C}[G]_q$ is a domain; see Proposition 2.46.

COROLLARY 2.203. *The algebra* $\mathrm{Pol}(\mathfrak{p}^-)_q$ *is a domain.*

REMARK 2.204. A number of statements from Subsections 2.4.1, 2.4.2, 2.4.3, 2.4.4 can be easily transferred to the compact quantum group case. Let us restrict ourselves to some definitions and statements. They are normally used in comparing quantum analogs of compact and noncompact symmetric spaces which are dual in the sense of Cartan. Consider the unital algebra $\mathrm{Fun}(U)_q \supset \mathrm{Pol}(U)_q$ generated by its subalgebra $\mathrm{Pol}(U)_q$ and an element f_0 which satisfies the defining relations

$$f_0^2 = f_0, \quad \psi f_0 = \langle\psi\rangle f_0, \ \psi \in \mathbb{C}[\mathfrak{p}^-]_q, \quad f_0\psi = \langle\psi\rangle f_0, \ \psi \in \mathbb{C}[\mathfrak{p}^+]_q.$$

Let us endow the algebra with an involution putting $f_0^\star = f_0$. There exists a unique extension of the $*$-representation $\mathcal{T}_F$ from $\mathrm{Pol}(U)_q$ to $\mathrm{Fun}(U)_q$ such that $\mathcal{T}_F(f_0)$ is an orthogonal projection onto the one-dimensional subspace $\mathbb{C}v_0$. The obtained representation of $\mathrm{Fun}(U)_q$ will be called a Fock representation also. The two-sided ideal $\mathcal{D}(U)_q$ of the algebra $\mathrm{Fun}(U)_q$ generated by f_0 will be called an algebra of finite functions. The formal series

$$f = \sum_{k,j=0}^{\infty} f_{k,j}, \qquad f_{k,j} \in \mathbb{C}[\mathfrak{p}^+]_{q,-k}\mathbb{C}[\mathfrak{p}^-]_{q,j},$$

will be called distributions. We endow the vector space $\mathcal{D}(U)_q'$ of all distributions with the topology of coefficientwise convergence. (Recall that all $\mathbb{C}[\mathfrak{p}^+]_{q,-k}\mathbb{C}[\mathfrak{p}^-]_{q,j}$ are finite-dimensional). The involution $\star$ and the action of $U_q\mathfrak{g}$ admit an extension by continuity to $\mathcal{D}(U)_q'$ and a subsequent restriction to $\mathcal{D}(U)_q$. The $*$-algebra $\mathcal{D}(U)_q$ is a $(U_q\mathfrak{g},\star)$-module. Now we endow $\mathcal{H}$ with a structure of $U_q\mathfrak{q}^-$-module and introduce the corresponding representation Γ of $U_q\mathfrak{q}^-$ via the bijection

$$\mathbb{C}[\mathfrak{p}^+]_q f_0 \stackrel{\approx}{\to} \mathcal{H}, \qquad \psi f_0 \mapsto \mathcal{T}_F(\psi)v_0.$$

There exists a unique (up to a scalar factor) positive $U_q\mathfrak{g}$-invariant integral on $\mathcal{D}(U)_q$, and

$$\int_{U_q} f d\nu = \mathrm{const}\cdot \mathrm{tr}_q\, \mathcal{T}_F(f),$$

where $\mathrm{tr}_q\,\mathcal{T}_F(f) = \mathrm{tr}(\mathcal{T}_F(f)\Gamma(K_{-2\rho}))$ and each $\mathcal{T}_F(f)$, $f \in \mathcal{D}(U)_q$, is a finite operator; see Subsection 2.4.1.

Evidently, an invariant integral on $\mathcal{D}(U)_q$ can be obtained via the natural embedding of $\mathcal{D}(U)_q$ into an algebra of continuous functions on a compact quantum group and the results from Subsection 2.2.5.

2.5.7. $U_q\mathfrak{k}$-invariant polynomials. We introduce pairwise commuting $U_q\mathfrak{k}$-invariants $y_1, \dots, y_r \in \mathbb{C}[\mathfrak{p}^-]_q$ to be used extensively in the sequel. Here r denotes the rank of an irreducible bounded symmetric domain $\mathbb{D}$; see Subsection 2.5.10.

LEMMA 2.205. *The set of all eigenvalues of the linear operator $H_{\mathbb{S}}$ in $L(\overline{\omega}_{l_1})$ is the set*

$$\{\overline{\omega}_{l_1}(H_{\mathbb{S}}), \overline{\omega}_{l_1}(H_{\mathbb{S}}) - 2, \ldots, -\overline{\omega}_{l_1}(H_{\mathbb{S}}) + 2, -\overline{\omega}_{l_0}(H_{\mathbb{S}})\}.$$

The number of eigenvalues is $r+1$, where r is the rank of the bounded symmetric domain $\mathbb{D}$.

PROOF. The highest weight of the $U_q\mathfrak{g}$-module $L(\overline{\omega}_{l_1})$ equals $\overline{\omega}_{l_1}$; the lowest one equals $-\overline{\omega}_{l_0}$. Hence the first statement follows from the simplicity of the $U_q\mathfrak{g}$-module $L(\overline{\omega}_{l_1})$ and the definition of $H_{\mathbb{S}} \in \mathfrak{h}$. The second statement is equivalent to

$$\overline{\omega}_{l_1}(H_{\mathbb{S}}) + \overline{\omega}_{l_0}(H_{\mathbb{S}}) = 2r. \tag{2.145}$$

The last equation can be obtained from the Cartan list of irreducible bounded symmetric domains [**123**]. In fact, the ranks of these domains are known from [**123**, Sec. 9.4.4], and the summands in the left-hand side of (2.145) are just the coefficients of α_{l_0} in the expansions of the fundamental weights $\overline{\omega}_{l_1}$, $\overline{\omega}_{l_0}$ into sums of simple roots of $\mathfrak{g}$. These coefficients are listed in [**44**]. □

Recall the notation

$$v_{\lambda,j}^{\overline{\omega}_{l_1}}, \qquad c_{\overline{\omega}_{l_1};\lambda,j}^{\overline{\omega}_{l_1}}$$

introduced in Subsection 2.2.3 for the elements of a special basis of $L(\overline{\omega}_{l_1})$ and the matrix elements of the corresponding fundamental representation in this basis.

LEMMA 2.206. *To each matrix element $c_{\overline{\omega}_{l_1};\lambda,j}^{\overline{\omega}_{l_1}}$ there is a unique element $z_{\lambda,j}$ of $\mathbb{C}[\mathfrak{p}^-]_q$ such that*

$$\mathcal{I}(z_{\lambda,j}) = t^{-1} \cdot c_{\overline{\omega}_{l_1};\lambda,j}^{\overline{\omega}_{l_1}}.$$

PROOF. The uniqueness follows from the injectivity of $\mathcal{I}$. To prove the existence of $z_{\lambda,j}$, consider the subspace $t \cdot \mathcal{I}\mathbb{C}[\mathfrak{p}^-]_q \subset \mathbb{C}[G]_{q,x}$. It contains t and is a $U_q\mathfrak{b}^+$-submodule of the $U_q\mathfrak{b}^+$-module $\mathbb{C}[G]_{q,x}$ since for each $j = 1, 2, \ldots, l$

$$K_j^{\pm 1}(t\mathcal{I}(f)) = (K_j^{\pm 1}t)\mathcal{I}(K_j^{\pm 1}f),$$
$$E_j(t\mathcal{I}(f)) = (E_jt)\mathcal{I}f + (K_jt)\mathcal{I}(E_jf), \qquad f \in \mathbb{C}[\mathfrak{p}^-]_q,$$

and $K_j^{\pm 1}t$, $E_jt \in t \cdot \mathcal{I}\mathbb{C}[\mathfrak{p}^-]_q$. Hence,

$$t \cdot \mathcal{I}\mathbb{C}[\mathfrak{p}^-]_q \supset U_q\mathfrak{b}^+ \cdot t,$$

and it now just remains to observe that $c_{\overline{\omega}_{l_1};\lambda,j}^{\overline{\omega}_{l_1}} \in U_q\mathfrak{b}^+ \cdot t$. □

For short, we use the notation z_λ instead of $z_{\lambda,1}$ if the weight subspace $L(\overline{\omega}_{l_1})_\lambda$ is one dimensional.

Lemmas 2.205 and 2.206 make it possible to introduce the elements

$$y_i = \sum_{\{(\lambda,j) \mid \lambda(H_{\mathbb{S}}) = -\overline{\omega}_{l_0}(H_{\mathbb{S}}) + 2i\}} z_{\lambda,j} \cdot z_{\lambda,j}^*, \qquad i = 1, \ldots, r. \tag{2.146}$$

Obviously, $y_i = y_i^*$ for each $i = 1, 2, \ldots, r$. Put $y_0 = 1$.

PROPOSITION 2.207. (1) *The elements $y_1, y_2, \ldots, y_r$ of $\mathrm{Pol}(\mathfrak{p}^-)_q$ are $U_q\mathfrak{k}$-invariant and do not depend on the choice of the orthonormal basis $\left\{v_{\lambda,j}^{\overline{\omega}_{l_1}}\right\}$ in $L(\overline{\omega}_{l_1})$.*

(2) *$y_iy_j = y_jy_i$ for all $i, j = 1, 2, \ldots, r$.*

PROOF. The first statement follows from the facts that $\mathrm{Pol}(\mathfrak{p}^-)_q$ is a $(U_q\mathfrak{k}, *)$-module algebra, the Hermitian form in $L(\overline{\omega}_{l_1})$ is $U_q\mathfrak{k}$-invariant, and the basis $\left\{v_{\lambda,j}^{\overline{\omega}_{l_1}}\right\}$ is orthonormal.

The second statement is due to the $U_q\mathfrak{k}$-invariance of y_i, together with the following result obtained by Shklyarov. □

LEMMA 2.208. *The subalgebra of $U_q\mathfrak{k}$-invariants in* $\mathrm{Pol}(\mathfrak{p}^-)_q$ *is commutative.*

PROOF. Consider the $U_q\mathfrak{k}$-submodule $\mathbb{C}[\mathfrak{p}^-]_q f_0 \subset \mathcal{D}(\mathbb{D})_q$. Using the quantum version of the Hua–Schmid theorem Proposition 2.228, one can prove that, as a $U_q\mathfrak{k}$-module,

$$\mathbb{C}[\mathfrak{p}^-]_q f_0 \approx \bigoplus_{\lambda\in\mathcal{A}_+} L(\mathfrak{k},\lambda),$$

where $\mathcal{A}_+ \subset P_+^{\mathbb{S}}$ is introduced in Subsection 2.5.10. Hence each endomorphism of the $U_q\mathfrak{k}$-module $\mathbb{C}[\mathfrak{p}^-]_q$ acts as a scalar multiplication in each $U_q\mathfrak{k}$-isotypic components of $\mathbb{C}[\mathfrak{p}^-]_q$. Therefore the algebra of endomorphisms of the $U_q\mathfrak{k}$-module $\mathbb{C}[\mathfrak{p}^-]_q$ is commutative.

In particular, the operators of multiplication by φ, ψ

$$f \mapsto \varphi f, \qquad f \mapsto \psi f,$$

commute if $\varphi, \psi \in \mathrm{Pol}(\mathfrak{p}^-)_q$ are $U_q\mathfrak{k}$-invariant. This means that the operators of the Fock representation $T_F(\varphi)$, $T_F(\psi)$ of $\mathrm{Pol}(\mathfrak{p}^-)_q$ in the pre-Hilbert space $\mathcal{H} \cong \mathbb{C}[\mathfrak{p}^-]_q f_0$ commute; see Subsection 2.4.3. The result then follows from Lemma 2.135 on the faithfulness of the Fock representation. □

REMARK 2.209. In the category of $U_q\mathfrak{k}$-modules,

$$\mathcal{H} = \bigoplus_{\lambda\in\mathcal{A}_+} \mathcal{H}_\lambda, \qquad \mathcal{H}_\lambda \approx L(\mathfrak{k},\lambda),$$

and each $U_q\mathfrak{k}$-invariant $\psi \in \mathrm{Pol}(\mathfrak{p}^-)_q$ is determined by its Fourier coefficients $\widehat{\psi}(\lambda) = T_F(\psi)|_{\mathcal{H}_\lambda}$; see the proof of Lemma 2.208.

2.5.8. Representations of the $*$-algebra $\mathrm{Pol}(\mathfrak{p}^-)_q$. In this subsection we finish the proof of the existence and uniqueness of a faithful irreducible $*$-representation of $\mathrm{Pol}(\mathfrak{p}^-)_q$ by bounded operators in a Hilbert space.

Use $y_1, y_2, \ldots, y_r$ introduced in the last subsection.

PROPOSITION 2.210. *There exists a unique element y in* $\mathrm{Pol}(\mathfrak{p}^-)_q$ *such that $\mathcal{I}y = x^{-1}$. It is $U_q\mathfrak{k}$-invariant and is given by*

$$y = 1 + \sum_{i=1}^{r} (-1)^i y_i. \tag{2.147}$$

PROOF. The uniqueness of y is obvious. Let us now deduce its existence and (2.147) from (2.48). In fact, (2.48) implies that

$$\sum_{(\lambda,j)} c_{\overline{\omega}_{l_1};\lambda,j}^{\overline{\omega}_{l_1}} \cdot \left(c_{\overline{\omega}_{l_1};\lambda,j}^{\overline{\omega}_{l_1}}\right)^{\star} = 1. \tag{2.148}$$

But

$$\left(c_{\overline{\omega}_{l_1};\lambda,j}^{\overline{\omega}_{l_1}}\right)^{\star} = (-1)^{\frac{1}{2}(\lambda(H_{\mathbb{S}})+\lambda(\overline{\omega}_{l_0}))} \left(c_{\overline{\omega}_{l_1};\lambda,j}^{\overline{\omega}_{l_1}}\right)^{*} \tag{2.149}$$

since

$$t^\star = t^*, \qquad F_j^\star = \begin{cases} F_j^*, & j \neq l_0, \\ -F_j^*, & j = l_0, \end{cases}$$

and

$$(\xi f)^\star = (S(\xi))^\star f^\star, \qquad (\xi f)^* = (S(\xi))^* f^*$$

for each $\xi \in U_q\mathfrak{g}$, $f \in \mathbb{C}[G]_q$. The definition of y_i and (2.148), (2.149) imply that

$$t\mathcal{I}\left(\sum_{i=0}^{r}(-1)^i y_i\right) t^* = 1.$$

Now it remains to multiply both sides of the last equality by t^{-1} and $(t^*)^{-1}$.

Due to (2.147) and Proposition 2.207, we have the $U_q\mathfrak{k}$-invariance of y. □

COROLLARY 2.211. *For each $z \in \mathbb{C}[\mathfrak{p}^-]_{q,1}$, one has*

$$zy = q_{l_0}^{-2} yz, \qquad z^* y = q_{l_0}^2 yz^*. \tag{2.150}$$

PROOF. In the special case $z = z_{\text{low}}$ the commutation relations (2.150) can be easily deduced from the equality $\mathcal{I}y = x^{-1}$ and Corollary 2.48. The general case reduces to the special one via the action of $U_q\mathfrak{k}$. □

Recall the notation T_F for the Fock representation of $\mathrm{Pol}(\mathfrak{p}^-)_q$ in the pre-Hilbert space $\mathcal{H}$ and its completion $\overline{\mathcal{H}}$. As it is proved in Subsection 2.4.6, the operators $T_F(f)$ are bounded, hence T_F admits an extension by continuity to a representation $\overline{T}_F$ in the Hilbert space $\overline{\mathcal{H}}$.

PROPOSITION 2.212. (1) *The representation $\overline{T}_F$ is a faithful irreducible $*$-representation of* $\mathrm{Pol}(\mathfrak{p}^-)_q$ *by bounded operators in a Hilbert space.*

(2) *A representation with these properties is unique up to a unitary equivalence.*

PROOF. 1. The faithfulness of $\overline{T}_F$ follows from the faithfulness of T_F that is proved in Subsection 2.4.1. Now let us turn to the proof of irreducibility.

According to (2.146), the self-adjoint linear operator $\overline{T}_F(y)$ is compact and its spectrum is the closure of a geometric series. More precisely,

$$\overline{T}_F(y)|_{\mathcal{H}_k} = q_{l_0}^{2k} \cdot 1, \qquad k \in \mathbb{Z}_+, \tag{2.151}$$

with $\mathcal{H}_k = \mathbb{C}[\mathfrak{p}^-]_{q,k} \cdot v_0$.

Let $\mathfrak{A}$ be the closure of the algebra of operators $\left\{\overline{T}_F(f) \middle|\, f \in \mathrm{Pol}(\mathfrak{p}^-)_q\right\}$ with respect to the operator norm. By (2.151), the algebra $\mathfrak{A}$ contains the orthogonal projections onto all the subspaces $\mathcal{H}_k$. These subspaces are pairwise orthogonal, finite-dimensional, and their sum is dense in $\overline{\mathcal{H}}$. Hence Lemma 2.134 implies that $\mathfrak{A}$ contains all compact linear operators in $\overline{\mathcal{H}}$. This leads to the irreducibility of $\overline{T}_F$.

Obviously, $\overline{T}_F$ is a $*$-representation.

2. Let T' be a faithful irreducible $*$-representation of $\mathrm{Pol}(\mathfrak{p}^-)_q$ by bounded linear operators in a Hilbert space. Let us now show that T' and $\overline{T}_F$ are unitary equivalent.

The faithfulness of T' implies that $T'(y) \neq 0$. This allows us to apply the commutation relations (2.150) to prove the fact that the spectrum of the self-adjoint linear operator $T'(y)$ is the closure of a geometric series; see Subsection 1.2.3.

Let v' be a normalized eigenvector with eigenvalue of maximum modulus. Then (2.150) implies $T'(z)^* v' = 0$ for each $z \in \mathbb{C}[\mathfrak{p}^-]_{q,1}$. Hence $(T'(f)v', v') = (\overline{T}_F(f)v_0, v_0)$ for each $f \in \mathrm{Pol}(\mathfrak{p}^-)_q$. Therefore the map $v_0 \mapsto v'$ admits an extension to an isometric linear operator that intertwines $\overline{T}_F$ and T'. The operator is surjective since T' is irreducible. □

REMARK 2.213. In the proof of uniqueness we only use the fact that $T'(y) \neq 0$, not the faithfulness of T'. This observation has an interesting application. Let $J = J^*$ be a nonzero two-sided ideal of $\mathrm{Pol}(\mathfrak{p}^-)_q$. If the quotient algebra $\mathrm{Pol}(\mathfrak{p}^-)_q/J$ has irreducible $*$-representations that separate its points, then $y \in J$. In the special case of the quantum matrix ball and the defining ideal J of its Shilov boundary [**318**, pp. 381–383], this implies the inclusion $(y) \subset J$, which means that the Shilov boundary belongs to the topological boundary of the ball. Here (y) is the two-sided ideal generated by y.

The multiplicative system $y^{\mathbb{Z}_+}$ is an Ore set of the domain $\mathrm{Pol}(\mathfrak{p}^-)_q$; see (2.150) and Corollary 2.203. Consider the localization $\mathrm{Pol}(\mathfrak{p}^-)_{q,y}$ of the algebra $\mathrm{Pol}(\mathfrak{p}^-)_q$ with respect to the set $y^{\mathbb{Z}_+}$. Evidently, $\mathrm{Pol}(\mathfrak{p}^-)_q \hookrightarrow \mathrm{Pol}(\mathfrak{p}^-)_{q,y}$. Moreover, the natural extension of T_F from $\mathrm{Pol}(\mathfrak{p}^-)_q$ to $\mathrm{Pol}(\mathfrak{p}^-)_{q,y}$ is a faithful representation of $\mathrm{Pol}(\mathfrak{p}^-)_{q,y}$ in $\mathcal{H}$. Similarly to the proof of Theorem 2.202, one has

PROPOSITION 2.214. *The natural extension of the canonical embedding*

$$\mathrm{Pol}(\mathfrak{p}^-)_q \hookrightarrow \mathbb{C}[\mathbb{X}_{\mathbb{S}}]_{q,x}$$

to the localization $\mathrm{Pol}(\mathfrak{p}^-)_{q,y}$ *is an embedding of algebras*

$$\mathrm{Pol}(\mathfrak{p}^-)_{q,y} \hookrightarrow \mathbb{C}[\mathbb{X}_{\mathbb{S}}]_{q,x}. \tag{2.152}$$

REMARK 2.215. As in the previous subsections, one can prove that the structure of $(U_q\mathfrak{g}, *)$-module algebra can be canonically extended from $\mathrm{Pol}(\mathfrak{p}^-)_q$ to $\mathrm{Pol}(\mathfrak{p}^-)_{q,y}$. Also, (2.152) is a morphism of $(U_q\mathfrak{g}, *)$-module algebras (we use the fact that the elements $\xi(fy^n)y^{-n}$, $\xi \in U_q\mathfrak{g}$, $f \in \mathrm{Pol}(\mathfrak{p}^-)_q$, are values of the corresponding Laurent polynomials on a geometric series).

We maintain the notation $\mathcal{I}$ for *the canonical embedding* (2.152) and now describe the image of $\mathrm{Pol}(\mathfrak{p}^-)_{q,y}$ under it.

Let K be the connected affine algebraic subgroup of G with Lie algebra $\mathfrak{k}$ (see Subsection 2.3.8), w_0 the longest element of W, and

$$K_1 = w_0 K w_0^{-1}. \tag{2.153}$$

Obviously, if $a, b \in G$, then $w_0 \cdot Ka = w_0 \cdot Kb$ is equivalent to $K_1(w_0 a) = K_1(w_0 b)$.

Solutions of the system of equations

$$L_{\mathrm{reg}}(E_i)f = L_{\mathrm{reg}}(F_i)f = 0, \qquad i \neq l_1,$$
$$L_{\mathrm{reg}}(K_j^{\pm 1})f = f, \qquad j = 1, 2, \ldots, l,$$

form a subalgebra in the $U_q\mathfrak{g}$-module algebra $\mathbb{C}[G]_q$. It can be treated as a quantum analog of the algebra $\mathbb{C}[K_1 \backslash G]$ of regular functions on the affine algebraic variety $K_1 \backslash G$; see Subsection 2.2.5 and Proposition 2.179. Denote by $\mathbb{C}[K_1 \backslash G]_q$ the latter $U_q\mathfrak{g}$-module algebra.

The algebra $\mathbb{C}[K_1 \backslash G]_q$ is generated by $c^{\overline{\omega}_{l_1}}_{\overline{\omega}_{l_1};\lambda,j}(c^{\overline{\omega}_{l_1}}_{\overline{\omega}_{l_1};\mu,k})^\star$; see Subsection 2.2.5. Hence it can be defined as the smallest $U_q\mathfrak{g}$-module subalgebra that contains x. This definition is used implicitly in the sequel.

In Subsection 2.5.6 the algebra $\mathbb{C}[\mathbb{X}_\mathbb{S}]_q$ is endowed with the grading. We now extend it to the localization $\mathbb{C}[\mathbb{X}_\mathbb{S}]_{q,x}$.

LEMMA 2.216. *There exists a unique $\mathbb{Z}$-grading of $\mathbb{C}[\mathbb{X}_\mathbb{S}]_{q,x}$ such that*

$$\deg\left(c^{\overline{\omega}_{l_1}}_{\overline{\omega}_{l_1};\lambda,j}\right)=1,\qquad \deg\left(c^{\overline{\omega}_{l_0}}_{-\overline{\omega}_{l_1};\mu,k}\right)=-1.$$

PROOF. The uniqueness is obvious. A construction of a grading with the required properties can be done as in Subsection 2.5.4. In fact,

$$L_{\mathrm{reg}}(H_{l_1})c^{\overline{\omega}_{l_1}}_{\overline{\omega}_{l_1};\lambda,j}=c^{\overline{\omega}_{l_1}}_{\overline{\omega}_{l_1};\lambda,j},\qquad L_{\mathrm{reg}}(H_{l_1})c^{\overline{\omega}_{l_0}}_{-\overline{\omega}_{l_1};\mu,k}=-c^{\overline{\omega}_{l_0}}_{-\overline{\omega}_{l_1};\mu,k},$$

and the differentiation $L_{\mathrm{reg}}(H_{l_1})$ on $\mathbb{C}[\mathbb{X}_\mathbb{S}]_q$ admits an extension to a differentiation on $\mathbb{C}[\mathbb{X}_\mathbb{S}]_{q,x}$. It remains to set $\deg f=j$ for each $f\in\mathbb{C}[\mathbb{X}_\mathbb{S}]_{q,x}$ such that $L_{\mathrm{reg}}(H_{l_1})f=jf$. □

Note that the localization $\mathbb{C}[K_1\backslash G]_{q,x}$ of $\mathbb{C}[K_1\backslash G]_q$ with respect to the multiplicative system $x^{\mathbb{Z}_+}$ is naturally embedded into $\mathbb{C}[\mathbb{X}_\mathbb{S}]_{q,x}$.

PROPOSITION 2.217. $\mathcal{I}\,\mathrm{Pol}(\mathfrak{p}^-)_{q,y}=\mathbb{C}[K_1\backslash G]_{q,x}$.

PROOF. The embedding of $U_q\mathfrak{g}$-module algebras $\mathcal{I}\,\mathrm{Pol}(\mathfrak{p}^-)_{q,y}\subset\mathbb{C}[K_1\backslash G]_{q,x}$ is obvious. To prove the opposite inclusion, consider the subspace

$$\left(\bigoplus_{j=1}^{\infty}t^j\cdot\mathcal{I}\,\mathrm{Pol}(\mathfrak{p}^-)_{q,y}\right)\bigoplus\mathcal{I}\,\mathrm{Pol}(\mathfrak{p}^-)_{q,y}\bigoplus\left(\bigoplus_{j=1}^{\infty}\mathcal{I}\,\mathrm{Pol}(\mathfrak{p}^-)_{q,y}\cdot t^{*j}\right).$$

It is a $U_q\mathfrak{g}$-module subalgebra; it contains t, t^*, x^{-1}. Hence it coincides with $\mathbb{C}[\mathbb{X}_\mathbb{S}]_{q,x}$. Therefore $\mathcal{I}\,\mathrm{Pol}(\mathfrak{p}^-)_{q,y}$ contains all elements of degree zero from $\mathbb{C}[\mathbb{X}_\mathbb{S}]_{q,x}$. It remains now just to use the fact that nonzero elements of $\mathbb{C}[K_1\backslash G]_{q,x}$ have zero degree. □

2.5.9. Appendix. A cone over $\partial\mathbb{D}$. Consider the two-sided ideal (y) of the algebra $\mathrm{Pol}(\mathfrak{p}^-)_q$ generated by $y=y^*$ and the corresponding quotient algebra

$$\mathbb{C}[\partial\mathbb{D}]_q=\mathrm{Pol}(\mathfrak{p}^-)_q/(y).$$

It is a q-analog for the algebra of regular functions on the affine algebraic variety $\partial\mathbb{D}$ which is the topological boundary of $\mathbb{D}$. It is important to note that the two-sided ideal (y) is a $U_q\mathfrak{g}$-module subalgebra of the $U_q\mathfrak{g}$-module algebra $\mathrm{Pol}(\mathfrak{p}^-)_q$. In fact,

$$E_jy=F_jy=0,\quad j\neq l_0,\qquad K_i^{\pm1}y=y,\quad i=1,2,\ldots,l,$$

and it remains to prove that $E_{l_0}y\in(y)$, $F_{l_0}y\in(y)$. The last claim can be verified easily by using the equality $\mathcal{I},y=t^{-1}(t^*)^{-1}$. Hence $\mathbb{C}[\partial\mathbb{D}]_q$ is a $(U_q\mathfrak{g},*)$-module algebra.

Consider the auxiliary $*$-subalgebra of $\mathbb{C}[\mathbb{X}]_{q,x}$ generated by the set $\mathcal{I}\mathbb{C}[\mathfrak{p}^-]_{q,1}$ and t. It is a $(U_q\mathfrak{g},*)$-module subalgebra since

$$E_jt=F_jt=(K_j^{\pm1}-1)t=0,\qquad j\neq l_0,$$
$$F_{l_0}t=(K_{l_0}^{\pm1}-1)t=0,\qquad E_{l_0}t=q_{l_0}^{-1/2}\,t\,(\mathcal{I}z_{\mathrm{low}}).$$

One can describe the latter $*$-algebra by its generators t, $z \in \mathbb{C}[\mathfrak{p}^-]_{q,1}$ and the relations

$$tt^* = t^*t, \qquad tt^*y = ytt^* = 1, \tag{2.154}$$

$$tz = q_{l_0}^{-1}zt, \qquad t^*z = q_{l_0}^{-1}zt, \qquad\qquad z \in \mathbb{C}[\mathfrak{p}^-]_{q,1}. \tag{2.155}$$

Moreover, the set of relations should include the defining relations of the $*$-algebra $\mathrm{Pol}(\mathfrak{p}^-)_q$.

To obtain (2.154) and (2.155), one can identify an element $z \in \mathbb{C}[\mathfrak{p}^-]_{q,1}$ with its image under the canonical embedding $\mathcal{I}$.

The action of $U_q\mathfrak{g}$ is given as before:

$$E_jt = F_jt = (K_j^{\pm 1} - 1)t = 0, \qquad j \neq l_0,$$
$$F_{l_0}t = (K_{l_0}^{\pm 1} - 1)t = 0, \qquad E_{l_0}t = q_{l_0}^{-1/2}t(\mathcal{I}z_{\mathrm{low}}).$$

2.5.10. Appendix on spherical $(\mathfrak{g}, \mathfrak{k})$-modules and the Hua–Schmidt theorem. In this Subsection we recall well-known and regularly used results from the theory of Hermitian symmetric spaces and bounded symmetric domains [**274**], [**304**].

Consider the Hermitian symmetric pair $(\mathfrak{g}, \mathfrak{k})$ corresponding to the set $\mathbb{S} = \{1, 2, \ldots, l\} \setminus \{l_0\}$, and $H_{\mathbb{S}} \in \mathfrak{h}$; see Subsections 2.3.4 and 2.3.1.

The set $\Phi_c = \{\beta \in \Phi | \, \beta(H_{\mathbb{S}}) = 0\}$ is called a set of compact roots, while the set $\Phi_n = \Phi \setminus \Phi_c$ is called a set of noncompact roots . The compact roots form a root system of the semisimple complex Lie algebra $\mathfrak{k}_{ss} \subset \mathfrak{k}$ generated by $\{E_i, F_i, H_i\}_{i \neq l_0}$. The Weyl group $W_{\mathbb{S}}$ of the root system is naturally embedded into W and acts on the set of positive noncompact roots.

Two linearly independent roots γ', γ'' are called strongly orthogonal if neither $\gamma' + \gamma''$ nor $\gamma' - \gamma''$ are roots. In this case γ' and γ'' are orthogonal, as can be proved easily by considering the set of roots $\{\gamma'' + j\gamma' \in \Phi \, | \, j \in \mathbb{Z}\}$; see [**44**, pp. 185–186].

Following [**89**, p. 572], we introduce a family of simple complex Lie algebras $\mathfrak{g}_1, \mathfrak{g}_2, \ldots, \mathfrak{g}_r$ (up to an isomorphism) with root systems embedded into Φ. Their maximal roots $\{\gamma_1, \gamma_2, \ldots, \gamma_r\}$ will be strongly orthogonal positive noncompact roots of $\mathfrak{g}$ and will play an important role in the theory of Hermitian symmetric spaces.

Consider the Dynkin diagram of $\mathfrak{g}$ and the node corresponding to the simple root α_{l_0}. The Lie algebras $\mathfrak{g}_1, \mathfrak{g}_2, \ldots, \mathfrak{g}_r$ should satisfy the following conditions: (i) their systems of simple roots are embedded into the system of simple roots of $\mathfrak{g}$; (ii) their systems of simple roots contain α_{l_0}.

The construction of these Lie algebras is given by recursion. The first one is $\mathfrak{g}_1 = \mathfrak{g}$. To pass from $\mathfrak{g}_n$ to $\mathfrak{g}_{n+1}$, consider the subgraph of the Dynkin diagram of $\mathfrak{g}_n$ that consists of roots γ such that γ is orthogonal to the maximal root γ_n of $\mathfrak{g}_n$. If this subgraph does not include the α_{l_0} node, the construction is finished. Otherwise there is a graph component that contains the α_{l_0} node. It is a Dynkin diagram and defines a simple complex Lie algebra $\mathfrak{g}_{n+1} \subset \mathfrak{g}_n$ whose root system can be embedded into the root system of $\mathfrak{g}_n$. As a result, we get a linearly ordered set of positive noncompact strongly orthogonal roots $\{\gamma_1, \gamma_2, \ldots, \gamma_r\}$, which is used in [**134**, p. 82], [**242**, pp. 3776–3777], [**304**, p. 443].

In this construction r is called the rank of the irreducible bounded symmetric domain related to the Hermitian symmetric pair $(\mathfrak{g}, \mathfrak{k})$.

EXAMPLE 2.218. The Lie algebras $\mathfrak{g}_1, \mathfrak{g}_2, \ldots, \mathfrak{g}_r$ can be easily found by using an explicit form of the expansion of the maximal root δ as a linear combination of fundamental weights $\overline{\omega}_1, \overline{\omega}_2, \ldots, \overline{\omega}_l$; see tables in [**44**]. For example, for the exceptional Lie algebras E_6, E_7 one has

$$E_6, \quad A_5; \qquad\qquad E_7, \quad D_6, \quad A_1.$$

In the first case the rank is 2, in the second it is 3.

REMARK 2.219. In [**274**, p. 334] the roots $\{\gamma_1, \gamma_2, \ldots, \gamma_r\}$ appear in the reverse order. Many authors [**114, 197, 266, 142, 121**] prefer the following system of strongly orthogonal noncompact positive roots $\{w_{0,\mathbb{S}}\gamma_1, \ w_{0,\mathbb{S}}\gamma_2, \ \ldots, \ w_{0,\mathbb{S}}\gamma_r\}$, where $w_{0,\mathbb{S}}$ is the longest element of the Weyl group $W_{\mathbb{S}}$.

Let us use the nondegenerate scalar product in $\mathfrak{h}^*$ chosen in Subsection 2.1.3 to identify $\mathfrak{h}^*$ with $\mathfrak{h}$. Each root $\beta \in \Phi$ is a linear functional on $\mathfrak{h}^*$, so it corresponds to an element H_β of the Cartan subalgebra $\mathfrak{h}$ under the isomorphism $\mathfrak{h} \cong \mathfrak{h}^*$. As mentioned in Subsection 2.1.6,

$$H_{\alpha_i} = d_i H_i, \qquad i = 1, 2, \ldots, l.$$

Following [**274**], introduce the linear span $\mathfrak{h}^-$ of $H_{\gamma_1}, H_{\gamma_2}, \ldots, H_{\gamma_r}$ and the subspace

$$\mathfrak{h}^+ = \bigcap_{j=1} \operatorname{Ker} \gamma_j.$$

It is easy to prove that $\mathfrak{h}^+$ is orthogonal to $\mathfrak{h}^-$ in $\mathfrak{h}$ with respect to the invariant scalar product in $\mathfrak{g}$, and

$$\mathfrak{h} = \mathfrak{h}^+ \oplus \mathfrak{h}^-.$$

EXAMPLE 2.220. Let $N = m+n$ and $n \geq m$. Consider the Hermitian symmetric pair $(\mathfrak{g}, \mathfrak{k})$:

$$\mathfrak{g} = \mathfrak{sl}_N, \qquad \mathfrak{k} = \mathfrak{s}(\mathfrak{gl}_n \times \mathfrak{gl}_m).$$

The positive noncompact roots are

$$\varepsilon_i - \varepsilon_j, \qquad i \in \{1, 2, \ldots, n\}, \ j \in \{n+1, n+2, \ldots, N\},$$

where

$$\varepsilon_i(H_k) = \begin{cases} 1, & i = k, \\ -1, & i = k+1, \\ 0, & i \notin \{k, k+1\}. \end{cases}$$

The group W acts in $\mathfrak{h}^*$ by permutations of ε_i. Obviously, $r = m$ and

$$\gamma_j = \varepsilon_j - \varepsilon_{N+1-j} = \sum_{i=j}^{N-j} \alpha_i, \qquad j = 1, 2, \ldots, m,$$

$$H_{\alpha_i} = H_i, \qquad H_{\gamma_j} = \sum_{i=j}^{N-j} H_i.$$

The subspace $\mathfrak{h}^-$ is spanned by

$$H_j + H_{N-j}, \quad j = 1, 2, \ldots, m-1; \qquad \sum_{i=m}^{N-m} H_i.$$

Consider the restriction of a linear functional on $\mathfrak{h}^-$:

$$\mathfrak{h}^* \mapsto (\mathfrak{h}^-)^*, \qquad f \mapsto f|_{\mathfrak{h}^-}. \tag{2.156}$$

The set of all nonzero elements in the image of Φ under the map (2.156) will be denoted by Φ_{res}, and the number of elements in the preimage of an element will be called the multiplicity of the element. It is known that the set $\Phi_{\mathrm{res}} \subset (\mathfrak{h}^-)^*$ is a root system. Its elements are called restricted roots, and its Weyl group W_{res} is called the restricted Weyl group. Nonzero restrictions of positive roots on $\mathfrak{h}^-$ span the set Φ^+_{res}, which is called the set of restricted positive roots.

REMARK 2.221. The maximal Abelian Lie subalgebra $\mathfrak{a}_{\mathbb{C}} \subset \mathfrak{p}^+ \oplus \mathfrak{p}^-$is widely used in representation theory and can be obtained from $\mathfrak{h}^-$ via the Caley transform [**197**, p. 212], [**274**, p. 334].

We identify the linear functionals $\{\gamma_1, \gamma_2, \dots, \gamma_r\}$ with their restrictions to $\mathfrak{h}^-$. It will not lead to ambiguities since the restrictions are linearly independent [**114**].

PROPOSITION 2.222. [**197**, p. 212] *The group W_{res} is generated by permutations of the basis elements $\{\gamma_1, \gamma_2, \dots, \gamma_r\}$ and multiplications of these by ± 1. The set Φ_{res} consists of the roots $\pm\gamma_j$ with multiplicity 1, the roots $\pm\frac{1}{2}\gamma_i \pm \frac{1}{2}\gamma_j$ with multiplicity a, and, perhaps, the roots $\pm\frac{1}{2}\gamma_j$ with even multiplicity $2b$. If $\frac{1}{2}\gamma_i \in \Phi_{\mathrm{res}}$, then Φ_{res} is a nonreduced root system of type BC; otherwise it is a reduced root system of type C.*

It is not difficult to describe the restrictions of the simple compact roots on $\mathfrak{h}^-$ [**134**, p.83]:

$$\{\tfrac{1}{2}(\gamma_1 - \gamma_2), \tfrac{1}{2}(\gamma_2 - \gamma_3), \dots, \tfrac{1}{2}(\gamma_{r-1} - \gamma_r), \tfrac{1}{2}\gamma_r\}, \qquad \text{for } BC,$$
$$\{\tfrac{1}{2}(\gamma_1 - \gamma_2), \tfrac{1}{2}(\gamma_2 - \gamma_3), \dots, \tfrac{1}{2}(\gamma_{r-1} - \gamma_r)\}, \qquad \text{for } C.$$

REMARK 2.223. Using the classification of irreducible bounded symmetric domains [**123**], one can show that such a domain (up to an isomorphism) is determined by the triple of numbers r, a, b [**197**, p. 213], [**9**, p. 16]. The number $p = (r-1)a + b + 2$ is called the genus of irreducible bounded symmetric domain $\mathbb{D}$ [**9**, p.16], [**197**, p. 239].

In the sequel we use the same notation for $\mathfrak{g}$-modules and $\mathfrak{k}$-modules as for $U_q\mathfrak{g}$-modules and $U_q\mathfrak{k}$-modules, just omitting the index q.

A weight $\lambda \in P_+$ is called spherical if the simple finite-dimensional $\mathfrak{g}$-module $L(\lambda)$ with highest weight λ is spherical, i.e., if it contains a nonzero $\mathfrak{k}$-invariant vector. In this case the subspace of $\mathfrak{k}$-invariant vectors is one dimensional [**123**, Sec. 10.4], and hence the corresponding matrix elements $\varphi_\lambda(g)$ are well defined. They are called *zonal spherical functions*.

Recall a well-known description of the set P_+^{spher} of spherical weights and the Abelian subgroup $P^{\mathrm{spher}} \subset P$ generated by P_+^{spher}.

PROPOSITION 2.224. [**124**, Sec. V.4.1] *A simple finite-dimensional $\mathfrak{g}$-module is spherical if and only if its highest weight λ satisfies the following conditions:*

(1) $\lambda|_{\mathfrak{h}^+} = 0$,

(2) $\dfrac{(\lambda, \gamma)}{(\gamma, \gamma)} \in \mathbb{Z}$ *for each restricted root* γ.

EXAMPLE 2.225. Let $n > 1$. Consider the Hermitian symmetric pair $(\mathfrak{sl}_{n+1}, \mathfrak{s}(\mathfrak{gl}_n \times \mathfrak{gl}_1))$. In this case $r = 1$,

$$\gamma_1 = \overline{\omega}_1 + \overline{\omega}_n = \sum_{i=1}^{n} \alpha_i, \qquad H_{\gamma_1} = \sum_{i=1}^{n} H_i = e_{1,1} - e_{n+1,n+1},$$

where $\{e_{i,j}\}$ is the standard basis of the vector space of matrices. Hence the restricted root system is $\Phi_{\mathrm{res}} = \left\{-\gamma_1, -\frac{1}{2}\gamma_1, \frac{1}{2}\gamma_1, \gamma_1\right\}$, and $\mathfrak{h}^+$ is spanned by

$$H_1 - H_n, \qquad H_j, \quad j = 2, 3, \ldots, n-1.$$

The conditions $\lambda \in P_+$, $\lambda|_{\mathfrak{h}^+} = 0$ mean that $\lambda \in \mathbb{Z}_+(\overline{\omega}_1 + \overline{\omega}_n)$. The second condition holds for all such λ since

$$(\gamma_1, \gamma_1) = 2, \qquad (\overline{\omega}_1 + \overline{\omega}_n, \gamma_1) = \left(\overline{\omega}_1 + \overline{\omega}_n, \sum_{i=1}^{n} \alpha_i\right) = 2,$$

which follows from (2.7) and (2.8).

It is known that

$$P_+^{\mathrm{spher}} = \bigoplus_{i=1}^{r} \mathbb{Z}_+ \mu_i, \qquad P^{\mathrm{spher}} = \bigoplus_{i=1}^{r} \mathbb{Z} \mu_i,$$

where $\mu_1, \mu_2, \ldots, \mu_r$ are the fundamental spherical weights; see [**126**, p. 6]. Their explicit form is given in, e.g., [**212**, p. 478].

Let us endow P_+^{spher} with the partial order by setting $\sum_{j=1}^{r} m_j' \mu_j \preceq \sum_{j=1}^{r} m_j'' \mu_j$ if $m_j' \le m_j''$ for each $j = 1, 2, \ldots, r$. It is easy to show that

$$\varphi_{\lambda'}(g) \varphi_{\lambda''}(g) = \sum_{\nu \preceq \lambda' + \lambda''} c_\nu(\lambda', \lambda'') \, \varphi_\nu(g),$$

where $c_\nu(\lambda', \lambda'')$ are nonnegative numbers and $c_{\lambda'+\lambda''}(\lambda', \lambda'') \neq 0$; see [**331**], [**323**, p. 297].

Consider the algebra $\mathrm{Pol}(\mathfrak{p}^-)$ of polynomials on the real vector space $\mathfrak{p}^-$ and the subalgebra $\mathrm{Pol}(\mathfrak{p}^-)^{\mathfrak{k}}$ of $\mathfrak{k}$-invariants. One can easily obtain a basis of $\mathrm{Pol}(\mathfrak{p}^-)^{\mathfrak{k}}$ as a vector space using the matrix elements of the finite-dimensional spherical $\mathfrak{g}$-modules $L(\lambda)$ corresponding to their $\mathfrak{k}$-invariant vectors. The passage to polynomials on $\mathfrak{p}^-$ can be done via the canonical embedding $\mathbb{C}[\mathfrak{p}^-] \hookrightarrow \mathbb{C}[G]_t$. The next result provides a description of $\mathrm{Pol}(\mathfrak{p}^-)^{\mathfrak{k}}$.

PROPOSITION 2.226. [**142**, p. 74] *There exist r homogeneous polynomials $P_1, P_2, \ldots, P_r \in \mathrm{Pol}(\mathfrak{p}^-)^{\mathfrak{k}}$ such that*

(1) $\deg P_i = 2i$, *$i = 1, 2, \ldots, r$;*
(2) $\mathrm{Pol}(\mathfrak{p}^-)^{\mathfrak{k}}$ *is a free commutative algebra with the generators $P_1, P_2, \ldots, P_r$.*

A similar result holds for the Lie subalgebra $\mathfrak{n}_{\mathfrak{k}}^+ \subset \mathfrak{k}$ generated by E_i, $i \in \mathbb{S}$. Let $\mathbb{C}[\mathfrak{p}^-]^{\mathrm{inv}}$ be the subalgebra of $\mathfrak{n}_{\mathfrak{k}}^+$-invariants in $\mathbb{C}[\mathfrak{p}^-]$.

PROPOSITION 2.227. [**142**, p. 73] *There exist r homogeneous polynomials $p_1, p_2, \ldots, p_r \in \mathbb{C}[\mathfrak{p}^-]^{\mathrm{inv}}$ such that*

(1) $\deg p_i = i$;
(2) *the weight of p_i equals $\sum_{k=1}^{i} \gamma_k$;*
(3) $\mathbb{C}[\mathfrak{p}^-]^{\mathrm{inv}}$ *is a free commutative algebra with the generators $p_1, p_2, \ldots, p_r$;*

(4) *the generators* $p_1, p_2, \ldots, p_r$ *of* $\mathbb{C}[\mathfrak{p}^-]^{\mathrm{inv}}$ *are determined by the conditions* (1)–(3) *(up to a scalar factor).*

In Propositions 2.226 and 2.227 r denotes the rank of $\mathbb{D}$.

The next result is called the Hua-Schmidt theorem. It follows from Proposition 2.227.

PROPOSITION 2.228. [**142**, p. 73], [**304**, p. 443] *We have*

$$\mathbb{C}[\mathfrak{p}^-] \approx \bigoplus_{\lambda \in \mathcal{A}_+} L(\mathfrak{k}, \lambda)$$

in the category of $\mathfrak{k}$*-modules, where*

$$\mathcal{A}_+ = \bigoplus_{j=1}^{r} \mathbb{Z}_+ \left(\sum_{i=1}^{j} \gamma_i \right). \tag{2.157}$$

COROLLARY 2.229. $\mathbb{C}[\mathfrak{p}^-]$ *is a multiplicity free* $\mathfrak{k}$*-module.*

2.6. Covariant differential calculi and invariant differential operators

2.6.1. First-order differential calculi. Let us introduce necessary definitions as in [**174**]. Consider an algebra F which will be assumed unital unless the contrary is stated explicitly.

A first-order differential calculus over F is a pair (M, d) where M is an F-bimodule and $d : F \to M$ is a linear map such that

(1) for all $f_1, f_2 \in F$ one has

$$d(f_1 \cdot f_2) = df_1 \cdot f_2 + f_1 \cdot df_2; \tag{2.158}$$

(2) M is a linear span of the vectors $f_1 \cdot df_2 \cdot f_3$, where $f_1, f_2, f_3 \in F$.

It follows from (2.158) that $d1 = 0$. The second axiom is equivalent to the assumption that M is a linear span of $f_1 \cdot df_2$, where $f_1, f_2 \in F$.

Let A be a Hopf algebra, and suppose that F is an A-module algebra. A first-order differential calculus (M, d) over F is said to be covariant if M is an A-covariant F-bimodule and d is a morphism of A-modules.

We use the abbreviation CFODC to denote a *covariant first order differential calculus.* Two CFODC (M', d') and (M'', d'') are said to be isomorphic if there exists an isomorphism of A-covariant F-bimodules $i : M' \to M''$ such that $d'' = i \cdot d'$.

We restrict ourselves to the Hermitian symmetric case and produce a CFODC over the $U_q\mathfrak{g}$-module algebra $\mathbb{C}[\mathfrak{p}^-]_q$.

It is well known [**26**, p. 82] that in the classical case $q = 1$ the differential is the dual map for the morphism of generalized Verma modules

$$d^* : N(\mathfrak{q}^+, -\alpha_{l_0}) \to N(\mathfrak{q}^+, 0), \qquad d^* : v(\mathfrak{q}^+, -\alpha_{l_0}) \mapsto F_{l_0} v(\mathfrak{q}^+, 0), \tag{2.159}$$

in the category of graded $U\mathfrak{g}$-modules. Let us turn to the quantum universal enveloping algebra.

Let d^* be the morphism of $U_q\mathfrak{g}$-modules defined by the same formula (2.159) as in the classical case. The existence and uniqueness of such a morphism is obvious; see Subsection 2.3.1. Recall that the vector space $N(\mathfrak{q}^+, -\alpha_{l_0})$ is graded,

$N(\mathfrak{q}^+, -\alpha_{l_0}) = \bigoplus_{j=1}^{\infty} N(\mathfrak{q}^+, -\alpha_{l_0})_{-j}$. Introduce the dual graded vector space

$$\Lambda^1(\mathfrak{p}^-)_q = \bigoplus_{j=1}^{\infty} \Lambda^1(\mathfrak{p}^-)_{q,j}, \qquad \Lambda^1(\mathfrak{p}^-)_{q,j} = N(\mathfrak{q}^+, -\alpha_{l_0})^*_{-j},$$

together with the dual linear map

$$d : \mathbb{C}[\mathfrak{p}^-]_q \to \Lambda^1(\mathfrak{p}^-)_q.$$

We will show that $\Lambda^1(\mathfrak{p}^-)_q$ is a $U_q\mathfrak{g}$-covariant $\mathbb{C}[\mathfrak{p}^-]_q$-bimodule.

Let $\lambda \in P_+^{\mathbb{S}}$. Consider the morphisms

$$\Delta^+_{\mathrm{left},\lambda} : N(\mathfrak{q}^+, \lambda) \to N(\mathfrak{q}^+, 0) \otimes N(\mathfrak{q}^+, \lambda),$$
$$\Delta^+_{\mathrm{right},\lambda} : N(\mathfrak{q}^+, \lambda) \to N(\mathfrak{q}^+, \lambda) \otimes N(\mathfrak{q}^+, 0)$$

in the category of $U_q\mathfrak{g}^{\mathrm{cop}}$-modules defined by their actions on the generators

$$\Delta^+_{\mathrm{left},\lambda} : v(\mathfrak{q}^+, \lambda) \mapsto v(\mathfrak{q}^+, 0) \otimes v(\mathfrak{q}^+, \lambda),$$
$$\Delta^+_{\mathrm{right},\lambda} : v(\mathfrak{q}^+, \lambda) \mapsto v(\mathfrak{q}^+, \lambda) \otimes v(\mathfrak{q}^+, 0).$$

The following relations can be deduced from the definitions:

$$(\mathrm{id} \otimes \Delta^+_{\mathrm{left},\lambda})\Delta^+_{\mathrm{left},\lambda} = (\Delta^+ \otimes \mathrm{id})\Delta^+_{\mathrm{left},\lambda}, \tag{2.160}$$

$$(\Delta^+_{\mathrm{right},\lambda} \otimes \mathrm{id})\Delta^+_{\mathrm{right},\lambda} = (\mathrm{id} \otimes \Delta^+)\Delta^+_{\mathrm{right},\lambda}, \tag{2.161}$$

$$(\varepsilon^+ \otimes \mathrm{id})\Delta^+_{\mathrm{left},\lambda} = (\mathrm{id} \otimes \varepsilon^+)\Delta^+_{\mathrm{right},\lambda} = \mathrm{id}, \tag{2.162}$$

$$(\mathrm{id} \otimes \Delta^+_{\mathrm{right},\lambda})\Delta^+_{\mathrm{left},\lambda} = (\Delta^+_{\mathrm{left},\lambda} \otimes \mathrm{id})\Delta^+_{\mathrm{right},\lambda} \tag{2.163}$$

(the coproduct Δ^+ and the counit ε^+ in $N(\mathfrak{q}^+, 0)$ were introduced in Subsection 2.3.2). For example, (2.163) follows from

$$(\mathrm{id} \otimes \Delta^+_{\mathrm{right},\lambda})\Delta^+_{\mathrm{left},\lambda} v(\mathfrak{q}^+, \lambda) = v(\mathfrak{q}^+, 0) \otimes v(\mathfrak{q}^+, \lambda) \otimes v(\mathfrak{q}^+, 0),$$
$$(\Delta^+_{\mathrm{left},\lambda} \otimes \mathrm{id})\Delta^+_{\mathrm{right},\lambda} v(\mathfrak{q}^+, \lambda) = v(\mathfrak{q}^+, 0) \otimes v(\mathfrak{q}^+, \lambda) \otimes v(\mathfrak{q}^+, 0).$$

Consider the category of graded vector spaces and the space $\Gamma(\mathfrak{p}^-, \lambda)_q$ dual to $N(\mathfrak{q}^+, \lambda)$ in this category. We endow it with the structure of a $U_q\mathfrak{g}$-module, together with left and right actions of $\mathbb{C}[\mathfrak{p}^-]_q$ given by

$$m_{\mathrm{left},\lambda} \stackrel{\mathrm{def}}{=} (\Delta^+_{\mathrm{left},\lambda})^*, \qquad m_{\mathrm{right},\lambda} \stackrel{\mathrm{def}}{=} (\Delta^+_{\mathrm{right},\lambda})^*.$$

From (2.160)–(2.163) we deduce the following result.

Proposition 2.230. *For each $\lambda \in P_+^{\mathbb{S}}$, the graded vector space $\Gamma(\mathfrak{p}^-, \lambda)_q$ is a $U_q\mathfrak{g}$-covariant $\mathbb{C}[\mathfrak{p}^-]_q$-bimodule.*

Remark 2.231. Evidently,

$$\Gamma(\mathfrak{p}^-, \lambda)_q \cong \left\{ f \in \mathrm{Hom}_{U_q\mathfrak{q}^+}\left(U_q\mathfrak{g}, L(\mathfrak{q}^+, -w_0\lambda)\right) \Big| \, \dim(U_q\mathfrak{h} \cdot f) < \infty \right\}.$$

Let $N(\mathfrak{q}^+, \lambda)_{\mathrm{highest}}$ be the highest homogeneous component of the graded vector space $N(\mathfrak{q}^+, \lambda)$ and P_{highest} the projection of $N(\mathfrak{q}^+, \lambda)$ onto the subspace $N(\mathfrak{q}^+, \lambda)_{\mathrm{highest}}$ parallel to the sum of the remaining homogeneous components.

LEMMA 2.232. *The linear maps*

$$(\mathrm{id}\otimes P_{\mathrm{highest}})\Delta^+_{\mathrm{left},\lambda}: N(\mathfrak{q}^+,\lambda)\to N(\mathfrak{q}^+,0)\otimes N(\mathfrak{q}^+,\lambda)_{\mathrm{highest}},$$
$$(P_{\mathrm{highest}}\otimes \mathrm{id})\Delta^+_{\mathrm{right},\lambda}: N(\mathfrak{q}^+,\lambda)\to N(\mathfrak{q}^+,\lambda)_{\mathrm{highest}}\otimes N(\mathfrak{q}^+,0)$$

are injective.

PROOF. We keep the notation of Proposition 2.76. The injectivity of the linear maps in question follows from that proposition and the isomorphism $N(\mathfrak{q}^+,\lambda)_{\mathrm{highest}}\cong L(\mathfrak{q}^+,\lambda)$, together with the additional observation that for each weight vector $v\in N(\mathfrak{q}^+,\lambda)_{\mathrm{highest}}$ one has

$$(\mathrm{id}\otimes P_{\mathrm{highest}})\Delta^+_{\mathrm{left},\lambda}\left(F^{j_M}_{\beta_M}F^{j_{M-1}}_{\beta_{M-1}}\cdots F^{j_{M'+1}}_{\beta_{M'+1}}v\right)$$
$$=\mathrm{const}'\cdot F^{j_M}_{\beta_M}F^{j_{M-1}}_{\beta_{M-1}}\cdots F^{j_{M'+1}}_{\beta_{M'+1}}v(\mathfrak{q}^+,0)\otimes v,$$

$$(P_{\mathrm{highest}}\otimes \mathrm{id})\Delta^+_{\mathrm{right},\lambda}\left(F^{j_M}_{\beta_M}F^{j_{M-1}}_{\beta_{M-1}}\cdots F^{j_{M'+1}}_{\beta_{M'+1}}v\right)$$
$$=\mathrm{const}''\cdot v\otimes F^{j_M}_{\beta_M}F^{j_{M-1}}_{\beta_{M-1}}\cdots F^{j_{M'+1}}_{\beta_{M'+1}}v(\mathfrak{q}^+,0),$$

where the scalar factors const′, const″ are nonzero. □

PROPOSITION 2.233. (1) *The $\mathbb{C}[\mathfrak{p}^-]_q$-bimodule $\Gamma(\mathfrak{p}^-,\lambda)_q$ is both a free left and a free right $\mathbb{C}[\mathfrak{p}^-]_q$-module.*
(2) *Let $\Gamma(\mathfrak{p}^-,\lambda)_{q,\mathrm{lowest}}$ be the lowest homogeneous component of $\Gamma(\mathfrak{p}^-,\lambda)_q$. Then the maps*

$$\mathbb{C}[\mathfrak{p}^-]_q\otimes\Gamma(\mathfrak{p}^-,\lambda)_{q,\mathrm{lowest}}\xrightarrow{\approx}\Gamma(\mathfrak{p}^-,\lambda)_q,\qquad f\otimes v\mapsto fv, \tag{2.164}$$
$$\Gamma(\mathfrak{p}^-,\lambda)_{q,\mathrm{lowest}}\otimes\mathbb{C}[\mathfrak{p}^-]_q\xrightarrow{\approx}\Gamma(\mathfrak{p}^-,\lambda)_q,\qquad v\otimes f\mapsto vf. \tag{2.165}$$

are isomorphisms of $U_q\mathfrak{k}$-modules.

PROOF. The first claim follows from the second one. The morphisms of $U_q\mathfrak{k}$-modules (2.164) and (2.165) are also morphisms of graded vector spaces. A duality argument allows one to deduce from Proposition 2.76 that the dimensions of corresponding homogeneous components are finite and equal. Hence, the bijectivity of the morphisms (2.164) and (2.165) follows from their surjectivity, and the latter property is due to Lemma 2.232. □

Of course, $\Gamma(\mathfrak{p}^-,\lambda)_q$ fails to be a free $\mathbb{C}[\mathfrak{p}^-]_q$-*bimodule*. Let us find the relations between the generators of $\Gamma(\mathfrak{p}^-,\lambda)_{q,\mathrm{lowest}}$. Since $\Gamma(\mathfrak{p}^-,\lambda)_q$ is a lowest weight $U_q\mathfrak{g}$-module, one has the well-defined morphism of $U_q\mathfrak{g}$-modules

$$\check{R}_{\mu,\lambda}\stackrel{\mathrm{def}}{=}\check{R}_{\Gamma(\mathfrak{p}^-,\mu)_q,\Gamma(\mathfrak{p}^-,\lambda)_q}:\Gamma(\mathfrak{p}^-,\mu)_q\otimes\Gamma(\mathfrak{p}^-,\lambda)_q\to\Gamma(\mathfrak{p}^-,\lambda)_q\otimes\Gamma(\mathfrak{p}^-,\mu)_q$$

introduced in Subsection 2.1.4 in terms of the universal R-matrix. In particular,

$$\check{R}_{0,\lambda}:\mathbb{C}[\mathfrak{p}^-]_q\otimes\Gamma(\mathfrak{p}^-,\lambda)_q\to\Gamma(\mathfrak{p}^-,\lambda)_q\otimes\mathbb{C}[\mathfrak{p}^-]_q,$$

because in the category of $U_q\mathfrak{g}$-covariant $\mathbb{C}[\mathfrak{p}^-]_q$-bimodules one has

$$\mathbb{C}[\mathfrak{p}^-]_q\cong\Gamma(\mathfrak{p}^-,0)_q.$$

The first claim of the following proposition can be reformulated in the sense that the $\mathbb{C}[\mathfrak{p}^-]_q$-bimodules in question are local in $\mathcal{C}^-$; see Subsection 1.3.8.

PROPOSITION 2.234. (1) *One has* $m_{\mathrm{left},\lambda} = m_{\mathrm{right},\lambda} \cdot \check{R}_{0,\lambda}$ *and* $m_{\mathrm{right},\mu} = m_{\mathrm{left},\mu} \cdot \check{R}_{\mu,0}$.

(2) $\check{R}_{0,\lambda}(\mathbb{C}[\mathfrak{p}^-]_{q,1} \otimes \Gamma(\mathfrak{p}^-,\lambda)_{q,\mathrm{lowest}}) \subset \Gamma(\mathfrak{p}^-,\lambda)_{q,\mathrm{lowest}} \otimes \mathbb{C}[\mathfrak{p}^-]_{q,1}$.

PROOF. We start with the first claim. The first relation can be proved by the passage to dual maps. In fact, it is equivalent to coincidence of the morphisms of $U_q\mathfrak{g}^{\mathrm{cop}}$-modules

$$(\check{R}_{0,\lambda})^* \Delta^+_{\mathrm{right},\lambda} = \Delta^+_{\mathrm{left},\lambda}. \tag{2.166}$$

In turn, (2.166) follows from the fact that the vector $v(\mathfrak{q}^+,\lambda)$ generates the $U_q\mathfrak{g}^{\mathrm{cop}}$-module $N(\mathfrak{q}^+,\lambda)$, together with the relations

$$\Delta^+_{\mathrm{right},\lambda} v(\mathfrak{q}^+,\lambda) = v(\mathfrak{q}^+,\lambda) \otimes v(\mathfrak{q}^+,0), \qquad \Delta^+_{\mathrm{left},\lambda} v(\mathfrak{q}^+,\lambda) = v(\mathfrak{q}^+,0) \otimes v(\mathfrak{q}^+,\lambda),$$
$$\check{R}^*_{0,\lambda} v(\mathfrak{q}^+,\lambda) \otimes v(\mathfrak{q}^+,0) = v(\mathfrak{q}^+,0) \otimes v(\mathfrak{q}^+,\lambda).$$

The second relation can be proved in a similar way.

The last claim of Proposition 2.234 can be proved by observing that $\check{R}_{0,\lambda}$ is a morphism of $U_q\mathfrak{g}$-modules, hence this linear map commutes with the action of $H_{\mathbb{S}} \otimes 1 + 1 \otimes H_{\mathbb{S}}$. □

COROLLARY 2.235. *Let* $\{z_i\}$ *be a basis of the finite-dimensional vector space* $\mathbb{C}[\mathfrak{p}^-]_{q,1}$ *and* $\{\gamma_j\}$ *a basis of the finite-dimensional vector space* $\Gamma(\mathfrak{p}^-,\lambda)_{q,\mathrm{lowest}}$. *There exists a unique matrix* $(\check{R}^{km}_{ij}(\lambda))$ *such that*

$$\check{R}_{0,\lambda}(z_i \otimes \gamma_j) = \sum_{k,m} \check{R}^{km}_{ij}(\lambda)\gamma_k \otimes z_m,$$

where $i,m = 1,2,\ldots,\dim \mathbb{C}[\mathfrak{p}^-]_{q,1}$, $j,k = 1,2,\ldots,\dim \Gamma(\mathfrak{p}^-,\lambda)_{q,\mathrm{lowest}}$.

REMARK 2.236. It might seem a hard problem to find the functions $\check{R}^{km}_{ij}(\lambda)$ because their definition involves an action of the universal R-matrix in tensor products of infinite-dimensional $U_q\mathfrak{g}$-modules. Let us show that in fact there is no essential obstacle here. Just as in Subsection 2.3.7, consider the Hopf subalgebra $U_q\mathfrak{k}_{\mathrm{ss}} \subset U_q\mathfrak{g}$ generated by $K_i^{\pm 1}$, E_i, F_i with $i \neq l_0$. The action of the universal R-matrix of $U_q\mathfrak{g}$ on vectors from $\mathbb{C}[\mathfrak{p}^-]_{q,1} \otimes \Gamma(\mathfrak{p}^-,\lambda)_{q,\mathrm{lowest}}$ differs from the action of the universal R-matrix of $U_q\mathfrak{k}_{\mathrm{ss}}$ only by the scalar factor

$$\mathrm{const} = q_{l_0}^{\frac{(\lambda,\overline{\omega}_{l_0})}{(\overline{\omega}_{l_0},\overline{\omega}_{l_0})}}. \tag{2.167}$$

To prove (2.167), it suffices to use (2.11), the reduced expression of w_0 as in Subsection 2.3.1, and the relations (2.12),

$$(\alpha_{l_0},\lambda) = (\alpha_{l_0|\mathfrak{h}\cap\mathfrak{k}}, \lambda_{|\mathfrak{h}\cap\mathfrak{k}}) + \frac{(\alpha_{l_0},\overline{\omega}_{l_0})(\overline{\omega}_{l_0},\lambda)}{(\overline{\omega}_{l_0},\overline{\omega}_{l_0})}, \qquad (\alpha_{l_0},\overline{\omega}_{l_0}) = d_{l_0},$$

which allow one to compare the Cartan factors (see (2.11), together with Remark 2.97). Since $\check{R}_{\mu,\lambda}$ is a morphism of $U_q\mathfrak{g}$-modules, one has

$$\check{R}_{\mu,\lambda} : \Gamma(\mathfrak{p}^-,\mu)_{q,\mathrm{lowest}} \otimes \Gamma(\mathfrak{p}^-,\lambda)_{q,\mathrm{lowest}} \to \Gamma(\mathfrak{p}^-,\lambda)_{q,\mathrm{lowest}} \otimes \Gamma(\mathfrak{p}^-,\mu)_{q,\mathrm{lowest}}.$$

Just as above, one can find the scalar factor which determines the correspondence between the actions of R-matrices of $U_q\mathfrak{g}$ and of $U_q\mathfrak{k}_{\mathrm{ss}}$:

$$\mathrm{const} = q^{-\frac{(\mu,\overline{\omega}_{l_0})(\lambda,\overline{\omega}_{l_0})}{(\overline{\omega}_{l_0},\overline{\omega}_{l_0})}}.$$

Propositions 2.233 and 2.234 imply the following claims.

PROPOSITION 2.237. *The set* $\{\gamma_j\}_{j=1,2,\ldots,\dim\Gamma(\mathfrak{p}^-,\lambda)_{q,\mathrm{lowest}}}$ *generates the* $\mathbb{C}[\mathfrak{p}^-]_q$*-bimodule* $\Gamma(\mathfrak{p}^-,\lambda)_q$*, and the relations*

$$z_i\gamma_j=\sum_{k,m}\check{R}_{ij}^{km}(\lambda)\gamma_k z_m \tag{2.168}$$

are defining.

By our construction, $\Lambda^1(\mathfrak{p}^-)_q=\Gamma(\mathfrak{p}^-,-\alpha_{l_0})_q$. Hence $\Lambda^1(\mathfrak{p}^-)_q$ is a $U_q\mathfrak{g}$-covariant $\mathbb{C}[\mathfrak{p}^-]_q$-bimodule.

COROLLARY 2.238. *For all* $i,j=1,2,\ldots,\dim\mathfrak{p}^-$*, one has*

$$z_i dz_j=\sum_{k,m=1}^{\dim\mathfrak{p}^-}\check{R}_{ij}^{km}dz_k\,z_m, \tag{2.169}$$

where $\check{R}_{ij}^{km}=\check{R}_{ij}^{km}(-\alpha_{l_0})$ *and* (2.169) *is a defining list of relations between the generators of the* $\mathbb{C}[\mathfrak{p}^-]_q$*-bimodule* $\Lambda^1(\mathfrak{p}^-)_q$.

REMARK 2.239. Let γ_{low} be the lowest weight vector of the $U_q\mathfrak{g}$-module $\Gamma(\mathfrak{p}^-,\lambda)_q$ and $\{z_i\}$ a basis of weight vectors of the $U_q\mathfrak{k}$-module $\mathbb{C}[\mathfrak{p}^-]_{q,1}$: $H_jz_i=\mu_i(H_j)z_i$. Then $z_i\gamma_{\mathrm{low}}=q^{-(\lambda,\mu_i)}\gamma_{\mathrm{low}}z_i$ due to (2.168), (2.11), and (2.13). In particular,

$$z_{\mathrm{low}}dz_{\mathrm{low}}=q_{l_0}^{-2}dz_{\mathrm{low}}\,z_{\mathrm{low}}.$$

This generalizes the relation (1.78).

Now we are about to prove that $(\Lambda^1(\mathfrak{p}^-)_q,d)$ is a first-order differential calculus over $\mathbb{C}[\mathfrak{p}^-]_q$. The Leibnitz rule (2.158) can be readily deduced via the passage to dual linear maps in the following equality of morphisms of $U_q\mathfrak{g}^{\mathrm{cop}}$-modules.

LEMMA 2.240. $\Delta^+d^*=(d^*\otimes\mathrm{id})\Delta^+_{\mathrm{right},-\alpha_{l_0}}+(\mathrm{id}\otimes d^*)\Delta^+_{\mathrm{left},-\alpha_{l_0}}$.

PROOF. It suffices to use the relations

$$\begin{aligned}\Delta^+d^*v(\mathfrak{q}^+,-\alpha_{l_0})&=\Delta^+(F_{l_0}v(\mathfrak{q}^+,0))=F_{l_0}(\Delta^+v(\mathfrak{q}^+,0))\\&=F_{l_0}(v(\mathfrak{q}^+,0)\otimes v(\mathfrak{q}^+,0))=F_{l_0}v(\mathfrak{q}^+,0)\otimes v(\mathfrak{q}^+,0)+v(\mathfrak{q}^+,0)\otimes F_{l_0}v(\mathfrak{q}^+,0),\end{aligned}$$

$$\begin{aligned}&((d^*\otimes\mathrm{id})\Delta^+_{\mathrm{right},-\alpha_{l_0}}+(\mathrm{id}\otimes d^*)\Delta^+_{\mathrm{left},-\alpha_{l_0}})v(\mathfrak{q}^+,-\alpha_{l_0})\\&\qquad=d^*v(\mathfrak{q}^+,-\alpha_{l_0})\otimes v(\mathfrak{q}^+,0)+v(\mathfrak{q}^+,0)\otimes d^*v(\mathfrak{q}^+,-\alpha_{l_0})\\&\qquad=F_{l_0}v(\mathfrak{q}^+,0)\otimes v(\mathfrak{q}^+,0)+v(\mathfrak{q}^+,0)\otimes F_{l_0}v(\mathfrak{q}^+,0).\ \square\end{aligned}$$

It remains to show that the elements df, $f\in\mathbb{C}[\mathfrak{p}^-]_q$, generate the $\mathbb{C}[\mathfrak{p}^-]_q$-bimodule $\Lambda^1(\mathfrak{p}^-)_q$. We are about to obtain an even stronger statement.

PROPOSITION 2.241. *If* $\{z_j\}$ *is a basis of the vector space* $\mathbb{C}[\mathfrak{p}^-]_{q,1}$*, then*

$$\Lambda^1(\mathfrak{p}^-)_q=\bigoplus_{j=1}^{\dim\mathfrak{p}^-}\mathbb{C}[\mathfrak{p}^-]_q dz_j=\bigoplus_{j=1}^{\dim\mathfrak{p}^-}dz_j\mathbb{C}[\mathfrak{p}^-]_q.$$

PROOF. By Proposition 2.233, it suffices to demonstrate that the linear map

$$d|_{\mathbb{C}[\mathfrak{p}^-]_{q,1}} : \mathbb{C}[\mathfrak{p}^-]_{q,1} \to \Lambda^1(\mathfrak{p}^-)_{q,1}$$

is one-to-one. To make this plausible, observe that the dual linear map is a morphism of simple weight $U_q\mathfrak{k}$-modules and is nonzero, because

$$d^*v(\mathfrak{q}^+, -\alpha_{l_0}) = v(\mathfrak{q}^+, 0) \neq 0.$$

□

2.6.2. Universal differential calculi. We use the notions introduced in Subsection 1.5.1. Recall that a differential calculus over an algebra F is a differential graded algebra (Ω, d) such that $\Omega_0 = F$ and Ω is generated by $F \oplus dF$.

It is well known [**174**, p. 463] that every first-order differential calculus (M, d) over F determines a differential calculus $(\Omega^{\mathrm{univ}}, d^{\mathrm{univ}})$ over F with the following properties:

(1) $\Omega_1^{\mathrm{univ}} = M$;
(2) $d^{\mathrm{univ}}|_F = d$;
(3) for each differential calculus (Ω', d') over F with the previous two properties ($\Omega'_1 = M$, $d'|_F = d$), there exists a morphism of differential graded algebras $\Omega^{\mathrm{univ}} \to \Omega'$ which is the identity on $F \oplus M$.

Such differential calculus is unique up to an isomorphism which is the identity on $F \oplus M$, and is called the *differential enveloping* or *universal differential calculus*. It exists and is covariant if the initial first-order differential calculus is covariant [**174**, pp. 463, 464].

REMARK 2.242. The work [**174**] assumes implicitly that all the $U_q\mathfrak{g}$-modules under consideration are locally finite dimensional. This does not affect the proof of covariance of the universal differential calculus [**174**, p. 464].

Below we present three examples of the construction of a universal differential calculus. Every passage to the next first-order differential calculus will be done by altering the previous one via *extending the list of defining relations.*

EXAMPLE 2.243. (No relations.) Consider a vector space V. We are about to produce a differential calculus over its tensor algebra $T(V)$. Choose a vector space V' isomorphic to V and an isomorphism $d : V \to V'$. Endow the tensor algebra $\Omega' = T(V \oplus V')$ with the grading as follows:

$$\deg v = 0, \quad v \in V; \qquad \deg v' = 1, \quad v' \in V'.$$

Obviously, $\Omega' = \bigoplus_{j \in \mathbb{Z}_+} \Omega'_j$, where $\Omega'_j = \{t \in \Omega' |\ \deg t = j\}$. Let us define a linear operator d' in Ω' by a recursion:

$$\begin{aligned} d'1 &= 0; \\ d'v &= dv, \quad v \in V; \qquad d'v' = 0, \quad v' \in V'; \\ d'(v \otimes t) &= dv \otimes t + v \otimes d't, \qquad v \in V, \quad t \in \Omega'; \\ d'(v' \otimes t) &= -v' \otimes d't, \qquad v' \in V', \ t \in \Omega'. \end{aligned}$$

The relations $(d')^2 = 0$ and (1.75) are easy consequences of the definitions. The pair $(\Omega'_1, d'|_{\Omega'_0})$ is a first-order differential calculus over the tensor algebra $T(V)$, and the pair (Ω', d') is its universal differential calculus.

EXAMPLE 2.244. (Relations between the coordinates are imposed.) Let J_0 be a two-sided ideal of the tensor algebra $T(V)$ and $J_0 = \bigoplus_{j\geq 2}(J_0 \cap V^{\otimes j})$. We intend to produce a differential calculus over the algebra $F = T(V)/J_0$.

Introduce the two-sided ideal J_F of Ω' generated by J_0 and $d'J_0$. The algebra $\Omega^F = \Omega'/J_F$ inherits the grading $\Omega^F = \bigoplus_{j\in\mathbb{Z}_+}\Omega_j^F$.

Obviously, $V \hookrightarrow F$. Since $d'J_F \subset J_F$, the action of the differential d' can be transferred to $\Omega^F = \Omega'/J_F$. Thus we get a linear operator d_F and the differential calculus (Ω^F, d^F) over the algebra $F = \Omega_0^F$. It is the universal differential calculus for the first-order differential calculus $(\Omega_1^F, d^F|_F)$ and is well known in quantum group theory [**272**, p. 462].

EXAMPLE 2.245. (Commutation relations between the coordinates and the differentials are imposed.) Consider the invertible linear operator $\check{\mathcal{R}} : V\otimes V' \to V'\otimes V$. Let J_1 be the subbimodule of the F-bimodule Ω_1^F generated by the subspace

$$\left\{vv' - v'v \,\middle|\, v\otimes v' = \check{\mathcal{R}}(v\otimes v'), \quad v\in V,\ v'\in V'\right\} \subset \Omega_1^F.$$

Factorization of $(\Omega_1^F, d^F|_F)$ by J_1 leads to a first-order differential calculus over F. To produce the associated universal differential calculus, consider the two-sided ideal J_M of Ω^F generated by J_1, $d^F J_1$, and then divide out Ω^F by J_M. The grading and the differential are transferred to Ω^F/J_M since the ideal J_M is homogeneous and $d^F J_M \subset J_M$.

Consider the first-order differential calculus $(\Lambda^1(\mathfrak{p}^-)_q, d)$ over the algebra $\mathbb{C}[\mathfrak{p}^-]_q$ introduced in Subsection 2.6.1. Let $(\Lambda(\mathfrak{p}^-)_q, d)$ be the associated universal differential calculus over $\mathbb{C}[\mathfrak{p}^-]_q$ and $\wedge$ the multiplication in $\Lambda(\mathfrak{p}^-)_q = \bigoplus_{j\in\mathbb{Z}_+}\Lambda^j(\mathfrak{p}^-)_q$.

It is easy to describe the differential algebra $\Lambda(\mathfrak{p}^-)_q$ in terms of generators and relations using the results of Subsection 2.6.1 and Example 2.245. For that, it suffices to complete the list of defining relations of the quadratic algebra $\mathbb{C}[\mathfrak{p}^-]_q$ by the relations (2.169) and the relations derived from (2.169) via differentiation. Namely,

$$dz_i \wedge dz_j = -\sum_{k,m=1}^{\dim\mathfrak{p}^-} \check{R}_{ij}^{km} dz_k \wedge dz_m.$$

REMARK 2.246. Let us use the results of Subsection 1.3.8 to obtain another description of the differential calculus $(\Lambda(\mathfrak{p}^-)_q, d)$. This will make clear the covariance of the differential calculus in question. First, the algebra $F = \mathbb{C}[\mathfrak{p}^-]_q$ is commutative in the category $\mathcal{C}^-$. This means that F is an object of $\mathcal{C}^-$, the multiplication $m : F\otimes F \to F$ is a morphism in $\mathcal{C}^-$, and $m = m\check{R}_{F,F}$. Second, $M = \Lambda^1(\mathfrak{p}^-)_q$ is a local bimodule over F, i.e., M is an object of $\mathcal{C}^-$, the actions $m_{\text{left}} : F\otimes M \to M$, $m_{\text{right}} : M\otimes F \to M$ are morphisms in $\mathcal{C}^-$, and

$$m_{\text{right}} = m_{\text{left}}\check{R}_{M,F}, \qquad m_{\text{left}} = m_{\text{right}}\check{R}_{F,M}.$$

Third, $d : F \to M$ is a morphism in $\mathcal{C}^-$ and (M, d) is a first-order differential calculus over F.

By Proposition 1.62, the automorphism $\check{R}_{M,M}$ of $M\otimes M$ induces an automorphism $\bar{R}_{M,M}$ of $M\otimes_F M$. Namely, $\bar{R}_{M,M}\cdot j = j\cdot\check{R}_{M,M}$, where j is the canonical projection morphism $M\otimes M \to M\otimes_F M$.

Consider the tensor algebra

$$T_F(M) = \bigoplus_{j\in\mathbb{Z}_+} T_F(M)_j, \qquad T_F(M)_j = \underbrace{M \otimes_F M \otimes_F \cdots \otimes_F M}_{j}.$$

Here $T_F(M)_0 = F$, $T_F(M)_1 = M$. Let J be the two-sided ideal of this algebra generated by

$$\{\omega_1 \otimes_F \omega_2 + \bar{R}_{M,M}(\omega_1 \otimes_F \omega_2) |\, \omega_1, \omega_2 \in M\}.$$

It follows from the definitions that $\Lambda(\mathfrak{p}^-)_q \cong T_F(M)/J$, hence $\Lambda(\mathfrak{p}^-)_q$ is a $U_q\mathfrak{g}$-module algebra. It remains to observe that, first, d is a morphism of $U_q\mathfrak{g}$-modules, because

$$d(f_0 \wedge df_1 \wedge df_2 \wedge \cdots \wedge df_n) = df_0 \wedge df_1 \wedge df_2 \wedge \cdots \wedge df_n$$

for all $n \in \mathbb{Z}_+$, $f_0, \ldots, f_n \in F$, and, second, the first-order differential calculus (M, d) is covariant.

Let us prove that $\Lambda^j(\mathfrak{p}^-)_q$ is a free left $\mathbb{C}[\mathfrak{p}^-]_q$-module of rank $\binom{\dim \mathfrak{p}^-}{j}$ and a free right $\mathbb{C}[\mathfrak{p}^-]_q$-module of rank $\binom{\dim \mathfrak{p}^-}{j}$, just as in the classical case $q = 1$.

Note that the linear span of j-forms with constant coefficients

$$dz_{i_1} \wedge dz_{i_2} \wedge \cdots \wedge dz_{i_j}, \qquad i_1, i_2, \ldots, i_j \in \{1, 2, \ldots, \dim \mathfrak{p}^-\},$$

is a $U_q\mathfrak{k}$-module. We denote it by $\Lambda^j(\mathfrak{p}^-)_q^{\mathrm{const}}$.

PROPOSITION 2.247. *There exist isomorphisms of $U_q\mathfrak{k}$-modules*

$$\mathbb{C}[\mathfrak{p}^-]_q \otimes \Lambda^j(\mathfrak{p}^-)_q^{\mathrm{const}} \xrightarrow{\sim} \Lambda^j(\mathfrak{p}^-)_q, \qquad f \otimes \omega \mapsto f\omega,$$
$$\Lambda^j(\mathfrak{p}^-)_q^{\mathrm{const}} \otimes \mathbb{C}[\mathfrak{p}^-]_q \xrightarrow{\sim} \Lambda^j(\mathfrak{p}^-)_q, \qquad \omega \otimes f \mapsto \omega f.$$

PROOF. It suffices to use Proposition 2.233 and the definition of $\Lambda(\mathfrak{p}^-)_q$. □

Let us turn to computing $\dim \Lambda^j(\mathfrak{p}^-)_q^{\mathrm{const}}$.

LEMMA 2.248. *The dimension of $\Lambda^2(\mathfrak{p}^-)_q^{\mathrm{const}}$ does not exceed its value in the classical case $q = 1$:*

$$\dim \Lambda^2(\mathfrak{p}^-)_q^{\mathrm{const}} \le \frac{\dim \mathfrak{p}^-(\dim \mathfrak{p}^- - 1)}{2}. \tag{2.170}$$

PROOF. Consider the linear operator

$$\widetilde{R} : dz_i \otimes dz_j \to \sum_{k,m=1}^{\dim \mathfrak{p}^-} \check{R}_{ij}^{km} dz_k \otimes dz_m$$

in $\Lambda^1(\mathfrak{p}^-)_q^{\mathrm{const}} \otimes \Lambda^1(\mathfrak{p}^-)_q^{\mathrm{const}}$. One has the natural isomorphism

$$\Lambda^2(\mathfrak{p}^-)_q^{\mathrm{const}} \cong \left\{ v \in \Lambda^1(\mathfrak{p}^-)_q^{\mathrm{const}} \otimes \Lambda^1(\mathfrak{p}^-)_q^{\mathrm{const}} \,\middle|\, \widetilde{R}v = -v \right\}.$$

Remark 2.236 and (2.31) easily imply that all the eigenvalues of this linear operator are real and nonzero. We reproduce the argument used to prove Lemma 2.115 in assuming that the functions $\check{R}_{ij}^{km}$ of q are analytic, hence continuous on $(0, 1]$. For all $q \in (0, 1]$ the number of negative eigenvalues (with multiplicities) of the operator $\widetilde{R}$ is $\frac{\dim p^-(\dim p^- - 1)}{2}$. Hence the dimension of the eigenspace corresponding to the eigenvalue -1 does not exceed $\frac{\dim p^-(\dim p^- - 1)}{2}$. □

The inequality opposite to (2.170) will be obtained when we succeed in proving that -1 is an eigenvalue of $\widetilde{R}$ whose multiplicity is at least $\frac{\dim p^-(\dim p^- -1)}{2}$. It suffices to produce such an estimate for the linear operator $(\widetilde{R}^*)^{-1}$. The latter can be identified with the restriction of $\check{R}_{N(\mathfrak{q}^+,-\alpha_{l_0}),N(\mathfrak{q}^+,-\alpha_{l_0})}$ to the tensor product of highest homogeneous components $N(\mathfrak{q}^+,-\alpha_{l_0})_{\text{highest}} \otimes N(\mathfrak{q}^+,-\alpha_{l_0})_{\text{highest}}$.

Let us turn to the classical case $q = 1$. The following result is due to Kostant [**201**, pp. 359–360], [**263**, p. 348].

LEMMA 2.249. *Consider the $U\mathfrak{k}$-module $(\mathfrak{p}^-)^{\wedge r}$, where $r = 1, 2, \ldots, \dim \mathfrak{p}^-$. Its isotypic components are simple $U\mathfrak{k}$-modules whose weight subspaces are one dimensional, and the set of weights has the form*

$$\{w\rho - \rho |\, w \in W^{\mathbb{S}} \quad \text{and} \quad l(w) = r\},$$

where $\mathbb{S} = \{1, 2, \ldots, l\} \setminus l_0$, $W^{\mathbb{S}} = ({}^{\mathbb{S}}W)^{-1}$, and ${}^{\mathbb{S}}W$ is introduced in Subsection 2.2.5.

LEMMA 2.250. *The dimension of $\Lambda^2(\mathfrak{p}^-)_q^{\text{const}}$ is at least its value at $q = 1$, that is*

$$\dim \Lambda^2(\mathfrak{p}^-)_q^{\text{const}} \geq \frac{\dim \mathfrak{p}^-(\dim \mathfrak{p}^- - 1)}{2}. \tag{2.171}$$

PROOF. For $q = 1$, the inequality (2.171) is obvious. By Lemma 2.249, this implies that the highest weights of simple $U\mathfrak{k}$-modules that admit an embedding to $\mathfrak{p}^- \wedge \mathfrak{p}^-$ belong to $W\rho - \rho$, and the sum of their multiplicities is $\frac{\dim p^-(\dim p^- -1)}{2}$.

The passage from the classical case to the quantum case does not change highest weights or dimensions of simple $U_q\mathfrak{k}$-submodules of the finite-dimensional $U_q\mathfrak{k}$-module $N(\mathfrak{q}^+,-\alpha_{l_0})_{\text{highest}} \otimes N(\mathfrak{q}^+,-\alpha_{l_0})_{\text{highest}}$. Also, the multiplicities of finite-dimensional simple $U_q\mathfrak{k}$-modules in decompositions of tensor products remain intact.

It remains to make sure that the restriction of $\check{R}_{N(\mathfrak{q}^+,-\alpha_{l_0}),N(\mathfrak{q}^+,-\alpha_{l_0})}$ to the image of every morphism of $U_q\mathfrak{g}$-modules of the form

$$N(\mathfrak{q}^+, w\rho - \rho) \to N(\mathfrak{q}^+,-\alpha_{l_0}) \otimes N(\mathfrak{q}^+,-\alpha_{l_0}), \qquad w \in W^{\mathbb{S}} \quad \text{and} \quad l(w) = 2,$$

is -1. In fact, it is enough to prove that it is ± 1, because (2.31) and continuity in q implies that we are inside the spectral subspace corresponding to the negative part of the spectrum.

Let us apply (2.31). It is clearly visible from the proof of this relation as in [**80**, p. 39] that the restriction of $\check{R}_{N(\mathfrak{q}^+,\mu),N(\mathfrak{q}^+,\lambda)}\check{R}_{N(\mathfrak{q}^+,\lambda),N(\mathfrak{q}^+,\mu)}$ to the image of the morphism $N(\mathfrak{q}^+,\nu) \to N(\mathfrak{q}^+,\lambda) \otimes N(\mathfrak{q}^+,\mu)$ reduces to multiplication by the scalar

$$q^{-(\mu,\mu+2\rho)-(\lambda,\lambda+2\rho)+(\nu,\nu+2\rho)} = q^{-(\mu+\rho,\mu+\rho)-(\lambda+\rho,\lambda+\rho)+(\nu+\rho,\nu+\rho)+(\rho,\rho)}.$$

We substitute $\lambda = \mu = -\alpha_{l_0}$, $\nu = w\rho - \rho$ and obtain 1 because the weights $-\alpha_{l_0} + \rho$, $\nu + \rho$, ρ are in the same W-orbit, hence have the same length. □

Let us turn to higher-order differential forms.

LEMMA 2.251. *For all $j \geq 3$, one has*

$$\dim \Lambda^j(\mathfrak{p}^-)_q^{\text{const}} \leq \binom{\dim \mathfrak{p}^-}{j}.$$

PROOF. We start with choosing the basis of weight vectors

$$\{dz_1, dz_2, \dots, dz_{\dim \mathfrak{p}^-}\} \subset \Lambda^1(\mathfrak{p}^-)_q^{\mathrm{const}}, \tag{2.172}$$

as in Remark 2.117. Let us also introduce the lexicographic order on the set $\{dz_i \otimes dz_k\}_{i,k=1,2,\dots,\dim \mathfrak{p}^-}$. The action of the universal R-matrix of $U_q\mathfrak{k}$ in this basis is described by a triangular matrix with positive entries at the principal diagonal; see (2.11). Hence the only nonzero contribution to the right-hand side of

$$dz_i \otimes dz_j = -\sum_{k,m=1}^{\dim \mathfrak{p}^-} \check{R}_{ij}^{km} dz_k \otimes dz_m, \qquad i \geq j, \tag{2.173}$$

is made by the terms $dz_k \otimes dz_m$ with $k \leq m$. Therefore, every element of $\Lambda^j(\mathfrak{p}^-)_q^{\mathrm{const}}$ belongs to the linear span of the monomials

$$dz_{i_1} \wedge dz_{i_2} \wedge \cdots \wedge dz_{i_j}, \qquad 1 \leq i_1 < i_2 < \cdots < i_j \leq \dim \mathfrak{p}^-. \qquad \square$$

REMARK 2.252. The distinguished basis (2.172) can be used to identify the tensor algebra $T((\mathfrak{p}^-)^*)$ with the free noncommutative algebra $\mathbb{C}\langle dz_1, dz_2, \dots, dz_{\dim \mathfrak{p}^-}\rangle$. Consider the two-sided ideal I of the free algebra generated by the set of left- and right-hand sides of (2.173). These relations work as substitution rules in the final part of the proof of Lemma 2.251: the left-hand side of a relation, wherever it appears, is replaced by its right-hand side.

The proof of the following lemma reproduces the proof of a similar result of Heckenberger and Kolb [**118**].

LEMMA 2.253. *For all* $j \geq 3$,

$$\dim \Lambda^j(\mathfrak{p}^-)_q^{\mathrm{const}} \geq \binom{\dim \mathfrak{p}^-}{j}. \tag{2.174}$$

PROOF. If the inequality (2.174) holds for $j = 3$, then it follows from the Diamond Lemma 2.128 that G is a Gröbner basis of the two-sided ideal I; see Remark 2.252. This already implies (2.174) for all $j > 3$. Hence while proving the proposition, one may restrict to the special case $j = 3$.

We identify $(\mathfrak{p}^-)^*$ with $\Lambda^1(\mathfrak{p}^-)_q^{\mathrm{const}}$ and introduce the notation $(\mathfrak{p}^-)^{*\wedge 2}$ for the subspace $\left\{v \in (\mathfrak{p}^-)^{*\otimes 2} \,\middle|\, \check{R}v = -v\right\}$, cf. (2.173). Lemmas 2.248 and 2.250 yield that

$$\dim((\mathfrak{p}^-)^{*\wedge 2}) = \frac{\dim \mathfrak{p}^-(\dim \mathfrak{p}^- - 1)}{2}. \tag{2.175}$$

Consider the subspaces $L_1 = (\mathfrak{p}^-)^{*\wedge 2} \otimes (\mathfrak{p}^-)^*$, $L_2 = (\mathfrak{p}^-)^* \otimes (\mathfrak{p}^-)^{*\wedge 2}$ of the vector space $(\mathfrak{p}^-)^{*\otimes 3}$, together with the complex

$$0 \to L_1 \cap L_2 \to L_1 \oplus L_2 \xrightarrow{j} (\mathfrak{p}^-)^{*\otimes 3} \to \mathbb{C}[\mathfrak{p}^-]_{q,3} \to 0,$$
$$j : v_1 \oplus v_2 \mapsto v_1 - v_2, \qquad v_j \in L_j \subset (\mathfrak{p}^-)^{*\otimes 3},$$

which is exact in all terms except, possibly, the term $(\mathfrak{p}^-)^{*\otimes 3}$. A computation of the Euler characteristic of this complex leads to

$$-\dim(L_1 \cap L_2) + \dim(L_1 \oplus L_2) - \dim\left((\mathfrak{p}^-)^{*\otimes 3}\right) + \dim \mathbb{C}[\mathfrak{p}^-]_{q,3} \leq 0;$$

see [**202**, p. 91] and [**211**, p. 119]. Then (2.175), (2.64) yield

$$\dim(L_1 \cap L_2) \geq n^2(n-1) - n^3 + \frac{n(n+1)(n+2)}{6} = \frac{n(n-1)(n-2)}{6},$$

where $n = \dim \mathfrak{p}^- = \dim(\mathfrak{p}^-)^*$. The description of universal differential calculus as in Example 2.244 implies

$$\dim \Lambda^3(\mathfrak{p}^-)_q^{\mathrm{const}} \geq \dim(L_1 \cap L_2) \geq \binom{\dim \mathfrak{p}^-}{3}. \qquad \square$$

The following statement can be deduced from the lemmas proved in this subsection.

PROPOSITION 2.254. *The homogeneous components $\Lambda^j(\mathfrak{p}^-)_q$ of the differential graded algebra $\Lambda(\mathfrak{p}^-)_q$ are nonzero for $j \leq \dim \mathfrak{p}^-$. Everyone of those is both free left and free right $\mathbb{C}[\mathfrak{p}^-]_q$-module of rank $\binom{\dim \mathfrak{p}^-}{j}$.*

In the important special case $\mathfrak{g} = \mathfrak{sl}_N$ this result is well known [**226**, p. 296].

REMARK 2.255. Assuming additionally that q is transcendental, one can easily prove exactness for the de Rham complex, together with the fact that the dual resolution of the trivial $U_q\mathfrak{g}$-module is isomorphic to the generalized Bernstein–Gelfand–Gelfand resolution; see [**286**]. The recent results of Heckenberger and Kolb demonstrate that one can get rid of the transcendence assumption on q; see [**120, 119**].

2.6.3. Differential forms with polynomial coefficients. This and the next subsections contain generalizations of the results of Subsections 1.5.1 and 1.5.2, respectively.

We start with producing CFODC over $\mathrm{Pol}(\mathfrak{p}^-)_q$. Recall that one has $\mathrm{Pol}(\mathfrak{p}^-)_q = \mathbb{C}[\mathfrak{p}^-]_q \otimes \mathbb{C}[\mathfrak{p}^+]_q$ in the category of $U_q\mathfrak{g}$-modules.

Consider the $U_q\mathfrak{g}$-module $\Omega^{1,0}(\mathfrak{p}^-)_q \stackrel{\mathrm{def}}{=} \Lambda^1(\mathfrak{p}^-)_q \otimes \mathbb{C}[\mathfrak{p}^+]_q$. Using the universal R-matrix just as in Subsection 2.3.3, we endow $\Omega^{1,0}(\mathfrak{p}^-)_q$ with the structure of $U_q\mathfrak{g}$-covariant $\mathrm{Pol}(\mathfrak{p}^-)_q$-bimodule

$$\begin{array}{c}
\mathrm{Pol}(\mathfrak{p}^-)_q \otimes \Omega^{1,0}(\mathfrak{p}^-)_q \\
\Big\downarrow \cong \\
\mathbb{C}[\mathfrak{p}^-]_q \otimes \big(\mathbb{C}[\mathfrak{p}^+]_q \otimes \Lambda^1(\mathfrak{p}^-)_q\big) \otimes \mathbb{C}[\mathfrak{p}^+]_q \\
\Big\downarrow \mathrm{id}\otimes\check{R}\otimes\mathrm{id} \\
\mathbb{C}[\mathfrak{p}^-]_q \otimes \big(\Lambda^1(\mathfrak{p}^-)_q \otimes \mathbb{C}[\mathfrak{p}^+]_q\big) \otimes \mathbb{C}[\mathfrak{p}^+]_q \\
\Big\downarrow \cong \\
\big(\mathbb{C}[\mathfrak{p}^-]_q \otimes \Lambda^1(\mathfrak{p}^-)_q\big) \otimes \big(\mathbb{C}[\mathfrak{p}^+]_q \otimes \mathbb{C}[\mathfrak{p}^+]_q\big) \\
\Big\downarrow \\
\Omega^{1,0}(\mathfrak{p}^-)_q.
\end{array}$$

We introduce the differential

$$\partial : \mathbb{C}[\mathfrak{p}^-]_q \otimes \mathbb{C}[\mathfrak{p}^+]_q \to \Lambda^1(\mathfrak{p}^-)_q \otimes \mathbb{C}[\mathfrak{p}^+]_q, \qquad \partial = d \otimes \mathrm{id}.$$

The results of Subsection 2.6.1 and Proposition 2.9 easily imply the following result.

PROPOSITION 2.256. (1) *$\Omega^{1,0}(\mathfrak{p}^-)_q$ is both a free left* $\mathrm{Pol}(\mathfrak{p}^-)_q$*-module and a free right* $\mathrm{Pol}(\mathfrak{p}^-)_q$*-module.*

(2) *$(\Omega^{1,0}(\mathfrak{p}^-)_q, \partial)$ is a CFODC over* $\mathrm{Pol}(\mathfrak{p}^-)_q$.

In a similar way one can introduce the CFODC $\left(\Omega^{0,1}(\mathfrak{p}^-)_q, \overline{\partial}\right)$ over $\mathrm{Pol}(\mathfrak{p}^-)_q$. Let us describe the corresponding constructions, starting with the algebra $\mathbb{C}[\mathfrak{p}^+]_q$. We are about to apply the antilinear map

$$* : \mathbb{C}[\mathfrak{p}^-]_q \to \mathbb{C}[\mathfrak{p}^+]_q, \qquad * : f \mapsto f^*,$$

together with the dual antilinear map

$$* : N(\mathfrak{q}^+, 0) \to N(\mathfrak{q}^-, 0), \qquad * : \xi v(\mathfrak{q}^+, 0) \mapsto (S^{-1}(\xi))^* v(\mathfrak{q}^-, 0).$$

Note that

$$(F_{l_0} v(\mathfrak{q}^+, 0))^* = (S^{-1}(F_{l_0}))^* v(\mathfrak{q}^-, 0) = (-K_{l_0} F_{l_0})^* v(\mathfrak{q}^-, 0) = E_{l_0} v(\mathfrak{q}^-, 0).$$

The $U_q\mathfrak{g}$-morphism

$$N(\mathfrak{q}^+, -\alpha_{l_0}) \to N(\mathfrak{q}^+, 0), \qquad v(\mathfrak{q}^+, -\alpha_{l_0}) \mapsto F_{l_0} v(\mathfrak{q}^+, 0),$$

as in Subsection 2.6.1, is to be replaced by the morphism of $U_q\mathfrak{g}$-modules

$$N(\mathfrak{q}^-, \alpha_{l_0}) \to N(\mathfrak{q}^-, 0), \qquad v(\mathfrak{q}^-, \alpha_{l_0}) \mapsto E_{l_0} v(\mathfrak{q}^-, 0),$$

in order to obtain a CFODC $\left(\Lambda^1(\mathfrak{p}^+)_q, d\right)$ over the $U_q\mathfrak{g}$-module algebra $\mathbb{C}[\mathfrak{p}^+]_q$. Similarly to Proposition 2.92, one can prove

PROPOSITION 2.257. *Let $* : \Lambda^1(\mathfrak{p}^-)_q \to \Lambda^1(\mathfrak{p}^+)_q$ be the antilinear map dual to the map*

$$* : N(\mathfrak{q}^-, \alpha_{l_0}) \to N(\mathfrak{q}^+, -\alpha_{l_0}), \qquad * : \xi v(\mathfrak{q}^-, \alpha_{l_0}) \mapsto (S^{-1}(\xi))^* v(\mathfrak{q}^+, -\alpha_{l_0}).$$

Then for all $f, f_1, f_2, f_3 \in \mathbb{C}[\mathfrak{p}^-]_q$, one has

$$(f_1 df_2\, f_3)^* = f_3^* (df_2)^* f_1^*, \qquad (df)^* = df^*.$$

If $\{z_i\}$ is a basis of the vector space $\mathbb{C}[\mathfrak{p}^-]_{q,1}$, then (2.169) and Proposition 2.257 imply that

$$dz_j^*\, z_i^* = \sum_{k,m=1}^{\dim \mathfrak{p}^-} \check{R}_{ij}^{km} z_m^* dz_k^*.$$

Consider the $U_q\mathfrak{g}$-module $\Omega^{0,1}(\mathfrak{p}^-)_q \stackrel{\text{def}}{=} \mathbb{C}[\mathfrak{p}^-]_q \otimes \Lambda^1(\mathfrak{p}^+)_q$. Just as before, we use the universal R-matrix to endow $\Omega^{0,1}(\mathfrak{p}^-)_q$ with a structure of $U_q\mathfrak{g}$-covariant $\mathrm{Pol}(\mathfrak{p}^-)_q$-bimodule. Impose the differential

$$\overline{\partial} : \mathbb{C}[\mathfrak{p}^-]_q \otimes \mathbb{C}[\mathfrak{p}^+]_q \to \mathbb{C}[\mathfrak{p}^-]_q \otimes \Lambda^1(\mathfrak{p}^+)_q, \qquad \overline{\partial} = \mathrm{id} \otimes d.$$

The following claim can be proved in the same way as Proposition 2.256.

PROPOSITION 2.258. (1) *$\Omega^{0,1}(\mathfrak{p}^-)_q$ is both a free left* $\mathrm{Pol}(\mathfrak{p}^-)_q$*-module and a free right* $\mathrm{Pol}(\mathfrak{p}^-)_q$*-module.*

(2) *$(\Omega^{0,1}(\mathfrak{p}^-)_q, \overline{\partial})$ is a CFODC over* $\mathrm{Pol}(\mathfrak{p}^-)_q$.

Consider the $U_q\mathfrak{g}$-covariant $\mathrm{Pol}(\mathfrak{p}^-)_q$-bimodule

$$\Omega^1(\mathfrak{p}^-)_q = \Omega^{1,0}(\mathfrak{p}^-)_q \bigoplus \Omega^{0,1}(\mathfrak{p}^-)_q$$

and the differential $d = \partial + \overline{\partial}$.

COROLLARY 2.259. (1) *$\Omega^1(\mathfrak{p}^-)_q$ is both a free left* $\mathrm{Pol}(\mathfrak{p}^-)_q$*-module and a free right* $\mathrm{Pol}(\mathfrak{p}^-)_q$*-module.*

(2) $(\Omega^1(\mathfrak{p}^-)_q, d)$ *is a CFODC over* $\mathrm{Pol}(\mathfrak{p}^-)_q$.

It is easy to show that $y\omega = \omega y$ for all $\omega \in \Lambda^1(\mathfrak{p}^-)_{q,1} \oplus \Lambda^1(\mathfrak{p}^+)_{q,-1}$; see the proof of Lemma 2.263 below. One can use the localization of $\mathrm{Pol}(\mathfrak{p}^-)_q$-bimodule $\Omega^1(\mathfrak{p}^-)_q$ with respect to the multiplicative system $y^{\mathbb{Z}_+}$, together with the argument described in Subsection 2.5.4, to produce a CFODC over the $U_q\mathfrak{g}$-module algebra $\mathrm{Pol}(\mathfrak{p}^-)_{q,y}$.

Let us make a passage from first-order differential calculi to full differential calculi over $\mathrm{Pol}(\mathfrak{p}^-)_q$. Consider the first-order differential calculus $(\Lambda^1(\mathfrak{p}^+)_q, d)$ over $\mathbb{C}[\mathfrak{p}^+]_q$, together with the associated universal differential calculus $(\Lambda(\mathfrak{p}^+)_q, d)$. It follows from the results of Subsection 2.6.2 that $\Lambda^j(\mathfrak{p}^+)_q$ is both a free left and a free right $\mathbb{C}[\mathfrak{p}^+]_q$-module of rank $\binom{\dim \mathfrak{p}^+}{j}$.

Let $(\Omega^{*,0}(\mathfrak{p}^-)_q, \partial)$, $(\Omega^{0,*}(\mathfrak{p}^-)_q, \overline{\partial})$, and $(\Omega(\mathfrak{p}^-)_q, d)$ be the universal differential calculi associated to the first-order differential calculi $(\Omega^{1,0}(\mathfrak{p}^-)_q, \partial)$, $(\Omega^{0,1}(\mathfrak{p}^-)_q, \overline{\partial})$ and $(\Omega^1(\mathfrak{p}^-)_q, d)$, respectively. In the category of $U_q\mathfrak{g}$-modules one has

$$\Omega^{*,0}(\mathfrak{p}^-)_q \cong \Lambda(\mathfrak{p}^-)_q \otimes \mathbb{C}[\mathfrak{p}^+]_q, \qquad \partial = d \otimes \mathrm{id},$$
$$\Omega^{0,*}(\mathfrak{p}^-)_q \cong \mathbb{C}[\mathfrak{p}^-]_q \otimes \Lambda(\mathfrak{p}^+)_q, \qquad \overline{\partial} = \mathrm{id} \otimes d,$$
$$\Omega(\mathfrak{p}^-)_q \cong \Omega^{*,0}(\mathfrak{p}^-)_q \otimes_{\mathrm{Pol}(\mathfrak{p}^-)_q} \Omega^{0,*}(\mathfrak{p}^-)_q \cong \Lambda(\mathfrak{p}^-)_q \otimes \Lambda(\mathfrak{p}^+)_q,$$

so that in the latter case the differential is given by $d \otimes \mathrm{id} \oplus \mathrm{id} \otimes d$.

We keep the notation ∂, $\overline{\partial}$ for the operators $d \otimes \mathrm{id}$, $\mathrm{id} \otimes d$ in $\Omega(\mathfrak{p}^-)_q$. The algebra $\Omega(\mathfrak{p}^-)_q$ is graded:

$$\Omega(\mathfrak{p}^-)_q = \bigoplus_{k=1}^{\dim \mathfrak{p}^-} \Omega^k(\mathfrak{p}^-)_q,$$

and

$$\Omega^k(\mathfrak{p}^-)_q = \bigoplus_{i+j=k} \Omega^{i,j}(\mathfrak{p}^-)_q, \qquad \Omega^{i,j}(\mathfrak{p}^-)_q \cong \Lambda^i(\mathfrak{p}^-)_q \otimes \Lambda^j(\mathfrak{p}^+)_q,$$
$$\partial : \Omega^{i,j}(\mathfrak{p}^-)_q \to \Omega^{i+1,j}(\mathfrak{p}^-)_q, \qquad \overline{\partial} : \Omega^{i,j}(\mathfrak{p}^-)_q \to \Omega^{i,j+1}(\mathfrak{p}^-)_q.$$

The relation $d^2 = 0$ implies

$$\partial^2 = \overline{\partial}^2 = 0, \qquad \partial\overline{\partial} = -\overline{\partial}\partial. \tag{2.176}$$

An argument similar to those in Subsection 2.3.3 allows one to endow $\Omega(\mathfrak{p}^-)_q$ with a structure of $(U_q\mathfrak{g}, *)$-module algebra using the involution

$$* : \omega \otimes 1 \mapsto 1 \otimes \omega^*, \qquad \omega \in \Lambda^1(\mathfrak{p}^-)_q.$$

It follows from the definitions that

$$\overline{\partial}\omega = (-1)^k(\partial\omega^*)^*, \qquad \omega \in \Omega^k(\mathfrak{p}^-)_q. \tag{2.177}$$

We are about to present some commutation relations. Let $\{z_i\}$ be a basis of the vector space $\mathbb{C}[\mathfrak{p}^-]_{q,1}$. The universal R-matrix of $U_q\mathfrak{g}$ is used to impose the operator of *quantum flip of tensor factors*

$$\mathbb{C}[\mathfrak{p}^+]_{q,-1} \otimes \Lambda^1(\mathfrak{p}^-)_{q,1} \to \Lambda^1(\mathfrak{p}^-)_{q,1} \otimes \mathbb{C}[\mathfrak{p}^+]_{q,-1},$$

$$z_i^* \otimes dz_j \mapsto \sum_{k,m=1}^{\dim \mathfrak{p}^-} \widehat{R}_{ij}^{km} dz_k \otimes z_m^*.$$

Now the definitions yield that

$$z_i^* dz_j = \sum_{k,m=1}^{\dim \mathfrak{p}^-} \widehat{R}_{ij}^{km} dz_k\, z_m^*.$$

In particular,

$$z_{\mathrm{low}}^* dz_{\mathrm{low}} = q_{l_0}^2 dz_{\mathrm{low}} z_{\mathrm{low}}^*.$$

An application of the involution $*$ allows us to obtain

$$dz_j^*\, z_i = \sum_{k,m=1}^{\dim \mathfrak{p}^-} \widehat{R}_{ij}^{km} z_m dz_k^*,$$

$$dz_{\mathrm{low}}^* z_{\mathrm{low}} = q_{l_0}^2 z_{\mathrm{low}} dz_{\mathrm{low}}^*.$$

Due to the definition of multiplication in $\mathrm{Pol}(\mathfrak{p}^-)_q$ and (2.11),

$$z_i^* z_j = \sum_{k,m=1}^{\dim \mathfrak{p}^-} \widehat{R}_{ij}^{km} z_k z_m^* + (z_j, z_i), \tag{2.178}$$

where $(\cdot,\cdot)$ is the standard Hermitian form in $\mathbb{C}[\mathfrak{p}^-]_q$; see Subsection 2.3.3.

REMARK 2.260. Let $(z_i, z_j) = \delta_{i,j}$. The commutation relations (2.178) are q-analogs of the canonical commutation relations and determine the Wick algebra [**146, 145**]. Besides that, in the formal limit as $q \to 1$, the elements $\{z_i\}_{i=1,2,\ldots,\dim \mathfrak{p}^-}$ pairwise commute. Hence $\mathrm{Pol}(\mathfrak{p}^-)_q$ is a q-analog of the Weyl algebra of differential operators with polynomial coefficients on $\mathfrak{p}^-$, and the scalar product $(\cdot,\cdot)$ is a q-analog of the Fisher–Fock scalar product in the vector space of holomorphic polynomials on $\mathfrak{p}^-$ [**9**, p. 24].

2.6.4. Differential forms with distribution coefficients. The algebras $\Lambda(\mathfrak{p}^-)_q$, $\Lambda(\mathfrak{p}^+)_q$ are weight $U_q\mathfrak{g}$-modules, hence graded vector spaces

$$\Lambda(\mathfrak{p}^-)_q = \bigoplus_{i=0}^{\infty} \Lambda(\mathfrak{p}^-)_{q,i}, \qquad \Lambda(\mathfrak{p}^+)_q = \bigoplus_{j=0}^{\infty} \Lambda(\mathfrak{p}^+)_{q,-j}.$$

Therefore the vector space $\Omega(\mathfrak{p}^-)_q$ is endowed with the bigrading

$$\Omega(\mathfrak{p}^-)_q = \bigoplus_{(i,j)\in\mathbb{Z}_+^2} \Omega(\mathfrak{p}^-)_{q,i,-j},$$

as in the previous subsection. The results of Subsection 2.6.3 yield that

$$\dim(\Omega(\mathfrak{p}^-)_{q,i,-j}) < \infty$$

for all $(i,j) \in \mathbb{Z}_+^2$.

Recall that the topological vector space $\mathcal{D}(\mathbb{D})'_q$ of distributions on a quantum bounded symmetric domain was obtained in Subsection 2.4.2 via the passage from finite sums (2.98) to formal series. Here we proceed in a similar way. Let

$$\Omega(\mathbb{D})'_q = \prod_{(i,j)\in\mathbb{Z}_+^2} \Omega(\mathfrak{p}^-)_{q,i,-j}$$

be the vector space of formal series with terms from $\Omega(\mathfrak{p}^-)_{q,i,-j}$. We endow $\Omega(\mathbb{D})'_q$ with the topology of termwise convergence. Obviously, $\Omega(\mathfrak{p}^-)_q$ is a dense linear subvariety of $\Omega(\mathbb{D})'_q$. Equivalently, $\Omega(\mathbb{D})'_q$ is a completion of the vector space $\Omega(\mathfrak{p}^-)_q$ endowed with the topology of termwise convergence. Of course, $\Omega(\mathbb{D})'_q = \bigoplus_{i,j=0}^{\dim\mathfrak{p}^-} \Omega^{i,j}(\mathbb{D})'_q$, where $\Omega^{i,j}(\mathbb{D})'_q$ is the closure of the linear variety $\Omega^{i,j}(\mathfrak{p}^-)_q \subset \Omega(\mathbb{D})'_q$. The elements $\omega \in \Omega(\mathbb{D})'_q$ will be called differential forms.

The linear operators ∂, $\overline{\partial}$, and the antilinear operator $*$ admit an extension by continuity from $\Omega(\mathfrak{p}^-)_q$ to $\Omega(\mathbb{D})'_q$. The relations (2.176) and (2.177) remain valid after the extension. The algebra $\Omega(\mathfrak{p}^-)_q$ is a $U_q\mathfrak{g}$-covariant $\Omega(\mathfrak{p}^-)_q$-bimodule. The structure of the $U_q\mathfrak{g}$-covariant $\Omega(\mathfrak{p}^-)_q$-bimodule extends from $\Omega(\mathfrak{p}^-)_q$ to $\Omega(\mathbb{D})'_q$ by continuity. In this new context the relation (1.75) is still valid if either of the differential forms ω', ω'' belongs to $\Omega(\mathfrak{p}^-)_q$.

A differential form $\omega \in \Omega(\mathbb{D})'_q$ is called finite if for some $N(\omega) \in \mathbb{N}$

$$\bigoplus_{j=N(\omega)}^{\infty} \mathbb{C}[\mathfrak{p}^+]_{q,-j} \cdot \omega = 0, \qquad \omega \cdot \bigoplus_{j=N(\omega)}^{\infty} \mathbb{C}[\mathfrak{p}^-]_{q,j} = 0$$

(cf. Proposition 2.142). The subspace of finite-differential forms $\omega \in \Omega(\mathbb{D})'_q$ will be denoted by $\Omega(\mathbb{D})_q$.

Evidently, $\Omega(\mathbb{D})_q = \bigoplus_{i,j=0}^{\dim\mathfrak{p}^-} \Omega^{i,j}(\mathbb{D})_q$, where $\Omega^{i,j}(\mathbb{D})_q = \Omega^{i,j}(\mathbb{D})'_q \cap \Omega(\mathbb{D})_q$.

PROPOSITION 2.261. *$\partial(\Omega(\mathbb{D})_q) \subset \Omega(\mathbb{D})_q$ and $\overline{\partial}(\Omega(\mathbb{D})_q) \subset \Omega(\mathbb{D})_q$.*

PROOF. The linear operators

$$\partial : \Lambda(\mathfrak{p}^-)_q \to \Lambda(\mathfrak{p}^-)_q, \qquad \overline{\partial} : \Lambda(\mathfrak{p}^+)_q \to \Lambda(\mathfrak{p}^+)_q$$

are morphisms of $U_q\mathfrak{g}$-modules; hence they preserve the degree of homogeneity. Therefore,

$$\partial\mathbb{C}[\mathfrak{p}^-]_{q,j} \subset \mathbb{C}[\mathfrak{p}^-]_{q,j-1} \cdot \Lambda^1(\mathfrak{p}^-)_{q,1}, \qquad \overline{\partial}\mathbb{C}[\mathfrak{p}^+]_{q,-j} \subset \Lambda^1(\mathfrak{p}^+)_{q,-1} \cdot \mathbb{C}[\mathfrak{p}^+]_{q,-(j-1)}$$

and the relations

$$\mathbb{C}[\mathfrak{p}^+]_{q,-j} \cdot \omega = \omega \cdot \mathbb{C}[\mathfrak{p}^-]_{q,j} = 0$$

imply

$$\mathbb{C}[\mathfrak{p}^+]_{q,-(j+1)} \cdot \partial\omega = \partial\omega \cdot \mathbb{C}[\mathfrak{p}^-]_{q,j+1} = 0,$$
$$\mathbb{C}[\mathfrak{p}^+]_{q,-(j+1)} \cdot \overline{\partial}\omega = \overline{\partial}\omega \cdot \mathbb{C}[\mathfrak{p}^-]_{q,j+1} = 0. \qquad \square$$

Let us show that the structure of the $U_q\mathfrak{g}$-covariant $\mathrm{Pol}(\mathfrak{p}^-)_q$-bimodule in $\Omega(\mathbb{D})_q$ and a continuity argument allow one to endow $\Omega(\mathbb{D})_q$ with a structure of the $U_q\mathfrak{g}$-covariant $\mathcal{D}(\mathbb{D})_q$-bimodule.

We restrict ourselves temporarily to finite type distributions, that is, the formal series

$$f=\sum_{\{i,j\mid |i-j|\le N(f)\}} f_{ij},\qquad f_{ij}\in\mathbb{C}[\mathfrak{p}^-]_{q,i}\cdot\mathbb{C}[\mathfrak{p}^+]_{q,-j}.$$

The sequence of distributions f_n, $n\in\mathbb{N}$, will be said to be convergent to zero if it converges to zero in $\mathcal{D}(\mathbb{D})_q'$ and $N(f_n)$ can be chosen so that $\sup_n N(f_n)<\infty$. As in the previous subsections, the structure of $U_q\mathfrak{g}$-module algebra in $\mathrm{Fun}(\mathbb{D})_q$ can be extended from $\mathrm{Pol}(\mathfrak{p}^-)_q$ by continuity. (In fact, if $f_n\to f$, $g_n\to g$, then $f_ng_n\to fg$.) One can impose in a similar way the forms ω of finite type, the convergence $\omega_n\to\omega$, and to extend by continuity the structure of $U_q\mathfrak{g}$-module algebra from $\Omega(\mathfrak{p}^-)_q$ to $\Omega(\mathfrak{p}^-)_q\oplus\Omega(\mathbb{D})_q$. After this procedure (1.75) is still valid.

It will be shown in the sequel (see Remark 2.266) that $\partial\mathcal{D}(\mathbb{D})_q$ generates the $\mathcal{D}(\mathbb{D})_q$-bimodule $\Omega^{1,0}(\mathbb{D})_q$, and $\overline{\partial}\mathcal{D}(\mathbb{D})_q$ generates the $\mathcal{D}(\mathbb{D})_q$-bimodule $\Omega^{0,1}(\mathbb{D})_q$. Thus one has

PROPOSITION 2.262. *The pairs* $\left(\Omega^{1,0}(\mathbb{D})_q,\partial\right)$ *and* $\left(\Omega^{0,1}(\mathbb{D})_q,\overline{\partial}\right)$ *are CFODC over the* $U_q\mathfrak{g}$*-module algebra* $\mathcal{D}(\mathbb{D})_q$.

Let us find an explicit form of the element $\overline{\partial}f_0$, where f_0 is the standard generator of the $\mathrm{Pol}(\mathfrak{p}^-)_q$-bimodule $\mathcal{D}(\mathbb{D})_q$; see Subsection 2.4.1.

LEMMA 2.263. *For each* $\omega\in\Lambda^1(\mathfrak{p}^-)_{q,1}$ *one has*

$$f_0\cdot\omega=\omega\cdot f_0.$$

PROOF. Let

$$\widetilde{R}:\mathcal{D}(\mathbb{D})_q'\otimes\Lambda^1(\mathfrak{p}^-)_q\to\Lambda^1(\mathfrak{p}^-)_q\otimes\mathcal{D}(\mathbb{D})_q'$$

be the morphism of $U_q\mathfrak{g}$-modules determined by the action of the universal R-matrix of $U_q\mathfrak{g}$. Then $\widetilde{R}(f_0\otimes\omega)=\omega\otimes f_0$, which can be deduced from the explicit formula (2.11), the relation $F_{l_0}\omega=0$, together with the $U_q\mathfrak{k}$-invariance of f_0:

$$E_jf_0=0,\quad j\ne l_0,\qquad H_if_0=0,\quad i=1,2,\ldots,l.$$

It remains to observe that the structure of $\Lambda^1(\mathfrak{p}^-)_q$-bimodule in $\Omega(\mathbb{D})_q'$, just as that in $\Omega(\mathfrak{p}^-)_q$, can be described in terms of the operator $\widetilde{R}$. □

An application of the involution $*$ yields the relation $\omega\cdot f_0=f_0\cdot\omega$ for all $\omega\in\Lambda^1(\mathfrak{p}^+)_{q,-1}$. Let us use the standard notation

$$dz=\partial z,\quad dz^*=\overline{\partial}z^*,\qquad z\in\mathbb{C}[\mathfrak{p}^-]_{q,1},$$

and the Hermitian form in $\mathbb{C}[\mathfrak{p}^-]_q$ introduced in Subsection 2.3.3.

PROPOSITION 2.264. *Let* $\{z_j\}$ *be an orthogonal basis of the Hermitian vector space* $\mathbb{C}[\mathfrak{p}^-]_{q,1}$. *Then one has*

$$\overline{\partial}f_0=-\sum_{j=1}^{\dim\mathfrak{p}^-}\frac{1}{(z_j,z_j)}z_j\cdot f_0\cdot dz_j^*. \tag{2.179}$$

PROOF. We are going to produce a system of equations which determines uniquely the differential form $\omega = \overline{\partial} f_0$. After that it will remain to verify that the right-hand side of (2.179) is a solution of that system of equations. Let

$$\Psi = \sum_{i=1}^{\dim \mathfrak{p}^-} \frac{z_i z_i^*}{(z_i, z_i)}.$$

Then $\overline{\partial}\Psi = \sum_i (z_i, z_i)^{-1} z_i dz_i^*$. An application of $\overline{\partial}$ to both sides of $\Psi f_0 = 0$ yields

$$\sum_i (z_i, z_i)^{-1} z_i dz_i^* f_0 + \sum_i (z_i, z_i)^{-1} z_i z_i^* \overline{\partial} f_0 = 0.$$

Displacing dz_i^* with f_0, we obtain the first equation for $\omega = \overline{\partial} f_0$:

$$\Psi\omega = -\sum_i (z_i, z_i)^{-1} z_i f_0 dz_i^*. \tag{2.180}$$

One can extend the linear operator $H_{\mathbb{S}}$ to $\Omega^{0,1}(\mathbb{D})'_q$ by continuity. The second equation is due to the fact that $\overline{\partial}$ commutes with $H_{\mathbb{S}}$:

$$H_{\mathbb{S}}\omega = 0. \tag{2.181}$$

The differential form $-\sum_i (z_i, z_i)^{-1} z_i f_0 dz_i^*$ is a solution of the system of equations (2.180) and (2.181) in $\Omega^{0,1}(\mathbb{D})'_q$. It remains to prove the uniquenes for a solution of this system, that is, to prove that the corresponding homogeneous system of equations

$$\Psi\omega = H_{\mathbb{S}}\omega = 0, \qquad \omega \in \Omega^{0,1}(\mathbb{D})'_q, \tag{2.182}$$

has only zero solution. By Corollary 2.141,

$$\Omega^{0,1}(\mathbb{D})'_q \cong \prod_{j,k} \mathbb{C}[\mathfrak{p}^-]_{q,j} f_0 \Lambda^1(\mathfrak{p}^+)_{q,-k}.$$

If the system (2.182) has a nonzero solution, then there exists a nonzero element

$$f \in \mathbb{C}[\mathfrak{p}^-]_{q,j} f_0, \qquad j \neq 0,$$

such that $\Psi f = 0$, hence $f^* \Psi f = 0$. This contradicts to positivity of the standard Hermitian form in $\mathbb{C}[\mathfrak{p}^-]_q$; see Proposition 2.98. □

COROLLARY 2.265. $\partial f_0 = -\sum_{j=1}^{\dim \mathfrak{p}^-} \frac{1}{(z_j, z_j)} dz_j \cdot f_0 \cdot z_j^*$.

REMARK 2.266. Due to (2.178),

$$\mathcal{D}(\mathbb{D})_q \cdot \partial f_0 \cdot \mathcal{D}(\mathbb{D})_q = \Omega^{1,0}(\mathbb{D})_q, \qquad \mathcal{D}(\mathbb{D})_q \cdot \overline{\partial} f_0 \cdot \mathcal{D}(\mathbb{D})_q = \Omega^{0,1}(\mathbb{D})_q.$$

The above results allow us to describe the universal differential calculi over $\mathrm{Pol}(\mathfrak{p}^-)_q \oplus \mathcal{D}(\mathbb{D})_q$ associated to the CFODC

$$\left(\Omega^{1,0}(\mathfrak{p}^-)_q \oplus \Omega^{1,0}(\mathbb{D})_q, \partial\right), \quad \left(\Omega^{0,1}(\mathfrak{p}^-)_q \oplus \Omega^{0,1}(\mathbb{D})_q, \overline{\partial}\right), \quad \left(\Omega^{1}(\mathfrak{p}^-)_q \oplus \Omega^{1}(\mathbb{D})_q, d\right).$$

2.6.5. Invariant differential operators on $K\backslash G$. Consider a Hermitian symmetric pair $(\mathfrak{g}, \mathfrak{k})$ and the $U_q\mathfrak{g}$-module algebra

$$\mathbb{C}[K\backslash G]_q = \{f \in \mathbb{C}[G]_q |\, L_{\mathrm{reg}}(\xi)f = \varepsilon(\xi)f, \quad \xi \in U_q\mathfrak{k}\}.$$

As was noted in Subsections 2.2.5 and 2.5.8, this algebra is a quantum analog of the algebra of regular functions on the affine algebraic variety $K\backslash G$.

Let us introduce an algebra of linear operators in $\mathbb{C}[K\backslash G]_q$ which is a quantum analog of the algebra of G-invariant differential operators on $K\backslash G$. In the classical case $q = 1$, the approach we use here is described in [**83, 198**].

We follow the terminology of Drinfeld [**79**] in calling $(U_q\mathfrak{g})^*$ the algebra of functions on the *formal quantum group* G. It is a $U_q\mathfrak{g}$-module algebra:

$$R_{\mathrm{reg}}(\xi)l(\eta) = l(\eta\xi), \qquad \xi, \eta \in U_q\mathfrak{g}, \quad l \in (U_q\mathfrak{g})^*.$$

Its $U_q\mathfrak{g}$-module subalgebra

$$\mathrm{Hom}_{U_q\mathfrak{k}}(U_q\mathfrak{g}, \mathbb{C}) \cong \left(\mathbb{C} \otimes_{U_q\mathfrak{k}} U_q\mathfrak{g}\right)^*$$

is called the algebra of functions on the *formal quantum homogeneous space* $K\backslash G$.

By definition, $\mathbb{C}[G]_q$ is the largest weight locally finite-dimensional submodule of the $U_q\mathfrak{g}$-module $(U_q\mathfrak{g})^*$. Hence $\mathbb{C}[K\backslash G]_q$ is the largest weight locally finite-dimensional submodule of the $U_q\mathfrak{g}$-module $\mathrm{Hom}_{U_q\mathfrak{k}}(U_q\mathfrak{g}, \mathbb{C})$.

Endomorphisms of the right $U_q\mathfrak{g}$-module $U_q\mathfrak{g}$ are indexed by the elements $a \in U_q\mathfrak{g}$ and have the form $\eta \mapsto a\eta$. The dual linear operators

$$L_{\mathrm{reg}}(a) : (U_q\mathfrak{g})^* \to (U_q\mathfrak{g})^*, \qquad L_{\mathrm{reg}}(a) : f(\eta) \mapsto f(a\eta),$$

will be treated as invariant differential operators on the formal quantum group G.

Similarly, endomorphisms of the right $U_q\mathfrak{g}$-module $\mathbb{C} \otimes_{U_q\mathfrak{k}} U_q\mathfrak{g}$ are indexed by its $U_q\mathfrak{k}$ invariant elements

$$\mathrm{End}_{U_q\mathfrak{g}}\left(\mathbb{C} \otimes_{U_q\mathfrak{k}} U_q\mathfrak{g}\right) \cong \mathrm{Hom}_{U_q\mathfrak{k}}\left(\mathbb{C}, \mathbb{C} \otimes_{U_q\mathfrak{k}} U_q\mathfrak{g}\right) \cong \left(\mathbb{C} \otimes_{U_q\mathfrak{k}} U_q\mathfrak{g}\right)^{U_q\mathfrak{k}}$$

and have the form $1 \otimes \eta \mapsto (1 \otimes a)(1 \otimes \eta)$, where $1 \otimes a \in \left(\mathbb{C} \otimes_{U_q\mathfrak{k}} U_q\mathfrak{g}\right)^{U_q\mathfrak{k}}$. The dual linear operators will be called invariant differential operators on the formal quantum homogeneous spaces $K\backslash G$. These form an algebra $\mathbf{D}(K\backslash G)_q \cong \mathrm{End}_{U_q\mathfrak{g}}\left(\mathbb{C} \otimes_{U_q\mathfrak{k}} U_q\mathfrak{g}\right)^{\mathrm{op}}$. An operator from $\mathbf{D}(K\backslash G)_q$ is an endomorphism of the $U_q\mathfrak{g}$-module $\mathrm{Hom}_{U_q\mathfrak{k}}(U_q\mathfrak{g}, \mathbb{C})$, hence it maps the subspace $\mathbb{C}[K\backslash G]_q$ into itself. It follows from the definitions that the *invariant differential operator* associated to an element

$$1 \otimes a \in \left(\mathbb{C} \otimes_{U_q\mathfrak{k}} U_q\mathfrak{g}\right)^{U_q\mathfrak{k}} \cong \mathrm{End}_{U_q\mathfrak{g}}\left(\mathbb{C} \otimes_{U_q\mathfrak{k}} U_q\mathfrak{g}\right),$$

is the restriction of $L_{\mathrm{reg}}(a)$ to the subspace $\mathbb{C}[K\backslash G]_q$.

REMARK 2.267. An application of the antipode S makes it easy to replace right $U_q\mathfrak{g}$-modules by left $U_q\mathfrak{g}$-modules. Namely,

$$\mathbf{D}(K\backslash G)_q \cong \mathrm{End}_{U_q\mathfrak{g}}\left(U_q\mathfrak{g} \otimes_{U_q\mathfrak{k}} \mathbb{C}\right)^{\mathrm{op}},$$

and the invariant differential operator associated to an element

$$a \otimes 1 \in \left(U_q\mathfrak{g} \otimes_{U_q\mathfrak{k}} \mathbb{C}\right)^{U_q\mathfrak{k}} \cong \mathrm{Hom}_{U_q\mathfrak{k}}\left(\mathbb{C}, U_q\mathfrak{g} \otimes_{U_q\mathfrak{k}} \mathbb{C}\right) \cong \mathrm{End}_{U_q\mathfrak{g}}\left(U_q\mathfrak{g} \otimes_{U_q\mathfrak{k}} \mathbb{C}\right)$$

is the restriction of $L_{\mathrm{reg}}(S(a))$ to the subspace $\mathbb{C}[K\backslash G]_q$.

Let us turn from spaces of functions to spaces of sections of homogeneous vector bundles on $K\backslash G$.

Consider a finite-dimensional weight $U_q\mathfrak{k}$-module V. We will call the vector space

$$\mathrm{Hom}_{U_q\mathfrak{k}}(U_q\mathfrak{g}, V) \cong \left(V^* \otimes_{U_q\mathfrak{k}} U_q\mathfrak{g}\right)^*$$

the space of sections of the homogeneous vector bundle on the formal quantum homogeneous space $K\backslash G$. It is endowed in a natural way with a structure of $U_q\mathfrak{g}$-covariant $\mathrm{Hom}_{U_q\mathfrak{k}}(U_q\mathfrak{g}, \mathbb{C})$-bimodule: it suffices to use the map

$$\mathrm{Hom}_{U_q\mathfrak{k}}(U_q\mathfrak{g}, V_1) \times \mathrm{Hom}_{U_q\mathfrak{k}}(U_q\mathfrak{g}, V_2) \to \mathrm{Hom}_{U_q\mathfrak{k}}(U_q\mathfrak{g}, V_1 \otimes V_2),$$

where $(f_1(\xi), f_2(\xi)) \mapsto (f_1 \otimes f_2)\Delta(\xi)$. For that, we set first $V_1 = V$, $V_2 = \mathbb{C}$, and then $V_1 = \mathbb{C}$, $V_2 = V$.

Let us introduce invariant differential operators just as in the special case $V_1 = V_2 = \mathbb{C}$. If the linear map $A : V_2^* \otimes_{U_q\mathfrak{k}} U_q\mathfrak{g} \to V_1^* \otimes_{U_q\mathfrak{k}} U_q\mathfrak{g}$ is a morphism of right $U_q\mathfrak{g}$-modules, then the dual linear map

$$A^* : \mathrm{Hom}_{U_q\mathfrak{k}}(U_q\mathfrak{g}, V_1) \to \mathrm{Hom}_{U_q\mathfrak{k}}(U_q\mathfrak{g}, V_2)$$

is also a morphism of $U_q\mathfrak{g}$-modules. Such linear maps serve as quantum analogs of invariant differential operators on $K\backslash G$. Thus invariant differential operators are indexed by the elements of $\mathrm{Hom}_{U_q\mathfrak{g}}\left(V_2^* \otimes_{U_q\mathfrak{k}} U_q\mathfrak{g}, V_1^* \otimes_{U_q\mathfrak{k}} U_q\mathfrak{g}\right)$, that is, by the elements of the space

$$\mathrm{Hom}_{U_q\mathfrak{k}}\left(V_2^*, V_1^* \otimes_{U_q\mathfrak{k}} U_q\mathfrak{g}\right) \cong \left(\left(V_1^* \otimes_{U_q\mathfrak{k}} U_q\mathfrak{g}\right) \otimes V_2\right)^{U_q\mathfrak{k}}.$$

To describe the action of these operators, we follow [**198**]. To every element $\sum_j l_j \otimes \xi_j \otimes v_j \in V_1^* \otimes U_q\mathfrak{g} \otimes V_2$, we assign the linear map from $\mathrm{Hom}(U_q\mathfrak{g}, V_1)$ to $\mathrm{Hom}(U_q\mathfrak{g}, V_2)$:

$$f(\eta) \mapsto \sum_j l_j(f(\xi_j\eta))v_j, \qquad \eta \in U_q\mathfrak{g}.$$

This linear map is obviously a quantum analog of an invariant differential operator on G.

Each element

$$A \in \left(V_1^* \otimes_{U_q\mathfrak{k}} U_q\mathfrak{g}\right) \otimes V_2 \cong \mathrm{Hom}\left(V_2^*, V_1^* \otimes_{U_q\mathfrak{k}} U_q\mathfrak{g}\right)$$

determines the linear map $D_A : \mathrm{Hom}_{U_q\mathfrak{k}}(U_q\mathfrak{g}, V_1) \to \mathrm{Hom}(U_q\mathfrak{g}, V_2)$, which is a morphism of $U_q\mathfrak{g}$-modules. Suppose that

$$A \in \mathrm{Hom}_{U_q\mathfrak{k}}\left(V_2^*, V_1^* \otimes_{U_q\mathfrak{k}} U_q\mathfrak{g}\right).$$

Then $D_A(\mathrm{Hom}_{U_q\mathfrak{k}}(U_q\mathfrak{g}, V_1)) \subset \mathrm{Hom}_{U_q\mathfrak{k}}(U_q\mathfrak{g}, V_2)$. In fact, if π_j are the representations corresponding to the $U_q\mathfrak{k}$-modules V_j, and $A = \sum_i l_i \otimes \xi_i \otimes v_i$, then for all $\zeta \in U_q\mathfrak{k}$, $\eta \in U_q\mathfrak{g}$ one has

$$D_A f(\zeta\eta) = \sum_i l_i(\pi_1(\zeta)f(\xi_i\eta))v_i = \sum_i l_i(f(\xi_i\eta))\pi_2(\zeta)v_i = \pi_2(\zeta)\, D_A f(\eta),$$

where $f \in \mathrm{Hom}_{U_q\mathfrak{k}}(U_q\mathfrak{g}, V_1)$.

Let us pass from formal sections to regular sections of homogeneous vector bundles in question and their quantum analogs. Let $\mathbb{V}_1$ and $\mathbb{V}_2$ be the largest weight locally finite-dimensional submodules of the $U_q\mathfrak{g}$-modules $\mathrm{Hom}_{U_q\mathfrak{k}}(U_q\mathfrak{g}, V_1)$ and $\mathrm{Hom}_{U_q\mathfrak{k}}(U_q\mathfrak{g}, V_2)$, respectively. These are $U_q\mathfrak{g}$-covariant $\mathbb{C}[K\backslash G]_q$-bimodules. They are quantum analogs of spaces of (regular) sections of homogeneous vector bundles on $K\backslash G$. Evidently, $D_A : \mathbb{V}_1 \to \mathbb{V}_2$, and we thus get quantum analogs of invariant differential operators in spaces of sections of homogeneous vector bundles.

REMARK 2.268. Just as in the special case $V_1 = V_2 = \mathbb{C}$, the antipode S can be used to replace right $U_q\mathfrak{g}$-modules by left $U_q\mathfrak{g}$-modules.

2.6.6. Invariant differential operators on $K\backslash G$: the sequel. It is well known [**124**, p. 329] that the algebra $\mathbf{D}(K\backslash G)$ of invariant differential operators on $K\backslash G$ is commutative. Let us obtain a q-analog of this result by proving commutativity of the algebra $\mathbf{D}(K\backslash G)_q$.

The natural representation of this algebra in the vector space $\mathbb{C}[K\backslash G]_q$ is faithful since the pairing

$$\mathbb{C}[K\backslash G]_q \times \left(\mathbb{C} \otimes_{U_q\mathfrak{k}} U_q\mathfrak{g}\right) \to \mathbb{C}, \qquad f \times (1 \otimes \xi) \mapsto f(\xi),$$

is nondegenerate (see a proof, e.g., in [**117**]). Hence the commutativity of $\mathbf{D}(K\backslash G)_q$ follows from the commutativity of endomorphism algebra of the $U_q\mathfrak{g}$-module $\mathbb{C}[K\backslash G]_q$. The latter claim, in turn, is due to the multiplicity free action of $U_q\mathfrak{g}$ in $\mathbb{C}[K\backslash G]_q$:

$$\mathbb{C}[K\backslash G]_q \approx \bigoplus_{\lambda \in P_+^{\mathrm{spher}}} L(\lambda); \tag{2.183}$$

see Proposition 2.40 and Subsection 2.1.8.

Here we present the description of $\mathbf{D}(K\backslash G)_q$ used by Baldoni and Frajria [**21**, p. 1187], and discuss the main result of their work.

Recall that $U_q\mathfrak{g}$ is a $U_q\mathfrak{k}$-module algebra with respect to the adjoint action, and that the morphism of $U_q\mathfrak{k}$-modules

$$m : U_q\mathfrak{k} \otimes U_q\mathfrak{p}^- \otimes U_q\mathfrak{p}^+, \qquad m : \xi \otimes \eta \otimes \zeta \mapsto \xi\eta\zeta,$$

is one-to-one; see Subsection 2.3.6. Hence,

$$\mathbf{D}(K\backslash G)_q \cong \left(\mathbb{C} \otimes_{U_q\mathfrak{k}} U_q\mathfrak{g}\right)^{U_q\mathfrak{k}} \cong (U_q\mathfrak{p}^- U_q\mathfrak{p}^+)^{U_q\mathfrak{k}},$$

and the invariant differential operator associated to $a \in (U_q\mathfrak{p}^- U_q\mathfrak{p}^+)^{U_q\mathfrak{k}}$ is the restriction of the linear operator $L_{\mathrm{reg}}(a)$ to the invariant subspace $\mathbb{C}[K\backslash G]_q$.

EXAMPLE 2.269. The homogeneous components $(U_q\mathfrak{p}^+)_1$ and $(U_q\mathfrak{p}^-)_{-1}$ are simple finite-dimensional weight $U_q\mathfrak{k}$-modules dual to each other. Hence,

$$\dim\left((U_q\mathfrak{p}^-)_{-1}(U_q\mathfrak{p}^+)_1\right)^{U_q\mathfrak{k}} = \dim\left((U_q\mathfrak{p}^-)_{-1} \otimes (U_q\mathfrak{p}^+)_1\right)^{U_q\mathfrak{k}} = 1.$$

Therefore there exists a unique up to a scalar factor element of $D(K\backslash G)_q$ associated to a nonzero $U_q\mathfrak{k}$-invariant element of $(U_q\mathfrak{p}^-)_{-1}(U_q\mathfrak{p}^+)_1$. It is a quantum analog of the invariant Laplacian. In the simplest case $\mathfrak{g} = \mathfrak{sl}_2$ the element FE belongs to $\left((U_q\mathfrak{p}^-)_{-1}(U_q\mathfrak{p}^+)_1\right)^{U_q\mathfrak{k}}$, and for all $f \in \mathbb{C}[K\backslash \mathrm{SL}_2]_q$, one has $L_{\mathrm{reg}}(FE)f = L_{\mathrm{reg}}(C_q)f = R_{\mathrm{reg}}(C_q)f$, where C_q is the quantum analog of the

Casimir element; see (1.46). Its links to the invariant Laplacian on the quantum disc are discussed in Subsections 1.5.4 and 1.7.1.

A similar argument used in [**21**] allows one to introduce a distinguished basis of the vector space $D(K\backslash G)_q$. Components of the decompositions

$$U_q\mathfrak{p}^+ = \bigoplus_{\lambda\in\mathcal{A}_+}(U_q\mathfrak{p}^+)_\lambda, \qquad U_q\mathfrak{p}^- = \bigoplus_{\lambda\in\mathcal{A}_+}(U_q\mathfrak{p}^-)_{-\lambda} \tag{2.184}$$

in the category of $U_q\mathfrak{k}$-modules substitute simple $U_q\mathfrak{k}$-modules $(U_q\mathfrak{p}^+)_1$, $(U_q\mathfrak{p}^-)_{-1}$. Here $(U_q\mathfrak{p}^+)_\lambda \approx L(\mathfrak{k},\lambda)$, $(U_q\mathfrak{p}^-)_{-\lambda} \approx L(\mathfrak{k},\lambda)^*$, and the set $\mathcal{A}_+$ is given by (2.157). The decompositions (2.184) can be easily produced using the isomorphism of $U_q\mathfrak{k}$-module algebras $U_q\mathfrak{p}^+ \cong \mathbb{C}[\mathfrak{p}^-]_q$ (see Subsection 2.3.5), the Hua–Schmid decompositions (see Proposition 2.228), and the results of Subsection 2.1.8.

REMARK 2.270. A replacement of the trivial $U_q\mathfrak{k}$-module $\mathbb{C}$ in the above argument by a general one-dimensional $U_q\mathfrak{k}$-module leads to quantum analogs of the Shimura operators [**275**].

It is proved in [**21**] that for nonexceptional Lie algebras $\mathfrak{g}$ all the invariant differential operators in $\mathbb{C}[K\backslash G]_q$ are given by the action of the elements of the center $Z(U_q\mathfrak{g})$ of $U_q\mathfrak{g}$. It is known from [**125**] that this restriction to $\mathfrak{g}$ cannot be discarded even in the case $q=1$. These are just nonexceptional Lie algebras that are subjects of the main result of [**21**]. Namely, that work contains a description of the image of $\mathbf{D}(K\backslash G)_q$ under the Harish-Chandra homomorphism, which differs by an inessential shift from the embedding of $\mathbf{D}(K\backslash G)_q$ to the function algebra on P_+^{spher} corresponding to the decomposition (2.183).

The constructions of Subsection 2.6.3 supply the first-order differential calculi $(\Omega^{1,0}(\mathfrak{p}^-)_q,\partial)$, $(\Omega^{0,1}(\mathfrak{p}^-)_q,\bar\partial)$ over the $U_q\mathfrak{g}$-module algebra $\mathrm{Pol}(\mathfrak{p}^-)_q$. Let us sketch a different approach to those differential calculi. To simplify matters, we restrict ourselves to the first of those and assume that $\mathbb{D}\subset\mathfrak{p}^-$ is a bounded symmetric domain of tube type; see Subsection 2.3.8.

Consider the invariant differential operator ∂' associated to the morphism of $U_q\mathfrak{k}$-modules

$$(U_q\mathfrak{p}^+)_1 \to \mathbb{C}\otimes_{U_q\mathfrak{k}} U_q\mathfrak{g}, \qquad E_{l_0}\mapsto 1\otimes E_{l_0}.$$

It was shown in [**117**] that

$$\partial'(f_1f_2) = \partial'(f_1)f_2 + f_1\partial'(f_2), \qquad f_1,f_1\in\mathbb{C}[K\backslash G]_q,$$

i.e., ∂' supplies a CFODC over $\mathbb{C}[K\backslash G]_q$. Let us extend it to the localization $\mathbb{C}[K\backslash G]_{q,x}$ of the algebra $\mathbb{C}[K\backslash G]_q$ by the multiplicative system $x^{\mathbb{Z}_+}$ (the existence of such an extension can be proved in the same way as Proposition 2.185, using the commutation relations between x and the matrix elements $c^\Lambda_{\lambda,i;\mu,j}$; see Subsection 2.2.2).

It remains to transfer the above CFODC from $\mathbb{C}[K\backslash G]_{q,x}$ to the localization $\mathrm{Pol}(\mathfrak{p}^-)_{q,y}$ of $\mathrm{Pol}(\mathfrak{p}^-)_q$ with respect to the multiplicative system $y^{\mathbb{Z}_+}$ using the canonical isomorphism $\mathcal{I}:\mathrm{Pol}(\mathfrak{p}^-)_{q,y}\xrightarrow{\approx}\mathbb{C}[K\backslash G]_{q,x}$; see Subsection 2.5.8.

The fact that the above CFODC over the subalgebra $\mathrm{Pol}(\mathfrak{p}^-)_q\subset\mathrm{Pol}(\mathfrak{p}^-)_{q,y}$ is isomorphic to $\left(\Omega^{1,0}(\mathfrak{p}^-)_q,\partial\right)$ can be easily derived from Proposition 2.179 by a duality argument; see Subsections 2.4.5, 2.6.5, 2.6.1, and 2.6.3.

Recall the notation $K_1 = w_0Kw_0^{-1}$; see (2.153). Proposition 2.217 hints that every invariant differential operator in $\mathbb{C}[K_1\backslash G]_q$ should determine an invariant differential operator on a quantum bounded symmetric domain $\mathbb{D}$. Let us describe this operator.

Consider the subalgebra $U_q\mathfrak{k}_1$ generated by $\{E_j, F_j\}_{j\neq l_1} \cup \{K_i^{\pm 1}\}_{i=1,2,\dots,l}$, where l_1 is determined by $\bar{\omega}_{l_1} = -w_0\bar{\omega}_{l_0}$; see Subsection 2.5.4. Of course, $U_q\mathfrak{k}_1$ is a quantum universal enveloping algebra corresponding to the group K_1.

Consider the algebra $\mathbb{C}[G]_q^*$. Since $\widetilde{w}_0 \in \mathbb{C}[G]_q^*$ and $U_q\mathfrak{g}$ is embedded into $\mathbb{C}[G]_q^*$, one has $\widetilde{w}_0 U_q\mathfrak{g} \subset \mathbb{C}[G]_q^*$. Recall that $\mathbb{C}[G]_q^*$ is a $U_q\mathfrak{g}$-bimodule, and $\widetilde{w}_0 U_q\mathfrak{g}$ is its subbimodule

$$\xi \times \eta : \widetilde{w}_0\zeta \mapsto \widetilde{w}_0(\widetilde{w}_0^{-1}\xi\widetilde{w}_0)\zeta\eta; \tag{2.185}$$

see Proposition 2.179.

PROPOSITION 2.271. *Let A be an invariant differential operator in $\mathbb{C}[K_1\backslash G]_q$ and A^* the dual linear operator in $\mathbb{C}[K_1\backslash G]_q^*$. Then*

$$A^*\left(\mathbb{C}\otimes_{U_q\mathfrak{k}_1}\widetilde{w}_0 U_q\mathfrak{g}\right) \subset \mathbb{C}\otimes_{U_q\mathfrak{k}_1}\widetilde{w}_0 U_q\mathfrak{g}. \tag{2.186}$$

PROOF. It follows from Proposition 2.179 that the inclusion (2.186) is equivalent to

$$A^*\left(\mathbb{C}\otimes_{U_q\mathfrak{k}_1} U_q\mathfrak{g}\widetilde{w}_0\right) \subset \mathbb{C}\otimes_{U_q\mathfrak{k}_1} U_q\mathfrak{g}\widetilde{w}_0,$$

and the latter inclusion, in turn, follows from the obvious inclusion

$$A^*\left(\mathbb{C}\otimes_{U_q\mathfrak{k}_1} U_q\mathfrak{g}\right) \subset \mathbb{C}\otimes_{U_q\mathfrak{k}_1} U_q\mathfrak{g},$$

together with the fact that A is a morphism of $U_q\mathfrak{g}$-modules. □

Using the antipode S, we endow $\mathcal{D}(\mathbb{D})_q$ with a structure of right $U_q\mathfrak{g}$-module isomorphic to $\mathbb{C}\otimes_{U_q\mathfrak{k}} U_q\mathfrak{g}$:

$$\mathbb{C}\otimes_{U_q\mathfrak{k}} U_q\mathfrak{g} \to \mathcal{D}(\mathbb{D})_q, \qquad 1\otimes 1 \mapsto f_0;$$

see Subsection 2.4.5. It is easy to transfer the linear operator described in Proposition 2.271 from the vector space $\mathbb{C}\otimes_{U_q\mathfrak{k}_1}(\widetilde{w}_0 U_q\mathfrak{g})$ to $\mathbb{C}\otimes_{U_q\mathfrak{k}} U_q\mathfrak{g}$ and, finally, to $\mathcal{D}(\mathbb{D})_q$. For that, it suffices to use the isomorphism of right $U_q\mathfrak{g}$-modules

$$\mathbb{C}\otimes_{U_q\mathfrak{k}} U_q\mathfrak{g} \to \mathbb{C}\otimes_{U_q\mathfrak{k}_1}(\widetilde{w}_0 U_q\mathfrak{g}), \qquad 1\otimes 1 \mapsto 1\otimes\widetilde{w}_0. \tag{2.187}$$

In fact, existence of the morphism (2.187) follows from the $U_q\mathfrak{k}$-invariance of $1\otimes\widetilde{w}_0 \in \mathbb{C}\otimes_{U_q\mathfrak{k}_1} U_q\mathfrak{g}$; see (2.185). The uniqueness of such a morphism is obvious, and its bijectivity follows from the results of Subsections 2.3.6 and 2.5.2.

We thus obtain an invariant differential operator in $\mathcal{D}(\mathbb{D})_q$ which corresponds to the invariant differential operator in $\mathbb{C}[K_1\backslash G]_q$. Note that it is easy to pass from an invariant differential operator in the space $\mathcal{D}(\mathbb{D})_q$ of finite functions to an invariant differential operator in the space $\mathcal{D}(\mathbb{D})_q'$ of distributions on a quantum bounded symmetric domain.

2.6.7. Invariant differential operators on the big cell of the generalized flag space. Consider the $\mathfrak{g}$-module $L(\overline{\omega}_{l_0})$ with highest weight $\overline{\omega}_{l_0}$ and the highest weight vector $v(\overline{\omega}_{l_0})$. It is well known that the G-orbit of the point $\mathbb{C}v(\overline{\omega}_{l_0})$ of the projective space $P(L(\overline{\omega}_{l_0}))$ is the generalized flag space and possesses the standard partition into Schubert cells X^w of codimension $l(w)$, indexed by $w \in W^{\mathbb{S}}$. The point $\mathbb{C}v(\overline{\omega}_{l_0})$ belongs to the big cell X^e of this partition.

Let us reproduce constructions of Subsection 2.6.5 with $\mathfrak{k}$ being replaced by the parabolic Lie subalgebra $\mathfrak{q}^+ \supset \mathfrak{k}$ generated by

$$\{E_i, H_i | \, i = 1, 2, \ldots, l\} \cup \{F_j | \, j \neq l_0\}.$$

The $U_q\mathfrak{g}$-module algebra

$$\mathbb{C}[U]_q = \{f \in \mathrm{Hom}_{U_q\mathfrak{q}^+}(U_q\mathfrak{g}, \mathbb{C}) | \, \dim(U_q\mathfrak{h} \cdot f) < \infty\},$$

is a quantum analog of the $U\mathfrak{g}$-module algebra $\mathbb{C}[X^e]$ of regular functions on X^e. The $U_q\mathfrak{g}$-module algebra $\mathrm{Hom}_{U_q\mathfrak{q}^+}(U_q\mathfrak{g}, \mathbb{C}) \cong \left(\mathbb{C} \otimes_{U_q\mathfrak{q}^+} U_q\mathfrak{g}\right)^*$ is a quantum analog of the algebra of formal series on X^e.

Let us pass from functions on X^e to sections of homogeneous vector bundles.

Every finite-dimensional weight $U_q\mathfrak{k}$-module V is endowed with a structure of $U_q\mathfrak{q}^+$-module using the natural projection morphism of algebras $U_q\mathfrak{q}^+ \to U_q\mathfrak{k}$. The weight $U_q\mathfrak{g}$-module

$$\mathbb{V} = \left\{ f \in \mathrm{Hom}_{U_q\mathfrak{q}^+}(U_q\mathfrak{g}, V) \middle| \, \dim(U_q\mathfrak{h} \cdot f) < \infty \right\},$$

is a quantum analog of the space of sections of a homogeneous vector bundle over X^e with fiber V, and the quantum analog for the corresponding space of formal series is the vector space $\mathrm{Hom}_{U_q\mathfrak{q}^+}(U_q\mathfrak{g}, V)$.

The approach to producing spaces of sections of homogeneous vector bundles described here is well known in the classical case $q = 1$ [**330**, p. 52].

In what follows we essentially use the fact that the natural pairing

$$\mathrm{Hom}_{U_q\mathfrak{q}^+}(U_q\mathfrak{g}, V) \times (V^* \otimes_{U_q\mathfrak{q}^+} U_q\mathfrak{g}) \to \mathbb{C}, \qquad f \times (l \otimes \xi) \mapsto l(f(\xi)),$$

supplies the canonical isomorphism of $U_q\mathfrak{g}$-modules

$$\mathrm{Hom}_{U_q\mathfrak{q}^+}(U_q\mathfrak{g}, V) \cong \left(V^* \otimes_{U_q\mathfrak{q}^+} U_q\mathfrak{g}\right)^*.$$

Here $V^* \otimes_{U_q\mathfrak{q}^+} U_q\mathfrak{g}$ is a right $U_q\mathfrak{g}$-module since, just as in Subsection 2.6.5, we refrain from using the antipode S under the passage to dual spaces.

Let us introduce quantum analogs of invariant differential operators on X^e. Consider the weight finite-dimensional $U_q\mathfrak{k}$-modules V_1, V_2, the associated $U_q\mathfrak{q}^+$-modules V_1, V_2, and the spaces of *sections of homogeneous vector bundles* $\mathbb{V}_1 \subset \mathrm{Hom}_{U_q\mathfrak{q}^+}(U_q\mathfrak{g}, V_1)$, $\mathbb{V}_2 \subset \mathrm{Hom}_{U_q\mathfrak{q}^+}(U_q\mathfrak{g}, V_2)$.

If the linear map $A : V_2^* \otimes_{U_q\mathfrak{q}^+} U_q\mathfrak{g} \to V_1^* \otimes_{U_q\mathfrak{q}^+} U_q\mathfrak{g}$ is a morphism of right $U_q\mathfrak{g}$-modules, then the dual linear map

$$A^* : \mathrm{Hom}_{U_q\mathfrak{q}^+}(U_q\mathfrak{g}, V_1) \to \mathrm{Hom}_{U_q\mathfrak{q}^+}(U_q\mathfrak{g}, V_2)$$

is a morphism of left $U_q\mathfrak{g}$-modules and maps $\mathbb{V}_1$ to $\mathbb{V}_2$. Such operators A^* will be called quantum invariant differential operators on the big cell X^e.

These are indexed by the elements of $\mathrm{Hom}_{U_q\mathfrak{g}}(V_2^* \otimes_{U_q\mathfrak{q}^+} U_q\mathfrak{g}, V_1^* \otimes_{U_q\mathfrak{q}^+} U_q\mathfrak{g})$, i.e., by the elements of $\mathrm{Hom}_{U_q\mathfrak{q}^+}\left(V_2, V_1^* \otimes_{U_q\mathfrak{q}^+} U_q\mathfrak{g}\right) \cong \left(\left(V_1^* \otimes_{U_q\mathfrak{q}^+} U_q\mathfrak{g}\right) \otimes V_2\right)^{U_q\mathfrak{q}^+}$.

REMARK 2.272. 1. In the classical limit $q = 1$, our approach to producing invariant differential operators agrees with the description in the works by Harris and Jakobsen [**115, 135**].

2. Invariant differential operators on the big cell X^e naturally arise in problems of representation theory [**68, 69**], mathematical physics [**26**], and algebraic geometry [**91, 88, 87**].

CHAPTER 3

Conclusion

3.1. Boundaries and spherical principal series

Let us summarize.

- Quantum bounded symmetric domains of a general form are introduced in Chapter 2; in that way we produce a list of subjects to be investigated.
- Results of Chapter 1 on the quantum disc allow us to form a list of questions in noncommutative complex analysis and noncommutative harmonic analysis that we are interested in.

A typical problem is to answer one of these questions for a certain set of quantum bounded symmetric domains.

Such an approach to formulating problems is quite naive because many substantial results on bounded symmetric domains become trivial once we turn to the particular case of real rank 1. This means that the list of problems should be essentially extended. Let us give examples of problems that do not arise in the quantum disc case.

We start with some preliminaries on the simplest series of quantum bounded symmetric domains of tube type.

Let $n \in \mathbb{N}$ and $N = 2n$. We endow the vector space of linear operators in $\mathbb{C}^n$ and the canonically isomorphic vector space Mat_n of complex $n \times n$-matrices with operator norms. It is known that the unit disc $\mathbb{D} = \{\mathbf{z} \in \mathrm{Mat}_n \mid \mathbf{z}\mathbf{z}^* < 1\}$ is a bounded symmetric domain of real rank n; see [**123**, p. 387]. The quantum bounded symmetric domain corresponding to this classical domain is called the quantum matrix ball. In this case

$$l = N-1, \quad l_0 = n, \quad U_q\mathfrak{g} = U_q\mathfrak{sl}_N, \quad U_q\mathfrak{k} = U_q\mathfrak{s}(\mathfrak{gl}_n \times \mathfrak{gl}_n), \quad \mathfrak{p}^- = \mathrm{Mat}_n,$$

and the notation $U_q\mathfrak{su}_{n,n} = (U_q\mathfrak{sl}_N, *)$ is normally used; see (2.65).

Descriptions of the $U_q\mathfrak{sl}_N$-module algebra $\mathbb{C}[\mathfrak{p}^-]_q = \mathbb{C}[\mathrm{Mat}_n]_q$ and the $U_q\mathfrak{su}_{n,n}$-module $*$-algebra $\mathrm{Pol}(\mathfrak{p}^-)_q = \mathrm{Pol}(\mathrm{Mat}_n)_q$ in terms of the generators

$$\{z_a^\alpha\}_{\alpha,\,a=1,2,\dots,n} \subset \mathbb{C}[\mathfrak{p}^-]_{q,1}, \qquad \{z_a^\alpha, (z_a^\alpha)^*\}_{\alpha,\,a=1,2,\dots,n} \subset \mathbb{C}[\mathfrak{p}^-]_{q,1} \bigoplus \mathbb{C}[\mathfrak{p}^+]_{q,-1}$$

and commutation relations [**280, 283**] are well known. The generators z_a^α are quantum analogs of matrix elements of $\mathbf{z} = (z_a^\alpha)$ that are viewed as linear functionals on Mat_n. The commutation relations between them are well known:

$$z_a^\alpha z_b^\beta = \begin{cases} q z_b^\beta z_a^\alpha, & a = b \text{ and } \alpha < \beta \quad \text{or} \quad a < b \text{ and } \alpha = \beta, \\ z_b^\beta z_a^\alpha, & a < b \text{ and } \alpha > \beta, \\ z_b^\beta z_a^\alpha + (q - q^{-1}) z_a^\beta z_b^\alpha, & a < b \text{ and } \alpha < \beta. \end{cases}$$

For $n = 1$, i.e., in the case of quantum disc $\mathbb{D}$, there exists a single spherical principal unitary series of Harish-Chandra modules; see Subsection 1.5.5. It admits a geometrical realization in the space of Laurent polynomials, that is in the space of functions on the unit circle $\overline{\mathbb{D}} \setminus \mathbb{D}$, and the scalar product making these representations unitary is produced using the positive invariant integral on $\overline{\mathbb{D}} \setminus \mathbb{D}$; see Subsection 1.6.3.

When passing from $n = 1$ to $n > 1$, the situation substantially changes even in the classical case $q = 1$. First, the number of principal spherical unitary series increases. Second, the action of the group $\mathrm{Aut}(\mathbb{D})$ on $\overline{\mathbb{D}} \setminus \mathbb{D}$ fails to be transitive, thus preventing the construction of geometrical realizations of principal spherical unitary series in spaces of functions on the topological boundary of the domain $\mathbb{D}$. One needs different boundaries, for example, the Shilov boundary $S(\mathbb{D})$, or the maximal Furstenberg–Satake boundary [**196, 195**].

A quantum analog of the latter one gives a geometrical realization of the *nondegenerate* principal spherical series of unitarizable $U_q\mathfrak{su}_{n,n}$-modules. This series arises in a decomposition of the quasi-regular representation of the $*$-algebra $U_q\mathfrak{su}_{n,n}$ in the pre-Hilbert space $\mathcal{D}(\mathbb{D})_q$ into a direct integral over irreducible $*$-representations. An explicit form of the Plancherel measure is known.

A quantum analog of the Shilov boundary $S(\mathbb{D})$ of the matrix ball and the corresponding *degenerate* principal spherical series of $U_q\mathfrak{su}_{n,n}$-modules was studied in [**33, 318**]. In the classical case $q = 1$, the Shilov boundary $S(\mathbb{D})$ is the set of unitary matrices U_n and, therefore, the real affine variety determined by the matrix equation $\mathbf{z}\mathbf{z}^* = I$. Let us give a similar description of the Shilov boundary of the quantum matrix ball.

Consider the two-sided ideal J of the algebra $\mathrm{Pol}(\mathrm{Mat}_n)_q$ generated by

$$\sum_{j=1}^{n} q^{2n-\alpha-\beta} z_j^\alpha (z_j^\beta)^* - \delta^{\alpha\beta}, \qquad \alpha, \beta = 1, 2, \ldots, n.$$

Obviously, $J = J^*$ and one can verify that this two-sided ideal is $U_q\mathfrak{sl}_N$-invariant. Thus the quotient algebra $\mathbb{C}[S(\mathbb{D})]_q = \mathrm{Pol}(\mathrm{Mat}_n)_q/J$ is a $U_q\mathfrak{su}_{n,n}$-module $*$-algebra. It is called the *algebra of regular functions on the Shilov boundary of a quantum matrix ball.* The canonical homomorphism $\mathrm{Pol}(\mathrm{Mat}_n)_q \to \mathbb{C}[S(\mathbb{D})]_q$ is a quantum analog of the homomorphism which maps a polynomial function to its restriction to the Shilov boundary.

REMARK 3.1. Geometrical approaches to investigation of representations of real semisimple Lie groups have an extended history. We restrict ourselves by references to the papers of M. Graev [**107**], Harish-Chandra [**114**], M. Narasimhan and K. Okamoto [**238**], A.W. Knapp and N.R. Wallach [**176**], J.A. Wolf [**336, 337, 338**], W. Schmid [**269, 270**], L. Barchini [**22, 23**], R. Zierau [**352, 351**], J. Arazy and H. Upmeier [**13**], H. Wong [**340**], J.-T. Chang [**60, 61**]. R.J. Baston and M.G. Eastwood [**26**], L. Barchini, M.G. Eastwood and A.R. Gover [**24**], L. Barchini, A.W. Knapp and R. Zierau [**25, 351**], J.A. Wolf and R. Zierau [**339**]. Normally, the results on real and complex manifolds used in these papers were known long before they found applications to producing the geometrical realizations of representations of real semisimple Lie groups. In the theory of quantum bounded symmetric spaces the situation is substantially different. The literature on noncommutative geometry contains practically no results that are required to produce the geometrical realizations of Harish-Chandra modules over the algebra $U_q\mathfrak{g}$; see Section 2.3. This forces us to develop the required geometrical notions while producing the geometrical realizations of Harish-Chandra modules.

Using the isomorphism of $U_q\mathfrak{s}(\mathfrak{gl}_n \times \mathfrak{gl}_n)$-module algebras

$$\mathbb{C}[S(\mathbb{D})]_q \overset{\approx}{\to} \mathbb{C}[U_n]_q, \qquad z_a^\alpha \mapsto q^{\alpha-n} z_a^\alpha, \quad \alpha, a = 1, 2, \dots, n,$$

one can prove the existence and uniqueness of a $U_q\mathfrak{s}(\mathfrak{gl}_n \times \mathfrak{gl}_n)$-invariant integral

$$\nu : \mathbb{C}[S(\mathbb{D})]_q \to \mathbb{C}, \qquad \nu : f \mapsto \int\limits_{S(\mathbb{D})_q} f d\nu,$$

such that $\int_{S(\mathbb{D})_q} 1 d\nu = 1$. This integral is positive and is used in a construction of integral operators from $\mathbb{C}[S(\mathbb{D})]_q$ to $\mathcal{D}(\mathbb{D})'_q$.

In the classical case $q = 1$, the function

$$P(\mathbf{z}, \zeta) = \frac{\det(1 - \mathbf{z}\mathbf{z}^*)^n}{|\det(1 - \zeta^* \mathbf{z})|^{2n}}, \qquad \mathbf{z} \in \mathbb{D},\ \zeta \in S(\mathbb{D}),$$

on the Cartesian product $\mathbb{D} \times S(\mathbb{D})$ is called the Poisson kernel; see [**129**, p. 98].

The determining property of the Poisson kernel is that the integral operator with this kernel maps unit to unit and is a morphism from the $SU_{n,n}$-module of functions on $S(\mathbb{D})$ to the $SU_{n,n}$-module of functions on $\mathbb{D}$. Such a definition of the Poisson kernel is transferred to the quantum case without any changes. For example, for the quantum disc, the Poisson kernel unessentially differs from the invariant kernel $k_{22}^{-1}\, k_{11}^{-1}$ (see Subsection 1.7.3) and equals

$$P = (1 - z^*\zeta)^{-1}\, (1 - z^* z)\, (1 - z\zeta^*)^{-1}.$$

Similar arguments allow us to introduce a Poisson kernel for the quantum matrix ball. The definitions show that the image of the integral operator with the Poisson kernel consists of solutions of the equation $\widetilde{\Box} u = 0$ both in the classical and the quantum case, where $\widetilde{\Box}$ is the invariant Laplacian.

In the classical case $q = 1$, a surprising effect appears when passing from the disc to the $n \times n$-matrix ball with $n > 1$; this effect was first observed by Hua. Namely, the Poisson integrals do satisfy an invariant system of differential equations, the so-called Hua system,

$$\sum_{a,b=1}^{n} \left(\delta_{a,b} - \sum_{c=1}^{n} \overline{z_c^\alpha} z_c^\beta \right) \frac{\partial^2 u}{\partial z_c^\beta\, \partial \overline{z_c^\alpha}} = 0, \qquad \alpha, \beta = 1, 2, \dots, n;$$

see [**143**]. The equation $\widetilde{\Box} u = 0$ can be derived from the Hua system. Similar results are valid for the quantum $n \times n$-matrix ball, see [**35**].

A quite general approach to the construction of invariant systems of differential equations whose solutions are all the Poisson integrals is given in [**30**, Proposition 4.2]. This approach can be transferred to the quantum case, as it follows from the results of Subsection 2.6.6, where the notion of the invariant system of differential equations in $\mathcal{D}(\mathbb{D})'_q$ is discussed.

3.2. The analytic continuation of holomorphic discrete series and the Penrose transform

Let $\lambda \in \mathbb{C}$. The canonical embedding described in Subsection 2.5 allows us to introduce a representation π_λ of the algebra $U_q\mathfrak{su}_{n,n}$ in the vector space $\mathbb{C}[\mathrm{Mat}_n]_q$.

Let us give an explicit form of the operators of representations corresponding to the generators $E_j, F_j, K_j^{\pm}$:

$$\pi_\lambda(E_j)f = \begin{cases} E_j f, & j \neq n, \\ E_n f - q^{1/2}\frac{1-q^{2\lambda}}{1-q^2}(K_n f)z_n^n, & j = n, \end{cases}$$

$$\pi_\lambda(F_j)f = \begin{cases} F_j f, & j \neq n, \\ q^{-\lambda}F_n f, & j = n, \end{cases}$$

$$\pi_\lambda(K_j^{\pm 1})f = \begin{cases} K_j^{\pm 1} f, & j \neq n, \\ q^{\pm\lambda}K_n^{\pm 1} f, & j = n. \end{cases}$$

We verify that this $U_q\mathfrak{su}_{n,n}$-module admits a unitarization for $\lambda > N-1$. Later we formulate a result that allows one to weaken the assumption on λ; see Proposition 3.3.

In Subsection 2.4.4, an explicit form of a positive $U_q\mathfrak{g}$-invariant integral on $\mathcal{D}(\mathbb{D})_q$ is given. For the quantum $n \times n$-matrix ball, one has

$$\int\limits_{\mathbb{D}_q} f d\nu = (1-q^2)^{n^2}\operatorname{tr}(T_F(f)\Gamma(K_{-2\rho})). \tag{3.1}$$

Here

$$K_{-2\rho} = \prod_{j=1}^{N-1} K_j^{j(N-j)}.$$

As is explained in Subsection 2.4.4, the trace in the right-hand side of (3.1) is well defined. One can easily prove the following statement.

PROPOSITION 3.2. *Let* $\lambda > N-1$ *and*

$$C(\lambda) = \prod_{j=0}^{n-1}\prod_{k=0}^{n-1}(1 - q^{2(\lambda+1-N)}\, q^{2(j+k)}).$$

The linear functional

$$\nu_\lambda : \mathcal{D}(\mathbb{D})_q \to \mathbb{C}, \qquad \nu_\lambda : f \mapsto C(\lambda)\operatorname{tr}(T_F(f)T_F(y)^\lambda\Gamma(K_{-2\rho})),$$

admits an extension by continuity to $\mathrm{Pol}(\mathrm{Mat}_n)_q$, *and*

$$\int\limits_{\mathbb{D}_q} f d\nu_\lambda = C(\lambda)\operatorname{tr}(T_F(f)T_F(y)^\lambda\Gamma(K_{-2\rho})), \quad f \in \mathrm{Pol}(\mathrm{Mat}_n)_q.$$ [1]

For $\lambda > N-1$, the representation π_λ of $U_q\mathfrak{su}_{n,m}$ in the pre-Hilbert space $\mathbb{C}[\mathrm{Mat}_n]_q$ with the scalar product

$$(f_1, f_2)_\lambda = \int\limits_{\mathbb{D}_q} f_2^* f_1 d\nu_\lambda, \qquad f_1, f_2 \in \mathbb{C}[\mathrm{Mat}_n]_q, \tag{3.2}$$

is a $*$-representation. Such representations constitute the so-called holomorphic discrete series.

The theory of bounded symmetric domains contains the well-known Ørsted result [**245**] on analytic continuation of the integral (3.2) in the parameter λ. Let us give a q-analog of this result. Using the Hua–Schmid decomposition (see Subsection

[1] We use the topology of $C(\overline{\mathbb{D}})_q$; see Subsection 2.4.6.

2.5.10), one can easily observe that the $U_q\mathfrak{s}(\mathfrak{gl}_n \times \mathfrak{gl}_n)$-module $\mathbb{C}[\mathrm{Mat}_n]_q$ is a sum of pairwise nonisomorphic simple weight $U_q\mathfrak{s}(\mathfrak{gl}_n \times \mathfrak{gl}_n)$-submodules $\mathbb{C}[\mathrm{Mat}_n]_q^{(\mathbf{k})}$, where $\mathbf{k} = (k_1, k_2, \ldots, k_n) \in \mathbb{Z}_+^n$ and $k_1 \geq k_2 \geq \cdots \geq k_n$.

The Hermitian forms

$$(f_1, f_2)_\lambda = \int\limits_{\mathbb{D}_q} f_2^* f_1 d\nu_\lambda, \qquad (f_1, f_2)_{S(\mathbb{D})_q} = \int\limits_{S(\mathbb{D})_q} f_2^* f_1 d\nu$$

in $\mathbb{C}[\mathrm{Mat}_n]_q$ are positive definite and $U_q\mathfrak{s}(\mathfrak{u}_n \times \mathfrak{u}_n)$-invariant. Thus they differ only by a scalar factor on each of the subspaces $\mathbb{C}[\mathrm{Mat}_n]_q^{(\mathbf{k})}$.[2] Equivalently, for $\lambda > N-1$, one has

$$\int\limits_{\mathbb{D}_q} f_2^* f_1 d\nu_\lambda = c(\mathbf{k}, \lambda) \int\limits_{S(\mathbb{D})_q} f_2^* f_1 d\nu, \qquad f_1, f_2 \in \mathbb{C}[\mathrm{Mat}_{m,n}]_q^{(\mathbf{k})}.$$

A quantum analog of the Ørsted result claims that each of the functions $c(\mathbf{k}, \lambda)$ in variable λ admits an analytic continuation to a meromorphic function. Precisely, it is proved in [**285**] that

PROPOSITION 3.3. *For all* $\mathbf{k} = (k_1, k_2, \ldots, k_n)$ *we have*

$$c(\mathbf{k}, \lambda) = \frac{\prod\limits_{j=1}^{n} (q^{2(n+1-j)}; q^2)_{k_j}}{\prod\limits_{j=1}^{n} (q^{2(\lambda+1-j)}; q^2)_{k_j}}.$$

The proof extensively uses Milne's result [**231**], which is a special case of the Okounkov binomial formula for Macdonald polynomials.

Let us turn to the Penrose transform. Denote the analytic continuation of the integral $\int_{\mathbb{D}_q} f_2^* f_1 d\nu_\lambda$ in parameter λ by $(f_1, f_2)_\lambda$ and consider the special case of 2×2-matrix ball. Then we have

$$c(\mathbf{k}, \lambda) = \frac{(q^4; q^2)_{k_1} (q^2; q^2)_{k_2}}{(q^{2\lambda}; q^2)_{k_1} (q^{2(\lambda-1)}; q^2)_{k_2}}, \qquad k_1 \geq k_2 \geq 0, \tag{3.3}$$

and the function $(f_1, f_2)_\lambda$ in λ is analytic for $\lambda > 1$. Now (3.3) shows that the kernel L of the sesquilinear form $\lim_{\lambda \to 1+0}(1 - q^{2(\lambda-1)})(f_1, f_2)_\lambda$ equals

$$L = \bigoplus_{k=0}^{\infty} \mathbb{C}[\mathrm{Mat}_2]_q^{(k,0)}.$$

It is clear that L is a subrepresentation of π_1. Moreover, the $U_q\mathfrak{su}_{2,2}$-invariant Hermitian form

$$(f_1, f_2) = \lim_{\lambda \to 1+0} (f_1, f_2)_\lambda,$$

is well defined on L and endows it with the structure of a pre-Hilbert space.

The $*$-representation of $U_q\mathfrak{su}_{2,2}$ in L we just constructed is the so-called *ladder* representation. One can easily show that L is a kernel of a quantum invariant differential operator, more precisely, a differential operator that intertwines the representations π_1 and π_3 of the algebra $U_q\mathfrak{su}_{2,2}$.

---

[2]This claim can be easily deduced from the Schur lemma, more precisely, the Burnside theorem [**65**, p. 177].

In the classical limit $q \to 1$, we obtain the known realization of the ladder representation of $U\mathfrak{sl}_4$ in the space

$$\left\{ \psi \in \mathbb{C}[\mathbf{z}] \;\middle|\; \frac{\partial^2 \psi}{\partial z_1^1 \, \partial z_2^2} - \frac{\partial^2 \psi}{\partial z_2^1 \, \partial z_1^2} = 0 \right\}.$$

There is another realization of this representation that is known in the classical case. Namely, consider the coherent sheaf $\mathcal{O}(-2)$ on $\mathbb{CP}^3$ and the Zariski open set

$$U = \{(u_1 : u_2 : u_3 : u_4) \in \mathbb{CP}^3 \mid u_3 \neq 0 \text{ or } u_4 \neq 0\}.$$

The action of $U\mathfrak{sl}_4$ in the cohomology $H^1(U, \mathcal{O}(-2))$ determines its representation that is equivalent to the ladder representation.

The isomorphism of $U\mathfrak{sl}_4$-modules

$$H^1(U, \mathcal{O}(-2)) \to \left\{ \psi \in \mathbb{C}[\mathbf{z}] \;\middle|\; \frac{\partial^2 \psi}{\partial z_1^1 \, \partial z_2^2} - \frac{\partial^2 \psi}{\partial z_2^1 \, \partial z_1^2} = 0 \right\}$$

is unique up to a scalar factor and equals (up to a scalar factor again) the Penrose transform [**26**]. The contructions described can be reproduced in the quantum case and provide a quantum analog of the Penrose transform [**277**].

In [**26**] Baston and Eastwood consider a series of examples related to the Penrose transform that are more complicated than the ladder representation of $U\mathfrak{sl}_4$ given above. Unfortunately, methods of [**277**] are inapplicable in these more tricky examples. The point is that in [**277**] the authors tried to imitate Čech cohomology, that is to use implicitly the methods of noncommutative algebraic geometry which are not developed enough by now.

An alternative approach consists in replacing Čech cohomology with Dolbeaut cohomology, more precisely, in using a quantum analog of the Vogan–Zuckerman method of cohomological induction [**175**]. Apparently, such a quantum analog can be developed once one restricts to facts that are necessary to produce quantum analogs of the results from [**26**] without trying to develop too general constructions. The first step was made in [**287**].

REMARK 3.4. Papers and monographs on classical bounded symmetric domains, for example the papers of J. Arazy, K. Zhu, A. Koranyi and H. Upmeier [**9, 10, 11, 8, 13, 14, 349, 350, 197, 196, 195, 315, 310, 311, 312, 313, 314**] may serve as an additional source of problems of noncommutative analysis in quantum bounded symmetric domains.

Bibliography

[1] H. Ahmedov and I. H. Duru, *Green function on the q-symmetric space $SU_q(2)/U(1)$*, J. Phys. A **31** (1998), 5741–5754.

[2] ———, *Summation formulae for the product of the q-Kummer functions from $E_q(2)$*, J. Phys. A **34** (2001), 1333–1340.

[3] N. I. Akhiezer and I. M. Glazman, *Theory of linear operators in Hilbert space*, "Nauka", Moscow, 1966; English trans, Dover, 1993.

[4] ———, *Theory of linear operators in Hilbert space*, vol. 2, Izd-vo KhGU, Kharkov, 1978; English transl. Dover, 1993.

[5] L. Alvares-Gaumé, C. Gomes, and G. Sierra, *Quantum group interpretation of some conformal field theory*, Phys. Lett. B **220** (1989), 142–152.

[6] G. Andrews, *The theory of partitions*, Cambridge Univ. Press, Cambridge, 1998.

[7] G. E. Andrews and R. Askey, *Enumeration of partitions: the role of Eulerian series and q-orthogonal polynomials*, Higher Combinatorics (Proc. NATO Advanced Study Inst., Berlin, 1976), Reidel, Boston, MA, 1977.

[8] J. Arazy, *Realization of the invariant inner products on the highest quotients of the composition series*, Ark. Mat. **30** (1992), 1–24.

[9] ———, *A survey of invariant Hilbert spaces of analytic functions on bounded symmetric domains*, Contemp. Math., vol. 185, Amer. Math. Soc., Providence, RI, 1995, pp. 7–65.

[10] J. Arazy and S. D. Fisher, *Invariant Hilbert spaces of analytic functions on bounded symmetric domains*, Operator Theory. Advances and applications, vol. 48, Birkhäuser, Basel, 1990, pp. 67–91.

[11] J. Arazy, S. D. Fisher, and J. Peetre, *Möbius invariant function spaces*, J. Reine Angew. Math. **363** (1985), 110–145.

[12] J. Arazy and B. Ørsted, *Asymptotic expansions of Berezin transforms*, Indiana Univ. Math. J. **49** (2000), no. 1, 7–30.

[13] J. Arazy and H. Upmeier, *Discrete series representations and integration over boundary orbits of symmetric domains*, Perspectives of Quantization (L. A. Coburn and M. A. Riffel, eds.), Contemp. Math., vol. 214, Amer. Math. Soc., Providence, RI, 1998, pp. 1–22.

[14] ———, *Boundary integration and the discrete Wallach points*, ESI-Preprint Series, 1999.

[15] ———, *Invariant symbolic calculi and eigenvalues of invariant operators on symmetric domains*, Function Spaces, Interpolation Theory and Related Topics (Lund, 2000) (M. Cwikel, M. Englis, A. Kufner, L.-E. Persson, and G. Sparr, eds.), Walter de Gruyter, Berlin, 2002, pp. 151–211.

[16] V. I. Arnold, *Mathematical methods of classical mechanics*, "Nauka", Moscow, 1989; English transl., Springer, Heidelberg, 1989.

[17] W. Arveson, *Subalgebras of C^*-algebras*, Acta Math. **123** (1969), 141–224.

[18] I. V. Arzhantsev, *Gröbner basis and systems of algebraic equations*, Maks-Press, M., 2002. (Russian)

[19] R. Askey and M. Ismail, *Recurrence relations, continued fractions and orthogonal polynomials*, Mem. Amer. Math. Soc. **49** (1984), no. 300.

[20] R. Askey and J. Wilson, *Some basic hypergeometric orthogonal polynomials that generalize Jacobi polynomials*, Mem. Amer. Math. Soc. **54** (1985), no. 319.

[21] W. Baldoni and P. M. Frajria, *The Harish-Chandra homomorphism for a quantized classical Hermitean symmetric pair*, Ann. Inst. Fourier (Grenoble) **49** (1999), 1179–1214.

[22] L. Barchini, *Szegö mapping, harmonic forms, and Dolbeault cohomology*, J. Funct. Anal. **118** (1993), 351–406.

[23] ———, *Szegö kernel associated with Zuckerman modules*, J. Funct. Anal. **131** (1995), 145–182.

[24] L. Barchini, M. G. Eastwood, and A. R. Gover, *An integral operators into Dolbeault cohomology*, J. Funct. Anal. **137** (1996), 364–380.

[25] L. Barchini, A. W. Knapp, and R. Zierau, *Intertwining operators into Dolbeault cohomology representations*, J. Funct. Anal. **107** (1992), 302–341.

[26] R. J. Baston and M. G. Eastwood, *The Penrose transform. Its interaction with representation theory*, Clarendon Press, Oxford, 1989.

[27] H. Bateman and A. Erdelyi, *Higher transcendental functions*, vol. 1, McGraw-Hill, New York–Toronto–London, 1953.

[28] ———, *Higher transcendental functions*, vol. 2, McGraw-Hill, New York–Toronto–London, 1953.

[29] M. Beals, C. Fefferman, and R. Grossman, *Strictly pseudoconvex domains in* $\mathbb{C}^n$, Bull. Amer. Math. Soc. (N.S.), **8** (1983), no. 2, 125–322.

[30] C. Benson, D. Buraczewski, E. Damek, and G. Ratcliff, *Differential systems of type* $(1,1)$ *of Hermitian symmetric spaces and their zeros*, J. Funct. Anal. **215** (2004), 427–475.

[31] F. Berezin, *General concept of quantization*, Comm. Math. Phys. **40** (1975), no. 2, 153–174.

[32] G. M. Bergman, *The diamond lemma for ring theory*, Adv. Math. **29** (1978), 178–218.

[33] O. Bershtein, *Degenerate principal series of quantum Harish-Chandra modules*, J. Math. Phys. **45** (2004), no. 10, 3800–3827.

[34] O. Bershtein, Ye. Kolisnyk, and L. Vaksman, *On a q-analog of the Wallach–Okounkov formula*, Lett. Math. Phys. **78** (2006), 97–109.

[35] O. Bershtein and S. Sinel'shchikov, *A q-analog of the Hua equations*, Zh. Mat. Fiz. Anal. Geom. **5** (2009), 219–244.

[36] O. Bershtein, A. Stolin, and L. Vaksman, *Spherical principal series of quantum Harish-Chandra modules*, Zh. Mat. Fiz. Anal. Geom. **3** (2007), 157–169.

[37] L. C. Biedenharn and M. A. Lohe, *Quantum group symmetry and q-tensor algebras*, World Scientific, Singapore, 1995.

[38] F. Bonechi, N. Ciccoli, R. Giachetti, E. Sorace, and M. Tarlini, *Free q-Schrodinger equation from homogeneous spaces of the* $2-\dim$ *Euclidean quantum group*, Comm. Math. Phys. **175** (1996), 161–176.

[39] Y. G. Borisovich, N. M. Bliznyakov, T. N. Fomenko, and Y. A. Izrailevich, *Introduction to topology*, "Vysshaya Shkola", Moscow, 1980; English transl., *Introduction to differential and algebraic topology*, Kluwer, Dorderecht, 1995.

[40] D. Borthwick, A. Lesniewski, and H. Upmeier, *Non-perturbative deformation quantization of Cartan domains*, J. Funct. Anal. **113** (1993), 153–176.

[41] N. Bourbaki, *Algebra I: Chapters 1–3*, Springer, Heidelberg, 1989.

[42] ———, *Theory of sets*, Springer, Heidelberg, 2004.

[43] ———, *Commutative algebra: Chapters 1–7*, Springer, Heidelberg, 1989.

[44] ———, *Lie groups and Lie algebras: Chapters 4–6*, Springer, Heidelberg, 2009.

[45] ———, *Théories spectrales: Chapitres 1–2*, Springer, Heidelberg, 2007.

[46] ———, *Lie groups and Lie algebras: Chapters 1–3*, Springer, Heidelberg, 1989.

[47] ———, *Lie groups and Lie algebras: Chapters 7–9*, Springer, Berlin, 2005.

[48] O. Bratteli and D. W. Robinson, *Operator algebras and quantum statistical mechanics*, vol. 1, Springer, Heidelberg, 2002.

[49] K. A. Brown and K. R. Goodearl, *Lectures on algebraic quantum groups*, Birkhäuser, Basel-Boston-Berlin, 2002.

[50] I. Bucur and A. Deleanu, *Introduction to the theory of categories and functors*, Wiley, New York, 1968).

[51] E. Buffenoir and Ph. Roche, *Harmonic analysis on the quantum Lorentz group*, Comm. Math. Phys. **207** (1999), 499–555.

[52] ———, *Tensor product of principal unitary representations of quantum Lorentz group and Askey–Wilson polynomials*, J. Math. Phys. **41** (2000), no. 11, 7715–7751.

[53] I. M. Burban and A. U. Klimyk, *Representations of the quantum algebra* $U_q(su_{1,1})$, J. Phys. A **26** (1993), 2139–2151.

[54] M. Cahen, S. Gutt, and J. Rawnsley, *Quantization of Kähler manifolds. III*, Lett. Math. Phys. **30** (1994), 291–305.

[55] ———, *Quantization on Kähler manifolds. IV*, Lett. Math. Phys. **34** (1995), 159–168.

[56] E. Celeghini, R. Giachetti, E. Sorace, and M. Tarlini, *Three-dimensional quantum groups from contractions of $SU(2)_q$*, J. Math. Phys. **31** (1990), 2548–2551.

[57] ———, *The quantum Heisenberg group $H(1)_q$*, J. Math. Phys. **32** (1991), 1159–1165.

[58] ———, *Contractions of quantum groups*, Quantum groups (P. P. Kulish, ed.), Lecture Notes in Math., vol. 1510, Springer, Berlin, 1992, pp. 221–244.

[59] A. Chakrabarty, *q-Analogs of $IU(n)$ and $U(n,1)$*, J. Math. Phys. **32** (1991), 1227–1334.

[60] J.-T. Chang, *Special K-types, tempered characters and the Beilnson–Bernstein realization*, Duke Math. J. **56** (1988), 345–383.

[61] J.-T. Chang, *Remarks on localization and standard modules: the duality theorem on a generalized flag varieties*, Proc. Amer. Math. Soc. **117** (1993), 585–591.

[62] V. Chari and A. Pressley, *A guide to quantum groups*, Cambridge Univ. Press, Cambridge, 1994.

[63] D. Cox, J. Little, and D. O'Shea, *Ideals, varieties, and algorithms: an introduction to computational algebraic geometry and commutative algebra*, Springer, New York, 1997.

[64] H. S. M. Coxeter and W. O. J. Moser, *Generators and relations for discrete groups*, Springer, Hedelberg, 1980.

[65] C. W. Curtis and I. Reiner, *Representation theory of finite groups and associative algebras*, Amer. Math. Soc., Providence, RI, 2006.

[66] A. Van Daele, *The Haar measure on a compact quantum group*, Proc. Amer. Math. Soc. **123** (1995), no. 10, 3125–3128.

[67] I. Damiani and C. De Concini, *Quantum groups and Poisson groups*, Representations of Lie groups and quantum groups (V. Baldoni and M. A. Picardello, eds.), New York, 1994, pp. 1–45.

[68] M. G. Davidson, T. J. Enright, and R. J. Stanke, *Covariant differential operators*, Math. Ann. **288** (1990), 731–739.

[69] ———, *Differential operators and highest weight repesentations*, Mem. Amer. Math. Soc., vol. 455, Amer. Math. Soc., Providence, R.I., 1991.

[70] C. De Concini and V. Kac, *Representations of quantum groups at roots of 1*, Operator Algebras, Unitary Representations, Envelopping Algebras and Invariant Theory (A. Connes, M. Duflo, and R. Rentchler, eds.), Birkhäuser, Boston-Basel-Berlin, 1990, pp. 472–506.

[71] C. De Concini and C. Procesi, *Quantum groups*, *D*-modules, Representation Theory, and Quantum Groups (L. Boutet De Monvel and C. De Conchini, C. Procesi, P. Shapira, and M. Vergne, eds.), Lecture Notes Math., vol. 1565, Springer, Berlin, 1993, pp. 31–140.

[72] M. S. Dijkhuizen and T. H. Koornwinder, *CQG algebras: a direct algebraic approach to quantum groups*, Lett. Math. Phys. **32** (1994), 315–330.

[73] J. Dixmier, *C^*-algebras*, North-Holland, Amsterdam, 1982.

[74] ———, *Enveloping algebras*, North-Holland, Amsterdam, 1977.

[75] V. K. Dobrev, *Canonical q-deformation of non-compact Lie (super-) algebras*, J. Phys. A **26** (1993), 1317–1334.

[76] R. G. Douglas, *Banach algebra techniques in operator theory*, Academic Press, New York–London, 1972.

[77] V. G. Drinfeld, *Constant quasi-classical solutions of the Yang–Baxter equations*, Dokl. Akad. Nauk SSSR **273** (1983), 531–536; English transl., Soviet Math. Dokl. **28** (1983), 667–671).

[78] ———, *Hopf algebras and the quantum Yang–Baxter equation*, Dokl. Akad. Nauk SSSR **282** (1985), 1060–1064; English transl., Soviet Math. Dokl. **32** (1985), 254–258.

[79] ———, *Quantum groups*, Proc. Intern. Congress of Mathematicians (Berkeley, 1986) (A. M. Gleason, ed.), Amer. Math. Soc., Providence, RI, 1987, pp. 798–820.

[80] ———, *On almost cocommutative Hopf algebras*, Algebra i Analiz **1** (1989), 30–46; English transl., Leningrad Math. J. **1** (1989), 321–342.

[81] ———, *Quasi-Hopf algebras and Knizhnik–Zamolodchikov equations*, Problems of Modern Quantum Field Theory (A. A. Belavin, A. U. Klimyk, and A. B. Zamolodchikov, eds.), Springer, Berlin, 1989, pp. 1–13.

[82] N. Dunford and J. T. Schwartz, *Linear Operators, Part 2. Spectral Theory, Self Adjoint Operators in Hilbert Space*, Wiley, New York, 1988.

[83] M. G. Eastwood and J. W. Rice, *Conformally invariant differential operators on Minkowski space and their curved analogues*, Comm. Math. Phys. **109** (1987), 207–228.

[84] R. Edwards, *Functional analysis: Theory and applications*, Dover, New York, 1995.

[85] M. Engliš, *Invariant operators and the Berezin transform on Cartan domains*, Math. Nachr. **195** (1998), 61–75.

[86] ———, *Berezin–Toeplitz quantization on the Schwartz space of bounded symmetric domains*, J. Lie Theory **15** (2005), 27–50.

[87] T. J. Enright and M. Hunziker, *Resolutions and Hilbert series of determinantal varieties and unitary highest weight modules*, J. Algebra **273** (2004), 608–639.

[88] ———, *Resolutions and Hilbert series of the unitary highest weight modules of the exceptional groups*, J. Represent. Theory **8** (2004), 15–51.

[89] T. J. Enright and A. Joseph, *An intrinsic analysis of unitarizable highest weight modules*, Math. Ann. **288** (1990), 571–594.

[90] T. J. Enright, R. Parthasarathy, N. R. Wallach, and J. A. Wolf, *Unitary derived functor modules with small spectrum*, Acta Math. **154** (1985), no. 1–2, 105–136.

[91] T. J. Enright and J. F. Willenbring, *Hilbert series, Howe duality and branching for classical groups*, Ann. of Math. **159** (2004), 337–375.

[92] P. Etingof and D. Kazhdan, *Quantization of Lie bialgebras, I*, Selecta Math. (N.S.) **1** (1996), 1–41.

[93] H. Exton, *q-Hypergeometric functions and applications*, Ellis Horwood, New York, 1983.

[94] L. D. Faddeev, N. Yu. Reshetikhin, and L. A. Takhtajan, *Quantization of Lie groups and Lie algebras*, Algebra i Analiz **1** (1989), 178–206; English transl., Leningrad Math. J. **1** (1990), 193–225.

[95] R. Fioresi, *A deformation of the big cell inside the Grassmannian manifold $G(r,n)$*, Rev. Math. Phys. **11** (1999), no. 1, 25–40.

[96] P. M. Frajria, *A note on the continuation of Borel–De Siebenthal discrete series out of the canonical chamber*, J. Algebra **154** (1993), 95–107.

[97] J. Fuchs and C. Schweigert, *Category theory for conformal boundary conditions*, Vertex Operator Algebras in Mathematics and Physics (S. Berman, Y. Billig, Y.-Z. Huang, and J. Lepowsky, eds.), Fields Inst. Commun., vol. 39, Amer. Math, Soc., Providence, RI, 2003, pp. 25–70.

[98] T. Gamelin, *Uniform algebras*, Amer. Math. Soc., Proidence, RI, 1984.

[99] G. Gasper and M. Rahman, *Basic hypergeometric series*, Cambridge Univ. Press, Cambridge, 2004.

[100] S. I. Gelfand and Yu. I. Manin, *Homological algebra*, Algebra-5 (R. V. Gamkrelidze, ed.), Sovrem. Probl. Mat. Fundam. Naprl., vol. 38, VINITI, Moscow, 1989, pp. 7–238; English transl, Algebra V, Encyclopedia Math. Sci., vol. 38, Springer, Heodelberg, 1994).

[101] J. E. Gilbert, R. A. Kunze, and P. A. Tomas, *Intertwining kernels and invariant differential operators in representation theory*, Probability Theory and Harmonic Analysis (J.-A. Chao and W. A. Woyczynsky, eds.), Pure Appl. Math. (N. Y.), vol. 98, Marcel Dekker, New York, 1983, pp. 91–112.

[102] I. M. Glazman and Ju. I. Ljubic, *Finite-dimensional linear analysis*, "Nauka", Moscow, 1969; English transl., *Finite-dimensional linear analysis: A systematic presentation in problem form*, Dover New York, 2006).

[103] I. C. Gohberg and M. G. Krein, *Introduction to the theory of linear non-selfadjoint operators in Hilbert space*, "Nauka", Moscow, 1965; English transl., Amer. Math. Soc., Providence, RI, 1969.

[104] ———, *Theory and applications of Volterra operators in Hilbert space*, "Nauka", Moscow, 1967; English transl., Amer. Math. Soc., Providence, RI, 2004.

[105] A. L. Gorodentsev, A. S. Khoroshkin, and A. N. Rudakov, *On syzygies of highest weight orbits*, Moscow Seminar on Mathematical Physics II., Amer. Mat. Soc. Transl. Ser, 2, vol. 221, Amer. Math. Soc., Providence, RI, 2007, pp. 79–120.

[106] M. Goto and F. Grosshans, *Semisimple Lie algebras*, Marcel Dekker, New York–Basel, 1978.

[107] M. I. Graev, *Unitary representations of real simple Lie groups*, Tr. Mosk. Mat. Obs. **7** (1958), 335–389; English transl., Amer. Math. Soc. Transl. Ser, 2, vol. 66, Amer. Math. Soc., Providence, RI, 1968, pp. 1–62).

[108] P. Griffiths and J. Harris, *Principles of algebraic geometry. I* Wiley, New York, 1978).

[109] W. Groenevelt, *Tensor product representations and special functions*, Thesis, DUP Science, Delft, 2004.

[110] D. P. Gupta, M. E. Ismail, and D. R. Masson, *Contiguous relations, basic hypergeometric functions, and orthogonal polynomials III. Associated dual q-Hahn polynomials*, J. Comput. Appl. Math. **68** (1996), 115–149.

[111] W. Hahn, *Über orthogonalpolynome die q-differenzengleichungen genügen*, Math. Nachr. **2** (1949), 4–34.

[112] P. H. Hái, *An embedding theorem for Abelian monoidal categories*, Compos. Math. **132** (2002), 27–48.

[113] P. Halmos, *A Hilbert space problem book*, Springer, Heidelberg, 1982.

[114] Harish-Chandra, *Representations of semisimple Lie groups VI*, Amer. J. Math. **78** (1956), 564–628.

[115] M. Harris and H. P. Jakobsen, *Covariant differential operators*, Group Theoretical Methods in Physics, Istanbul 1982 (M. Serdaroğlu and E. Inönü, eds.), Lecture Notes Phys., vol. 180, Springer, Berlin, 1983, pp. 16–41.

[116] H. Hecht and W. Schmid, *Characters, asymptotics and* $\mathfrak{n}$*-homology of Harish-Chandra modules*, Acta Math. **151** (1983), no. 1–2, 49–151.

[117] I. Heckenberger and S. Kolb, *The locally finite part of the dual coalgebra of quantized irreducible flag manifolds*, Proc. Lond. Math. Soc. **89** (2004), 457–484.

[118] ———, *De Rham complex for quantized irreducible flag manifolds*, J. Algebra **305** (2006), 704–741.

[119] ———, *Differential forms via the Bernstein–Gelfand–Gelfand resolution for quantized irreducible flag manifolds*, J. Geom. Phys. **57** (2007), 2316–2344.

[120] ———, *On the Bernstein–Gelfand–Gelfand resolution for Kac–Moody algebras and quantized enveloping algebras*, Transform. Groups **12** (2007), no. 4, 647–655.

[121] G. Heckman and H. Schlichtkrull, *Harmonic analysis and special functions on symmetric spaces*, Academic Press, New York–Boston, 1994.

[122] H. Hedenmalm, B. Korenblum, and K. Zhu, *Theory of Bergman spaces*, Springer, New York–Berlin–Heidelberg, 2000.

[123] S. Helgason, *Differential geometry, Lie groups, and symmetric spaces*, Amer. Math. Soc., Providence, RI, 2001).

[124] ———, *Groups and geometric analysis: integral geometry, invariant differential operators, and spherical functions*, Amer. Math. Soc., Providence, RI, 2000.

[125] ———, *Invariant differential operators and Weyl group invariants*, Harmonic Analysis on Reductive Groups (W. Barker and P. Sally, eds.), Birkhäuser, Boston, 1991, pp. 193–200.

[126] B. Hoogenboom, *Intertwining functions on compact Lie groups*, vol. 5, Mathematisch Centrum, Amsterdam, 1984.

[127] L. Hörmander, *Introduction to the theory of functions of several complex variables*, Van Nostrand, Princeton, NJ, 1966.

[128] ———, *The analysis of linear partial differential operators I. Distribution theory and Fourier analysis*, Classics in Math., Springer, Berlin, 2003.

[129] L. K. Hua, *Harmonic analysis of functions of several complex variables in the classical domains*, Izd-vo In. Lit., Moscow, 1959: English transl., Amer. Math. Soc., Providence, RI, 1963.

[130] J. E. Humphreys, *Linear algebraic groups*, Springer, Heidelberg, 1975.

[131] ———, *Reflection groups and Coxeter groups*, Cambridge Univ. Press, Cambridge, 1990.

[132] ———, *Conjugacy classes in semisimple algebraic groups*, Amer. Math. Soc., Providence, RI, 1995.

[133] ———, *Introduction to Lie algebras and representation theory*, Birkhäuser, Boston, 1994.

[134] T. Inoue, *Unitary representations and kernel functions associated with boundaries of a bounded symmetric domain*, Hiroshima Math. J. **10** (1980), 75–140.

[135] H. P. Jakobsen, *Basic covariant differential operators on Hermitean symmetric spaces*, Ann. Sci. Éc. Norm. Supér. **18** (1985), 421–436.

[136] ———, *Quantized Hermitian symmetric spaces*, Lie Theory and Its Applications in Physics (Clausthal, 1995), World Sci. Publ., River Edge, NJ, 1996, pp. 105–116.

[137] ———, *Unitarity of highest-weight modules for quantum groups*, Lett. Math. Phys. **41** (1997), 119–133.

[138] ———, *A class of quadratic matrix algebras arising from the quantized enveloping algebra* $U_q(A_{2n-1})$, J. Math. Phys. **41** (2310–2336), no. 4, 119–133.

[139] J. C. Jantzen, *Lectures on quantum groups*, Amer. Math. Soc., Providence, RI, 1996.

[140] M. Jimbo, *A q-difference analogue of $U(\mathfrak{g})$ and the Yang–Baxter equation*, Lett. Math. Phys. **10** (1985), 63–69.

[141] ———, *Quantum R-matrix for the generalized Toda system: an algebraic approach*, Field theory, Quantum Gravity and Strings (H. J. De Vega and N. Sanchez, eds.), Lecture Notes Phys., vol. 246, Springer, Berlin, 1986, pp. 335–361.

[142] K. Johnson, *On a ring of invariant polynomials on a Hermitean symmetric space*, J. Algebra **67** (1980), 72–81.

[143] K. Johnson and A. Korányi, *The Hua operators on bounded symmetric domains of tube type*, Ann. of Math. **111** (1980), 589–608.

[144] K. D. Johnson and N. R. Wallach, *Composition series and intertwining operators for the spherical principal series. I*, Trans. Amer. Math. Soc. **229** (1977), 137–173.

[145] P. E. T. Jørgensen, D. P. Proskurin, and Yu. S. Samoilenko, *The kernel of Fock representations of Wick algebras with braided operator of coefficients*, Pacific J. Math. **198** (2001), no. 1, 109–122.

[146] P. E. T. Jørgensen, L. M. Schmitt, and R. F. Werner, *Positive representations of general commutation relations allowing Wick ordering*, J. Funct. Anal. **134** (1995), 33–99.

[147] A. Joseph, *Faithfully flat embeddings for minimal primitive quotients of quantized enveloping algebras*, Quantum Deformations of Algebras and their Representations (A. Joseph and S. Shnider, eds.), Israel Math. Conf. Proc., vol. 7, Berlin, 1993, pp. 79–106.

[148] ———, *Quantum groups and their primitive ideals*, Springer, Berlin-Heidelberg, 1995.

[149] ———, *Some remarks on the R-matrix*, J. Algebra **180** (1996), 412–430.

[150] A. Joyal and R. Street, *Braided tensor categories*, Adv. Math. **102** (1993), 20–78.

[151] V. G. Kac, *Infinite-dimensional Lie algebras*, Birkhäuser, Boston, MA, 1983.

[152] I. I. Kachurik and A. U. Klimyk, *On Racah coefficients of the quantum algebra $U_q(su_2)$*, J. Phys. A **23** (1990), 2717–2728.

[153] ———, *General recurrence relations for Clebsh–Gordan coefficients of the quantum algebra $U_q(su_2)$*, J. Phys. A **24** (1991), 4009–4015.

[154] A. Kamita, *Quantum deformations of certain prehomogeneous vector spaces III*, Hiroshima Math. J. **30** (2000), no. 1, 79–115.

[155] ———, *Quantum b-functions of prehomogeneous vector spaces of commutative parabolic type*, J. Algebra **244** (2001), 581–603.

[156] A. Kamita, Y. Morita, and T. Tanisaki, *Quantum deformations of certain prehomogeneous vector spaces I*, Hiroshima Math. J. **28** (1998), no. 3, 527–540.

[157] A. V. Karabegov, *Deformation quantization with separation of variables on Kähler manifolds*, Comm. Math. Phys. **180** (1996), 745–755.

[158] M. Kashiwara and W. Schmid, *Quasi-equivariant $\mathcal{D}$-modules, equivariant derived category, and representations of reductive Lie groups*, Lie Theory and Geometry (J.-L. Brylinski and R. Brylinski, ed.), Progr. Math., vol. 123, Birkhäuser, Boston, 1994, pp. 457–488.

[159] C. Kassel, M. Rosso, and V. Turaev, *Quantum groups and knot invariants*, Institut komp'uternyh issledovaniy, Moscow, 2002 (in Russian; English version: Panor. Synthèses, vol. 5, Soc. Math. France, Paris, 1997).

[160] C. Kassel and V. Turaev, *Double construction for monoidal categories*, Acta Math. **175** (1995), 1–48.

[161] Ch. Kassel, *Quantum groups*, Springer, New York, 1995.

[162] T. Kato, *Perturbation theory for linear operators*, Springer, Heidelberg, 1966).

[163] M. Kébé, *$\mathcal{O}$-algèbres quantiques*, C. R. Math. Acad. Sci. Paris, Série I **322** (1996), 1–4.

[164] ———, *Sur la classification des $\mathcal{O}$-algèbres quantiques*, J. Algebra **212** (1999), 626–659.

[165] S. M. Khoroshkin and V. N. Tolstoy, *Universal R-matrix for quantized (Super) algebras*, Comm. Math. Phys. **141** (1991), 599–617.

[166] ———, *The uniqueness theorem for the universal R-matrix*, Lett. Math. Phys. **24** (1992), 231–244.

[167] T. Kimura, *A classification theory of prehomogeneous vector spaces*, Representations of Lie Groups, Kyoto, Hiroshima, 1986 (K. Okamoto and T. Oshima, eds.), Adv. Stud. Pure Math., vol. 14, Academic Press, Boston, MA, 1988, pp. 223–256.

[168] A. A. Kirillov, *Elements of the theory of representations*, "Nauka", Moscow, 1978; English transl., Springer, New York, 1976.

[169] A. A. Kirillov, Jr., and V. Ostrik, *On q-analog of McKay correspondence and ADE classification of $\widehat{\mathfrak{sl}}(2)$ conformal field theory*, Adv. Math. **171** (2002), 183–227.

[170] A. N. Kirillov, *Clebsch–Gordan quantum coefficients*, Zap. nauchnyh seminarov LOMI (L. D. Faddeev, ed.), vol. 170, "Nauka", Leningrad, 1988, pp. 67–84; English transl., J. Math. Sci. (N. Y.) **53** (1991), no. 3, 264–276.

[171] A. N. Kirillov and N. Yu. Reshetikhin, *q-Weyl group and a multiplicative formula for universal R-matrices*, Comm. Math. Phys. **134** (1990), 421–431.

[172] S. Klimek and A. Lesniewski, *Quantum Riemann surfaces I. The unit disc*, Comm. Math. Phys. **146** (1992), 103–122.

[173] ———, *A two-parameter quantum deformation of the unit disc*, J. Funct. Anal. **115** (1993), 1–23.

[174] A. Klimyk and K. Schmüdgen, *Quantum groups and their representations*, Springer, Berlin, 1997.

[175] A. Knapp and D. A. Vogan, *Cohomological induction and unitary representations*, Princeton Univ. Press, Princeton, NJ, 1995.

[176] A. W. Knapp and N. R. Wallach, *Szegö kernels associated with discrete series*, Invent. Math. **34** (1976), 163–200.

[177] S. Kobayashi and K. Nomizu, *Foundations of differential geometry*. Vol. II, Wiley, New York, 1996.

[178] R. Koekoek and R. F. Swattouw, *The Askey-scheme of hypergeometric orthogonal polynomials and its q-analogue*, Tech. Report 98-17, Delft University of Technology, Delft, 1998.

[179] E. Koelink and J. Kustermans, *A locally compact quantum group analogue of the normalizer of $SU(1,1)$ in $SL(2,\mathbb{C})$*, Comm. Math. Phys. **233** (2003), no. 2, 231–296.

[180] H. T. Koelink, *The addition formula for continuous q-Legendre polynomials and associated spherical elements on the $SU(2)$ quantum group related to Askey–Wilson polynomials*, SIAM J. Math. Anal. **25** (1994), no. 1, 197–217.

[181] ———, *The quantum group of plane motions and the Hahn–Exton q-Bessel function*, Duke Math. J. **76** (1994), no. 2, 483–508.

[182] ———, *Addition formula for big q-Legendre polynomials from the quantum $SU(2)$ group*, Canad. J. Math. **47** (1995), 436–448.

[183] ———, *The quantum group of plane motions and basic Bessel functions*, Indag. Math. (N.S.) **6** (1995), no. 2, 191–211.

[184] ———, *Askey–Wilson polynomials and the quantum $SU(2)$ group: survey and applications*, Acta Appl. Math. **44** (1996), 295–3520.

[185] ———, *Yet another basic analogue of Graf's addition formula*, J. Comput. Appl. Math. **68** (1996), 209–220.

[186] H. T. Koelink and W. Van Assche, *Orthogonal polynomials related to the Hahn–Exton q-Bessel function*, Constr. Approx. **11** (1995), 477–512.

[187] H. T. Koelink and R. F. Swarttouw, *On the zeros of the Hahn–Exton q-Bessel functions and associated q-Lommel polynomials*, J. Math. Anal. Appl. **186** (1994), no. 3, 690–710.

[188] ———, *On q-analogue of Graf's addition formula for the Hahn–Exton q-Bessel function*, J. Approx. Theory **81** (1995), no. 2, 260–273.

[189] H. T. Koelink and J. Verding, *Spectral analysis and the Haar functional on the quantum $SU(2)$ group*, Comm. Math. Phys. **177** (1996), 399–415.

[190] A. N. Kolmogorov and S. V. Fomin, *Elements of the theory of functions and functional analysis*, "Nauka", Moscow, 1972; English transl., Dover, New York, 1999.

[191] T. H. Koornwinder, *Representations of the twisted $SU(2)$ quantum group and some q-hypergeometric orthogonal polynomials*, Indag. Math. (N.S.) **51** (1989), 97–117.

[192] ———, *Askey–Wilson polynomials for root systems of type BC*, Hypergeometric Functions on Domains of Positivity, Jack Polynomials, and Applications (D. Richards, ed.), Contemp. Math., vol. 138, 1992, pp. 189–204.

[193] ———, *Askey–Wilson polynomials as zonal spherical functions on the $SU(2)$ quantum group*, SIAM J. Math. Anal. **24** (1993), no. 3, 795–813.

[194] T. H. Koornwinder and R. Swarttouw, *On q-analogue of the Fourier and Hankel transforms*, Trans. Amer. Math. Soc. **333** (1992), 445–461.

[195] A. Korányi, *Harmonic functions on symmetric spaces*, Symmetric spaces. Short courses presented at Washington University (W. M. Boothby and G. L. Weiss, eds.), Pure Appl. Math. (N. Y.), vol. 8, Marcel Dekker, New York, 1972, pp. 379–412.

[196] ———, *Poisson integrals and boundary components of symmetric spaces*, Invent. Math. **34** (1976), 19–35.

[197] ______, *Function spaces on bounded symmetric domains*, Analysis and Geometry on Complex Homogeneous Domains (J. Faraut, S. Kaneyuki, Lu Qikeng, and G. Roos, eds.), Progr. Math., vol. 185, Birkhäuser, Boston-Basel-Berlin, 1999, pp. 183–281.

[198] A. Korányi and H. M. Reimann, *Equivariant first order differential operators on boundaries of symmetric spaces*, Invent. Math. **139** (2000), 371–390.

[199] L. I. Korogodski and Yan S. Soibelman, *Algebras of functions on quantum groups: Part* I, Amer. Math. Soc., Providence, RI, 1998.

[200] L. I. Korogodsky, *Quantum group* $SU(1,1) \rtimes \mathbb{Z}_2$ *and "super-tensor" products*, Comm. Math. Phys. **163** (1994), 433–460.

[201] B. Kostant, *Lie algebra cohomology and the generalized Borel–Weil theorem*, Ann. of Math. **74** (1961), no. 2, 329–387.

[202] A. I. Kostrikin and Yu. I. Manin, *Linear algebra and geometry*, "Nauka", Moscow, 1986; English transl., Gordon and Breach, New York, 1989.

[203] S. G. Krein, Ju. I. Petunin, and E. M. Semenov, *Interpolation of linear operators*, "Nauka", Moscow, 1978; English transl., Amer. Math. Soc., Providence, RI, 1982.

[204] P. P. Kulish and N. Yu. Reshetikhin, *Quantum linear problem for the sine-Gordon equation and higher representations*, Zap. nauchnyh seminarov LOMI (L. D. Faddeev, ed.), vol. 101, "Nauka", Leningrad, 1981, pp. 101–110; English transl., J. Math. Sci. (N. Y.) **23** (1983), no. 4, 2435–2441.

[205] P. P. Kulish, N. Yu. Reshrtikhin, and E. K. Sklyanin, *Yang–Baxter equation and representation theory. I*, Lett. Math. Phys. **5** (1985), 393–403.

[206] P. P. Kulish and E. K. Sklyanin, *Solutions of the Yang–Baxter equation*, Zap. nauchnyh seminarov LOMI (L. D. Faddeev, ed.), vol. 95, "Nauka", Leningrad, 1980, pp. 129–160; English transl., J. Math. Sci. (N. Y.) **19** (1982), no. 5, 1596–1620.

[207] J. Kustermans and S. Vaes, *Locally compact quantum groups*, Ann. Sci. Éc. Norm. Supér. **33** (2000), 837–934.

[208] ______, *Locally compact quantum groups in the von Neumann algebraic setting*, Math. Scand. **92** (2003), no. 1, 68–92.

[209] V. Lakshmibai and N. Reshetikhin, *Quantum flag and Schubert schemes*, Contemp. Math., vol. 134, Amer. Math. Soc., Providence, RI, 1992, pp. 145–181.

[210] T. Y. Lam, *Lectures on modules and rings*, Springer, New York, 1999.

[211] S. Lang, *Algebra*, Addison-Wesley, Reading, MA, 1965.

[212] J. Lepowsky, *Generalized Verma modules, the Cartan–Helgason theorem, and the Harish-Chandra homomorphism*, J. Algebra **49** (1977), 470–495.

[213] A. Lesniewski and M. Rinaldi, *Tensor products of representations of* $C(SU_q(2))$, J. Math. Phys. **34** (1993), no. 1, 305–314.

[214] G. Letzter, *Representation theory for quantized enveloping algebra*, Proc. Symp. Pure Math., vol. 56, Amer. Math. Soc., Providence, RI, 1994, pp. 63–80.

[215] S. Z. Levendorskii and Ya. S. Soibelman, *Some applications of quantum Weyl groups*, J. Geom. Phys. **7** (1990), 241–254.

[216] ______, *Algebras of functions on compact quantum groups, Schubert cells and quantum tori*, Comm. Math. Phys. **139** (1991), 141 –170.

[217] J. Lukierski and A. Nowicky, *Quantum deformations of* $D = 4$ *Poincare algebra*, Quantum Groups and Related Topics (R. Gierlak and et al., eds.), Kluwer, Dordrecht, 1992, pp. 33–44.

[218] V. A. Lunts and A. L. Rosenberg, *Differential operators on noncommutative rings*, Selecta Math. (N.S.) **3** (1997), 335–359.

[219] ______, *Localization for quantum groups*, Selecta Math. (N.S.) **5** (1999), 123–159.

[220] G. Lusztig, *Quantum groups at roots of 1*, Geom. Dedicata **35** (1990), 89–114.

[221] V. Lychagin and A. V. Prasolov, *Products and quantization in* K*-theory of braiding-commutative algebras*, Algebraic K-theory and Its Applications, TCTP Trieste 1997 (H. Bass, A. Kuku, and C. Pedrini, eds.), vol. 132, World Scientific, Singapore, 1999, pp. 450–489.

[222] V. Lyubashenko, *Modular transformations for tensor categories*, J. Pure Appl. Algebra **98** (1995), 279–327.

[223] ______, *Tangles and Hopf algebras in braided categories*, J. Pure Appl. Algebra **98** (1995), 245–278.

[224] I. G. Macdonald, *Symmetric functions and Hall polynomials*, Clarendon Press, Oxford, 1995.

[225] S. Majid, *Foundations of quantum group theory*, Cambridge Univ. Press, Cambridge, 1995.
[226] G. Maltsiniotis, *Le langage des espaces et des groupes quantiques*, Comm. Math. Phys. **151** (1993), 275–302.
[227] Yu. I. Manin, *Quantum groups and noncommutative geometry*, Université de Montréal, CRM, Montreal, 1988.
[228] T. Masuda, K. Mimachi, Y. Nakagami, M. Noumi, and K. Ueno, *Representations of quantum groups and a q-analogue of orthogonal polynomials*, C. R. Math. Acad. Sci. Paris, Série I **307** (1988), 559–564.
[229] ———, *Representations of the quantum group $SU(2)$ and the little q-Jacobi polynomials*, J. Funct. Anal. **99** (1991), 357–386.
[230] O. Mathieu, *Classification of irreducible weight modules*, Ann. Inst. Fourier (Grenoble) **50** (2000), no. 2, 537–592.
[231] S. C. Milne, *A triple product identity for Schur functions*, J. Math. Anal. Appl. **160** (1991), no. 2, 446–458.
[232] C. Moreno and P. Ortega-Navarro, *$*$-Products on $D^1(\mathbb{C})$, S^2 and related spectral analysis*, Lett. Math. Phys. **7** (1983), 181–193.
[233] Y. Morita, *Quantum deformation of certain prehomogeneous vector spaces II*, Osaka J. Math. **37** (2000), no. 2, 385–403.
[234] I. Muller, H. Rubenthaler, and G. Shiffmann, *Structures des espaces préhomogènes associes à certaines algèbres de Lie graduées*, Math. Ann. **274** (1986), 95–123.
[235] M. F. Müller-Bahns and N. Neumaier, *Invariant star products of Wick type: classification and quantum moment mapping*, Lett. Math. Phys. **70** (2004), 1–15.
[236] G. Nagy and A. Nica, *On the "quantum disc" and a "non-commutative circle"*, Birkhäuser, Boston, 1994, pp. 277–290.
[237] M. Naimark, *Normed rings*, "Nauka", Moscow, 1968; English transl., Noordhoff, Groningen, 1964.
[238] M. S. Narasimhan and K. Okamoto, *An analogue of the Borel–Weil–Bott theorem for Hermitian symmetric pairs of non-compact type*, Ann. of Math. **91** (1970), no. 2, 486–511.
[239] M. Noumi and K. Mimachi, *Big q-Jacobi polynomials, q-Hahn polynomials and a family of quantum 3-spheres*, Lett. Math. Phys. **19** (1990), 299–305.
[240] ———, *Quantum 2-spheres and big q-Jacobi polynomials*, Comm. Math. Phys. **128** (1990), 521–531.
[241] C. Năstăcescu and F. Van Oystaeyen, *Graded ring theory*, North-Holland, Amsterdam, 1982.
[242] G. Ólafsson and B. Ørsted, *Causal compactification and Hardy spaces*, Trans. Amer. Math. Soc. **351** (1999), 3771–3792.
[243] M. A. Olshanetsky and V. B. K. Rogov, *Liouville quantum mechanics on a lattice from geometry of quantum Lorentz group*, J. Phys. A **27** (1994), 4669–4683.
[244] ———, *Unitary representations of quantum Lorentz group and quantum relativistic Toda chain*, Teoret. Mat. Fiz. **130** (2002), 355–382; arXiv: math/110182, 2001).
[245] B. Ørsted, *Composition series for analytic continuations of holomorphic discrete series representations of $SU(n,n)$*, Trans. Amer. Math. Soc. **260** (1980), no. 2, 563–573.
[246] B. Pareigis, *On braiding and dyslexia*, J. Algebra **171** (1995), 413–425.
[247] V. Pasquier and H. Saleur, *Common structures between finite systems and conformal field theories through quantum groups*, Nuclear Phys. B **330** (1990), 523–556.
[248] I. Penkov and V. Serganova, *Generalized Harish-Chandra modules*, Mosc. Math. J. **2** (2002), no. 4, 753–767.
[249] I. Penkov and G. Zuckerman, *Generalized Harish-Chandra modules: a new direction in structure theory of representations*, Acta Appl. Math. **81** (2004), 311–326.
[250] ———, *Generalized Harish-Chandra modules with generic minimal $\mathfrak{k}$-type*, Asian J. Math. **8** (2004), no. 4, 795–812.
[251] P. Podleš, *Quantum spheres*, Lett. Math. Phys. **14** (1987), 193–202.
[252] ———, *Differential calculus on quantum spheres*, Lett. Math. Phys. **18** (1989), 107–119.
[253] ———, *Complex quantum groups and their real representations*, Publ. Res. Inst. Math. Sci. **28** (1992), 709–745.
[254] P. Podleš, *Symmetries of quantum spaces, subgroups and quotient spaces of quantum $SU(2)$ and $SO(3)$ group*, Comm. Math. Phys. **170** (1995), 1–20.
[255] P. Podleš and S. Woronovicz, *Quantum deformation of Lorentz group*, Comm. Math. Phys. **130** (1990), 381–431.

[256] V. L. Popov, *Generators and relations of the affine coordinate rings of connected semisimple algebraic groups*, Preprint ESI **972**, 2000.

[257] V. V. Prasolov, *Polynomials*, MCNMO, Moscow, 2001; English transl., Springer, New York, 2004.

[258] D. Proskurin and L. Turowska, *On the C^*-algebra associated with* $\mathrm{Pol}(\mathrm{Mat}_{2,2})_q$, Methods Funct. Anal. Topology **7** (2001), 88–92.

[259] W. Pusz, *Irreducible unitary representations of quantum Lorentz group*, Comm. Math. Phys. **152** (1993), 591–626.

[260] M. Reed and B. Simon, *Methods of modern mathematical physics I: Functional analysis*, Academic Press, London–New York, 1980).

[261] N. Reshetikhin and M. Yakimov, *Quantum invariant measures*, Comm. Math. Phys. **224** (2001), no. 2, 399–426.

[262] N. Yu. Reshetikhin, *Quasitriangular Hopf algebras and invariants of tangles*, Algebra i Analiz **1** (1989), no. 2, 169–188; English transl., Leningrad Math. J. **1** (1990), no. 2, 491–513.

[263] A. Rocha-Caridi, *Splitting criteria for $\mathfrak{g}$-modules induced from a parabolic and Bernstein–Gelfand–Gelfand resolution of a finite dimensional irreducible $\mathfrak{g}$-module*, Trans. Amer. Math. Soc. **262** (1980), 335–366.

[264] M. Rosso, *Représentations des groupes quantiques*, Séminaire Bourbaki, Astérisque, vol. 201–203, Soc. Math. France, Paris, 1992, pp. 443–483.

[265] H. Rubenthaler, *Algèbres de Lie et espaces préhomogènes*, Hermann, Paris, 1992.

[266] H. Rubenthaler and G. Shiffmann, *Opérateurs différentiels de Shimura et espaces préhomogènes*, Invent. Math. **90** (1987), 409–442.

[267] I. Satake, *Algebraic structure of symmetric domains*, Iwanami Shoten, Tokyo; Princeton Univ. Press, Princeton, NJ, 1980.

[268] M. Sato, M. Kashiwara, T. Kimura, and T. Oshima, *Micro-local analysis of prehomogeneous vector spaces*, Invet. Math. **62** (1980), 117–179.

[269] W. Schmid, *Homogeneous complex manifolds and representations of semisimple Lie groups (thesis, Berkeley*, 1967), Representation Theory and Harmonic Analysis on Semisimple Lie Groups, Math. Surveys Monogr., vol. 31, Amer. Math. Soc., Providence, RI, 1989, pp. 223–286.

[270] ———, *Construction and classification of irreducible Harish-Chandra modules*, Harmonic Analysis on Reductive Groups (W. Baker and P. Sally, eds.), vol. 201–203, Birkhäuser, Boston MA, 1991, pp. 235–275.

[271] K. Schmüdgen and A. Schüler, *Covariant differential calculi on quantum spaces and on quantum groups*, C. R. Math. Acad. Sci. Paris, Série I **316** (1993), 1155–1160.

[272] A. Schüler, *Differential Hopf algebras on quantum groups of type A*, J. Algebra **214** (1999), 479–518.

[273] N. N. Shapovalov, *On a bilinear form on the universal enveloping algebra of a complex semisimple Lie algebra*, Funktsional. Anal. i ego Prilozhen. **6** (1972), 65–70; English transl., Functional Anal. Appl., **6** (1972), 307–312.

[274] N. Shimeno, *The Plancherel formula for spherical functions with a one-dimensional K-type on a simply connected simple Lie group of Hermitian type*, J. Funct. Anal. **121** (1994), 330–388.

[275] G. Shimura, *Invariant differential operators on Hermitian symmetric spaces*, Ann. of Math. **132** (1990), 237–272.

[276] D. Shklyarov, *On q-analogues of certain prehomogeneous vector spaces: comparison of several approaches*, Math. Phys. Anal. Geom. **8** (2001), no. 3, 325–345.

[277] D. Shklyarov, S. Sinel'shchikov, A. Stolin, and L. Vaksman, *On a q-analogue of the Penrose transform*, Ukrain. Fiz. Zh. **47** (2002), no. 3, 288–292.

[278] D. Shklyarov, S. Sinel'shchikov, and L. Vaksman, *On function theory in quantum disc: q-differential equations and Fourier transform*, Preprint arXiv:math/9909036, 1999.

[279] ———, *A q-analogue of the Berezin quantization method*, Lett. Math. Phys. **49** (1999), 253–261.

[280] ———, *q-Analogues of some bounded symmetric domains*, Czechoslovak J. Phys. **50** (2000), no. 1, 175–180.

[281] ———, *Geometric realizations for some series of representations of the quantum group* $SU_{2,2}$, Math. Phys. Anal. Geom. **8** (2001), no. 1, 90–110.

[282] ———, *Hidden symmetry of some algebras of q-differential operators*, Noncommutative Structures in Mathematics and Physics (S. Duplij and J. Wess, eds.), NATO Sci. Ser., vol. 22, Kluwer, Dordrecht-Boston-London, 2001, pp. 309–320.

[283] ———, *Fock representations and quantum matrices*, Int. J. Math. **15** (2004), no. 9, 1–40.

[284] D. Shklyarov and G. Zhang, *A q-analogue of the Berezin transform on the unit ball*, J. Math. Phys. (2004).

[285] ———, *Covariant q-differential operators and unitary highest weight representations for* $U_q\mathfrak{su}_{n,n}$, J. Math. Phys. **46** (2005), 062307 1–24.

[286] S. Sinel'shchikov, A. Stolin, and L. Vaksman, *Differential calculi on some quantum prehomogeneous vector spaces*, Preprint arXiv: math/0606499, 1998.

[287] ———, *A quantum analogue of the Bernstein functor*, J. Lie Theory **17** (2007), no. 1, 73–89.

[288] S. Sinel'shchikov and L. Vaksman, *Hidden symmetry of the differential calculus on the quantum matrix space*, J. Phys. A **30** (1997), L23–L26.

[289] ———, *On q-analogues of bounded symmetric domains and Dolbeault complexes*, Math. Phys. Anal. Geom. **1** (1998), no. 1, 75–100.

[290] ———, *Quantum groups and bounded symmetric domains*, Contemp. Math., vol. 391, Amer. Math. Soc, Providence, RI, 2005, pp. 323–330.

[291] E. K. Sklyanin, *Some algebraic structures connected with the Yang–Baxter equation*, Funktsional. Anal. i ego Prilozhen. **16** (1982), 27–34; English transl., Functional Anal. Appl. **16** (1982), 263–270).

[292] ———, *On certain algebra generated by quadratic relations*, Uspekhi Mat. Nauk **40** (1985), 214. (Russian)

[293] E. K. Sklyanin and L. D. Faddeev, *Quantum mechanical approach to completely integrable field theory models*, Dokl. Akad. Nauk SSSR **243** (1978), 1430–1433; English transl., Soviet Phys. Dokl. (1978), no. 23, 902–904.

[294] E. K. Sklyanin, L. A. Takhtadzhyan, and L. D. Faddeev, *Quantum inverse problem method. I*, Teoret. Mat. Fiz. **40** (1979), no. 2, 194–220; English transl., Theoret. Math. Phys. **40** (1979), no. 2, 688–706.

[295] Ya. S. Soibelman and L. L. Vaksman, *On some problems in the theory of quantum groups*, Representation Theory and Dynamical Systems (A. M. Vershik, ed.), Adv. Soviet Math., vol. 9, Amer. Math. Soc., Providence, RI, 1992, pp. 3–55.

[296] T. Springer, *Jordan algebras and algebraic groups*, Springer, Berlin–Heidelberg–New York, 1973.

[297] ———, *Linear algebraic groups*, Birkhäuser, Boston, 1998.

[298] R. Steinberg, *Lectures on Chevalley groups*, Dept. Mathematics, Yale Univ., 1968.

[299] J. V. Stokman, *The quantum orbit method for generalized flag manifolds*, Math. Res. Lett. **10** (2003), no. 4, 469–481.

[300] J. V. Stokman and M. S. Dijkhuizen, *Quantized flag manifolds and irreducible *-representations*, Comm. Math. Phys. **203** (1999), 297—324.

[301] R. F. Swarttouw, *An addition theorem and some product formulas for the Hahn–Exton q-Bessel functions*, Canad. J. Math. **44** (1992), 867–879.

[302] B. Sz.-Nagy and C. Foias, *Harmonic analysis of operators on a Hilbert space*, Elsevier, New York, 1970.

[303] M. Takesaki, *Theory of operator algebras I*, Springer, Berlin–New York, 1979.

[304] M. Takeuchi, *Polynomial representations associated with symmetric bounded domains*, Osaka J. Math. **10** (1973), 441–475.

[305] L. A. Takhtadzhyan and L. D. Faddeev, *The quantum method of the inverse problem and the Heisenberg* XYZ*-model*, Uspekhi Mat. Nauk **34** (1979), no. 5, 13–63. (Russian)

[306] T. Tanisaki, *Killing forms, Harish-Chandra homomorphisms and universal R-matrices for quantum algebras*, Infinite Analysis (A. Tsuchiya, T. Eguchi, and M. Jimbo, eds.), Internat. J. Modern Phys. A, Suppl 1B, vol. 7, World Scientific, Singapore, 1992, pp. 941–962.

[307] V. G. Turaev, *Quantum invariants of knots and* 3*-manifolds*, Walter de Gruyter, Berlin, 1994.

[308] L. Turowska, *Representations of a q-analogue of the *-algebra* $\mathrm{Pol}(\mathrm{Mat}_{2,2})$, J. Phys. A **34** (2001), 2063–2070.

[309] V. A. Ufnarovskii, *Combinatorial and asymptotic methods in algebra*, Algebra – 6, Itogi Nauki i Tekhniki. Ser. Sovrem. Probl. Mat. Fundam. Naprl., vol. 57, VINITI, Moscow, 1990, pp. 5–177. (Russian)

[310] A. Unterberger and H. Upmeier, *The Berezin transform and invariant differential operators*, Comm. Math. Phys. **164** (1994), 563–597.

[311] H. Upmeier, *Toeplitz operators on bounded symmetric domains*, Trans. Amer. Math. Soc. **280** (1983), no. 1, 221–237.

[312] ______, *Toeplitz C^*-algebras on bounded symmetric domains*, Ann. of Math. **119** (1984), 549–576.

[313] ______, *An index theorem for multivariable Toeplitz operators*, Integral Equations Operator Theory **9** (1986), 355–386.

[314] ______, *Fredholm indices for Toeplitz operators on bounded symmetric domains*, Amer. J. Math. **110** (1988), 811–832.

[315] ______, *Toeplitz operators and index theory in several complex variables*, Proc. Sympos. Pure Math., vol. 51, Amer. Math. Soc., Providence, RI, 1990, pp. 585–598.

[316] L. Vaksman, *q-Analogues of the Clebsch–Gordan coefficients and algebra of functions on the quantum group $SU(2)$*, Dokl. Akad. Nauk SSSR **306** (1989), no. 2, 269–271; English transl., Soviet Math. Dokl. 39 (1989), no. 3, 467–470.

[317] ______, *Lectures on q-analogs of Cartan domains and associated Harish-Chandra modules*, arXiv:math/0109198, 2001.

[318] L. L. Vaksman, *Quantum matrix ball: the Cauchy–Szegö kernel and the Shilov boundary*, Math. Phys. Anal. Geom. **8** (2001), no. 4, 366–384.

[319] L. L. Vaksman and L. I. Korogodskij, *An algebra of bounded functions on the quantum group of the motions of the plane, and q-analogues of Bessel functions*, Dokl. Akad. Nauk SSSR **304** (1989), 1036–1040; English transl., Soviet Math. Dokl. **39** (1989), no. 1, 173–177.

[320] L. L. Vaksman and D. L. Shklyarov, *Integral representations of functions in the quantum disc. I*, Math. Phys. Anal. Geom. **4** (1997), no. 3, 286–308. (Russian)

[321] L. L. Vaksman and Ya. S. Soibel'man, *Algebra of functions on the quantum group $SU(2)$*, Funktsional. Anal. i ego Prilozhen. **22** (1988), no. 3, 1–14; English transl., Functional Anal. Appl. **22** (1988), no. 3, 170–181.

[322] N. J. Vilenkin, *Special functions and the theory of group representations*, "Nauka", Moscow, 1965; English transl., Amer. Math. Soc., Providence, RI, 1968.

[323] N. Ja. Vilenkin and A. U. Klimyk, *Representations of Lie groups and special functions. Vol 3*, Kluwer, Dordrecht, 1992.

[324] E. Vinberg, *Classification of homogeneous nilpotent elements of a semisimple graded Lie algebra*, Trudy Sem. Vektor. Tenzor. Anal. (P. K. Rashevskiy, ed.), vol. XIX, Izd. MGU, Moscow, 1979, pp. 155–177; English transl., Selecta Math. Soviet. (1987), 15–35.

[325] E. B. Vinberg, *Linear representations of groups*, "Nauka", Moscow, 1985; English transl., Birkhäuser, Boston, 1989.

[326] E. B. Vinberg, V. V. Gorbatsevich, and A. L. Onishchik, *Structure of Lie groups and Lie algebras*, Itogi Nauki i Tekhniki. Ser. Sovrem. Probl. Mat. Fundam. Naprl., vol. 41, VINITI, Moscow, 1989; English transl., Encyclopaedia Math. Sci., 20, Springer, Berlin, 1993.

[327] E. B. Vinberg and A. L. Onishchik, *A seminar on algebraic groups and Lie groups*, "Nauka", Moscow, 1988; English transl., Springer, Berlin, 1990.

[328] E. B. Vinberg and V. L. Popov, *Invariant theory*, Algebraic geometry IV (D. I. Panyushev, ed.), Sovrem. Probl. Mat. Fundam. Naprl., vol. 55, VINITI, Moscow, 1989, pp. 137–309; English transl., Encyclopedia Math. Sci., vol. 55, Springer, Heidelberg, 1994, pp. 123–278.

[329] V. S. Vladimirov, *Equations of mathematical physics*, "Nauka", Moscow, 1971; English transl., Marcel Dekker, New York, 1971.

[330] D. A. Vogan, *Unipotent representations and cohomological induction*, The Penrose Transform and Analytic cohomology in Representation Theory (R. Zierau M. Eastwood, J. Wolf, ed.), Contemp. Math., vol. 154, Amer. Math. Soc., Providence, RI, 1993, pp. 47–69.

[331] L. Vretare, *Elementary spherical functions on symmetric spaces*, Math. Scand. **39** (1976), 343–358.

[332] N. R. Wallach, *Real reductive groups I*, Academic Press, Boston, 1988.

[333] ______, *Real reductive groups II*, Academic Press, Boston, 1992.

[334] N. E. Wegge-Olsen, *K-Theory and C^*-algebras. A friendly approach*, Oxford Univ. Press, Oxford, 1993.

[335] R. O. Wells, *Differential analysis on complex manifolds*, Springer, Heidelberg, 1980.

[336] J. Wolf, *The action of a real semisimple Lie group on a complex flag manifold. I: Orbit structure and holomorphic arc components*, Bull. Amer. Math. Soc. (N.S.) **75** (1969), 1121–1237.

[337] ———, *The action of a real semisimple Lie group on a complex flag manifold. II: Unitary representations on partially holomorphic cohomology spaces*, Mem. Amer. Math. Soc., vol. 138, Amer. Math. Soc., Providence, RI, 1974.

[338] ———, *Admissible representations and geometry of flag manifolds*, The Penrose Transform and Analytic Cohomology in Representation Theory (M. Eastwood, J. Wolf, and R. Zierau, eds.), Contemp. Math., vol. 154, Amer. Math. Soc., Providence, RI, 1993, pp. 21–45.

[339] J. A. Wolf and R. Zierau, *Holomorphic double fibration transform*, The Mathematical Legacy of Harish-Chandra (R. S. Doran and V. S. Varadarajan, eds.), Proc. Sympos. Pure Math., vol. 68, Amer. Math. Soc., Providence, RI, 2000, pp. 527–551.

[340] H. Wong, *Dolbeault cohomological realization of Zuckerman modules associated with finite rank representations*, J. Funct. Anal. **129** (1995), 428–454.

[341] S. L. Woronowicz, *Pseudospaces, pseudogroups, and Pontryagin duality*, Proceedings of the International Conference on Mathematics and Physics, Lausanne 1979 (K. Osterwalder, ed.), Lecture Notes Phys., vol. 116, Springer, Berlin, 1980, pp. 407–412.

[342] ———, *Compact matrix pseudogroups*, Comm. Math. Phys. **111** (1987), no. 4, 613–665.

[343] ———, *Twisted SU(2) group. An example of noncommutative differential calculus*, Publ. Res. Inst. Math. Sci. **23** (1987), 117–181.

[344] ———, *A remark on compact matrix quantum groups*, Lett. Math. Phys. **21** (1991), 35–39.

[345] ———, *Unbounded elements affiliated with C^*-algebras and noncompact quantum groups*, Comm. Math. Phys. **136** (1991), 399–432.

[346] ———, *Quantum $E(2)$-group and its Pontryagin dual*, Lett. Math. Phys. **23** (1992), 251–263.

[347] ———, *Quantum $SU(2)$ and $E(2)$ groups. Contraction procedure*, Comm. Math. Phys. **149** (1992), 637–652.

[348] D.P. Zhelobenko, *Compact Lie groups and their representations*, "Nauka", Moscow, 1970; English transl., Amer. Math. Soc., Providence, RI, 1973.

[349] K. Zhu, *Operator theory in functional spaces*, Marcel Dekker, New York, Basel, 1990.

[350] ———, *Spaces of holomorphic functions in the unit ball*, Springer, New York, 2005.

[351] R. Zierau, *Representations in Dolbeault cohomology*, Representation theory of Lie groups (Park City, UT, 1998), 91–146, IAS/Park City Math. Ser., 8, Amer. Math. Soc., Providence, RI, 2000.

[352] ———, *Geometric construction of certain highest weight modules*, Proc. Amer. Math. Soc. **95** (1985), no. 4, 631–635.

Index

Titles in This Series

For a complete list of titles in this series, visit the
AMS Bookstore at **www.ams.org/bookstore/**.